AND
THE EMERGENCE OF
EVOLUTIONARY THEORIES
OF MIND AND BEHAVIOR

Bishop Samuel Wilberforce, 1805–1873, portrait done in 1864.

AND
THE EMERGENCE OF
EVOLUTIONARY THEORIES
OF MIND AND BEHAVIOR

ROBERT J. RICHARDS

Science and Its Conceptual Foundations
David L. Hull, Editor

The University of Chicago Press
Chicago and London

The University of Chicago Press, Chicago 60637
The University of Chicago Press, Ltd., London

© 1987 by The University of Chicago
All rights reserved. Published 1987
Paperback edition 1989
Printed in the United States of America

98 97 96 95   7 6 5 4

Library of Congress Cataloging in Publication Data

Richards, Robert John
  Darwin and the emergence of evolutionary theories
of mind and behavior.

  (Science and its conceptual foundations)
  Bibliography: p.
  Includes index.
  1. Genetic psychology. 2. Darwin, Charles,
1809–1882—Views on genetic psychology. 3. Psychobiology.
4. Human evolution. I. Title. II. Series.
BF711.R53 1987   155.7   87-10891

ISBN 0-226-71200-1 (pbk.)

To My Parents

The recesses of feeling, the darker, blinder strata of character are the only places in the world in which we catch real fact in the making, and directly perceive how events happen, and how work is actually done.

—William James, *Varieties of Religious Experience*

Contents

Illustrations

Preface

This book traces the development of evolutionary theories of mind and behavior from their first distinct appearance in the eighteenth century to their controverted state in the present. The focus, however, tightens on their evolution during the nineteenth century and on the various factors in the conceptual environment that gave them shape. Darwin's ideas about instinct, reason, and morality—considered against the background of his personality, training, scientific and cultural concerns, and intellectual community—control the course of my reconstruction.

This history has explanation as a principal aim. It attempts to specify the antecedent causal matrix which, I argue, makes intelligible the consequent evolution of ideas. I believe that historical narratives provide understanding to the degree that they show how one set of factors impinged upon a subsequent set, influenced it, caused it. The ideal history would recover and make explicit those causal lines that necessarily gave rise to the events of interest. Actual histories, of course, can only nod toward the ideal. In practice, historians must assume certain states of affairs and causal actions that can only be vaguely suggested, and then they highlight some few factors believed to have had considerable weight in producing the events. Not all historians and philosophers of science would endorse this conception of historical narrative. And fewer would be immediately receptive to the particular historiographic theory I use to support the conception—namely, a natural selection approach to ideas. I do not believe, though, that this historiographic framework cuts through the surface of the narrative so as to distress most historians of science. My formal justification for the theory behind the practice will be found sketched in the introduction and elaborated in the first appendix. My informal justification is the narrative itself.

I have moved beyond strict narrative history in another and, I hope, equally unobtrusive way. I draw certain philosophical conclusions con-

cerning ethics from this history. But again the argument is confined to
an appendix. I do not think my narrative need collapse if the reader
rejects either my notions about the evolution of ideas or my particular
theory of evolutionary ethics, any more than, say, certain anatomical
descriptions would need be abandoned should current theory in evolution be substantially revised. A higher-level theory might lead to refined observations, be justified by those observations, yet subsequently
be replaced while the data remain intact. I do believe, though, that my
theory of the evolution of ideas is sustained by the practices of sensitive
historians.

In this book, I have attempted to advance several connected arguments in history and philosophy of science. The introduction and conclusion epitomize these arguments. I realize, however, that not all readers would care to pursue the more general themes through a book of
this size, so I have tried to make individual chapters and groups of
chapters fairly self-contained. Thus chapters 1, 2–5, 6–7, 8, 9, and 10, as
well as the two appendixes, can be read independently.

I owe many debts for the support of this project. Small grants from
the American Philosophical Society, the Fishbein Center for the History of Science and Medicine, the National Institutes of Health, the
National Science Foundation, and the Spencer Foundation enabled me
to travel to archives in the United States and Europe and to make copies
of needed materials. An Andrew Mellon Faculty Fellowship at Harvard
University during the academic year 1982–1983 provided an uninterrupted period to draft several chapters, and the masters of Lowell
House, William and Mary Bossert, supplied living quarters and generous hospitality. The conclusion was written during the summer of 1986
while participating in a seminar on the history of the human sciences
sponsored by the Center for Advanced Study in the Behavioral
Sciences.

Librarians in charge of special collections and manuscripts at several
institutions made research much easier. I am grateful for the aid received from staffs at the following libraries and archives: the library of
the American Philosophical Society; the Bibliothèque Nationale, Paris;
the Boston Public Library; Bristol University Library; the library of
the British Museum; the University Library, Cambridge University;
Haeckel Haus, Jena; Houghton Library, Harvard University; the library of the Institut de France, Paris; the Johns Hopkins Library; the
Library of Congress; Maastricht Natural History Museum, the Neth-
erlands; the Joseph Regenstein Library, the University of Chicago; Senate House Library, University of London; University College Library,
University of London; and the Wellcome Institute Library, London.

Where necessary, I have obtained permission to quote from unpublished papers and correspondence.

This book incorporates portions of some previously published essays, which originally appeared in *Biology and Philosophy, British Journal for the History of Science, Harvard Library Bulletin, Journal of the History of Biology,* and the book *Scientific Inquiry and the Social Sciences,* ed. M. Brewer and B. Collins (Josey-Bass).

Many of my ideas have been tried out (and some have expired) under auspices of several organizations at the University of Chicago. The Fishbein Center for the History of Science and Medicine, the Chicago Group for the History of the Human Sciences, and the Committee on the Conceptual Foundations of Science created an intellectual environment that would test the mettle of the hardiest of ideas. These groups formed for me a kind of Tierra del Fuego of the mind.

I am fortunate that friends who read drafts of various chapters did not dull their critical faculties on my account. I acknowledge with pleasure the help of Mitchell Ash, Keith Baker, Len Berk, Richard Blackwell, John Cornell, Kurt Danziger, Allen Debus, Robert DiSalle, Alan Gewirth, Sophie Haroutunian, David Hollinger, Malcolm Kottler, William Kruskal, David Leary, Gerald Myers, Peter Novick, Philip Pauly, Eugene Taylor, Stephen Toulmin, William Wimsatt, William Woodward, and Robert Wozniak. Richard Burkhardt, David Hull, and George Stocking patiently read the entire manuscript; I am especially grateful for their effort. These scholars saved me from many mistakes of fact and argument. They gave their advice thoughtfully and generously, and the reader may wish I had more often heeded it. My wife Barbara encouraged and supported me in all the ways possible—*eis aiona tui sum.*

Darwin
AND
THE EMERGENCE OF
EVOLUTIONARY THEORIES
OF MIND AND BEHAVIOR

Thomas Henry Huxley, 1825–1895, photograph from ca. 1860.

Introduction

The scene is familiar and emblematic of the Darwinian revolution. Thomas Henry Huxley confronts Bishop Samuel Wilberforce at the Oxford meeting of the British Association in 1860, a few months after publication of the *Origin of Species*. Wilberforce, armored in righteousness and crammed in biology by Richard Owen, represents orthodoxy both in religion and science. Huxley, Darwin's partisan and an intellect of dazzling agility, stands for the new scientific order. Wilberforce spoke first. An auditor recalled much later, in 1898, that the Bishop intoned "for full half an hour with inimitable spirit, emptiness, and unfairness." The recollection remained warm for almost forty years:

> In a light, scoffing tone, florid and fluent, he assured us there was nothing in the idea of evolution; rock-pigeons were what rock-pigeons had always been. Then, turning to his antagonist with a smiling insolence, he begged to know, was it through his grandfather or his grandmother that he claimed his descent from a monkey?[1]

Wilberforce strode into disaster. Huxley slapped his knee and whispered to his companion, "The Lord hath delivered him into mine hands." "On this," the correspondent continued,

> Huxley slowly and deliberately arose. A slight tall figure, stern and pale, very quiet and very grave, he stood before us and spoke those tremendous words—words which no one seems sure of now, nor, I think, could remember just after they were spoken, for their meaning took away our breath, though it left

---

1. "Grandmother's Tale," *Macmillan's Magazine* 78 (1898): 425–35. Leonard Huxley relied on these recollections in recounting his father's debate with Wilberforce. See Leonard Huxley, *Life and Letters of Thomas H. Huxley* (New York: D. Appleton, 1900), 1:197–98.

us in no doubt as to what it was. He was not ashamed to have
a monkey for his ancestor; but he would be ashamed to be
connected with a man who used great gifts to obscure the
truth. No one doubted his meaning, and the effect was tremendous. One lady fainted and had to be carried out; I, for one,
jumped out of my seat.[2]

Huxley conquered, there is no doubt, and Darwinism remains triumphant in our own time. Our usual assessment of the revolution in
thought carried out by Darwin and other evolutionists of the period
has, however, much the same texture as that remembrance of the Oxford debate. We are not sure of what actually happened, but we do
know the meaning of the event. That meaning has become the received
view, and almost every historian has been seduced by the clear vantage
it offers on nineteenth-century science, as well as by its significance for
contemporary biology.

Components of the Received View

Historians taking the long perspective—frequently sighting from just
over theory in the last part of the twentieth century (particularly in
genetics and systematics)—have reached some consensus about the significant components in that pattern of thought ascribed to Darwin and
his followers. Darwinism, Mayr insists, introduced "population thinking" into biology and so banished essentialism—the idea that species
members instantiated immutable types.[3] Concomitantly, that intellectual movement undermined the teleological approach to nature, showing, Hull concludes, "it was of no relevance to science."[4] As the evolutionary constructions of chance and necessity arose, they pushed back
the sea of faith, which submissively retreated from the scientific shore.
"Deism verging toward agnosticism," according to Greene, was the trajectory of Darwinism.[5] Lewontin, Rose, and Kamin, and most other
scholars, agree that "Natural-selection theory and physiological reductionism were explosive and powerful enough statements of a research
program to occasion the replacement of one ideology—of God—by
another: a mechanical, materialist science."[6] An evolutionary process

2. "Grandmother's Tale," pp. 433–34.

3. Ernst Mayr, "The Nature of the Darwinian Revolution," *Evolution and the Diversity of Life* (Cambridge: Harvard University Press, 1976), p. 293.

4. David Hull, *Darwin and His Critics* (Cambridge: Harvard University Press, 1973), p. 66.

5. John Greene, "Darwinism as a World View," *Science, Ideology, and World View* (Berkeley: University of California Press, 1981), p. 151.

6. R. C. Lewontin, Steven Rose, and Leon Kamin, *Not in Our Genes: Biology, Ideology, and Human Nature* (New York: Pantheon, 1984), p. 51.

guided by purely material forces cannot guarantee progress. Neglect of this deep logical feature of Darwinism, in Mayr's estimation, is richly displayed in the work of men like Bergson and Teilhard de Chardin.[7] Cold-blooded Darwinism, by contrast, has come to regard nature as "morally meaningless."[8]

The historians who have thus described the Darwinian revolution as denying the biological reality of transindividual objects, as rejecting the operation of nonmaterial forces, and as destroying the doctrine of the purposeful development of life—these scholars have focused their analyses generally on the larger structural features of the theory. When they have cause to consider that one special part of Darwin's conception—the citadel itself, man—they often move quickly to make the reasonable inferences. The Darwinian image of man is that of a competitively isolated individual, a completely material creature, subject to its evolutionary history, to its consequent biological form, and to its immediate natural environment. Darwinian man has a brain that requires no guiding mind; reason that cannot transcend its animal origins; religious aspirations that have become barren in the sterile light of science; and morals that are subjective and Benthamitic.

Some historians and philosophers, while acknowledging this as the image of man forged by nineteenth-century evolutionary theory, are uneasy. Though Darwin mused that "he who understands baboon would do more toward metaphysics than Locke,"[9] these modern Darwinian scholars retain a nostalgia for Locke. They refuse to go the whole orang and instead insist, like Eiseley, that "man was not Darwin's best subject."[10] Himmelfarb, with less reserve, simply maintains that Darwin's efforts in biopsychology display his "failures of logic and crudities of imagination."[11] Even Darwin's enthusiasts defend him only obliquely. They often exculpate him by settling the human evolutionary debt on Spencer—urging that what goes under the rubric of social Darwinism ought really to be called "social Spencerianism."[12]

The received view has taken on a life of its own, irrespective of how adequately it represents what early evolutionists actually believed,

7. Mayr, "The Nature of the Darwinian Revolution," p. 292.

8. Susan Cannon, *Science in Culture* (New York: Science History Publications, 1978), p. 276.

9. Charles Darwin, *Notebook M*, MS p. 84, transcribed by Paul Barrett, in Howard Gruber, *Darwin on Man* (New York: Dutton, 1974), p. 281.

10. Loren Eiseley, *Darwin and the Mysterious Mr X* (New York: Dutton, 1979), p. 202.

11. Gertrude Himmelfarb, *Darwin and the Darwinian Revolution* (New York: Norton, 1968), p. 375.

12. Derek Freeman, "The Evolutionary Theories of Charles Darwin and Herbert Spencer," *Current Anthropology* 15 (1974): 211–21; Stephen Jay Gould, *Ever Since Darwin* (New York: Norton, 1977), pp. 36–38; Lewontin, Rose, and Kamin, *Not in Our Genes*, p. 242.

wrote, and accomplished. But the historian must continually test such errant creatures of collective memory by reconstructing the past and pressing them against its sharpened edges. I do not wish to deny that the received view incorporates aspects of scientific thought in the last century. I do reject, though, the presumption that it embodies the essence of nineteenth-century Darwinism. We get a very different perspective on this intellectual movement if we examine subjects more immediately relevant for assessing the cultural shift the new biological thought produced, namely those sets of problems and proffered solutions forming the matrix of emerging evolutionary theories of mind and behavior. I believe a historical analysis with this focus will show, first, that what is called "population thinking" owes its origins to deeper philosophical foundations, which made Darwin's own approach to species and to mind possible. Further, a redirected historical investigation reveals that endemic to evolutionary thought in the last century was the conviction that organisms were not simply passive products of their histories and immediate environments, but that they took an active part in their own transformation. Early evolutionists, such as Lamarck and Cabanis, through Darwin, Wallace, Spencer, and later Darwinians—all proposed, though in a variety of ways, that behavior and mind drove the evolutionary process. Many of these thinkers supported this scientific diagnosis with a carefully worked out metaphysics that stood completely opposed to mechanistic materialism. They philosophically dissected nature and found mind at its core. And with a sensitivity heightened by their novel examination, they detected the faint pulse of divinity yet animating the whole.

The most significant reorientation produced by a study fixing on emerging evolutionary theories of mind and behavior occurs in respect to morals. The received view holds that the Darwinian revolution eviscerated nature of moral purpose and ethically neutered the human animal. Selfishness bred by competitive selection, so the tale goes, has seeped into the very marrow of man's being, rendering all ethical behavior a pretext for individual advantage. Darwinism, according to Himmelfarb, "de-moralized man" in replacing "moral man by amoral nature."[13] But a closer historical and philosophical analysis demonstrates that Darwin, Spencer, and their disciples had a very different conception of the implications of evolutionary theory for man. They believed that their evolutionary constructions reanimated moral life, that the evolutionary process gave heart to unselfish, altruistic behavior.

13. Gertrude Himmelfarb, "Social Darwinism, Sociobiology, and the Two Cultures," in *Marriage and Morals among the Victorians* (New York: Knopf, 1986), p. 79.

Man, I hope to show, was indeed Darwin's best subject—and Spencer's as well. Their scientific and philosophic considerations were penetrating and sophisticated. And while a contemporary philosopher might reasonably fault some features of their formulations, he or she cannot justly impugn the logical foundations or the basic thrust of their theorizing. I believe nineteenth-century scientists, particularly Darwin, disclosed a way to establish an evolutionary ethics unimpeachable by today's standards. In the second appendix to this volume, I attempt to make the necessary alterations and defend an ethics based on evolutionary theory.

Scope and Justification of This Study

I did not originally intend to suggest revisions in the received view of Darwinian man. My interest was to explore those areas of evolutionary thought that joined science and philosophy. I wished particularly to understand how philosophical conceptions of mind, reason, and moral sense took on biological and psychological form within a scientific theory. No previous investigation has attempted to sound the several main currents of nineteenth-century evolutionary theories of mind and behavior—though some important studies have begun to recover aspects of that history; and virtually no effort has gone into restoring evolutionary theories of morals to their context. Almost every consideration of Darwinian or Spencerian ethics in the recent past has been curt and dismissive; historically contextualizing and philosophically impartial evaluations of these evolutionary moral theories are vanishingly few. Yet theories of mind and morals occupied most every major evolutionary thinker of the past century, and their reincarnation in the last quarter of ours rarely fails to transmogrify taciturn and reserved scientists into impassioned advocates or vitriolic critics. So the time for a thorough examination of evolutionary theories of instinct, reason, and morals appeared historically and philosophically ripe. What I take to be the needed revisions in the image of Darwinian man simply came as natural consequences of this original intention.

I chose instinct, reason (including intelligence, from which it was sometimes distinguished), and morals as the subjects to instantiate more precisely my concerns with mind and behavior. I had several reasons for doing this. First, and most importantly, these topics played dominant roles in the general evolutionary theories of the scientists whom I consider. Second, these subjects were conceptually and logically conjoined by the theorists. They regarded instinct as the paradigm of evolved behavior, the model for weaving other aspects of mind into their evolutionary schemes. Depending on the thinker, instinct was understood to have given rise to intelligence or, for many, to be the inherited outcome of "lapsed intelligence." Reason, as a more abstractive and generalizing faculty—sometimes assigned even to lower animals, sometimes reserved for man alone—formed the next step in phylogenetic development. And moral behavior came to be construed by Darwin, Spencer, and their disciples as a species of instinct brought under the guidance of reason. Instinct thus formed the evolutionary hinge linking the minds of lower animals with that of man. But I have focused on these three topics for an added reason. They require a historical analysis that more directly connects an individual's scientific work with the broader philosophical, religious, social, and psychological features of his intellectual environment. Recently historians and particularly sociologists of science have tried to drop evolutionary theory into these general contexts. But their adhesion to the generic aspects of evolutionary theory has usually produced only suggestive rather than persuasive contextualizations, typically of the sort that portrays natural selection theory as merely the biologized version of capitalistic economy. But focus on ideas about the evolution of reason and morals in particular allows the historian to specify more exactly and, I think, convincingly the causal connections of evolutionary thought with its intellectual and cultural surroundings. Finally, the kinds of links among these three subjects wrought by scientists in the last century simply have their own fascinations.

Reconstructions in history by their nature require boundaries. Even the conjectured ideal of a complete history of a period lets slip epistemological conundrums. The need for constraints, then, requires decisions about themes to be worked through, individuals to be considered, and depth and scope of analysis to be attempted. I have already mentioned the themes of this study—evolutionary theories of mind and behavior, especially as articulated by the concepts of instinct, reason, and morals. But in a history of science such as this, what thinkers should be chosen and what levels of analysis tried? Concerning this last decision, the historian might fabricate a broad survey, including as many relevant thinkers and their leading ideas as possible; alternatively he or she might stick to the development of only one or a few figures, mentioning others as seems important for illuminating the thought of the central characters. Broad surveys give an impression of a period but only superficial knowledge about the rise of and interaction among even the important theories. Surveys in intellectual history tend either to vulgar Platonism—by hypostatizing trains of disembodied and ill defined ideas—or to even more vulgar Marxism—by reducing vague ideas to vaguer yet social conditions. In contrast, studies of individual figures can provide intimate knowledge of the development of significant theory by situating ideas in their immediate causal contexts and allowing deeper logical analyses of their content—something especially important in the history of science. But scientists live in social and intellectual environments that alter the shape of their thought. They also come into conceptual legacies. To understand Darwin's theory of instinct, for example, demands consideration of the views he derived from Erasmus Darwin (his grandfather) and Lamarck. Likewise the theories of his contemporaries must be taken into account—certain natural theologians, whose work he studied, and Spencer, some of whose notions he adopted and others he opposed. Moreover, to understand Darwin's developing theory of instinct also requires, I am convinced, an appreciation of what it became in the hands of his disciples. Aristotle was right. To understand a historical entity, which a theory surely is, means that we must construe not only its origins, but its subsequent phases as well. To comprehend an idea or theory is also to discriminate those incipient structures that later become more manifest; but this can only be done retrogressively, by historically retracing the later florescence to the seminal elements. Consequently, though squeezing and concentrating all the intellectual juices from one major thinker can be profitable and satisfying, there are real advantages in treating several figures whose ideas intertwine. This history attempts to hit the just measure between broad survey and individual intellectual biography, though inclining toward the latter more than historians of science might think wise.

My plan was to select major biologists and psychologists in the nineteenth century for consideration. While these professional categories are flexible and permeable, especially for sifting through scientific theories of the last century, they at least have enough stability to justify excluding philosophers such as Pierce and (fortunately) Hegel, on the one hand, and anthropologists such as Lubbock and Tylor, on the other. More coherent stories can be told of evolutionary philosophers and anthropologists by historians of these disciplines;[14] and in any case the enduring contributions to evolutionary science have been made, I believe, by those who comfortably fit the two professional designations mentioned. I had intended to include, beyond England and America,

14. Philip Wiener's *Evolution and the Founders of Pragmatism* (Cambridge: Harvard University Press, 1949) is still the most useful account of the response of American philosophers to evolutionary theory. George Stocking splendidly reconstructs the history of nineteenth-century British evolutionary anthropology in *Victorian Anthropology* (New York: Free Press, 1987). See also Stocking's *Race, Culture, and Evolution*, 2d ed. (Chicago: University of Chicago Press, 1981).

biologists and psychologists of France and Germany as well. But I have only imperfectly realized that aim: pre-Darwinian evolutionary theory in France occupies the first chapter, and German developments are worked in along the way.[15] I found that remaining primarily with British and American thinkers permitted a detailed and historically tight narrative. A cooled ambition anyway will lessen the strain on the patience and eyes of readers.

The first chapter examines the origins of evolutionary biology of behavior, especially as it was fostered by the epistemology of Enlightenment sensationalists and achieved definition in the works of Erasmus Darwin, Pierre-Jean Cabanis, Jean-Baptiste de Lamarck, and Frédéric Cuvier. Chapter 2 describes the various roles of behavior in Darwin's developing evolutionary views and considers particularly his early theories of instinct, reason, and moral sense. Chapter 3 discusses the work of certain theologians who brought Darwin to understand how natural selection might operate on behavior, but who also suggested a conceptual problem that threatened to undermine his entire evolutionary theory. Darwin's several attempts to resolve this difficulty and his eventual success are detailed. In chapter 4, the ideas about human evolution advanced by Lyell, Huxley, Galton, Greg, and Wallace are assessed. Chapter 5 concentrates on Darwin's own mature ideas about the evolution of man, especially human rational and moral faculties. That chapter also examines the reactions to Darwin's conception of man and analyzes the connection of his moral theory with utilitarianism. The next two chapters take up Herbert Spencer's achievements in evolutionary theory, with special attention given to his proposals about the evolution of moral sentiment and to Huxley's rejection of his friend's notions. Chapter 8 explores the metaphysical and religious dimensions of Darwinism as they are played out in the intellectual development of George Romanes, Darwin's anointed, of St. George Mivart, Darwin's great antagonist, and of Conwy Lloyd Morgan, who injected the genetical ideas of Weismann into evolutionary biopsychology. The two most influential purveyors of Darwinism in psychology on the American scene, William James and James Mark Baldwin, are the subjects of chapters 9 and 10. The last chapter attempts to account for the decline of evolutionary theorizing about mind and behavior during the first half of this century and then to sketch its gradual resurgence since the 1960s. Finally, there are two appendixes. The first explains the historiographic

15. A preliminary study of the German phase in nineteenth-century evolutionary biopsychology can be found in Robert J. Richards, "Wundt's Early Theories of Unconscious Inference and Cognitive Evolution in their relation to Darwinian Biopsychology," in W. Bringmann and R. Tweney, eds., *Wundt Studies* (Toronto: Hogrefe, 1980).

model that I have employed in this history, and the second defends a conception of evolutionary ethics, the elements of which have been drawn from the historical body of this volume.

Let me add a few words about the much abused Herbert Spencer. In recent historical work, he is often neglected, sometimes purposefully: among the almost one thousand pages of his *Growth of Biological Thought*, which is devoted mostly to evolutionary history, Mayr reluctantly expends three paragraphs on Spencer. Other treatments, which allot proportionally more space, usually smother his ideas in invective. This historiographic practice makes it perfectly unintelligible why so many major evolutionary thinkers placed great value on his work, and why even those who reacted hostilely nonetheless felt compelled to confront his theories. A good deal of what came to be known in the later part of the century as Darwinism included equal parts of Spencer. Indeed, in some respects—which Spencer himself would be loath to acknowledge—the "philosopher of the doctrine of Development," as Bain called him, might also be considered a Darwinian, so common a vision did the two major evolutionists of the period share. I have tried to adjust the historiographic balance a bit by taking Spencer as seriously as his contemporaries did.

Internalistic and Externalistic Temper in History of Science

While many historians reject Collingwood's thesis that "All history is the history of thought,"[16] few historians or philosophers of science would blanch at the emendation: all history of science is the history of thought. Historians of science have as their special subject the most refined and, in a perfectly straightforward way, the most successful kind of thought—that codified in the theories, hypotheses, observations, communities, and institutions of science. History has no argument, but historians do. In arguing the course of scientific thought, historians, with some interesting recent exceptions, have usually taken one of two basic approaches—that of 'internalism' or of 'externalism.' These categories creak a bit with age and no longer easily capture the practices of historians of science, yet they still serve remarkably well. Internalists focus on the development of scientific ideas and theories, tracing their internal logic and conceptual linkages, on the one hand, and the degree of their evidentiary support, on the other. In the extreme, internalists treat the historical transition from one set of ideas to another much as Platonic philosophers, weaving together the logical forms of ideas and

16. R. G. Collingwood, *The Idea of History* (Oxford: Oxford University Press, 1956), p. 215.

evidence while ignoring their psychological and social embodiments. Externalists, by contrast, embed scientific ideas and theories in the human world, in the minds of scientists whose psychological interests and social concerns mediate logical implication and empirical confirmation. Extreme externalists cloak themselves in Freud or Marx (or less dangerously, stretch the point of the *Annales* school); they suppose that scientific ideas reflect only psychological complexes or social relationships, or perhaps, merely bob on the surface of the *longue durée*.

Mayr's *Growth of Biological Thought,* for instance, displays the internalist's approach. Mayr concentrates "on the history of scientific problems and concepts," recommending that the *Dictionary of Scientific Biography* and Nordenskiöld's *History of Biology* be consulted for the "biographical and sociological aspects of the history of biology."[17] He executes his history with considerable historical and philosophical skill, and he brings a kind of sensitivity to the history of biology that perhaps only a great scientist who has made major contributions to his field can. Yet as an internalist, he forgets that ideas alone are causally impotent—that one idea cannot beget another. He fixes on "the insights and concepts of biology," while prescinding from the fact that it is human biologists who produce those insights and concepts. But ideas become historically linked only by passing through minds trapped in flesh, human minds which respond to logical implications, evidentiary support, and the legacy of scientific problems and concepts, though also to high emotion, religious feeling, class attitudes, and even perhaps Oedipal anxieties. To deal adequately with their subject—the growth of scientific ideas—historians of science cannot neglect the explanatory strategies of social, political, and cultural historians. And they must take special heed, I believe, of the observation of William James that serves as the admonitory epigraph for this history.

Orthodox internalists at times manifest externalist tendencies. This has been especially true of historians of evolutionary theory. Nordenskiöld, for instance, spun precisely in this direction when, in the early part of our century, he had to explain the currency of a conception that he knew had no logical or scientific merit:

> From the beginning Darwin's theory was an obvious ally to liberalism; it was at once a means of elevating the doctrine of free competition, which had been one of the most vital cornerstones of the movement of progress, to the rank of a natural

17. Ernst Mayr, *The Growth of Biological Thought* (Cambridge: Harvard University Press, 1982), pp. 6–7.

law, and similarly the leading principle of liberalism, progress, was confirmed by the new theory—the deeper down the origin of human culture was placed, the higher were the hopes that could be entertained for its future possibilities. It was no wonder, then, that the liberal-minded were enthusiastic; Darwinism must be true, nothing else was possible.[18]

Internalist historians often appeal to external causes when earlier scientific ideas seem logically incompatible, not so much with the original scientific matrix as with the historian's own scientific predilections—the "real" science of laboratory genetics in Nordenskiöld's case. It would be foolish to deny, though, that definable features of earlier science arose from definite social or psychological determinants. The mistake to avoid is the general presumption that scientific and logical factors can always be neatly separated from other cultural notions. Extreme externalists do not err in this way. But in their constructions one can hear the ticking message of the intellectual anarchist.

Extreme externalists, say of the Marxian or Durkheimian variety, explain scientific conception as they would any cultural artifact: they interpret it as totally determined by social and economic substructures. The more influential lately have been plying their trade out of Edinburgh. Bloor, for example, attempts to demonstrate that the apparently most objective, nonrelative, and certain of the sciences, mathematics, can be understood as a reflection of social practices that are conventional, varying, and epistemologically secure only in the sense that the community generally approves of its techniques. A culture radically different in relevant respects would, he argues, institute a valid mathematics, but one deviating from our own.[19] While extreme externalists have forced important lessons on historians of science, their "strong program" can, I believe, be terminally infected by a simple tu quoque argument: their thesis of social determinism must also be determined; but why should we listen to those whose views only reflect their dour Scottish culture? Simpler still: if these historians reserve "the word 'knowledge' for what is collectively endorsed, leaving the individual and idiosyncratic to count as mere belief,"[20] certainly we ought to regard their idosyncratic thesis as mere belief.

18. Erik Nordenskiold, *The History of Biology,* 2d ed. (New York: Tudor, [1920–1924] 1936), p. 477.

19. David Bloor, *Knowledge and Social Imagery* (London: Routledge & Kegan Paul, 1976), chaps. 5 and 6.

20. Ibid., p. 3. The strong program to explain scientific ideas by social forces is defended by Bloor's colleagues: Barry Barnes, "On the Conventional Character of Knowl-

Extreme internalists and extreme externalists stand by definition at the poles; the middle lattitudes support a larger population, among whom are some of the surest hands in the Darwin industry. These historians during the last decade and a half have mined the store of the Darwin papers at Cambridge to produce a richly detailed picture of the origins and early development of Darwin's ideas. The studies of many of the best of these have recently appeared in the volume *The Darwinian Heritage*.[21] My own debt to the real insights they provide will be made clear in the early chapters of this volume. What especially characterizes their meticulous efforts at reconstruction is the hybrid sentiment they express about internal and external factors. They recognize, for instance, that Darwin's ideas about natural law contain what we might for convenience label logical, scientific, philosophic, social, and religious elements, but which he himself regarded as the deposit of coherent scientific principle. Yet this capacious attitude about the complexion of science in the nineteenth century, as expressed in the work of these sensitive historians, does require more careful formalization and justification. This it can attain only when a satisfactory historiographic model has been constructed and measured against models venting other sentiments.

Models in the History of Science

Historians disposed toward internalism or externalism, or toward the hybrid view, specify their tendency by adopting—usually unreflectively—a historiographic model in light of which they articulate their subject. In this respect they function much like scientists. For historians, after all, do formulate theories, construct hypotheses, gather evidence, and, of necessity, employ models. Historiographic models comprise sets of assumptions concerning the nature of science, its developmental character, and the modes of scientific knowing. That historians must use models can be argued a priori: without antecedent conceptions about the character of science, they would have no clue about where to look for their subject matter, nor could they define its limits or determine what evidence would be relevant. That models have in fact been used can be established by an empirical survey of histories of science since the Renaissance. So, for instance, a model familiar to

———
edge and Cognition," *Philosophy of the Social Sciences* 11 (1981): 303–33; Andrew Pickering, *Constructing Quarks: A Sociological History of Particle Physics* (Chicago: University of Chicago Press, 1984).

21. David Kohn, ed., *The Darwinian Heritage* (Princeton: Princeton University Press, 1985).

most is Kuhn's paradigm model (what might also be called a Gestalt model) of science. Gillispie, in his *Edge of Objectivity,* more traditionally employs a revolutionary model (not to be confused with Kuhn's conception of scientific revolutions).[22] This model, introduced by historians in the eighteenth century, assumes that a discipline must undergo a fundamental upheaval to put it on the road to modern science—before the revolution there was not science; afterward scientists gradually laid a path of truth leading right up to the modern age. This model has been used to forestall external stimulants to the heart of science. Gillispie, for instance, refused Lamarck the title of scientist because his ideas derived from romantic ideology, while he awarded Darwin the crown as the Newton of biology, since the Englishman had introduced a quantificational spirit into the discipline, transforming it into authentic science.[23]

The several other models to be found in the writings of historians of science likewise express internalistic or externalistic attitudes. I describe the more prominent of these models in the first appendix to this volume. These descriptions furnish comparative standards for justifying the model that guides my own historical account of evolutionary theories of mind and behavior. The model I favor incorporates, of course, the resourceful features of the other models, while avoiding, of course, their misdirections. It is a model, I believe, that captures the practice of sensitive historians, and like other such heuristic devices, it offers recommendations for the further reflective conduct of the historian's craft. The first appendix also details this model and attempts to demonstrate its advantages. Here I will simply broadly sketch its features.

The Natural Selection Model

I have adopted what might appear initially as an implausible model of scientific development; it requires a moment of willing suspension of disbelief. I regard scientific conceptual systems as analogous to evolving biological species and the mechanisms of conceptual change to be formally similar to those available to the neo-Darwinian biologist.[24] The gene pool constituting such a specieslike entity consists of the ideas or concepts that form the genomic individuals—for example, the ideas making up Spencer's theory of the moral sense. The genetic elements of

22. Charles Coulston Gillispie, *The Edge of Objectivity* (Princeton: Princeton University Press, 1959).

23. Ibid., pp. 276 and 339.

24. David Hull has been working toward a similar historiographic conception. See his "Darwinism as a Historical Entity: A Historiographic Proposal," in Kohn, *Darwinian Heritage,* pp. 773–812.

conceptual systems are united by logical bonds of inclusion, implication, and coherence, as well as by the historical ties of a common evolutionary history. In biology the genome gives rise to different phenotypic expressions, depending on the environment; so too in the case of scientific thought. The individually identical system of ideas, those constituting James's theory of mind, for instance, will be differently expressed in his textbook, in his lectures, and in his conversation. The genotype-phenotype distinction in conceptual evolution permits the historian, therefore, to ascribe the same theoretic system, *individually* the same, to a given scientist, who may express it differently in different circumstances. Or the historian may ascribe the same theoretic system, *specifically* the same, to a group of scientists (e.g., the Aristotelians, the creationists, the Darwinians), though their individual descriptions of the system may vary more significantly.

The conceptual systems entertained by a scientist come to inhabit three distinguishable but continuous environments: his own mind, the scientific community, and the general culture. First, as the scientist begins to formulate his theory, selective pressures will be exerted by the private environment of his immediate conceptual concerns. This first intellectual niche may exhibit pressing problems from the scientific literature as well as from particular religious and philosophic ideas, psychological needs, or social interests. The selective fitness of the system will be a function of its heterogeneous terrain. A new theoretical formulation consequently will either survive in its intellectual niche because it is compatible with and supported by other resident conceptual systems, or if it is incompatible with the more firmly entrenched systems, it will be rejected by the scientist and perish. If the system of thought remains robust, it might then migrate into the wider and more challenging environment constituted by the mind of the scientific community. Finally, a conceptual system, such as Darwin's own evolutionary theory, might eventually become a vital element in the larger culture, actually changing the ecological relationships and perhaps causing the extinction or greatly reducing the vitality of other sorts of systematic thought.

This natural selection model of the evolution of conceptual systems raises to intelligibility, I believe, certain hard configurations in the history of science that other models attempt to grind away. It suggests, for instance, that science advances neither by reason of gradual increments of new truth, implied by the growth model of science (an internalistic model), nor by saltations of incommensurable paradigms, as required by Kuhn's Gestalt model. Rather the natural selection model indicates that the elements of scientific evolution—its ideas—will be discrete but genetically related to prior conceptual states; that the usual sources of variability will be the recombination of ideas rather than novel mutations; and that in general, conceptual systems change more slowly in some climates, more rapidly in others, but never in profoundly discontinuous fashion. The model also demands that the historian attend not only to the logic of scientific theory and the evidence supporting it but also to the psychological, political, economic, and religious factors determining its shape. The model thus forbids the historian to assume a priori that any one feature of the ecology will necessarily dominate in determining the evolution of a particular system of ideas; it rather encourages the historian to assess differentially the pressures exerted by the variety of factors constituting the interlocking niches in which conceptual systems come to dwell. Growth and revolutionary models exclude 'extrascientific' causes of scientific thought, while Gestalt and psychosocial models, tending toward extreme externalism, refuse all else.

Narrative Explanation in the History of Science

This natural selection model of conceptual evolution aids in the formation of a narrative structure that preserves the power found in the best examples of historical explanation. This is no accident, for evolutionary biology is itself a historical science whose typical explanatory form is the narrative. Evolutionary biologists explain the development, adaptational features, or extinction of a species by telling a story, based on paleontological evidence and neo-Darwinian theory, of the circumstances in light of which we can understand such phenomena. In biology the circumstances include the presumed genetic heritage of the animal group, various entrenched morphological structures, and the environmentally produced selective pressures. Similarly the historian of science, guided by a natural selection model, will account for the fate of conceptual systems and explain their evolution by appeal to the legacy of past developments in science, to established organizational structures, and to the shaping forces of the variegated mental environments which these systems inhabit.

Narratives in evolutionary biology and evolutionary history exhibit the sense of an ending. For in formulating the narrative, the researcher must begin with the terminus ad quem. The biologist who wishes to account for the rise of the mammals, for instance, already knows about the characteristics of contemporary mammals and therefore what needs to be explained. In the reconstruction the biologist may well discover new things about mammals of the past, may locate those peculiar transitional forms of the mammal-like reptiles, and as a result may conceive mammals of the present differently. But the researcher structures the story so that carefully isolated antecedent events make the subsequent events intelligible, so that they all contribute to the necessary conditions for explaining the phenomena of concern. Without this teleological structuring, narrative would degenerate into chronicle. There would be no sense of the connectedness of events, no way of weighting the various contingencies that produced explananda; explanatory force would be dissipated.

I have attempted to structure the narrative that follows with a comparable sense of ending, without, I hope, reading back the triumphs and failures of the future into the past. Historians of science have been rightly wary of the dangers of presentism, which has infected a good deal of earlier writing in history of science. The retrospective abuse heaped on Herbert Spencer should stand as a warning against this historiographic sin. But virtue does not lie in a return to Eden. Some historians, attracted by the promise of new ethnographic methods in sociology of science, argue that the processes of knowledge acquisition can best be understood by going epistemologically native, by the historian divesting himself or herself of knowledge about what lies in the future of the actors in the narrative. But if the historian attempts to construct a narrative strictly from the point of view of the historical subjects, to make the tale "rigorously nonretrospective," in Rudwick's terms[25]—this confuses the protagonists' awareness with the historian's. The plotting of a narrative demands selecting events, forming descriptions, and setting the scenes, none of which can be intelligently done without anticipating the next stages in the story. Darwin, of course, could not describe his own work on species in 1837 as "pre-Malthusian," but the historian can—and quite properly does. Indeed, without the sense of an ending, the historian would not even have reason to mention Darwin's reflections on species, rather, than, say, his worries about his railroad stocks. In the narrative, the actors' thoughts and activities occur within the plotted frames, which carry the story toward its conclusion; the actors themselves should remain in the dark about the future. The historian, not the characters, portrays the sense of an ending.

Because narratives exhibit a sense of an ending, they also incorporate either implicitly or explicitly value judgments about their subjects. The historian of evolutionary theory needs to sift out what was important—what prior experiences or beliefs had cognitive value, for

25. Martin Rudwick, *The Great Devonian Controversy* (Chicago: University of Chicago Press, 1985). I believe Rudwick's crafty practice delivers a narrative more solid and rich than his theoretical principles would ever permit.

example, in Wallace's argument (after the late 1860s) that natural selection failed to operate in the case of man. The historian, however, must give account not only of what Wallace believed was important for his argument, the way he estimated it, but what actually was important—insofar as the evidence supports such judgments. This latter sort of historical evaluation constitutes a deep structure of narrative explanation and should not be thought another example of Whiggism or presentism. Even to describe Wallace as "arguing," as opposed to "saying," "believing," or "naively supposing," implies that the historian has tacitly invoked certain standards or norms (in this case logical ones), that is, certain values, to render the description. The danger to avoid, of course, is the application of evaluational standards that make a scientist either too wise or too benighted before his time. To condemn Wallace for foolish credulity because he believed in the spirit world, while praising Darwin since he thought spiritualism codswallop, again confuses the historian's awareness with that of the actors. There may be reason for judging Wallace credulous, but it should be done on nineteenth-century grounds, not on ours. This danger should not, however, inhibit reflective evaluations in history of science. The historian has privileged knowledge denied the actors, sometimes even knowledge about motives the actors never consciously admit; narrative explanations must take advantage of that knowledge to give the best, that is, the truest account of the past.

In what follows, I have employed considerations drawn from a natural selection model, without, I hope, allowing its framework to poke through the fabric of the narrative. Insofar as the model pretends to formalize sound historiographic practice, its stranger devices properly should not obtrude into the story. Some readers will likely feel that, as in the case of Ptolemaic models in astronomy, the course of events can be mapped onto a great many implausible formal models, and that it would have been better had I rendered my account unvarnished, without glossing it in the higher historiography. Even after considering the defense in the first appendix, they may remain unconvinced. I do not think this demur should seriously bias their judgment of the soundness of the history offered in the next eleven chapters. The deficiencies of the narrative will likely have more prosaic causes. But to the degree that the tale appears to capture the actual texture of the past and unravel its knotty complexes, to that degree some empirical confirmation will have to be conceded the model which guided it.

1

Origins of Evolutionary
Biology of Behavior

The sciences of ethology and sociobiology have as premises that certain dispositions and behavioral patterns have evolved with species and that the acts of individual animals and men must therefore be viewed in light of innate determinants. These ideas are much older than the now burgeoning disciplines of the biology of behavior. Their elements were fused in the early constructions of evolutionary theory, and they became integral parts of the developing conception of species transformation.[1] Historians, however, have usually neglected close examination of the role behavior has played in the rise of evolutionary thought.[2]

Yet behavior has been an important consideration virtually from the beginnings of systematic biological theorizing. Aristotle devoted generous portions of his *Historia animalium* to discussion of species-typical habits and relative grades of animal intelligence. Galen conducted a set of elegant experiments to show that certain actions of animals were

1. Though the terms "evolutionary" and "evolution" are most closely associated with the views of Charles Darwin, I have used them to refer generally to theories proposing the modification of species over generations. The words "transformism," "transmutation," and the like are meant also to convey the same meaning. The use of "evolution" to describe early nineteenth-century theories of species change might seem anachronistic, but it is not. Charles Lyell used it in referring to Lamarck's arguments "in favour of the fancied evolution of one species out of another." See Lyell's *Principles of Geology* (London: Murray, 1830–1833), 2:60.

2. This is less the case for Darwin's early views about evolution. See, for example, Sandra Herbert, "The Place of Man in the Development of Darwin's Theory of Transmutation," *Journal of the History of Biology* 10 (1977): 155–227; Howard Gruber, *Darwin on Man* (New York: Dutton, 1974); Edward Manier, *The Young Darwin and His Cultural Circle* (Dordrecht: D. Reidel, 1978); Charles Swisher, "Charles Darwin on the Origins of Behavior," *Bulletin of the History of Medicine* 41 (1967): 24–43; Richard Burkhardt, "Darwin on Animal Behavior and Evolution," in *The Darwinian Heritage,* ed. David Kohn (Princeton: Princeton University Press, 1985), pp. 327–65.

innate and not learned.[3] In the seventeenth and eighteenth centuries, naturalists often disputed violently over interpretations of animal behavior, contending whether the activities of brutes were to be regarded as congenitally fixed or as the consequences of reasoned choice. These debates formed the immediate environment for the emergence of evolutionary theories at the turn of the eighteenth century. In this chapter, I wish to focus on the problems that animal instinct and intelligence posed for early evolutionary theorists. A chief difficulty stemmed from their commitment to the doctrines of sensationalism. Adherents of this persuasion in the seventeenth and eighteenth centuries generally argued that all ideas merely imaged sensations and that rational behavior, of which even animals were capable, derived from habitually associated ideas. Sensationalists thus tended to deny the existence of innate and mechanically expressed instincts, the most likely candidates for evolutionary treatment. My intention here is to explain how evolutionists accommodated to sensationalist doctrine the conviction that behavior did evolve. For this purpose, I will examine the roles of instinct, intelligence, reason, and particularly the mediating construct of habit in the theories of four early evolutionists: Erasmus Darwin (1731–1802), Pierre-Jean Cabanis (1757–1808), Jean-Baptiste de Lamarck (1744–1829), and Frédéric Cuvier (1773–1838). Since I hold scientific theories to be analogous to evolving biological species, I will emphasize the conceptual legacies inherited by these early thinkers, the intellectual and cultural environments that shaped their ideas, and the competitive reactions they evoked from rivals.

Let me briefly indicate some of the conclusions toward which this chapter arches. The history examined reveals, I believe, that evolutionary ideas developed in response, not only to narrowly conceived problems in zoology, but also to critical difficulties in the epistemological, psychological, and social-political doctrines of sensationalism. More specifically, it makes clear the central importance of conceptions of

3. In *De locis affectus* (vol. 8 of *Opera omnia,* ed. C. Kuhn [Lipsiae: in Libraria Cnoblochii, 1824]), Galen reports that he and his associates reared in isolation a young goat, which was taken by cesarean section "so that it would never see the one who bore it." Just after removal from its mother's womb, the kid was placed in a room in which there were several bowls with different nutriments—wine, oil, honey, milk, grains, and fruits. He describes their observations: "We observed that kid take its first steps as if it were hearing that it had legs; then, it shook off the moisture from its mother; the third thing it did was to scratch its side with its foot; next we saw it sniff each of the bowls in the room, and then from among all of these, it smelled the milk and lapped it up. And with this everyone gave a yell, seeing realized what Hippocrates had said: 'The natures of animals are untutored.' "

habit, instinct, intelligence, and reason for the first formulations of evolutionary principles. Finally, it shows that behavior was originally regarded, not merely as an outcome of the evolutionary process, but also as its principal agent.

Controversies over Animal Instinct and Intelligence in the Seventeenth and Eighteenth Centuries[4]

Aristotelians, Cartesians, and Sensationalists

During the seventeenth and eighteenth centuries, disputes over the nature and capacities of human mind were frequently waged on foreign territory—in the field of animal psychology. Fresh evidence from natural history, whose practitioners increased in number during the period, was brought to bear on metaphysical and epistemological issues. This new evidence, however, did not so much test ideas in contention as open the battle on another front. The disputants grouped themselves into three camps, within which, of course, factional differences often arose. The Aristotelians distinguished the human soul, with its rational abilities, from the animal soul, which could be guided only by sensory cognition. The Cartesians also separated man from animals, though more decisively. For Descartes and his disciples, animals mimicked intelligent action, but operated as mere machines: brutes consisted of extended matter alone and functioned according to the laws of physics. Finally, the sensationalists (who adopted the basic tenets of Locke's epistemology) held that human knowing drew exclusively on the same resources available to animals—sensations. The epistemology of sensationalism seemed to be confirmed by the successes of experimental methods in the various sciences and technologies; but toward the end of the eighteenth century, careful observations of animal behavior began to undermine the assumptions of sensationalist epistemology. Naturalists committed to sensationalism thus faced a critical problem, which had implications for their conception not only of animal psychology, but of human psychology as well. The disputes over animal abilities and the dilemma confronted by sensationalists centered on the problems of brute instinct and intelligence.

Aristotelians and Cartesians differed profoundly on the ultimate principles of animal psychology. They nonetheless agreed that complex animal behavior (e.g., birds' building their nests and bees' their cells)

---

4. I have discussed the pre-nineteenth-century debates over animal instinct and intelligence in more detail in "Influence of Sensationalist Tradition on Early Theories of the Evolution of Behavior," *Journal of the History of Ideas* 40 (1979): 85–105.

should be explained by appeal to instincts, which they understood as blind, innate urges instilled by the Creator for the welfare of his creatures.[5] Pierre Gassendi (1592–1655) forcefully opposed this interpretation of animal behavior. In his *Syntagma philosophicum* (posthumous, 1658), he undertook a comparative study of animal and human cognitive abilities and discovered they were logically similar: both human and animal souls operated on sensory images to yield reasoned actions.[6] Marin Cureau de La Chambre (1594–1669), an associate of Gassendi, concurred in his friend's conclusions;[7] and through the next century

5. From the resources of Aristotle's *De anima*, Avicenna developed a theory of instinct in *Kitāb al shifā*, the *Sufficientiae* of the medieval translation. The distinctive and skillful behaviors of different species evinced to him that the estimative faculty, that internal sense which detected *intentiones* not available in the immediate data of the external sense, was infused with a divine "inspiration" (*ilhām*, rendered by the Latin translator variously as *cautela naturalis* and *instinctus insitus*). See S. Van Riet's critically edited *Avicenna Latinus, Liber de anima,* 4.3 and 5 1 (Leiden: Brill, 1968–1972), 2:37, 73. This elaboration of Aristotelian psychology was adopted by Thomas Aquinas in *Summa theologica,* I, Quest 78, art. 4, resp., *Opera omnia* (Romae: Ex Typographia Polyglotta, 1891), 6:99; Francis Suarez in *De anima,* 3.30, n. 7, *Opera omnia* (Paris: Vives, 1856–1878), 3:705; the Jesuit Fathers at Coimbra in *In octo libros Physicorum Aristotelis Stagirita,* prima pars, 2.9, quest. 3 and quest. 4 (Coloniae: Zetznerius, 1602), cols. 420–29; Pierre Chanet in *Considerations sur la sagesse de Charron* (Paris: Le Groult, 1643); Gaston Pardies in part 2 of *Discours de la connoissance des bestes* (Paris: Delaulne, 1672). These are only the more prominent among the many thinkers who contributed to and preserved the Aristotelian interpretation of animal instinct. In the eighteenth and nineteenth centuries, the Aristotelian tradition was kept alive in the works of Hermann Samuel Reimarus, *Allgemeine Betrachtungen uber die Triebe der Thiere,* 3d ed. (Hamburg: Bohn, [1760] 1773); Erich Wasmann, *Instinct und Intelligenz im Thierreich* (Freiburg im Breisgau: Herder'sche Verlagshandlung, 1897).

Descartes discussed the theory of animal behavior in his letter (1646) to the Marquess of Newcastle, *Oeuvres de Descartes,* ed. C. Adam and P. Tannery (Paris: Cerf, 1897–1913), 4:573–75, and variously in other of his works. Descartes was followed by Antoine Dilly in *Traitté de l'ame et de la connoissance des bêtes* (Amsterdam: Gallet, 1676); F. B., an anonymous Englishman, in *A Letter Concerning the Soul and Knowledge of Brutes; Wherein is shewn They are Void of One, and Incapable of the Other* (London: Roberts, 1721); and a host of other disciples. Of particular interest is Thomas Willis's intelligent treatment in *De anima brutorum quae hominis vitalis ac sensitiva est* (1672), *Opera omnia,* ed. Geradus Blasius (Amsterdam: Westen, 1682); see especially chaps. 6 and 7 Lenora Rosenfield traces the development of the Cartesian conception of animals in *From Beast-Machine to Man-Machine* (New York: Octagon Books, 1968). See also my "Influence of Sensationalist Tradition."

6. Pierre Gassendi, "De functionibus phantasiae," *Physicae,* 3.2.8.4, in *Syntagma Philosophicum, Opera Omnia* (Lugduni: Anisson and Devenet, 1658), 2:409–14.

7. In "Quelle est la connoissance des bestes," an addition to his *Les characters des passions,* 2d ed. (Amsterdam: Michel, [1640] 1685), Cureau de La Chambre agreed with Gassendi that just as the human understanding composes and divides, so in the beast "the imagination does nothing else but unite and separate images of objects which the senses furnish in order to judge what is good or bad for the animal" (p. 544).

French sensationalists continued to be chary of the use of instinct in the account of animal behavior. The attitude of Jean-Antoine Guer (1713– 1764), a historian of animal psychology writing in mid-century, is representative. Guer believed the ascription of instinct to animals confounded any real attempt at scientific explanation, for "nothing is easier to say about whatever animals do than they do it from instinct."[8] Rather than attempt to uncover the reasons for animal activities, the Aristotelians, according to Guer, invoked the myth of substantial forms (in the guise of brute souls), which were to serve as repositories for instincts that supposedly predetermined behavior. Guer held Descartes and his disciples in no higher regard. The Cartesians refused beasts even the low-level, sensitive cognition granted by Aristotelians. Instead, they presumed all animal behavior to tick off like a clock. Guer delighted in this absurdity of Cartesian animal psychology: "Take your dog. Let us wind up that clock and set it, say, for six o'clock; that is, let us suppose a certain disposition in the organs of the animal, a certain arrangement, a certain sort of heat in its heart and stomach. Behold, the clock runs!"[9] The Cartesian beast machine, by sensationalists' lights, could be neither living beast nor preset machine. It was not truly an animal, since animals obviously did perceive, feel, and act intelligently, nor a machine, since Cartesian matter lacked any active principles that could explain these qualities.

Sensationalists easily exploited the dilemma of the Cartesians, who wished to explain behavior on simple, natural principles and yet to capture the complexities it revealed. The Abbé de Condillac (1715–1780), for instance, found an exemplar of this Cartesian problem in the theory of animal automatism formulated by the great natural historian Georges Leclerc, Comte de Buffon (1707–1788) in his *Histoire naturelle* (1749– 1804). Condillac pointed out that Buffon's insistence that animals were unthinking, instinctive machines clashed with his attribution to them of perception and feelings of pleasure and pain. In Condillac's sensationalist epistemology, such predications implied that a creature could think and make rational determinations, which were merely the results of complex associations of sensory images.[10]

8. Jean-Antoine Guer, *Histoire critique de l'ame des bêtes* (Amsterdam: Changuion, 1749), 2.191–92.

9. Ibid., pp. 242–43.

10. Georges Louis Leclerc, Comte de Buffon was convinced that uniformity in the behavior of animals provided strong "proof that their actions are only mechanical and purely material responses." See "De la nature de l'homme" (1749), *Histoire naturelle*, in *Oeuvres complètes de Buffon*, ed. Pierre Flourens (Paris: Garnier, 1853–1855), 2:7. Etienne-Bonnot de Condillac delighted in pointing out the liabilities of mechanistic interpretations of animal behavior. See his *Traité des animaux* (1755), 1.2, *Oeuvres complètes de Condillac* (Paris. Houel, 1798), 3:458–59.

The sensationalists resolved the Cartesian dilemma by reformulating both physical and psychological theory. First, they admitted that animals—and men—were machines, though not composed of inert matter. Thus Julien Offray de La Mettrie's (1709–1751) *L'homme machine* (1748) did not merely extend a Cartesian mechanistic analysis to the human mind, but reconstructed the very idea of matter. According to La Mettrie, matter harbored active properties of motion and sensation, which were expressed when it became organized in living beings. This new conception of matter allowed La Mettrie and other sensationalists to refer intricate and complex behaviors to a medium plastic enough to produce them, but these thinkers could still maintain the ideal of simple, natural principles of explanation.[11]

The sensationalists also introduced important epistemological and psychological reformations to the account of animal behavior. They argued that ideas were only copies of impressions received by sensory machines. Rational intelligence, they claimed, was not the product of an immaterial mind but of refined habit and complex processes of sensory association. Animals, then, might entertain ideas, which were more or less detailed representations of their environments. Through memory and imaginative associations, their behavior could thus be guided by *reasonable* considerations. Condillac insisted that a careful examination of animal activities would discover, contrary to Cartesian opinion, that supposedly blind instincts were really intelligently acquired habits.[12] Therefore we should not, his expositor Le Roy declared in the *Encyclopédie,* use "instinct" to refer to animal behavior "except that that word becomes synonymous with 'intelligence.'"[13]

Though Charles-Georges Le Roy (1723–1789) affirmed Condillac's sensationalist interpretation of animal action in the *Encyclopédie,* he remained a bit more conservative in his own diagnoses of mammalian

11. For further discussion of the sensationalists' analysis of matter, see Aram Vartanian, "Trembley's Polyp, La Mettrie, and Eighteenth-Century French Materialism," *Journal of the History of Ideas* 11 (1950): 259–86; and my "Influence of Sensationalist Tradition." In his "From *Homme Machine* to *Homme Sensible,*" *Journal of the History of Ideas* 39 (1978): 45–60, Sergio Maravia assumes that physiologists in the seventeenth and eighteenth centuries can fairly easily be distinguished into mechanists, like Descartes and Boerhaave, who regarded organic bodies as composed of inert and statically related parts, and vitalists, like Bordeu, who endowed organisms with extrinsic, vital forces. Maravia fails to emphasize that sensationalists, such as La Mettrie and Condillac, took a middle road, granting nonliving matter intrinsic, active powers which would express themselves when properly organized. This latter conception measurably influenced late-eighteenth-century theories of *homme sensible.*

12. Condillac, *Traité des animaux,* 1.3, p. 534.

13. [Charles-Georges Le Roy], "Instinct," *Encyclopédie ou dictionnaire raisonné des sciences, des arts et des métiers,* ed. Denis Diderot (Paris: Faulche, 1751–1765), 8:796.

behavior. In his *Lettres sur les animaux* (1768), a work later admired by Charles Darwin, Le Roy traced the development of intelligence in young wolves, foxes, and deer against the background of their social and natural circumstances. In his analyses, he preserved the notion of instinct to refer to basic, physiologically determined desires—the need for certain foods, shelter, and acceptable climate. But he refused instinct any role in directing behavior designed to satisfy those needs. He rather believed this was accomplished by sensory experience and the applications of wakening intelligence.[14]

To argue successfully that individual intelligence shaped animal behavior, the sensationalists had to deny that innate images, which many instinct theorists postulated,[15] played any role in guiding actions. Connate images, though, were not so much argued against as simply dismissed as repugnant to the accepted Lockean conviction that all ideas ultimately derived from sense experience. This same epistemological tenet also weakened the support for a common feature of traditional theories of instinct, the assumption that instinctive activities were rigidly uniform in a particular species. For instance, René-Antoine de Réaumur (1683–1757), who detected the stirrings of intelligence even among insects, protested the Cartesians' presumption of machinelike, predetermined fixity in the conduct of animals.[16] His objection conformed to the sensationalists' insistence that ideas (including those directing behavior) were not universals but particulars, fainter copies of sensations. Particular ideas could well account for variability in animal activity. And uniformity of action displayed by members of the same species could be attributed, according to Condillac and Le Roy, to the community of fundamental needs and the similarity of environments in

14. Charles-Georges Le Roy, *Lettres sur les animaux,* new ed. (Nuremberg: Saugrain, [1768] 1781), pp. 68–69. Though Le Roy deprecated Cartesian notions of instinct, he nevertheless suggested that habits might, after cultivation for several generations, become hereditary, so that the constitution of animals might be continually reformed and perfected. I owe a debt to Marc Swetlitz for bringing this feature of Le Roy's analysis to my attention.

15. Even Descartes referred to "images" and "ideas" of animal corporeal imagination. See, for example, Descartes, *L'homme,* in *Oeuvres de Descartes* 11:177. Thomas Willis described the cerebral dispositions determining animal instinct as "innate notions" (*notitia ingenita*) in his *De anima brutorum,* p. 32. Hermann Samuel Reimarus spoke of the animal having an inborn "idea or image" (*eine Idee order ein Denkbild*) to guide instinctive behavior. See Reimarus, *Abhandlungen von den vornehmsten Wahrheiten der natürlichen Religion,* 5th ed. (Tübingen: Frank und Schramm, [1754] 1782) p. 405.

16. René-Antoine de Réaumur, *Mémoires pour servir à l'histoire des insectes* (Paris: L'Imprimerie Royale, 1734–1742), 1:22–23.

which the young were reared. There was no cause to postulate of animals innate, universal ideas to explain their behavior.[17]

Social Implications of Sensationalism

An evolutionary model of scientific change recommends that the historian assess the complex environments against which scientific ideas were selected. The various overlapping intellectual ecologies for ideas about the evolution of behavior consisted of theological, metaphysical, psychological, and epistemological conceptions, aspects of which I have already touched on. I wish now, rather briefly, to consider how certain social views gave support to one strain of development—that culminating in Cabanis's behavioral evolutionism.[18] In a moment, I will indicate how a related set of social and professional constraints forced the convergence of Erasmus Darwin's biopsychological thought with that of Cabanis and Lamarck.

Sensationalist psychological theory supported the social doctrines of Enlightenment thinkers, particularly the French philosophes and ideologues.[19] Jean d'Alembert (1717–1783), in his *Discours préliminaire* (1751) to the *Encyclopédie,* explained that refined sensory experience naturally generated those clear ideas that lay at the foundation of exact science. He attempted to demonstrate this by tracing the progressive development of scientific knowledge from the first experiential generalizations of early men to the accomplishments recorded in the *Encyclopédie.* The editors of that massive work, of which d'Alembert was one, regarded

17. Condillac, *Traité des animaux,* 1.3, p. 534; Le Roy, *Lettres sur les animaux,* pp. 73–74.

18. Martin Staum explores the broad social implications of sensationalism in his *Cabanis: Enlightenment and Medical Philosophy in the French Revolution* (Princeton: Princeton University Press, 1980), especially chaps. 3–8. See also the extensive consideration of this problem in Keith Baker's *Condorcet: From Natural Philosophy to Social Mathematics* (Chicago: University of Chicago Press, 1975), especially chaps. 4–6.

19. Antoine-Louis Destutt de Tracy coined the term "ideology" to refer to a science of the formation, expression, and organization of ideas. The ideologues adopted the sensationalist interpretation of ideas and in accord with that view made systematic recommendations for social and educational reform. In addition to Destutt de Tracy, the immediate circle of ideologues included Cabanis, the great mathematician and educational reformer the Marquis de Condorcet, the historian of antiquity Constantin Francois Volney, and the zoologist the Comte de Lacépède. This group, considerably enlarged by many other French intellectuals (e.g., Condillac, Holbach, Diderot, and Turgot), held council in the salon of Madame Helvetius in the village of Auteuil during the last two decades of the eighteenth century. For an extended account of the ideologues and the variety of their views, see Jay Stein, *The Mind and the Sword* (New York: Twayne, 1961), and Staum, *Cabanis.*

its volumes as the repository of scientific and technical achievements grounded in direct observation and practice. They believed that steady application of the principles derived from these sources promised continued development of socially useful innovations.[20] D'Alembert's younger colleague, the Marquis de Condorcet (1743–1794), also uncovered the roots of scientific progress in the psychological ability of men to receive and compare sensations, to order them by means of signs, and to use them in various combinations. In his unfinished *Esquisse d'un tableau historique des progrès de l'esprit humain* (1795), Condorcet sought to demonstrate historically how the rational analysis of society, based on the methods of advancing science, had begun to release men from the tyranny of both nature and their own superstitions and to disclose a future of material and social progress.

During the Revolution, Condorcet's progressive social views could not disguise his noble lineage. He was hidden from the Terror by his ideologue friend, the physician Pierre-Jean Cabanis, who shared his conviction that the empirical sciences, especially medicine, might continue to promote social progress. Cabanis based his program for the moral use of medicine on his belief that ideas, from the most speculative to those guiding practice, bubbled up from sensations, and thus became subject to all the physical influences that operated on the external and internal organs of sense. Medicine, then, might aspire "to perfect human nature generally."[21] Cabanis held out this hope in his *Rapports du physique et du moral de l'homme* (1802):

> Without doubt, it is possible, by a plan of life, wisely conceived and faithfully followed, to alter the very habits of our constitution to an appreciable degree. It is thus possible to improve the particular nature of each individual; and this goal, so worthy of the attention of moralists and philanthropists, requires that all the discoveries of the physiologist and physician be considered. But if we are able usefully to modify each temperament, one at a time, then we can influence, extensively and profoundly, the character of the species, and can produce

20. As the editors wrote in the introduction to volume 3 of the *Encyclopédie*, p. vi: "It [the *Encyclopédie*] will exhibit the history of the riches of our century in this area; it will do so for this age which is ignorant of this history and for the ages to come, enabling them greatly to advance. These arts, inestimable monuments of human industry, will no longer have to fear being lost in oblivion; their deeds will no longer be hidden in the workshop and in the hands of the artists. They will be discovered to the philosopher, and reflection will finally be able to enlighten and simplify blind practice."

21. Pierre-Jean Cabanis, *Rapports du physique et du moral de l'homme,* sixième mémoire, in *Oeuvres complètes de Cabanis* (Paris: Bossange Frères, 1823–1825), 3:433.

an effect, systematically and continuously, on succeeding generations.[22]

Cabanis found empirical support for his plan of perfecting the human species in the experience of stockbreeders, who demonstrated effective natural methods for racial improvements.[23] His theoretical support lay in the animal psychology worked out by La Mettrie, Condillac, Le Roy, and similarly disposed sensationalists.

Thus for instance Le Roy, in his investigations of the psychological growth of young animals in their natural environments, isolated those social factors that led to the improvement or retardation of their intellectual faculties. La Mettrie showed that adoption of sensationalist epistemology led to the presumption that so-called inferior creatures might be improved, if given the same social advantages as enjoyed by men. So an orangutan, if taught to communicate by using techniques designed for the deaf, might "no longer be a wild man, nor a defective man; he would be a perfect man, a little gentleman, with as much substance or muscle as we have for thinking and profiting from his education."[24] La Mettrie, like Cabanis who read him, intended not only to demonstrate psychological continuity between men and animals, but also to use evidence of the connection to argue that moral and social progress could be promoted by controlling the physical sources of and influences on intellectual faculties.

Sensationalist psychological doctrines, especially the theory of animal perfectibility, thus both advanced and shaped the social goals of the philosophes. Theories asserting perfectibility assumed, however, that an animal's intricate behavior, which best manifested a progressive adap-

22. Ibid., p. 434. Compare Condorcet's similar considerations in his *Esquisse d'un tableau historique des progrès de l'esprit humain*, dixième epoque (Paris: Vrin, 1970), p. 238: "But are not the physical faculties, the strength, refinement, and acuteness of the senses among those qualities whose perfection the individual is able to transmit? Observation of the several races of domestic animals should convince us of this, and we may confirm it through direct observations of the human species. And, finally, can we not hold out this same hope for the intellectual and moral faculties? May not our parents, who transmit to us the advantages and defects of their characters, toward which we ourselves tend, and the distinctive features of our face and the disposition to certain physical ills, also transmit that part of their physical organization on which our intelligence, strength of mind, energy of soul and moral sensibility all depend? Is it not reasonable that education, in perfecting these qualities, influences that same organization, modifies and perfects it? Analogy, the analysis of the development of the human faculties and certain facts all seem to prove the reality of these conjectures, which again expand the bounds of our hopes."

23. Cabanis, *Rapports*, sixième mémoire, *Oeuvres* 3:434.

24. Julien Offray de La Mettrie, *L'homme machine*, ed. Aram Vartanian (Princeton: Princeton University Press, 1960), p. 162.

tation to the natural and social environments, was not the consequence of blind instinct, rather of intelligent consideration. But this interpretation of complex animal behavior came under increasing scrutiny by natural historians in the latter part of the eighteenth century.

The Challenge of Innate Behavior

More traditional theorists of instinct had hard evidence on their side. Hermann Samuel Reimarus (1694–1768), in *Allgemeine Betrachtungen über die Triebe der Thiere* (1760), pressed the fact that animals often exhibited completely formed and adaptive behaviors before they had any opportunity to learn them: chicks immediately after emerging from the egg began to peck at grain with coordinated movements; caterpillars that had never seen a cocoon skillfully wove one in the same design as that of their ancestors.[25] Further, though instinctive behavior might vary—Reimarus admitted this—it still retained an essential pattern in a variety of circumstances. In this respect instinct did not seem to differ from anatomical structures, and like them it gave evidence of being strongly inherited.[26] Reimarus used these properties of instinctive behavior in his argument against phylogenetic alteration of species in either habit or form. Behavior and anatomy were closely adapted to their circumstances; if they varied in essentials, the species could not survive. This argument alone sufficed for him (and later for Georges Cuvier) to reject the incipient evolutionary theories of La Mettrie, Maupertuis, and Buffon.[27]

Transformists of the late eighteenth and early nineteenth centuries—particularly Erasmus Darwin, Cabanis, and Lamarck—exhibited in their theories the force of the sensationalists' discussions of instinct. Under this influence they acknowledged and indeed insisted on the role

25. Reimarus, *Allgemeine Betrachtungen uber die Triebe der There*, pp. 157–60. See also Reimarus, *Abhandlungen von der vornehmsten Wahrheiten der naturlichen Religion*, p. 378. For an account of Reimarus's intellectual development, see Julian Jaynes and William Woodward, "In the Shadow of the Enlightenment," *Journal of the History of the Behavioral Sciences* 10 (1974): 3–15, 144–59.

26. Reimarus, *Allgemeine Betrachtungen uber die Triebe der There*, p. 160.

27. Reimarus, *Abhandlungen von der vornehmsten Wahrheiten der natürlichen Religion*, p. 392: "Everything conforms to the certain kind of life which each animal lives and for which it is determined. Give to one the size, build, and organs of another and it will not be able to lead its own kind of life; indeed it must be undone or suffer greatly." Reimarus's idea was similar to Georges Cuvier's key methodological principle, which is discussed below. Cuvier was a friend of Reimarus's son, J. A. H. Reimarus, who edited his father's works. For a description of the protoevolutionary ideas of La Mettrie, Maupertuis, and Buffon, see Bentley Glass, ed., *Forerunners of Darwin* (Baltimore: Johns Hopkins University Press, 1968), chaps. 3–5.

of intelligence in guiding animal actions. But they recognized the incontestable evidence of species-specific behaviors that appeared in animals before relevant environmental experience. To explain this they could not simply acquiesce in Aristotelian or Cartesian theories of preestablished integration of innate behavior and structure with the environment. The transformists had to show, not that behavior and structures were adapted to the environment, but that they were *adaptable,* while yet admitting that behavior had innate components. This problem was approached variously by the early transformists, and their several solutions suggested to the young Charles Darwin ways to conceptualize instinct and its part in organic evolution.

Erasmus Darwin's Sensationalist Interpretation of Instinct and Evolution

Darwin's Sensationalism

Erasmus Darwin, the grandfather of Charles, studied medicine at Edinburgh, then (in the 1750s) the preeminent medical school in the British Isles. The school had been founded in 1726; and the faculty, even in Darwin's time, still received their training largely at Leyden, where the great polymath Hermann Boerhaave (1668–1738) had taught generations of students a mechanistic physiology and a Cartesian mind-body dualism. In a letter (1802) to Darwin's son Robert, James Keir, who had studied with Darwin at Edinburgh, wondered how his friend had escaped the faculty's "Boerhaavian yoke." He marveled that Darwin's medical theory, detailed in his two volume *Zoonomia* (1794–1796), regarded man as a "living being" rather than a crude mechanism whose pipes occasionally became clogged with disease.[28]

The express aim of Darwin's *Zoonomia* was "to unravel the theory of diseases."[29] For this purpose he thought it was necessary to examine the structural and physiological principles governing the organization of the animal system. He adopted the framework of Albrecht von Haller's physiological theory, through which he wove a sensationalist psychology. Haller argued that muscular fibers had an intrinsic *vis insita* that responded to irritation.[30] Darwin, building on this idea, proposed that

28. Keir's letter is quoted in Desmond King-Hele, *Doctor of Revolution: the Life and Genius of Erasmus Darwin* (London: Faber & Faber, 1977), pp. 30–31.

29. Erasmus Darwin, *Zoonomia; or the Laws of Organic Life,* 2d ed. (London: Johnson, [1794] 1796), 1:1. (The second edition does not differ essentially from the first.)

30. Albrecht von Haller, *First Lines of Physiology,* notes by H. A. Wrisberg, trans. of 4th German ed. (Edinburgh: Elliot, 1786), pp. 231–33. Haller's *Primae lineae physiologiae* was first published in 1747.

Figure 1.1   Erasmus Darwin, 1731–1802, portrait done in 1774.

the most primitive sort of animal motion was irritative contraction of the fibers of muscles and sense organs. He believed this response was proximately produced by the *spirit of animation,* a subtle, material substance that flowed through the brain and nerves, and finally washed the fibers of muscles and organs. This tenuous liquid, he supposed, reacted immediately to internal and external stimuli, and caused fiber contractions. These contractions obeyed the laws of association, so that "all animal motions which have occurred at the same time, or in immediate succession, become so connected, that when one of them is reproduced the other has a tendency to accompany or succeed it."[31] Due to associative processes, then, one set of fibrous reactions could stimulate other

31. E. Darwin, *Zoonomia,* 9.7., vol. 1, p. 320.

sets; and so the sequential contractions required to play a piano or 'in-stinctively' to build a nest could be initiated without continued sensitive or volitional guidance.[32] But regardless of the exact types of connection forged, Darwin stipulated that all were "produced by habit, that is, by frequent repetition."[33]

Darwin's idea of an active spirit of animation derived from Haller's more mechanistically formulated "nervous liquor" and ultimately from Descartes's hydraulic theory of "animal spirits."[34] As in the older con-ceptions, the spirit of animation had not only motor functions but cog-nitive functions as well, namely, to produce ideas within the brain. In accord with sensationalist epistemology, Darwin understood these ideas to be the less perfect copies of immediate sense impressions. As such they could only be particulars. He thus acceded to the sensation-alists' rejection of Locke's general or abstract ideas, which had "no existence in nature, not even in the mind of their inventor."[35] For Dar-win, this meant that brutes were capable of all the basic mental opera-tions enjoyed by men: reasoning—that is, the concatenation of sensory ideas according to the laws of association—could be construed, there-fore, not as a barrier separating men from animals but as a link uniting them.[36] Darwin did recognize an important psychological distinction between men and animals, however. Provident, middle-class physician that he was, he believed that men more than animals worried about future happiness and the means to attain it, whether through "praying to the Deity" or "labouring for money."[37]

In his study, Darwin confined his attention to the material processes of the animal body, particularly to that "spirit of animation or sensorial power" which men shared with beasts.[38] He preferred to leave the "im-

32. Ibid., 6.1, pp. 34–35.
33. Ibid., 4.7, p. 31.
34. Descartes's animal spirits and Haller's nervous liquor were essentially inert, mate-rial fluids governed by the laws of mechanics. But in Darwin's theory, animal contraction, caused by the fluid spirit of animation, "is governed by laws of its own, and not by those of mechanics, chemistry, magnetism, or electricity" (ibid., 12.1, p. 65).
35. Ibid., 16.17, p. 188.
36. Ibid., 15.3, pp. 132–33.
37. Ibid., 16.17, p. 188.
38. Ibid., 14.2, p. 109. The sensorial power was characteristic not only of men and animals but also, acording to Darwin, of those less perfect animals, the plants (13.5, pp. 105–7). Because plants reacted differentially to heat and cold, light and darkness, and because in many plants the female parts approached the male parts, Darwin attributed to plants a sensorium, the passion of love and even "ideas of so many of the properties of the external world and of their own existence." If Condillac's statue could be brought to

mortal part" to the ruminations of theologians. One of these latter, William Paley—whose *Natural Theology* (1802) contains many allusions to *Zoonomia*—regarded Darwin's demur with suspicion. For Darwin's physiological analyses also included investigation of psychological activities—sensation, emotion, reason—which theologians had traditionally assigned to a separate, immaterial principle. Paley was especially concerned with Darwin's theories of instinct and generation, for good reason. Darwin's naturalistic analysis of instinct robbed natural theology of privileged evidence of God's design. And his theory of species generation hardly conformed to Biblical accounts of creation. Beyond the spare but firm evidence that the *Ens Entium* existed, Darwin found no suggestion of a special providence that interfered with the lawful operations of the universe.[39] His attempt to sidle away from theological confrontation was, however, hardly more successful than that of his grandson.

Rejection of Preestablished Instinct

Aristotelian and Cartesian instinct theorists had maintained that many animal behaviors, particularly those which seemed the most intelligent, were unlearned and independent of experience, due to the power of instinct, which, as Darwin irritably observed, "has been explained to be a *divine something,* a kind of inspiration; whilst the poor animal, that possesses it, has been thought little better than a machine!"[40] Darwin believed otherwise. He asserted that close inspection would show instinct "to have been acquired like all other animal actions, that are attended with consciousness, *by the repeated efforts of our*

---

think by the faintest titillation of the sense of smell, then surely one ought not be less generous with creatures manifestly more animate. Even Darwin's sober grandson mused on the cognitive abilities of plants. See Charles Darwin's *N Notebook,* MS pp 12–13, transcribed by Paul Barrett, in Gruber's *Darwin on Man,* p. 332.

39. See Darwin's profession of Deism to his Cambridge undergraduate friend Thomas Okes (November 1754). The letter is contained in *The Letters of Erasmus Darwin,* ed. Desmond King-Hele (Cambridge: Cambridge University Press, 1981), pp. 8–9.

40. E. Darwin, *Zoonomia,* 16.1, vol. I, p. 137. There are two theories of instinct to which Darwin was possibly referring, David Hartley's or Hermann Samuel Reimarus's, or both. Hartley, in *Observations on Man* (London: Leake, Frederick, Hitch, &Austen, 1749), 1 : 412, associated his view with the mechanical theory of Descartes and defined instinct as "a Kind of Inspiration to Brutes, mixing itself with, and helping out, that Part of their Faculties which corresponds to Reason in us, and which is extremely imperfect in them." Darwin may also have had in mind Reimarus's characterization of instinct (discussed above). Darwin nowhere directly refered to Reimarus, but he was a fellow student of Reimarus's son at Edinburgh and kept up a correspondence with his old friend.

muscles under the conduct of our sensations or desires."[41] He was convinced that even the behavior of insects did not display the uniformity commonly supposed and that their arts and improvements "arose in the same manner from experience and tradition, as the arts of our own species; though their reasoning is from fewer ideas, is busied about fewer objects, and is exerted with less energy."[42]

Réaumur, the great insectologist, had earlier arrived at a similar conclusion, but his was supported by exhaustive, first-hand observation. Darwin's was not. Like many early evolutionists and other naturalists, Darwin relied heavily on illustrative anecdotes and reasonable inferences drawn from sensationalist principles. For those working within this tradition, the very concept of behavior—as that of instrumental activity appropriate in a variety of circumstances—brought with it an assumption of intelligent control. Darwin thus argued that since the flesh fly often mistook the carrion flower for putrid flesh, it could not be guided by necessary instinct; for machines did not err. To make a mistake implied the possibility of correct reasoning.[43]

In his long chapter on instinct in *Zoonomia,* Darwin examined several kinds of activity usually attributed to instinct, with the purpose of suggesting how these might be construed as arising from experience and learning.[44] For instance, he deemed nest building in birds to result from observation and "their knowledge of those things, that are most agreeable to their touch in respect to warmth, cleanliness, and stability."[45] The variations in nest construction even among birds of the same species, he attributed to their intelligent accommodation to contingent circumstances.[46] In considering the cuckoo's habit of laying its eggs in other birds' nests, he rejected the belief that it was guided by instinct; he felt that anyone acquainted with the facts of animal life (apparently including its moral design) would "have very little reason himself, if he could imagine this neglect of her young to be a necessary instinct!"[47] Ironically, Darwin's grandson would treat this bit of moral delinquency as a paradigm of instinct, for which he argued that natural selection was the most likely explanation.[48] Behind the appearances of other supposed instincts—the swallow's migration, the turkey's danger

41. E. Darwin, *Zoonomia,* 16.2, vol. 1, p. 138.
42. Ibid., 16.16, p. 183.
43. Ibid., 16.11, p. 163.
44. Ibid., 16, pp. 136–88.
45. Ibid., 16.13, p. 172.
46. Ibid., 16.13, pp. 171–73.
47. Ibid., 16.13, p. 177.
48. Charles Darwin, *On the Origin of Species* (London: Murray, 1859), pp. 216–19.

alarms, and the rest—Darwin uncovered the operations of sensory experience and intelligence rather than the urgings of blind impulse.

But one feature of many putative instincts required more ingenuity to explain. Most animals manifested patterned activity immediately after birth. This was precisely the kind of evidence Reimarus, whom Darwin likely read,[49] used to demonstrate the innate and fixed character of complex animal behavior. Since Darwin could not argue that such behavior had a postnatal source, he located its origin in the fetal environment. He explained the coordinated pecking and swallowing of newly hatched chicks, for instance, as the consequence of the fetuses' continually gaping and gulping embryonic fluid. Their ability to walk just after birth could also be ascribed to habits acquired while ensconced in the egg: the struggles and swimming movements of fetal chicks resembled "their manner of walking, which they have thus in part acquired before their nativity, and hence accomplish it afterwards with very few efforts."[50] What appeared, then, to be phyletically constant instincts, Darwin found explicable through principles of embryogenesis. And there was a lesson in this: the account of an individual's behavioral patterns by appeal to prenatal habits could be generalized to explain the very structure of species in phylogenesis.

From Embryogenesis to Phylogenesis

Embryogenesis provided Darwin with a model for the developmental sequence that seemed to have given rise to all living species. Specifically, he isolated several kinds of change that occured in the embryo and developing organism (including the inheritance of traits acquired by parents, either naturally or from efforts of breeders) that, along with structural similarities of different animal species, indicated the likelihood of phylogenetic transformation. From this evidence he concluded:

> all warm-blooded animals have arisen from one living filament, which THE GREAT FIRST CAUSE endued with animality, with the power of acquiring new parts, attended with new propensities, directed by irritations, sensations, volitions, and associations; and thus possessing the faculty of continuing to improve by its own inherent activity, and of delivering down those improvements by generation to its posterity, world without end![51]

In Darwin's opinion, an individual embryo began as a single living

49. See note 40.
50. E. Darwin, *Zoonomia*, 16.2–3, vol. 1, pp. 138–39.
51. Ibid., 39.4, p. 509.

filament, which because of its sensitivity and contractility acquired certain associative modes of behavior and differential response. He believed these mechanisms would gradually shape the developing embryo into an articulated organism. He extended these same principles to that original living filament from which all life forms evolved. But after the first stirrings of this progenitor of animal life, when creatures organized in some primitive fashion made their initial appearance, he thought another factor of sensitive life began to dominate the course of species improvement—that of *need.* Adopting the same basic sensationalist framework as Cabanis and Lamarck, Darwin thus anticipated them in using the concept of sensitive need to explain species transformation. According to Darwin, the basic needs that drove animals were those of "lust, hunger, and security." For instance, male animals, impelled by their desires to possess females, developed horns, protective skin, spurs, and other accouterments in order to gain and defend their prizes: "The final cause of this contest amongst males seems to be, that the strongest and most active animal should propagate the species, which thence become improved."[52] But Darwin judged the impulse from needs to have still further consequences. He fancied that in the elephant's search for food, its trunk had been lengthened as it continued to reach for higher tree branches; similarly, he believed that in their attempts at self-preservation, some animals produced wings, while others strove for swiftness of foot or formed protective shells.[53] He was also willing to entertain the Linnaean and Buffonian hypothesis that some original species cross-mated to produce new hybrid species. For the most part, however, he emphasized mechanisms of association and need as chiefly responsible for introducing modifications into species.

The Inheritance of Acquired Characteristics

The doctrine of the inheritance of acquired characteristics, to which Darwin subscribed, was hardly novel, even in his own time. It can be traced back as far as the Hippocratic writers of *De aere, aquis et locis, De morbo sacro,* and *De generatione,* and forward well into the twentieth century. There were, however, two versions of the doctrine. The more venerable assumed that passively received modifications were inherited—diseases, scars, mutilations, and the direct effects of the environment. Buffon, for instance, pointed to dog breeders who cropped the ears and tails of puppies, a practice which would, in time, "transfer those defects, either whole or in part, to their descendants."[54] (Toward

52. Ibid., 39.4, pp. 507–8.
53. Ibid.
54. Buffon, "De la dégénération des animaux" (1766), *Histoire naturelle,* in *Oeuvres*

the end of his career, Buffon adopted a modified theory of evolution, according to which some few original types of animal developed through hybridization and the direct effects of the environment into all those species now populating the globe.)[55] This version of the inheritance of acquired characteristics, however, was decisively checked by August Weismann at the end of the nineteenth century, when he snipped the tails of generations of mice without producing any noticeable shortening of tails in the progeny.[56] But Weismann's experiments did not tell against the other version of the hypothesis, which supposed that what got primarily inherited were active modifications, that is, alterations induced by the habitual exercise of structures in the satisfaction of needs. This proposal, which reflected sensationalist notions of habit, became known in the nineteenth century as the theory of use inheritance, and seems to have first been systematically explored by Charles-Georges Le Roy and Erasmus Darwin. It was a powerful idea. Use inheritance became for Cabanis and Lamarck the principal mechanism of anatomical adaptation; it also functioned as the engine of evolution for the young Charles Darwin and persisted through later developments of his theory.

Erasmus Darwin rejected the notion that animal behavior was directed by innate ideas or compelled by blind impulses, even if divinely imposed. Behavior, in his view, required an intelligent (i.e., resulting from associative learning) comprehension of circumstances; consequently the traditional concept of instinct could not be used to describe the animal economy. But a short time after Darwin published *Zoonomia*, Cabanis constructed a theory of instinct—upon which Lamarck would further build—that did not compromise sensationalist principles and was remarkably consistent with Darwin's basic position. Because of the meaning instinct had acquired at the hands of more orthodox theorists, Darwin was disinclined to find in it any residual value. Cabanis and Lamarck merely abandoned more traditional interpretations of innate behavior and adopted precisely the theory that instincts were modes of phylogenetically acquired associative response,

---

complètes 4:116. Conway Zirkle has compiled passages from works that make use of the idea of inheritance of acquired characters. See his "The Early History of the Idea of the Inheritance of Acquired Characters and of Pangenesis," *Transactions of the American Philosophical Society*, n.s. 35 (1946): 91–151.

55. Buffon, "Dégénération des animaux," *Histoire naturelle*, in *Oeuvres complètes* 4:110–44.

56. August Weismann, "The Supposed Transmission of Mutilations" (1888), in his *Essays upon Heredity*, ed. and trans. E. Poulton et al., 2d ed. (Oxford: Clarendon Press, [1889] 1891), 1:444–45.

a theory in pleasing harmony with Darwin's principles of species generation.

The Historical Relation of Darwin's Evolutionism to Cabanis's and Lamarck's

To the historian, Darwin's theory of evolution strikes a resonant cord, since it presents the same basic structure as found in the ideas of Cabanis and Lamarck. Samuel Butler, in his *Evolution Old and New* (1879), also perceived the similarities and concluded that Lamarck must have known of Darwin's theory.[57] (Butler was unaware of Cabanis's work.) It now seems fairly clear that, while there may have been direct interaction between Cabanis and Lamarck on the species question, neither of them knew anything of Erasmus Darwin's ideas. But another kind of relation connected them, or so I think. Processes of conceptual descent and selection brought their ideas into convergence.

The similarities between the two French thinkers and Darwin can be understood as resulting from a common theoretical inheritance that had been shaped by like intellectual and social environments. Consider, first, their shared legacy regarding the stability of species. All three had read Linnaeus and Buffon, so they were quite at home with the idea that ecological factors (e.g., climate, terrain, food, etc.) might operate to alter species within some generous limits. Moreover, from scores of common sources (from Hippocrates to Buffon), including the folk wisdom of animal breeders, they were persuaded of the inheritance of acquired characteristics. Now this basic view, that species were alterable and could transmit modifications to descendants, became molded against their respective intellectual and social environments, which exerted remarkably similar pressures. Each of the three adopted a sensationalist metaphysics and epistemology, which implied that species as abstract entities or individuated essences could not be real: a species could only be a group of like individuals represented by a complex idea (or associated train of ideas) consisting of particular images. The sensationalist interpretation of ideas also convinced them that the cognitive functions of animals were logically similar to those of men; thus no insuperable metaphysical or psychological barriers separated men from animals. Furthermore, as sensationalists, Darwin, Cabanis, and Lamarck recognized the effectiveness of intelligently acquired habit in accommodating animals to their surroundings. Yet they were also aware, prompted by such diverse authors as Galen and Reimarus, that animals

57. Samuel Butler, *Evolution, Old and New* (London: Hardwicke and Bogue, 1879), pp. 258–60.

often exhibited fixed behaviors immediately after birth.[58] Finally, their social and professional concerns directed them to regard individuals and groups of individuals as improvable. Each of the three was trained as a physician. Indeed, Darwin and Cabanis developed their evolutionary theories in medical treatises designed to explain how to improve the physical and moral health of men. And most importantly, each of them believed that organic and mental progress flowed from perfectly natural causes that could be scientifically comprehended. Lamarck, for instance, rejected Cuvier's account of species extinction because it appealed to divinely ordained geological catastrophes. Darwin, expressing a similar sentiment, allowed the Creator only the privilege of igniting into life that first primitive speck; thereafter the *Causa causarum* rested.

Butler was right to suspect some link between Darwin and Lamarck. But he looked in the wrong place. He ignored the evolutionary connections, the common inheritance of these thinkers and the matching elements of their respective intellectual and social environments. These connections did not, of course, lead them ineluctably to similar theories of species transformation. Yet the processes of conceptual evolution did make the close resemblance of their ideas something more than mere accident.

Cabanis's Revision of Sensationalism

Rejection of Environmentalism

Cabanis and Lamarck both held positions in the Institut national des sciences et arts, which the French constitution established in 1795. The Institut embodied the still vital spirit of the encyclopedists, while it firmly aligned scientific and cultural activities with the interests of the state.[59] Cabanis was elected to the Class of Moral and Political Science, Section on the Analysis of Sensations and Ideas, in December of 1795, while Lamarck, who had been advanced by Buffon to membership in the suppressed Académie des sciences, was a member of the Class of

58. See the description of Galen's experiment on animal instinct described above in note 3. This experiment was cited by Erasmus Darwin in *Zoonomia,* 16.6, p. 143, and by Cabanis in *Rapports,* second mémoire, *Oeuvres complètes* 3:139. Lamarck, if he did not know of Galen's experiment directly, read of it in Cabanis. As explained in note 40, Darwin undoubtedly knew of Reimarus's work. Though Cabanis and Lamarck did not cite Reimarus directly, they likely read the article on "Instinct" in the *Supplement à l'Encyclopédie* (Amsterdam: Rey, 1776–1777), 3:608–11. This article provided a lengthy account of Reimarus's views.

59. Roger Hahn describes the structure of the Institut in his *The Anatomy of a Scientific Institution: the Paris Academy of Sciences, 1666–1803* (Berkeley: University of California Press, 1971), pp. 286–12.

Figure 1.2 Pierre-Jean Cabanis, 1757–1808.

Mathematics and Physics, Section on Botany and Plant Physiology, at the inception of the Institut.

Cabanis brought to the analysis of sensations and ideas the physician's concern with internal, physiological causes of behavior and a sensationalist epistemology, which he had cultivated as a member of the circle of ideologues.[60] Cabanis's medical theory and experience, however, led him to modify significant aspects of the received assumptions of sensationalism. The new synthesis produced a theory that won for him the admiration of his fellow Institut member Lamarck.

Lamarck showed keen interest in Cabanis's *Rapports du physique et du*

60. Staum's *Cabanis* situates the French physician's ideas among those of the ideologues.

moral de l'homme, referring to it more often than any other work in his own *Philosophie zoologique* (1809). Cabanis had read the first six memoirs of the *Rapports* to the Institut in the sessions of 1796–1797, and published the completed work of twelve memoirs in 1802. Lamarck undoubtedly sensed in the efforts of Cabanis the spirit of his mentor Buffon and recognized a development of thought congenial to his own.[61] Cabanis had, after all, denied the fixity of species, in express opposition to Georges Cuvier,[62] and had argued for an evolutionary conception, which supposed the origin of species to lie in the spontaneous generation of animaculae that gradually became adapted to their circumstances through habit.[63] This was essentially the theory that Lamarck himself promoted in his courses at the Muséum national d'histoire naturelle during the first two decades of the nineteenth century and that he described in some detail in a series of works beginning in 1800.

The evolutionary theory that Cabanis advanced in the completed *Rapports* served principally to support his major thesis: that man's physical constitution determined his understanding and behavior and that, consequently, these faculties bore the mark of the individual's age, sex, temperament, health, regimen, and environment. Cabanis's audience had become familiar with this thesis, especially through La Mettrie's portrayal of man as machine. Like La Mettrie, Cabanis illustrated his view with a perfectly outrageous metaphor, that just as the stomach digested food, so "the brain, after a fashion, digests impressions and organically secrets thought."[64]

In arguing his thesis, Cabanis did not wish to deny the Lockean axiom that all ideas ultimately derived from sensory experience.[65] He did dispute, however, the Abbé de Condillac's radical interpretation of that axiom. Condillac rejected the innate and zealously promoted the faith that external experience alone produced not just ideas but the very faculties of thought.[66] And if, as he dramatized in the *Traité des sensations* (1745), a statue could come cognitively alive through the reception of the merest impression, then animals were surely capable of intelligent action guided only by sensation, as he in fact argued in *Traité des ani-*

61. Louis Roule, *Lamarck et l'interprétation de la nature* (Paris: Flammarion, 1927), pp. 118–19.

62. Cabanis, *Rapports,* dixième mémoire, *Oeuvres complètes* 4:249–50. Cabanis acknowledged that Cuvier had discovered the remains of several extinct species, which, however, led him to suspect, contrary to the view of Cuvier, that "many races first made their appearance in a very different form from what they are today."

63. Ibid., pp. 249–53.

64. Ibid., seconde mémoire, *Oeuvres complètes* 3:160–61.

65. Ibid., préface, *Oeuvres complètes* 3:11.

66. This is the theme of Condillac's *Traité des sensations* (1754).

maux (1755). Cabanis hardly exaggerated, then, when he characterized Condillac as yielding to the empirically uncontrolled presumption "that all the determinations of an animal are the product of reasoned decision and a consequence of experience."[67] Were the Abbé a physician of the corporeal man as well as of the spiritual, he might have recognized that impressions also had internal sources in the viscera and that the innate dispositions of the nervous system molded patterns of experience. These "unconscious" influences Cabanis took to be of signal importance for a correct appraisal of human mental and moral character and for reassessing the elements of continuity linking men and animals. Both man and beast were endowed with essentially similar physical constitutions, which internally affected their thoughts and behavior. Cabanis found the paradigm of such internal conditioning in animal instincts. Thus to Condillac's strict environmentalism, he objected, "with observers of all times, that many of those determinations of animals are not able to be attributed to any sort of reasoning, and that, although they have their source in physical sensibility, they form themselves usually without any voluntary act on the part of the individual—except to better execute them. This group of determinations is designated by the name instinct."[68] In affirming the observations of more orthodox naturalists, Cabanis had not recanted the sensationalist faith in sensory experience as the ultimate source of mental activity. Rather, he showed that cognition had more proximate, internal causes, which had to be conceived as historical consequences of phylogenetic development. Such development itself, however, he interpreted as a product of sensory response to the environment.

Theory of Instinct

Cabanis's analysis of the ontogenetic origins of instinct agreed with that of Erasmus Darwin. Like Darwin, Cabanis recognized that animals and men displayed specific tendencies and patterns of behavior shortly after birth. Since these could not have been directly acquired from experience of external circumstances (as Condillac supposed), they must have arisen in the womb from what we would call proprioceptive sensations: "in the impressions received internally, in their simultaneous concourse, in their sympathetic combinations, in their continual repetition during the time of gestation, one necessarily finds the source of those tendencies that infants express at the moment of birth."[69] The

67. Cabanis, *Rapports,* seconde mémoire, *Oeuvres complètes* 3 : 105.
68. Ibid., pp. 105–6.
69. Ibid., p. 137.

infant comes into the world, therefore, not with a *tabula rasa,* but with a *tabula inscripta.* The infant has already acquired elemental ideas (e.g., the idea of resistance from the weight of its limbs and the effort perceived in moving them) and a fund of habits that furnishes a continuous interactive network affecting its development and use of reason.[70] Arguing in this way, Cabanis attempted to provide basic support for his thesis that "the operations of thought are all necessarily modified by such determinations and by the particular and general habits of instinct."[71]

Cabanis postulated two classes of instinct.[72] The first comprised those behaviors and tendencies displayed virtually from the moment of birth. Some of these were species-specific (e.g., the pecking of baby chicks) and arose from the particular character of the fetus's organic structure, its felt needs (e.g., to stretch), and the range of its intrauterine motions. Others were more general and resulted from structures, needs, and motions common to all organic beings: efforts at self-preservation (originating in impressions of continuous blood circulation and the want of oxygen at the time of birth), food-seeking behavior (from repeated swallowing of amniotic fluid), and movement (from struggles and felt resistance in the womb).[73] The second class of instincts depended upon organs which appeared during later stages of maturation (e.g., mating instincts). These too, according to Cabanis, were internally induced activities, not behaviors directed solely by an independent mind in contact with external nature.

Cabanis's theory of the internal sources of behavior responded to the deficiencies of Condillac's exaggerated environmentalism. In turn, Lamarck thought he detected in Cabanis an oversimplified distinction between instinct and reason. Lamarck seems to have based his criticism on the neat summary Cabanis's disciple Anthelm Richerand (1779–1840) gave his teacher's position: "As the philosopher [Cabanis] whom we just cited has well observed, instinct arises from impressions received through the interior organs, while reason is the product of external sensations."[74] Lamarck objected that all sensations were in fact internal

70. Ibid., dixième mémoire, *Oeuvres complètes* 4 : 294–96.
71. Ibid., p. 307.
72. Ibid., pp. 322–26.
73. Ibid., pp. 317–26.
74. Anthelme Richerand, *Nouveaux élémens de physiologie,* 5th ed. (Paris: Caille et Ravier, [1802] 1811), 2:157. Lamarck directly cites Cabanis's *Rapports* and Richerand's *Nouveaux élémens* throughout the third part of his *Philosophie zoologique,* the part devoted to the psychological faculties of animals and man. In the section on instinct, however, he refers to Cabanis as the author of the theory to which he takes some exception, but he only cites Richerand's summary of Cabanis's position.

and that the impressions of external objects could only be registered within the sensorium or *sentiment intérieur*.[75] The challenge was consistent with the sensationalist framework both naturalists had adopted; yet it was misleading, suggesting as it did that Cabanis had sealed off the operations of reason from instinct. He had not. Cabanis, as we have seen, insisted on the internal and instinctual conditioning of reason. He contended, for example, that the fetus's experience of its own movements within the womb induced ideas of resistance and force, and the sense of "self."[76] Moreover, in his epitome of Cabanis's theory, Richerand clearly stated that "these two parts of understanding, reason and instinct, unite, blend, and mix to produce our intellectual system and the diverse determinations to which we are susceptible."[77] Lamarck seems to have emphasized a small difference in order to preserve the independence of his own theory of instinct.

Evolutionary Theory

But aside from this nice objection to Cabanisean psychology, there was essential consonance between Lamarck and Cabanis on the question of the phylogenetic origins of instincts and other faculties. In his early memoirs, Cabanis postulated, as Lamarck would later, an interactive relationship between organically based *needs* and the response of *habits*, which together could alter the structure of an organism to expand its faculties and stimulate new needs.[78] Further, Cabanis did not confine this reciprocating engine of development to ontogeny:

> But the rule of habit is not limited to the profound and ineffaceable imprints made on each individual; these, at least in part, are able to be transmitted by way of generation. The abilities of certain organs to act, to produce certain movements, to execute certain functions which are considerable, in a word—those particular faculties which are developed in great degree—these are able to be propagated from generation to generation.[79]

Following Buffon's lead, Cabanis was ready to ascribe heritable modifications to the direct influence of the environment;[80] for "particu-

75. Jean-Baptiste de Lamarck, *Philosophie zoologique,* 3.5, (Paris: Dentu, 1809), 2:322–23.
76. Cabanis, *Rapports,* dixième mémoire, *Oeuvres complètes* 4:294–96.
77. Richerand, *Nouveaux élémens* 2:157.
78. Cabanis *Rapports,* second mémoire, *Oeuvres complètes* 3:110.
79. Ibid., neuvième mémoire, *Oeuvres complètes* 4:146–47.
80. See Buffon's "De la dégénération des animaux," *Histoire naturelle,* in *Oeuvres complètes* 4:110–44.

lar impressions that are constant and always the same are thus capable of modifying organic dispositions and rendering the modifications fixed in the races."[81] But like Lamarck, he was especially concerned to promote habit as the principal means by which animals accommodated themselves to their environments. In the estimation of both naturalists, enduring habits, developed in response to needs stimulated by a changing environment, had the effect of gradually transforming animal species over generations. For, as Cabanis claimed, "these causes of the initial habitual action do not cease to operate on many successive generations, they thus form a newly acquired nature, which, in its turn, is able to be changed only insofar as these same causes cease to operate for a long period and particularly when different causes begin to impress on the animal economy another set of determinations."[82]

It is tempting to see in Cabanis's *Rapport* the source of many of Lamarck's own evolutionary notions. Cabanis, like Lamarck, believed in the spontaneous generation of animaculae from the interaction of gases, heat, and water; and he cited experiments to demonstrate the phenomenon.[83] He differed from Lamarck in admitting that geological catastrophes had extinguished some animal lineages; in this view he more closely resembled Cuvier. But Cabanis also argued, quite congenially to Lamarck and contrary to Cuvier, that geological changes very probably compelled modifications in other animal lineages. Species were thus not unalterably fixed but might become transformed in response to environmental alterations:

> But if one forms a just idea of that sequence of circumstances in which the living races, having escaped destruction, must have successively yielded and conformed—whence, very likely, in each epoch new races were born from the preceding, being more appropriate to the new order of things; if one accepts them as given, the one supposition certain and the other very probable, then it does not seem so very impossible to conceive the initial production of the great animals from that of microscopic animaculae.[84]

Lamarck may have been directly influenced by Cabanis; the similarities in theory are unmistakable. In the *Philosophie zoologique,* Lamarck cited Cabanis often, though not on questions of species change; but his

81. Cabanis, *Rapports,* neuvième mémoire, *Oeuvres complètes* 4 : 147.
82. Ibid.
83. Ibid., dixième mémoire, *Oeuvres complètes* 4 : 246, n. 1. The note describing these experiments was added in the second edition of the *Rapports* (1805).
84. Ibid., dixième mémoire, *Oeuvres complètes* 4: 252.

own name is not to be found among the many that Cabanis mentioned in his *Rapports*. However, since Cabanis's evolutionary theory appeared only in the memoirs of the *Rapports* published in 1802—at the same time as Lamarck's extended evolutionary considerations in the *Recherches sur l'organisation des corps vivans* (1802) and after the brief sketch of his new theory presented in his course on invertebrates at the Muséum in 1800[85]—the influence may have flowed in the opposite direction. Or yet it may have been, as suggested above, the result of a common conceptual inheritance that took shape in similar intellectual circumstances.

The lines of dependence stand out more clearly on the question of instinct. Lamarck discussed Cabanis's theories at some length when working out matters of animal and human psychology; and though he was often critical of Cabanis and Richerand, his own approach to animal instinct and intelligence did not stray from the confines fixed by Cabanis's revision of sensationalist doctrine.

Lamarck: Behavior as Product and Instrument of Species Transformation

Evolution and the Mechanisms of Species Change

Historians have offered different reasons for Lamarck's conviction, first expressed in the "Discours d'ouverture" of 1800, that the scale of progression in the animal kingdom was due to nature's temporal creations, to her incessant redesigning the worm to become a man. Roule has suggested the influence of Cabanis.[86] Gillispie has made a case for seepage from Lamarck's chemical theories and belief in mineral mutability.[87] Greene has also pointed to the weight of his earlier geological ideas.[88] And Burkhardt has isolated the problem of species extinction and the acceptance of spontaneous generation as the inspiration for Lamarck's belief in evolution.[89] Only Burkhardt, however, places what he takes to be the principal precipitating causes of Lamarck's new theory

85. Lamarck's "Discours de ouverture" of 1800 was published as the preface to his *Système des animaux sans vertèbres* (Paris: Lamarck et Deterville, 1801).

86. Roule, *Lamarck et l'interprétation de la nature,* pp. 116–21.

87. Charles Gillispie, "The formation of Lamarck's Evolutionary Theory," *Archives internationales d'histoire des sciences* 9 (1956): 323–38

88. John Greene, "The Kuhnian Paradigm and the Darwinian Revolution in Natural History," in *Perspectives in the History of Science and Technology,* ed. D. Roller (Norman: University of Oklahoma Press, 1971), pp. 3–25.

89. Richard Burkhardt, "The Inspiration of Lamarck's Belief in Evolution," *Journal of the History of Biology* 5 (1972): 413–38, and his *The Spirit of System* (Cambridge: Harvard University Press, 1977), pp. 127–42.

Figure 1.3   Jean-Baptiste de Lamarck, 1774–1829, engraving in 1821.

within the larger conceptual environment of the zoologist's concerns: for instance, his classificatory studies, his insistence on natural explanations, and his ideas about the balance of species.[90] This, I believe, is the proper historiographic attitude. It is mistaken, though cognitively satisfying, to explain the origin of an important idea by appeal to one or two sources. Creative thought in science, as in other disciplines, usually emerges from a context of interwoven beliefs, suspicions, and observations. Unique causation of ideas is only a convenient fiction of historians. Undoubtedly all of the severally noted influences and others (e.g., sensationalist theories of psychological continuity between animals and man, sensationalist denial of abstract entities, etc.) operated to mold Lamarck's belief.

90. Ibid.

Lamarck proposed two chief mechanisms to account for the fact of evolution. These shaped the explanatory focus of his *Philosophie zoologique* and the introductory book (1815) of his seven-volume *Histoire naturelle des animaux sans vertèbres* (1815–1822). He conceived these evolutionary forces, however, during the period in which he initially formulated his theory; they were, I believe, salient elements of the original environment in which his belief in evolution grew. In other words, they gave rise to his conception of species transformation as much as they were required by it.[91] The first principle appeared in the "Discours d'ouverture" of 1800. It was based on two common observations, both enforced by the tradition of sensationalism: that animals adjusted their behaviors to alterations in their circumstances, and that the continued use of an organ strengthened it. From these premises, Lamarck inferred that as the environment slowly changed over long stretches of time, animals had to adopt new habits and lay old ones aside. Like Erasmus Darwin and Cabanis, he believed these new habits could produce heritable changes in organs and functions. He worked out the second device, which he joined to the first, in an unpublished discourse, "Sur l'origine des êtres vivans"[92] (1801–1803?), and in his *Recherches*.[93] It was anchored in his presumption that fluids moving through body parts could open passages in tissues, trench out canals, modify cellular masses, and gradually form new organs.

This second mechanism became for Lamarck the explanation of spontaneous generation and of the persistent tendency toward greater complexity surging through the animal kingdom. He believed the simplest infusoria were the products of very subtle fluids (principally caloric and electricity) that carved out vacuoles and cells in mucilaginous materials and thus vivified them. Following the logic of that claim, he imagined that the imponderable fluids, which controlled the flux of elemental life, became encapsulated and altered, and that they finally took

91. The problems of extinction and spontaneous generation undoubtedly influenced the formulation of Lamarck's theory, as Burkhardt suggests. But the immediate conceptual environment against which his theory took shape surely also included notions made popular by the sensationalists—especially the idea that animals became adapted to environmental change through habit. If this latter idea is linked to the theory that alterations of animal organs are heritable, as Lamarck's mentor Buffon insisted, species evolution follows as a natural consequence. I have developed this argument a bit further in the text below.

92. Lamarck's discourse, composed sometime between 1801 and 1803, has been published in *Inédits de Lamarck,* ed. M. Vachon, G. Rousseau, and Y. Laissus (Paris: Masson et Cie Editeurs, 1972), pp. 179–85.

93. Jean-Baptiste de Lamarck, *Recherches sur l'organisation des corps vivans* (Paris: Maillard, 1802), p. 105.

over the internal functions of independent organisms. Trapped caloric warmed the animal, and "animalized" electricity became its nervous fluid.[94] Had the environment become completely stable, the irregular branching of species would have disappeared; the liquors coursing through individuals would, over time, have created a smooth passage from less complex to more complex organisms. But, according to Lamarck, the environment did continually change, usually gradually, sometimes more dramatically, though never by the cataclysms Cuvier conjured up.[95] These changing conditions modified the needs and habits of animals; and new habits began restructuring their anatomy to fit in the irregular niches of nature. In this way, Lamarck explained the appearance of distinct species within the great upsurge of organic development.

In the "Discours d'ouverture" of 1800, Lamarck had not yet proposed the mechanism of the subtle fluids to explain species development; he merely supposed the environment worked both in direct (though unspecified) ways and in conjunction with the constructive power of habit to modify species. As he described them in the discourse, the principal conditions issuing new species were:

> the influence of climate, the variations of atmospheric temperature and surrounding environment, the diversity of locations, the varieties of habits, movements, actions, the means of living, and the ways of self-preservation, defense, and propagation. But as a result of these diverse influences, the faculties become extended and strengthen themselves through use; and they become diversified through new habits which have been practiced for a long time. Insensibly the structure, the firmness, in a

94. Lamarck, *Philosophie zoologique*, 2, introduction, vol 1, pp. 369–74; 2.3, vol. 2, pp. 4–19; 2.6, vol. 2, pp. 61–90; 3.2, vol. 2, pp. 235–51; and *Histoire naturelle des animaux san vertèbres* (Paris: Verdiere, 1815–1822), 1:177–85. The identification of nervous fluid with a kind of modified electrical juice was a common enough assumption (though denied by Erasmus Darwin; see note 34). The physiologist H. A. Wrisberg, annotator of Haller's *Liniae primae physiologiae,* believed the nervous fluid was of the same species as the imponderable fluids of electricity and magnetism, and claimed priority to Mesmer in this regard. See his note in Albrecht von Haller, *First Lines of Physiology,* pp. 221–22. Cabanis, in his *Rapports,* sixième mémoire, *Oeuvres complètes* 3:382, asserted that "the nervous organ is a kind of condenser, or rather a veritable reservoir of electricity." In the 1780s and 1790s, Lamarck himself was already proposing that the subtle fluids be identified as agents of vital activity. See Burkhardt, *Spirit of System,* pp. 63–68.

95. Lamarck was a uniformitarian in geology and rejected Cuvier's assumption of general upheavals as "a rather convenient means for those naturalists who wish to explain everything and who do not take any trouble to observe and study the course which nature follows in regard to her productions and all that constitute her domain" (*Système des animaux sans vertèbres,* p. 407).

word, the nature and state of the parts, as well as the organs, participate in the consequences of all of these influences, retain them, and propagate them through generation.[96]

In his *Recherches,* Lamarck reiterated these same ideas, though he now specified the imponderable fluids as the agents of environmental influence.[97] But in the *Philosophie zoologique,* he altered his view, or at least attempted to correct what he thought might have been a mistaken impression. There he permitted the environment no direct impact on the evolution of the higher animals; he allowed it only the indirect effect of modifying needs and forcing the adoption of different habits to satisfy those needs. The immediate source of adaptations among animals capable of independent behavior was, he maintained, the acquisition of new habits, with the consequences of habits passed to the progeny. Lamarck thus it made clear that the principal medium for transformation of species, from insects to elephants, was behavior, the habits acquired to meet the challenges of a shifting environment.[98]

   Lamarck's explanation of the way needs and habits accomplished their alterations should have scotched cartoons of the giraffe willing itself a longer neck. Will simply had no immediate role in evolution. Lamarck's contemporaries (e.g., Virey and Georges Cuvier), his later readers (Darwin conspicuously among them), and recent commentators have, nonetheless, persisted in regarding will as the Lamarckian mechanism of evolutionary change.[99] Lamarck, on the contrary, held that ani-

96. Lamarck, "Discours de ouverture" of 1800, *Système des animaux sans vertèbres,* p. 13.

97. Lamarck, *Recherches,* p. 52.

98. Lamarck, *Philosophie zoologique,* 1.7, vol. 1, pp. 221–24. Lamarck was not entirely consistent in elaborating his theory. In his account of the transformation of plant species, which could not employ habits, he admitted the direct effects of the environment and changed diet as agents of heritable change. In exemplifying this principle, he mentioned that altered diet could also modify animals, causing profound modifications of species (pp. 224–25).

99. Julien Virey derided the hypothesis of the "celebrated naturalist" as requiring that "the insect, the animal, and even the plant—and generally every organized body— dispose and voluntarily arrange their interior and exterior structure through a simple effect of their will, with that marvelous harmony we find, so that they are in accord with the circumstances in which they find themselves." See Virey's article "Instinct" (1817), *Nouveau dictionnaire d'histoire naturelle* (Paris: Deterville, 1803–1819), 16:311. See also Virey's *Histoire des moeurs et de l'instinct des animaux* (Paris: Deterville, 1822), 1:495–96. I will describe Cuvier's interpretation of Lamarck's mechanism of habit in the text below.

   Charles Darwin, Alfred Wallace, and even Herbert Spencer decried Lamarck's mechanism precisely on the grounds that it involved an animal's will efforts. Spencer—in the *Principles of Biology* (New York: D. Appleton, [1866–1867] 1884), 1:406—complained that since desires implied mental representation, "to assume that in the course of evolution there from time to time arose new kinds of action dictated by new desires, is simply to

mals lowest in the phylogenetic scale, since they had no nervous system, lacked the necessary instrument of willful behavior. The actions and evolution of infusoria and polyps, in his view, were entirely the result of impingements by external, imponderable fluids. Insects, the first creatures with proper nervous systems, did possess the requisite physiology for sensation and feeling; and this enabled them to act habitually and instinctively. Yet insects and the other invertebrates, since they had no cerebral cortex, could not formulate ideas. Lamarck believed that without thought these lower creatures were quite incapable of acts of will.[100] But even for the higher vertebrates, he did not regard evolution through stimulation of needs and habits to be a phenomenon of will.[101]

In Lamarck's theory, having felt needs usually meant that an animal found certain kinds of behavior agreeable, particularly the essential activities of survival: hunting, feeding, drinking, mating, and the rest. Some needs, however, he assumed to be merely consequences of use. Echoing Cabanis's theory of habit, Lamarck stipulated a reciprocal re-

---

remove the difficulty a step back." Wallace, in his celebrated 1858 paper "On the Tendency of Varieties to Depart Indefinitely from the Original Type," reprinted in *Natural Selection and Tropical Nature* (London: Macmillan, 1891), pp. 31–32, objected that "the powerful retractile talons of falcon and cat tribes have not been produced or increased by the volition of those animals." And in a letter to Hooker (1844), Darwin bid "Heaven forfend me from Lamarck's nonsense of a 'tendency to progression,' 'adaptations from slow willing of animals,' etc.!" However, Darwin did confess to Hooker that "the conclusions I am led to are not widely different from his; though the means of change are wholly so." The letter is published in *More Letters of Charles Darwin*, ed. F. Darwin (New York: D. Appleton, 1903), 1:41. Darwin made similar disparaging remarks about Lamarck's "absurd hypothesis of will" in his notebooks; see, for example, his *First Transmutation Notebook*, MS p. 21, transcribed by Gavin de Beer, *Bulletin of the British Museum (Natural History), Historical Series* 2 (1960): 43. Darwin may have been overly sensitive, for his own early mechanism of evolution was not very different from Lamarck's, as will be indicated in the next chapter. Virey, Cuvier, Spencer, Wallace, and Darwin are followed in supposing that Lamarck relied on animal will by Howard Gruber, *Darwin on Man*, p. 307; Gertrude Himmelfarb, *Darwin and the Darwinian Revolution* (New York: Norton, 1968), p. 317; and Michael Ruse, *The Darwinian Revolution* (Chicago: University of Chicago Press, 1979), p. 7. Burkhardt and other Lamarck scholars have striven to correct this entrenched conception of Lamarckian theory.

100. Jean-Baptiste de Lamarck, "Discours d'ouverture de 1814," *Inédits Lamarck*, p. 235: "It is certain and recognized that the *will* (*la volonté*) is a determination through thought, which is able to take place only when the being who wills is able not to will. That determination results from acts of intelligence, that is to say, from transactions among ideas; and in general it occurs in consequence of a comparison, of a choice, of a judgment, and always of premeditation. But as every premeditation is an employment of ideas, it supposes not only a faculty of acquiring ideas, but additionally, acts of intelligence to employ and form them." See also Lamarck, *Philosophie zoologique*, 3.6, vol. 2, pp. 330–45.

101. Ibid., 3.1–3, vol. 2, pp. 180–275.

lationship between needs and habits. He thought needs were often as much a product of the facilitation of constant behavior as they were its motivation.[102] After successive generations laid down such habit-need connections, continuous use could then operate slowly to alter heritable structures through further excavations of the animal body. According to Lamarck's geologically conceived physiology, "in every frequently repeated action, especially those that become habitual, the subtle fluids producing it carve out and progressively enlarge, by the repetition of particular displacements that they undergo, the routes that they have to pass through, and render them more and more easy."[103] These acquired modifications could then be passed on to the progeny as inherited traits—provided both parents had been similarly transformed.[104] And all of this would occur without any intervention of will.

The Ascent of Mind

Lamarck agreed with Cabanis in identifying the mental realm with the physical.[105] In his *Recherches sur l'organisation des corps vivans,* he represented the faculty of intelligence as graded throughout the animal kingdom, rising in perfection with the increasing complexity of physiological organization:

> It is true that one observes a kind of gradation in intelligence of animals, as it exists in the increased perfection of their organization, and one notes that they have ideas and memories; that they think, choose, love, hate; that they are susceptible of jealousy; and that by diverse inflexions of their voice and by signs they communicate and understand one another.[106]

Yet in the *Recherches* Lamarck hesitated. It appeared to him that "man alone is endowed with reason and that by that consideration he is well distinguished from all other productions of nature."[107] The smooth progression of intellect up the scale of nature met an invisible barrier

102. Lamarck, *Histoire naturelle* 1:248: "an action become entirely habitual, having modified the interior organization of the individual to facilitate its execution, is then so agreeable that the action becomes a need for the individual."

103. Ibid.

104. Lamarck, like Charles Darwin, was sensitive to the problem of small modifications being swamped out by one animal mating with another that lacked the trait in question. He thus assumed that an acquired characteristic would be passed on only if both parents possessed it—something to be expected when both were exposed to similar environmental pressures. See Lamarck, *Philosophie zoologique,* 1.7, vol. 1, pp. 261–62.

105. Ibid., 3, introduction, vol. 1, p. 364.

106. Lamarck, *Recherches,* p. 124.

107. Ibid.

and began again as something rather different on the other side, or so Lamarck at first believed. But in light of the concentrated investigations of the *Philosophie zoologique,* he was finally convinced that no division actually separated brute from human intelligence. He achieved the new perspective thanks to an objective instrument of measure. In the *Philosophie zoologique,* he introduced the empirical test of nervous development. Since insects first exhibited the fundaments of a common nerve center, he attributed to them obscure feelings and internally determined instincts. And because a primitive cortex appeared in fish, the first of the vertebrate orders, he assumed that ideas and thought began in the sea.[108] But use of this measure did not reveal to him any qualitative saltations among the higher animals. So in the *Philosophie zoologique,* he concluded that all animals having intelligence, made possible by cerebral hemispheres, were also capable of some degree of reason, which, as Cabanis helped him to see, was only the guide of experience in the correct use of judgment.[109]

Sentiment Intérieur

Lamarck supposed that the gradual internalization of subtle fluids and the evolution of a nervous system released animals from the domination of external forces. When a common nerve center evolved in the insects, it permitted them the experience of a *sentiment intérieur* and therewith an internal motive force for habitual and instinctive behavior. In the higher animals, this inner feeling served a necessary function in generating thought and acts of will. Lamarck's hypothesis of the *sentiment intérieur* was thus central to this theory of the evolution of behavior.

According to the hypothesis, the *sentiment intérieur* arises from the

108. Julien Virey (see note 99) may have been the source for Lamarck's criterion of nervous development. In his article "Animal" (1803), *Nouveau dictionnaire d'histoire naturelle* 1: 419–66, Virey distinguished animals having both a cerebral nervous system and sympathetic system (i.e., the vertebrates) from those having merely the latter (i.e., molluscs to worms), and these two groups from animals with only nervous "molecules" (e.g., echinoderms and sponges).

Though Lamarck believed particular instincts and modes of thought depended on inherited nervous structures, he rejected the suggestion that actual ideas came fully formed in the neonate. He was strongly influenced by Condillac's psychology, particularly as filtered through Cabanis and Richerand. He was consequently chary of the doctrine of innate ideas, even of the theory of innate intellectual faculties. He believed animals inherited organizational structures, like the cortex; but he contended that for these structures to yield ideas and thought, the animal had to exercise the faculties and strengthen them through use. See Lamarck's *Philosophie zoologique,* 3.7, vol. 2, pp. 268–69.

109. Ibid., 3.8, vol. 2, pp. 441–47.

confused assembly of internal sensations of organic activity, blood flow, and muscle movement, all of which produce a general agitation of nervous fluid in the common medullary reservoir.[110] This vibratory disturbance provokes in sentient animals a feeling of existence, more obscure in the lower animals, more vivid in the higher. External impingements also excite the *sentiment intérieur,* via the five senses, both to motivate behavior in sentient creatures and to arouse ideas in the vertebrates. In the higher animals, according to the theory, ideas are produced through a rather complicated hydraulic process of fluids rebounding between the *sentiment intérieur* (i.e., its medullary reservoir) and the cortex.[111] In all animals having sensation, the environment is able to stimulate more or less obscure feelings of unease and need, which only appropriate action can relieve.[112] If the animal is intelligent and the environmental impact moderate, then it has opportunity to exercise judgment and exert willpower to quell its desires. But if the animal is merely sentient or, as with vertebrates, the environmental pressures are immediate and severe, then fluids released to motor nerves gush into channels well grooved by prior habit, and instincts result.

Theory of Instinct

Like Charles Darwin, Lamarck believed that the wonderful instincts of animals presented the strongest evidence against a theory of species transformation.[113] Only the direct hand of God, it would seem, could have fashioned behavior so expressive of intelligent forethought as the caterpillar's spontaneous construction of its cocoon. Yet also like Darwin, Lamarck argued that a perfectly naturalistic account was possible. He set out to explain these wonderful instincts, not, however, by empirical study of their progressive complication in related species—one of Darwin's occupations[114]—but by proposing mechanisms that might

110. Lamarck's doctrine of the *sentiment intérieur* was elaborated in ibid., 3.4, vol. 2, pp. 276–301.

111. In Lamarck's theory, sensation occurs when a motion is excited in the fluid of a sensory nerve, transmitted to the medullary reservoir, reflected throughout the system, and then rebounds along the original nerve. For animals with a cerebral cortex, ideas arise when the nervous fluid, agitated from sensory impingements, moves to the cortex, where it leaves an impression of the object perceived and then returns to the inner feeling. The final return to inner feeling makes conscious those impressions deposited in the cortex. Lamarck discussed this theory in ibid., 3.7, vol. 2, pp. 374–79.

112. Lamarck, *Histoire naturelle* 1 : 267.

113. Lamarck, *Philosophie zoologique,* 1.3, vol. 1, pp. 66–67.

114. Charles Darwin, *Charles Darwin's Natural Selection, being the Second Part of his Big Species Book Written from 1856–1858,* ed. R. Stauffer (Cambridge: Cambridge University Press, 1975), chap. 10.

transform the habits of previous generations into the innate behavior of the present. He appears to have assumed that an appreciation of these mechanisms alone would be sufficient to demonstrate the validity of the explanation. And this was, perhaps, not an unreasonable expectation; for he had laid a modestly firm empirical foundation for his general evolutionary theory by carefully detailing relations of progressive development in his elaborate classificational studies (especially of invertebrate species). Further, since Condillac and the naturalists Réaumur and Le Roy had maintained that instincts were really only acquired habits, he could simply build on that to show how the idea of modifications acquired through habit could be extended to species. This, as I explained above, was basically Cabanis's theory of instinct. And if Cabanis did not initially suggest this interpretation of instinct to Lamarck, he at least confirmed it.

In his "Discours d'ouverture" of 1814, Lamarck defined instinct as "that singular power which operates without premeditation and from the consequences of felt emotion."[115] He used this definition to distinguish instinct from voluntary behavior on the one hand and from mere inherited habit on the other. Despite the sensationalist tradition, whence many of his psychological beliefs derived, Lamarck recognized that instincts were not disguises for intelligent behavior. He enumerated several criteria by which instincts might be separated from actions performed intelligently. Intelligent acts, he observed, are "often improper, sometimes deceitful, and do not always attain their desired end; while those which are executed through instinct are never deceitful, are adapted directly to their end, and are always the most proper to satisfy a felt need."[116] In addition, Lamarck emphasized a property which, from the time of Descartes, had come to be firmly associated with instinct: instinctive behaviors, unlike intelligent actions, were uniform in the race or species. Even the marvelous and seemingly intelligent behaviors of the social insects, if they were carefully examined, would be found to be stereotyped in each variety. These fixed patterns could be produced, he urged, by the mechanisms of environmentally induced needs and responding mutagenic habits: "But all of these actions, as complicated as they might be, are always the same, without variation in the individuals of each race; this is because they are among the habits which have modified the organization of these individuals, compelling them to execute only these actions, and because it is instinct alone

115. Lamarck, "Discours d'ouverture de 1814," *Inédits de Lamarck*, p. 236.
116. Jean-Baptiste de Lamarck, "Instinct" (1817), *Nouveau dictionnaire d'histoire naturelle* 16 : 339.

which, in consequence of felt needs, excites their performance."[117] Finally, Lamarck's own investigations convinced him that animals lacking a cerebral cortex could not exercise intelligence, though, as in the case of the social insects, they often displayed some of the most wonderful instincts.[118] Thus, despite the presumptions of Condillac, Le Roy and other sensationalists, Lamarck held that instinct was not merely intelligent behavior become habitual.[119]

Nor were instincts, in Lamarck's theory, simply inherited habits. The apathetic animals (i.e., those without nervous center and internal sense) often showed patterned activity that had to be ascribed to the excavations of imponderable fluids and the heritable effects of their flow. These animals were devoid of any internal driving force; they had no feelings of unease or need which might prompt them to exercise their inherited habits. By contrast, instincts were, in Lamarck's view, internally motivated acts; they were "the consequences of emotions excited in the *sentiment intérieur* by each felt need."[120] He was persuaded that complex displays of instinct were decidedly different from the simpler tropic reactions of lower animals and that the persistence of higher animals in carrying out instinctive performances indicated a motive force of subjective needs.

Reactions at the Muséum National d'Histoire Naturelle

Lacépède and Geoffroy

During the first part of the nineteenth century, Lamarck was not the only member of the Muséum national d'histoire naturelle to suggest that species might have become transformed over time. As Burkhardt has pointed out, two other prominent zoologists at the Muséum also argued for species mutability: Bernard-Etienne de La Ville, Comte de Lacépède (1756–1825) and Etienne Geoffroy Saint-Hilaire (1772–1844).[121] Lacépède, who had been Buffon's young colleague before the Revolution, survived the Terror and continued on at the Muséum and in various governmental capacities. He was a nobleman of great humility and deference, certainly assets during those precarious times. He

117. Jean-Baptiste de Lamarck, "Habitudes" (1817), *Nouveau dictionnaire d'histoire naturelle* 14 : 134–35.

118. Lamarck, *Philosophie zoologique*, 3.7, vol. 2, pp. 327–29.

119. Lamarck did not deny that the higher animals often guided their instinctive tendencies by intelligent consideration, for example, in the adaptation of nest construction to local conditions. See ibid., 3.5, vol. 2, p. 329.

120. Lamarck, "Instinct," *Nouveau dictionnaire d'histoire naturelle,* vol. 16, p. 334.

121. Burkhardt, *Spirit of System,* p. 202

adopted the basic ideas of his mentor Buffon. In the introductory "Discours sur la durée des espèces" of the second volume (1800) of his *Histoire naturelle des poissons* (1798–1803),[122] he declared, in the spirit of
Buffon, that taxonomic categories served merely as convenient labels
for grouping similar individuals, though one could suppose they did
capture some real differences in nature.[123] Like individuals, species suffered the vicissitudes of physical existence and the inexorable processes
of time. Lacépède thought they might go extinct from either sudden
geological cataclysms or more insidiously: perhaps by succumbing to
senescence, with the organs of species members simply wearing out
over generations; or, just the opposite, by a too rapid development of
their organs, so that their anatomy would become deranged and un-
viable. But instead of perishing, one species might possibly be altered,
so that "it is thus metamorphosized into a new species."[124]

Lacépède proposed that such species mutation would pass through
some twelve stages: from slight alterations of skin texture and color at
the beginning of the process, to more profound internal changes in
organs, and finally to modifications of habit and behavior.[125] He listed
a number of ways, borrowed largely from Buffon, by which man and
nature might transform species—for instance, through changes of climate, food, and water. In advancing these different means, he supposed,
of course, that the environment could directly produce heritable
changes in organisms. Lacépède recognized, however, that modifications in species would be propagated only when individuals having
similar traits mated—something man could arrange more efficiently
than nature. A human stand-in for nature could forcibly join individuals
who displayed "the first outlines of the new species that he desired to
see produced."[126]

Though Lacépède argued that species could be transformed over
generations, he failed to attend to the problem that occupied other early
evolutionists, as well as Georges Cuvier: the intimate relation of an organism to its environment. The problem of adaptation did not arise for
Lacépède, nor consequently, the need to provide an adequate explanation for the phenomenon.

During the Terror, Lacépède retired to the provinces. He was absent

122. Bernard-Germain-Etienne de la Ville-sur-Illon, Comte de Lacépède, *Histoire naturelle des poissons* (Paris: Plassan, 1798–1803).
123. Ibid., 2:32–33.
124. Ibid., p. 35.
125. Ibid., pp. 35–37.
126. Ibid., p. 41.

from Paris when the Convention reorganized the Jardin du Roi as the Muséum national d'histoire naturelle in 1793.[127] The Muséum's chair of zoology, which he might have expected to get, instead fell to Geoffroy Saint-Hilaire, the very young protégé of the great anatomist Louis Daubenton. After the death of Robespierre in 1794, Lacépède returned to the capital; and the chair of zoology was divided. Geoffroy retained the section on birds and mammals, while Lacépède, the more seasoned naturalist, received charge of fish and reptiles. Lacépède pursued with detached pleasure the art of classifying, as his five volumes on fish testify; but the younger scientist became passionately consumed by morphological research.

In the late 1790s, Geoffroy had formulated an initial version of a principle that would continue to guide his work—the idea that fundamental analogies united the animal kingdom. Though each species displayed parts that varied in size, shape, and even location, he yet found constant relations or "connections" among them.[128] All vertebrates, he concluded, expressed the same plan of organization. This was a theory that Georges Cuvier (1769–1832), Geoffroy's colleague and early collaborator, could endorse, since in his own *Règne animal* (1817), he represented all animals as constructed according to one of four basic plans: that of the radiata (e.g., jellyfish and starfish), mollusca (e.g., clams and octopuses), articulata (e.g., bees and lobsters), or vertebrata (e.g., fish and men).[129] Geoffroy, however, quickly reduced Cuvier's divisions to two. In 1820, he announced his discovery of analogies between radiata and mollusca, on the one hand, and articulata and vertebrata on the other. This discovery demonstrated that nature was more thoroughly unified than most morphologists suspected.[130] It required some creativity, perhaps, to perceive more than gossamer similarities joining the ant with the elephant, but an imagination of grand proportions to find, as Geoffroy claimed to in 1830, the same pattern to include the clam.[131] When

127. Camille Limoges describes the foundation and institutional evolution of the Muséum in "The Development of the Muséum d'Histoire Naturelle of Paris," in *The Organization of Science and Technology in France, 1808–1914,* ed. Robert Fox and George Weisz (Cambridge: Cambridge University Press, 1980), pp. 211–40.

128. The most thorough and sympathetic discussion of Geoffroy's morphology is found in E. S. Russell, *Form and Function: A Contribution to the History of Animal Morphology* (Chicago: University of Chicago Press, [1916] 1982).

129. Georges Cuvier, *Le règne animal,* 2d ed. (Paris: Deterville, 1829–1830), 1:48–51.

130. Etienne Geoffroy Saint-Hilaire, "Sur une colonne vertébrale et ses côtes dans les insectes apiropodes," *Isis* 2 (1820): 527.

131. Geoffroy presented his "theorie des analogues" in the preliminary discourse to his *Principes de philosophie zoologique* (Paris: Didier, 1830).

he declared that a single plan united the entire animal realm, Cuvier finally protested and ignited a dispute that badly injured Geoffroy's reputation as an exact scientist.[132]

Geoffroy did not originally attempt to explain morphological similarities by the evolutionary hypothesis. Only gradually and cautiously did he adopt that explanation. In a memoir on extinct crocodiles in 1825, he initially suggested that contemporary animals might be descendants of creatures that had lived before the flood. He ventured that the geological upheavals which had convulsed the earth in early times were "of a nature to have acted on the organs [of animals] . . . and to have done so precisely according to the two laws proposed by M. de Lamarck in his *Philosophie zoologique*."[133] Though he was not prepared in his memoir to advance the transformation hypothesis, he yet wished to show that "it is not repugnant to reason, that is, to physiological principles, that the crocodiles of the present age have descended by an uninterrupted succession from antediluvian species whose remains are found fossilized in our country."[134] In 1828, he again broached the problem of species mutation, still without attempting a definitive resolution.[135] He wanted, however, to provide a number of considerations that supported the hypothesis. He urged, for instance, that cases of monstrous births had evolutionary implications. Such cases weighed against the theory of preformation (accepted by Cuvier), according to which successive generations of a species were already formed in miniature and encapsulated in the germs, as in so many Chinese boxes. By contrast, the theory of epigenesis, holding that the embryo developed its parts out of an undifferentiated mass, allowed for the possibility that interfering causes could produce unusual progeny. But if the theory of epigenesis were true, then by analogy, might not species develop successively over the ages? Geoffroy thought the formative tendency (*nisus formativus*), which conserved the structure of species, must surely bend to the influences of the environment, particularly to those great changes spoken of

132. Toby Appel provides a thorough examination of the dispute in "The Cuvier-Geoffroy Debate and the Structure of Nineteenth-Century-French Zoology," (Ph.D. diss., Princeton University, 1975).

133. Etienne Geoffroy Saint-Hilaire, "Recherches sur l'organization des Gavials," *Mémoires du Muséum d'histoire naturelle* 21 (1825): 95–155; the sections referred to are on pp. 149–58. Lamarck's two laws, which Geoffroy mentioned, are the law that use and disuse modify anatomical structures and the law that these modifications can be inherited. See Lamarck, *Philosophie zoologique* 1:235.

134. Geoffroy, "Recherches sur l'organization des Gavials," pp. 152–53.

135. Etienne Geoffroy Saint-Hilaire, "Rapport fait a l'Académie royale des sciences sur un mémoire de M. Roulin," *Mémoires du Muséum d'histoire naturelle* 17 (1828): 201–29.

by geologists. While offering these considerations, he also felt compelled to defend his colleague Lamarck against the objections voiced by Georges Cuvier, who treated the question of species mutability "only according to the science of the present moment."[136]

Geoffroy believed that zoology had passed beyond the stage represented by Cuvier—a rash judgment, no doubt. It would not have been sustained by many of Geoffroy's contemporaries, or by scientists of succeeding periods, since Cuvier was generally acknowledged to be the very embodiment of the nineteenth-century scientist.[137] Yet Geoffroy claimed, in a memoir read to the Académie royale des sciences in 1831, that science had already demonstrated the principle of "the unity of organic composition" and had to go on to investigate how accidents of the environment modified the unity of organization to produce the multitude of species.[138] In this memoir, his most forthright advancement of the evolutionary hypothesis, he again recommended Lamarck's *Philosophie zoologique* as breaking new ground. But he differed from his late colleague on the mechanisms of transformation. Though he had referred favorably to Lamarck's mechanism of habit in his 1825 memoir, he now placed the burden of species change on the direct effects of the environment. Geoffroy had occupied himself throughout his career with tracing analogical similarities over the expanse of the animal kingdom. He now followed them back through the temporal depths of the kingdom. The problem of accounting for the differences among animals, particularly their finely determined adaptations to different environments, simply did not carry significant weight for him. Both Lacépède and Geoffroy virtually ignored what constituted for other evolutionists, as well as for Cuvier, a central fact of the animal economy.

Two points stand clear in the reaction of French scientists to theories of species transformation. The first is that, contrary to the usual assumption, such theories did not meet unanimous opposition. Opinions were divided concerning species mutation. Even Henri de Blainville (1777–1850), a nonevolutionist who succeeded to Lamarck's chair at the Muséum in 1829, had a strong sympathy for his predecessor, regarding him as "the man who contributed most to the progress of science" in

136. Ibid., p. 217.
137. John Theodore Merz, the great historian of nineteenth-century thought, judged that the scientific spirit of the modern age "knows no greater figure than Cuvier." See Merz, *A History of European Scientific Thought in the Nineteenth Century* (New York: Dover, [1904–1912] 1965), 1: 132–33.
138. Geoffroy Saint-Hilaire, "Le degré d'influence du monde ambiant pour modifier les formes animales," *Mémoires de l'Académie royale des sciences* 12 (1833): 63–92.

the modern era.[139] Blainville's estimation of Lamarck stemmed, to be sure, also from personal animus against Cuvier. Nonetheless, Blainville, Lacépède, and Geoffroy testify that sentiment at the Muséum during the first part of the nineteenth century was not entirely hostile to the idea of species transformation. The second point is more germane to the subject of this chapter: while Lacépède and Geoffroy adopted evolutionary theory, they allotted virtually no role to behavior as a mechanism of adaptation. Indeed, neither showed any keen sense of the need to consider the fit between an animal and its surroundings. But Lamarck, and to a lesser extent Cabanis, employed the mechanism of habit precisely because they recognized the importance of adaptation. Georges Cuvier, of course, regarded the articulation of a creature with its environment as an essential consideration for the zoologist. It is no accident, I believe, that Cuvier reacted strongly to the transformism of Lamarck, while virtually ignoring that of Lacépède and Geoffroy: Cuvier sensed the challenge of the behavioral mechanism of adaptation. Even so, he probably never seriously doubted his own conception of species, inclined as he was to accept the popular estimate of his scientific authority and merit. But his denunciation of transformism and his parody and misrepresentation of Lamarck's theory of adaptation bespeak some fear that other, less knowledgeable souls might be led astray.

Georges Cuvier's Criticism of Lamarckian Evolutionism

In the spring of 1832, Georges Cuvier began drafting a sustained attack on the theory of species evolution; but he died suddenly in May of that year after completing only a few manuscript pages.[140] Likely he would have developed objections he had previously made. His first extended analysis of transformism came in the introductory volume of his *Recherches sur les ossements fossiles* (1812). There he pursued his criticism along three lines: first, no remains of transitional species are found in the rocks, though mutability theory implies they should be; second, species vary only within narrow boundaries and then merely in nonessential characters, thus precluding radical alterability of animals over time; finally, three-thousand-year-old mummified animals from ancient

139. See Toby Appel, "Henri de Blainville and the Animal Series: A Nineteenth-Century Chain of Being," *Journal of the History of Biology* 13 (1980): 291–319.

140. The treatise was to be called "Sur la variété de composition des animaux." The first few pages of the introduction are in Fonds Cuvier, MS. 65, Institut de France. William Coleman refers to this unfinished work in *Georges Cuvier, Zoologist* (Cambridge: Harvard University Press, 1964), p. 204, n. 1.

Egypt are recognizably the same as found in modern Egypt, which suggests that transmutational variability is a myth.[141] Cuvier reiterated the second of these criticisms in the general introduction to his *Règne animal*(1817),[142] and with obvious allusion to Lamarck further argued that alterations of animal organization through the agency of vital fluids would "necessarily halt the general activity of life."[143] Cuvier's most devastating attack on Lamarck, however, came in his *éloge* (certainly a misnomer) for his dead colleague.[144]

After some faint praise for Lamarck's efforts in botany and invertebrate zoology, and then passing over his unfortunate attempts in chemistry and geology, Cuvier arrived at the infamous theory. He chose to reduce it to rubble by undermining its central support, the adaptational mechanism of habit. In Cuvier's caricature: "It is the power of the desire to swim that produces the membranes between the toes of aquatic birds; it is by reason of going in the water but wishing not to get wet that river birds have their legs lengthened; it is the power of desiring to fly that has changed their arms into wings and has developed their hair and scales into feathers."[145] In these and similar misrepresentations, Cuvier made Lamarck's intricate mechanism of habit into an absurdity: he transmogrified objective need, determined by the environment, and habitual responses into the wishes and fancies of dumb animals. These certainly could be dismissed as "something that might amuse the imagination of a poet."[146]

Cuvier's fundamental motive for opposing transformism seems clear enough: species mutation would be difficult (though not impossible) to reconcile with his key methodological principle that all the parts of an animal were finely coordinated with each other and with the environment in which the creature lived. He presumed that any basic changes in an animal would wreck internal organization and upset the

141. Georges Cuvier, *Recherches sur les ossements fossiles*, 4th ed. (Paris: D'Ocagne, [1812] 1834), 1:198–218. These passages do not differ from those of the first edition.

142. Cuvier, *Règne animal* 1:16.

143. Ibid., p. 14.

144. Lamarck died in December 1829. Cuvier's eulogy of Lamarck was read to the Académie des science in November, 1832. Cuvier himself had died suddenly the previous May, so the eulogy was read by a colleague. It was published with corrections and additions by his brother Frédéric in 1835. See the fair copy of the eulogy in Fonds Cuvier, MS. 3156. It was deposited by Georges Cuvier, Frédéric Cuvier's son. It carries the inscription: "Copie avec des corrections de mon père." It was published as "Eloge de M. De Lamarck," *Mémoires de l'Académie des sciences*, 2d series, 13 (1835): 1–xxxi.

145. Ibid., p. xix.

146. Ibid., p. xx.

Figure 1.4   Georges Cuvier, 1769–1832, portrait done ca. 1834.

delicate balance established by the Creator between an animal and its milieu.[147] But Lamarck's actual mechanism of adaptation did not really overturn these principles. It merely insulted the authority of Cuvier's static morphology and the omniscience that his position in French science accorded him. Certainly several later evolutionists who accepted the principle of morphological coordination—such as Charles Darwin, Herbert Spencer, and Ernst Haeckel—did not find the idea that habit produced heritable changes in species quite so implausible. Rather, they saw in that mechanism a means by which species might remain adapted to an ever-changing environment. The habit-mechanism of species change, however, had another advocate, one who nonetheless sided with Georges Cuvier. This was Frédéric Cuvier, his brother.

147. Cuvier developed his methodological theory of the correlation of parts at some length in *Recherches sur les ossements fossiles* 1:176–89.

Frédéric Cuvier's Theory of the Evolution of Behavior

The legend on his tomb in Strasbourg reads simply: "Frédéric Cuvier, frère de Georges Cuvier." In death as in life, Frédéric Cuvier was best known as the brother of the great zoologist. Yet Frédéric was an innovative scientist, who conducted exact experimental studies of mammalian behavior. Pierre Flourens (1794–1867), the permanent secretary of the Académie des sciences and a neurophysiologist of repute, wrote a monographic tribute to Frédéric's "positive science" of animal instinct.[148] Cuvier's work was known to an appreciative if small audience of later investigators, including Dugal Stewart, Charles Lyell, and Charles Darwin.[149]

Georges Cuvier brought his brother to the Muséum national d'histoire naturelle in 1797, initially to help arrange the exhibits in the hall of comparative anatomy. In 1804 Frédéric took charge of the menagerie of the Muséum and continued in that post until his death. Under his direction the menagerie flourished and grew considerably in size. Surviving early letters of Frédéric to Georges describe his activities in enlarging the number of specimens, and later letters detail the new building projects to house the growing collection.[150] He used the opportunities he had as director of the menagerie to undertake several studies of the behavior and psychological faculties of higher animals. It is these studies that make his position on the species question so very strange. His work brought him to conclude: that the rational abilities of higher animals were comparable to man's; that habits became hereditary and contributed to the perfection of animal groups; and that moral conscience, a seemingly distinctive human trait, was rooted in the animal instinct of sociability. He admitted that the psychological faculties and instincts of species might be transformed over time. But he remained faithful to the fundamental position of his brother that the anatomical features of species were unchanging.

Early in his career at the Muséum, Cuvier formed the plan of studying the intellectual faculties of one or two species in each genus of mammal, in order "to estimate the laws that operate in the entire class, and to understand the successive degradations (*dégradations*) which each

148. Pierre Flourens, *Résumé analytique des observations de Frédéric Cuvier sur l'instinct et l'intelligence des animaux* (Paris: Langloiset Leclercq, 1841). Flourens's monograph went through four editions between 1841 and 1861; with each Flourens added more of his own considerations on animal psychology and its relation to human psychology.

149. See, for example: Dugal Stewart, *Elements of the Philosophy of the Human Mind*, vol. 3 (Philadelphia: Carey, Lea & Carey, 1827), pp. 220, 326–27; Charles Lyell, *Principles of Geology* 2:38–44; and Charles Darwin, *On the Origin of Species*, p. 208.

150. The letters are in Fonds Cuvier, MS. 3342, Institut de France.

Figure 1.5   Frédéric Cuvier, 1773–1838, engraving in 1826.

class represents . . . in a word, to provide the foundations for that interesting part of natural history which till now has been constituted by an
imaginary system and obscure facts."[151] Cuvier did not advance very far
in his program, but he did produce some extraordinary studies of the
orangutan, seal, and wild dog, as well as important theories of animal
sociability and instinct.[152] His experimental observations of the orang
are of particular interest.

Travelers to the East Indies brought back wild tales of the very human activities of the orang, of its hissing language, its planned attacks

151. Frédéric Cuvier, "Description d'un orang-outang, et observations sur ses facultés
intellectuelles," *Annales du Muséum d'histoire naturelle* 16 (1810): 65.

152. Flourens describes Cuvier's studies of particular animals in his *Résumé analytique*,
pp. 87–118.

on men, and its dalliances with native women. Buffon had secured a supposed variety of the animal, a Jocko, who, however, failed to live up to its reputation. Buffon concluded that the *homo sylvestris* was really no relative of man, even though it preferred to take tea with human friends.[153] But fascination with the creature did not abate. Cuvier got his orang, a female of about fifteen months, from Borneo in 1808. On several occasions the animal did show humanlike ingenuity in manipulating a bolt lock and escaping from a room in which it had been kept. After a few such escapes, Cuvier made the problem more difficult, yet the animal quickly came to the solution. These and other displays led him to regard his orang as possessed of the generalizing faculty of reason: "it will hardly be possible," he observed, "not to see the result [of its action] as stemming from a combination of rather abstract ideas and not to recognize in the animal that is so able a faculty of generalizing."[154]

Though Cuvier did not hesitate in 1810 to claim that his orang exhibited an "extensive use of reason (*raisonnement*),"[155] he later grew more cautious. In 1822, in his article on instinct for the *Dictionnaire des sciences naturelles,*[156] he warned that human beings alone enjoyed the full abstractive and generalizing power of reason, which permitted them "to acquire pure ideas, to construct a conception of justice, of beauty, of truth, and to work to achieve their perfection."[157] Only poverty of psychological vocabulary, he confessed, sometimes led him to suggest otherwise.[158] He believed, however, that the term "intelligence" (*l'intelligence*), had its proper use in characterizing animal behavior, since it signified something less than reason. It indicated a faculty for appropriately modifying behavior in changed circumstances. Through the 1820s and 1830s, Cuvier took careful steps to avoid explicitly implying that there was a perfect continuity of mental development between animals and man. But his demur held weakly against the strong current of his psychological theories of animal behavior, especially his conception of instinct.

Cuvier's theory of the structure and origin of instinct bears strong resemblance to the views of Cabanis and Lamarck. To elucidate the

153. Buffon, "Les orangs-outangs, ou le pongo et le jocko" (1766), *Histoire naturelle,* in *Oeuvres complètes* 4 : 23–38.

154. F. Cuvier, "Description d'un orang-outang," p. 58.

155. Ibid., p. 62.

156. Frédéric Cuvier, "Instinct" (1822), *Dictionnaire des sciences naturelles* (Strasbourg: Levrault, 1816–1843), 23 : 528–44.

157. Ibid., p. 543.

158. Frédéric Cuvier, "De la sociabilité des animaux," *Mémoires du Muséum d'histoire naturelle* 13 (1825): 1, n. 3.

phenomenon of instinct, Cuvier chose the model of habit, which he understood as an organic disposition activated by some environmental situation and motivated by need. Practices that became habitual usually began with conscious intention, but passed into something like a mechanical response. Such fixed behavior, according to Cuvier, "establishes an immediate dependence between the organs of an animal and the natural needs, appetites, tendencies, and ideas, without any mediation of mind."[159] But this, he noted, was precisely the relation that instinct established between natural needs and behavioral dispositions. The principal distinction between habit and instinct was that an animal acquired habits, but instincts were innate—at least for an individual animal. But the case was different for a generational lineage. For also like Cabanis and Lamarck, Cuvier believed that firmly ingrained habits could be inherited as instincts by succeeding generations. In Cuvier's view, environmental needs would control the acquisition of habits in one generation; and if these needs persisted over subsequent generations, continued practice would render them innate:

> It is true that some of those qualities we regard as due to instinct in the mammals are subsumed by the laws which depend on education and that those which finally become instinctive or hereditary are ones which have been exercised for sufficient numbers of successive generations and that these are wiped out and removed to some degree when exercise ceases to strengthen and maintain them.[160]

Thus with Cabanis and Lamarck, Cuvier agreed that the innate behavioral patterns of species might originate in and be maintained by environmentally induced needs.

The conceptual bonds joining Cuvier with the evolutionists became even more tightly drawn by his proposal that the habit-mechanism of instinct would lead to the perfection of animal and human societies. Social groups of animals were capable, in Cuvier's estimation, of improving their accommodation to the environment by passing down through generations the acquired experience of predecessors.[161] The same was true, he thought, for the human animal. Our ancestors, for instance, lacked the ability directly to perceive objects as being at a determinate distance. But gradually as they came to associate sensations of touch with particular changes in the visual image, the judgment of distance became innate. Succeeding generations have thus come imme-

159. F. Cuvier, "Instinct," p. 540.
160. Frédéric Cuvier, "Observations sur le chien des habitants de la Nouvelle-Hollande," *Annales du Muséum d'histoire naturelle* 11 (1808): 462.
161. Ibid., p. 464.

diately to see objects at a particular distance. Similar adaptations, Cuvier believed, originally formed the different races of men and the varieties of animal species.[162]

In his early studies, Cuvier discovered in the higher animals rational abilities and social feelings comparable to man's own. Later he continued to detect close resemblances between animal and human societies; indeed the instinct of sociability displayed by animals, when illuminated by reason, became, he thought, the distinctively human trait of conscience.[163] But Cuvier drew the line between animals and men increasingly more sharply. In studies during the 1820s and 1830s, he granted that animals might be perfectible beyond their original station, but concluded that they never achieved the rational power of human mind.

Though he adopted the same habit-mechanism as Cabanis and Lamarck, and used it to explain psychological modifications of species over generations, Cuvier could not follow the transformist path to the end. In a set of preliminary observations to the posthumous fourth edition (1834) of his brother's *Recherches sur les ossements fossiles,* he reiterated Georges's arguments against fundamental anatomical changes in species,[164] neglecting to mention, however, his brother's similar arguments against instinctive changes.[165] Undoubtedly Frédéric's position was quite sincere: he did not believe that the anatomical patterns of species were modified over time (though he did admit they changed in nonessential ways through the inheritance of acquired characteristics).[166] His own close observations yet convinced him that the innate behavioral patterns fundamental to species survival did change, and by the very mechanism that the early evolutionists had advanced. Despite his fraternal devotion, Cuvier considered behavior as a powerful instrument for giving shape to the heritable constitutions of men and animals. He was a behavioral evolutionist, if a modest one.

Conclusion

Naturalists in the sensationalist tradition perceived the theory of animal instinct to be wedded to an outmoded philosophy. Réaumur, Guer, Condillac, and Le Roy contemned the theory that invoked innate images and employed a *machina ex Deo* to explain natural phenomena.

162. Frédéric Cuvier, "Examen de quelques observations de M. Dugal-Stewart, qui tendent à détruire l'analogie des phénomenes de l'instinct avec ceux de l'habitude," *Mémoires du Muséum d'histoire naturelle* 12 (1823): 256–57.

163. G. Cuvier, *Recherches sur les ossements fossiles* 1:179.

164. Frédéric Cuvier, "Observations préliminaires," to Georges Cuvier, *Recherches sur les ossements fossiles* 1:XI–XIV.

165. G. Cuvier, *Recherches sur les ossements fossiles* 1:179.

166. F. Cuvier, "Observations préliminaires," pp. xv–xvi.

Patient empirical research, like Réaumur's, seemed to show that even insects were capable of intelligent action. To Condillac and Le Roy, this meant that instincts could be interpreted as intelligent conduct become habitual. Yet the evidence of writers like Reimarus could not easily be ignored: animals did exhibit wonderfully complex patterns of behavior before they had any opportunity to learn them. Early evolutionists admitted evidence of this kind, but they attempted explanations that would be consistent with sensationalist ideas. Erasmus Darwin, true to the sensationalist faith, argued that congenital instincts were really learned behaviors, though acquired in the womb. Cabanis, while of similar opinion, yet stressed that instincts were the consequences of nonintelligent habits molded by inherited anatomical structures. Lamarck refined Cabanisean theory by reference to specific nervous complexes and a *sentiment intérieur*. Cabanis, Lamarck, and to some degree Erasmus Darwin all proposed a mechanism, under the inspiration of sensationalism, for the evolution of both anatomical and behavioral structures: the inheritance of the effects of habitual practices. As I will discuss in following chapters, this became a central principle of species mutation in the early theory of Charles Darwin as well as in the theories of a host of later evolutionists, such as Herbert Spencer, Douglas Spalding, George Romanes, Ernst Haeckel, and William McDougall.[167] Even such staunch neo-Darwinists as Conwy Lloyd Morgan and James Mark Baldwin contrived a means by which behavior might function as an engine of heritable alteration. Some early evolutionists, such as Lacépède and Geoffroy, did not, however, employ the mechanism of habit to explain species transformation. In confronting the doctrine of evolution, Georges Cuvier virtually ignored the arguments of Lacépède and Geoffroy. Instead he concentrated on the theory that had a plausible mechanism of change, Lamarck's. So powerful was the conception that behavior might alter the heritable constitution of animals, that Frédéric Cuvier, despite his fraternal allegiance, advanced it as an explanation of the instincts of animals and the perfectibility of their societies. It was thus a prevailing, if not universal, belief of early writers on evolution, and one which crept well into the twentieth century, that not only did behavior evolve, but it also functioned as a principal instrument of species modification.

---

167. The views of these authors concerning the evolutionary function of behavior are discussed in the chapters that follow.

2

Behavior and Mind in Evolution: Charles Darwin's Early Theories of Instinct, Reason, and Morality

Loren Eiseley once remarked that "man was not Darwin's best subject."[1] Eiseley saw Darwin as constantly trying to escape the bedevilment of metaphysicians, with their talk of reason, morals, and the nature of man. It is true, Darwin often enough deprecated his own abilities in the profounder sciences, so one might be persuaded to take him at his word.[2] The opinion is easily formed that, to adapt Hobbes's characterization of Descartes, had Darwin kept to beetles, worms, and orchids, he would have been the best biologist in England, but his head did not lie for philosophy. This opinion is abetted by the estimate frequently made of Darwin's general conceptual abilities. Jacques Barzun, displaying a Frenchman's taste for the nuances of argument and distaste for British empiricism, contended that "Darwin was a great assembler of facts and a poor joiner of ideas."[3] The Shrewsbury biologist, it seems, simply lacked the quick and penetrating genius required to deal with the complexities of human mental and moral development. He might be able to explain the shape of a finch's bill, but not the form of a moral judgment.

Darwin was not possessed of the genius, say, of Huxley, whose swiftness of insight often made the older man uncomfortable. But genius has its varieties. Darwin's own definition, which he offered in the *De-*

1. Loren Eiseley, *Darwin and the Mysterious Mr. X* (New York: Dutton, 1979), p. 202.

2. In his *Autobiography,* ed. Nora Barlow (New York: Norton, 1969), Darwin credited himself with the ability to sustain a line of thought—"for the *Origin of Species* is one long argument from beginning to end" (p. 140)—but not with a focusing intellect that could resolve remote concepts: "My power to follow a long and purely abstract train of thought is very limited; I should, moreover, never have succeeded with metaphysics or mathematics" (p. 140).

3. Jacques Barzun, *Darwin, Marx, Wagner,* 2d ed. (New York: Doubleday Anchor, 1958), p. 74. Barzun's opinion is shared by Gertrude Himmelfarb. See her *Darwin and the Darwinian Revolution* (New York: Norton, 1968), chaps. 15–17.

scent of Man, suggests another kind: "genius has been declared by a great authority to be patience; and patience, in this sense, means un-flinching, undaunted perseverance."[4] This describes, not accidentally, Darwin's own mental character. Yet he recognized that English dogged-ness was not enough. "But this view of genius," he allowed, "is perhaps deficient; for without the higher powers of the imagination and reason, no eminent success in many subjects can be gained."[5] Darwin's genius also encompassed these additional qualities. This becomes especially evident, I believe, in his theories of behavioral and mental evolution.

Aside from reservations about Darwin's intellectual abilities, his crit-ics, both in the nineteenth century and in our time, have taken excep-tion to his treatment of behavioral and mental evolution for further, substantive reasons. At the beginning of our century, for example, George Bernard Shaw, though an evolutionist, balked at the Darwinian biologist's entering the domain of the moral philosopher. In the preface to his play *Back to Methuselah,* Shaw observed:

> you cannot understand Moses without imagination nor Spur-geon [a famous preacher of the day] without metaphysics; but you can be a thorough-going Neo-Darwinian without imagi-nation, metaphysics, poetry, conscience, or decency. For "Natural Selection" has no moral significance: it deals with that part of evolution which has no purpose, no intelligence, and might more appropriately be called accidental selection, or bet-ter still, Un-natural Selection, since nothing is more unnatural than an accident. If it could be proved that the whole universe had been produced by such Selection, only fools and rascals could bear to live.[6]

The presumption is, of course, that when natural selection is brought to explain mind and morals, it voids both of significance. Mind becomes blind mechanism, and morals are eviscerated of higher purpose. Even among those generally persuaded of neo-Darwinian theory—Shaw sided with Lamarck—many are not prepared to go the "whole orang" with Darwin. They grant Darwinian mechanisms are capable of explain-ing man's general frame, but believe the resources of evolutionary biol-ogy cannot render intelligible the features of cultural behavior, certainly not the nature of ethical judgment. The anthropologist Marshall Sah-

4. Charles Darwin, *The Descent of Man, and Selection in Relation to Sex* (London: Mur-ray, 1871), 2:328.

5. Ibid.

6. George Bernard Shaw, "Preface," *Back to Methuselah* (London: Penguin Books, [1921] 1961), p. 44.

lins, for one, depicts social behavior as cut adrift from biological anchors, and so, unable to be captured in evolutionary terms. He thinks only a semiotic analysis will do.[7] And those of a more traditionally philosophical bent, from G. E. Moore to Anthony Flew, concur, if about little else, that evolutionary interpretations of ethical behavior produce a logically vulgar inference—the derivation of moral imperatives from scientifically factual statements, a lubricious slide from an *is* to an *ought.*[8] These latter two demurs usually lead to a third reason for objecting to an evolutionary construction of the higher faculties. It is the historical objection that Darwin unwittingly infused his theory with the political assumptions of laissez-faire English liberalism and the hedonistic selfishness of Benthamite utilitarianism and that this has hopelessly infected any evolutionary analysis of mind and morals.[9] It is argued that conceptions of rational ability and ethical choice which find root in the ideology of a particular culture—for example, nineteenth-century Victorian society—must be inherently defective or at least circumscribed thereby. Sahlins maintains that "Darwinism, at first appropriated to society as 'social Darwinism,' has returned to biology as a genetic capitalism." He warns us that the current incarnation of 'social Darwinism,' that is, sociobiology, "contributes primarily to the final translation of natural selection into social exploitation."[10]

A historical scrutiny of Darwin's theories of the evolution of instinct, reason, and morality will, I believe, discover a thinker of extraordinary

7. This is the theme of Marshall Sahlins's *The Use and Abuse of Biology* (Ann Arbor: The University of Michigan Press, 1976).

8. G. E. Moore analyzed evolutionary ethics in his *Principia Ethica* (London: Cambridge University Press, 1903), chap. 3; see my discussion below in chapter 7 and in appendix 2. For contemporary treatments of evolutionary ethics see Anthony Quinton, "Ethics and the Theory of Evolution," in *Biology and Personality,* ed. I. T. Ramsey (Oxford: Blackwell, 1966), pp. 107–130; and especially Anthony Flew, *Evolutionary Ethics* (New York: St. Martin's, 1967).

9. Ashley Montague—in *Darwin, Competition and Cooperation* (New York: Schuman, 1952), p. 32—gives expression to what has become a commonplace: "The truth is that Darwinian biology was largely influenced by the social and political thought of the first half of the nineteenth century, and that its own influence took the form of giving scientific support in terms of natural law for what had hitherto been factitiously imposed social law." For similar interpretations of Darwin's theory see: Bertrand Russell, *Religion and Science* (New York: Holt, 1935), pp. 72–73; Eric Nordenskiold, *The History of Biology* (New York: Tudor, 1935), p. 477; Marvin Harris, *The Rise of Anthropological Theory* (London: Routledge and Kegan Paul, 1968), pp. 108–25; Robert M. Young, "The Impact of Darwin on Conventional Thought," in his *Darwin's Metaphor* (Cambridge: Cambridge University Press, 1985), pp. 1–22; and Marshall Sahlins, in the passage quoted in the text and cited in the following note.

10. Sahlins, *Use and Abuse of Biology,* pp. 72–3.

intellectual power and sophistication, not the caricature of the good-natured but bumbling biologist. Moreover, it will demonstrate the centrality of certain concepts of mind and behavior to the development of Darwin's general theory of evolution. Such an examination will also, I think, weaken the objections mentioned in the preceding paragraph. This chapter and the following three will not, however, portray a scientist hermetically sealed from his culture—that could hardly be the consequence of an evolutionary historiography; yet they will reveal a thinker whose theories escaped the cultural and philosophical constraints generally assumed. I intend in these chapters also to indicate how Darwin strove to preserve the dignity of his principal subject, man as a moral creature. Finally, I will explore in particular his treatment of moral judgment. Darwin's theory of conscience remains philosophically attractive, as I believe historical analysis will show. Whether an evolutionary ethics must sin against moral logic is a question broached in chapter 7 but more thoroughly discussed in the second appendix to this book.

Preparations of an Evolutionary Thinker

Early Life

Born on 12 February 1809, the son of Robert and the grandson of Erasmus, Charles Darwin came into a family already consisting of a brother, Erasmus, and three sisters, Marianne, Susan, and Caroline; the last of Dr. Robert and Susannah Wedgwood Darwin's children, Catherine, was born a year after Charles. Charles's mother, a daughter of Josiah Wedgwood (who founded the famous pottery firm), died when he was eight years old. His sisters assumed domestic responsibilities. Life in the Darwin household appears to have been rather pleasant, with all the necessities and many of the luxuries of the upper middle class provided by Dr. Robert's ample income. The father's relations with his son suffered the usual strains that a successful professional has with his children. Charles remembered some fifty years later that his father had chastised him because of schoolboy idleness: "You care for nothing but shooting, dogs, and rat-catching," Dr. Darwin had admonished, "and you will be a disgrace to yourself and all your family."[11] Though just after recalling his father's words, Charles added in his *Autobiography:* "But my father, who was the kindest man I ever knew, and whose memory I love with all my heart, must have been angry and somewhat

---

11. Darwin, *Autobiography,* p. 28.

unjust when he used such words."[12] Charles's sentiments seem to have
been genuine and adequately recalled. There is no evidence that his
relationship with his father was infected by the extreme Oedipal anxieties that some psychoanalysts believe they have uncovered.[13]

The young Darwin spent seven years, from 1818 to 1825, at the Shrewsbury School, a prestigious public school run by Dr. Samuel Butler,
whose own son became one of Darwin's most acid critics in later years.
Charles felt that "nothing could have been worse for the development
of my mind than Dr. Butler's school."[14] The curriculum was strictly
classical, and the young Darwin's linguistic abilities were modest, which
caused his masters and his own father to consider him "a very ordinary
boy, rather below the common standard in intellect."[15] After leaving
Shrewsbury, Darwin matriculated at Edinburgh University in the fall of
1825, with the intention of following in the footsteps of his grandfather,
father, and brother. He recalled his pursuit of a medical education with
some chagrin:

> The instruction at Edinburgh was altogether by Lectures, and
> these were intolerably dull, with the exception of those on
> chemistry by Hope. . . . Dr. Duncan's lectures on Materia
> Medica at 8 o'clock on a winter's morning are something fear-
> ful to remember. Dr. Munro made his lectures on human
> anatomy as dull, as he was himself, and the subject disgusted
> me. It has proved one of the greatest evils in my life that I was
> not urged to practised dissection, for I should soon have got
> over my disgust; and the practice would have been invaluable
> for all my future work.[16]

It is doubtful how far along Darwin would have gotten with dissection,
since on the two memorable occasions when he witnessed bloody operations, he ran from the theater in revulsion.[17]

Though the medical curriculum failed to ignite his scientific imagination, Darwin did become interested in collateral disciplines that
would later become central to his concerns. He fell in with students and

---

12. Ibid.
13. The analyst Rankin Good, in his "Life of the Shawl," *The Lancet* (9 January 1954):
106–107, is one who has found the bond of paternity rotten. Peter Brent, however, has
shown to be wishful thinking the notions that Darwin hated his father and that the father
rejected the son. See Brent's thorough biography *Charles Darwin, a Man of Enlarged
Curiosity* (New York: Harper & Row, 1981).
14. Darwin, *Autobiography*, p. 27.
15. Ibid., p. 28.
16. Ibid., pp. 46–7.
17. Ibid., p. 48.

faculty active in the natural sciences. One of these, William Ainsworth, was a disciple of Abraham Werner, the great German geologist who advanced the notion that all rock formations were deposited as precipitates from several universal floods. Another, Robert Grant, spoke with enthusiasm about Lamarck. Grant was a physician and sometime lecturer at the university. He specialized in invertebrates and drew heavily from Lamarck's work on these lower creatures. Darwin, with his older friend's help, undertook investigation of marine invertebrates—an interest that would command a great deal of his research time during the *Beagle* voyage.[18] He judged that conversations with Grant, along with study of his grandfather's *Zoonomia,* may well have favored his pursuit of evolutionary ideas in later years.[19]

Darwin's introduction to the community of natural scientists inclined him further away from his intended profession. He became a member of the Plinian Society, a largely student group (though with some professors and and adjunct lecturers) before which, with the encouragement of Grant, he reported a discovery he had made in marine biology. At a meeting of this same society he likely heard a paper by one of his fellows, a Mr. Browne, arguing mind to be material, the result of matter more perfectly organized—a view Darwin himself would adopt in the early development of his theory of mental evolution.[20] He also attended meetings of the Wernerian Society, and became so enticed by the papers he heard read that in his second year at Edinburgh he enrolled in the geological and zoological lectures of the society's founder, Robert Jameson. Jameson preached catastrophe in geology and exemplified it in his lectures. As a result of a dreadful experience in the course, Darwin determined "never as long as I lived to read a book on Geology or in

18. Phillip Sloan has skillfully traced out Darwin's early preoccupation with invertebrate biology and assessed its preformative contribution to the development of his theory of evolution. See Sloan's "Darwin's Invertebrate Program, 1831–1836: Preconditions for Transformism," in *The Darwinian Heritage,* ed. David Kohn (Princeton: Princeton University Press, 1985), pp. 71–120.

19. Darwin, *Autobiography,* p. 49. Adrian Desmond describes Grant's evolutionary views and the intellectual atmosphere at Edinburgh during Darwin's time there in his "Robert E. Grant: The Social Predicament of a Pre-Darwinian Evolutionist," *Journal of the History of Biology* 17 (1984): 189–224.

20. The notes from the Plinian Society Minutes Book that describes Mr. Browne's paper on "Life and Mind" are transcribed in Howard Gruber, *Darwin on Man* (New York: Dutton, 1974), p. 479. Darwin discussed the merits of one form of materialism—phrenology—with his second cousin, William Darwin Fox. He admitted to Fox, however, that Sir James Mackintosh had "entirely battered down the very little belief of it that I picked up at Osmaston" [the Fox family home]. See Darwin to W. D. Fox (January 1830), in *The Correspondence of Charles Darwin,* vol. 1: *1821–1836,* ed. Frederick Burkhardt et al. (Cambridge: Cambridge University Press, 1985), pp. 96–7.

any way to study the science."[21] The vow, happily for his later fame, was often breached in succeeding years.

Darwin had initially excited hope in his father that he would become a successful physician, since his efforts at attending the poor of Shrewsbury prior to going up to Edinburgh indicated a deft and solicitous manner, which won the gratitude of his "patients." But the doctor's own success, along with his son's growing distaste for a physician's studies, foreclosed a medical career. "I became convinced from various small circumstances," he confessed in his *Autobiography,* "that my father would leave me property enough to subsist on with some comfort, though I never imagined that I should be so rich a man as I am; but my belief was sufficient to check any strenuous effort to learn medicine."[22] He came down from Edinburgh in summer of 1827, never to return.

Dr. Darwin concluded that the mental character of his son, particularly the avidity of Charles's pursuit of quail and beetles, precluded a serious professional life. The elder Darwin suggested that his son take religious orders after completing an education at Cambridge. The choice was dictated by the social possibilities for a young son of the gentry, not by any firm religious conviction of Dr. Darwin, who wore lightly the Deism of his own father, nor of Charles, who rarely scrupled over religious doctrine. At the time, Charles accepted a literal (but uninformed) rendering of the Bible;[23] and since the Anglican Creed presented no insuperable conflicts, he accepted it too. Only in the twilight of his recollections did he marvel at his own credulity: "It never struck me how illogical it was to say that I believed in what I could not understand and what is in fact unintelligible."[24] A Darwin brought to reflect more carefully about his mental habits, however, would not have been surprised. For his own thought, both in science and religion, moved slowly and gradually. It never heaved up and turned as though under the iron discipline of strict logic or the fevered hold of inflamed passion; his thought pulsed steadily, not like that, say, of William James, which jumped and raced in pursuit of the strong sentiment of rationality. After he went down from Cambridge, Darwin's religious convictions began slowly to slip away. By the time he wrote the *Origin* only a cautious Deism remained; and in the decade thereafter that too slid under, leav-

21. Darwin, *Autobiography,* p. 52

22. Ibid., p. 46.

23. Darwin had assumed that Bishop Ussher's calculation of the date of creation, 4004 B.C., was to be found in the text of Genesis itself. Only in 1861 did he discover his confusion. See Francis Darwin, ed., *More Letters of Charles Darwin* (London: Murray, 1903), 2:31.

24. Darwin, *Autobiography,* p. 57.

ing but a wariness of the power of religion to promote or smite a scientific theory, a reverence for his wife's beliefs, and a polite agnosticism.[25]

Darwin judged his three-year (1828–1831) course of studies at Cambridge hardly more useful than his time at Edinburgh. He read a little mathematics and classics, neither returning any investment. Strangely, only the books of the utilitarian moral philosopher and natural theologian William Paley produced a lasting effect. In order to pass his bachelor's exam, Darwin had to get up Paley's *Evidences of Christianity* and *Moral and Political Philosophy*. His reading notes on the *Evidences* carefully follow the theologian's argument for Christ's divinity to its ordained conclusion.[26] In his *Autobiography,* he still recalled how the logic of these works, as well as Paley's *Natural Theology,* "charmed and convinced" him.[27] This experience likely encouraged Darwin to remain open-minded about such literature, at least in the near term. During the critical years after his *Beagle* voyage until the mid-1840s, he avidly read the books of natural theologians. Some of their considerations became instrumental in important developments of his evolutionary theory, as I will try to show in the next chapter.

Though Darwin spent most of his time at Cambridge in his usual occupations—dining, hunting, and, as always, beetle collecting—he made personal acquaintances that gave form to his nebulous interests. He enjoyed the lectures of John Henslow on botany and relished the hospitality and encouragement this gracious man offered him. Through his friendship with Henslow, he came to know William Whewell, then a tutor at Trinity, and Adam Sedgwick, who rekindled his interest in geology by taking him on a field expedition in August of 1831. It was also Henslow who secured for Darwin the position as naturalist on HMS *Beagle*.

The Voyage of the *Beagle* and Darwin's Evolutionary Hypothesis

The *Beagle* sailed from Plymouth harbor on 27 December 1831. Its principal mission was to chart the coast of South America and a passage

25. Neil Gillespie, in *Charles Darwin and the Problem of Creation* (Chicago: University of Chicago Press, 1979), chap. 8, traces the slow decline of Darwin's religious beliefs.

26. Darwin's reading notes on Paley's *A View of the Evidences of Christianity* are preserved in container book 91, in the manuscript room of Cambridge University Library. Henceforth, I will refer to these container books holding the Darwin manuscripts by the standard designation "DAR"; e.g., DAR 91.

27. Darwin, *Autobiography,* p. 59.

through the Pacific to Australia. Captain Robert FitzRoy wished a gentleman companion, as much as a scientific naturalist, to accompany him, and in Darwin he got both.[28] Darwin's official assignment was to perform a geological survey and to secure as many plant and animal types as possible for periodic shipment back to London. Among the specimens he collected were the remains of many extinct creatures. For instance, in September 1832, he visited Punta Alta in Patagonia (Argentina), and there discovered the fossil *Megatherium*, a find that quickened the interest of the scientific community in his journey. Both Darwin's letters to Henslow, which his mentor shared with a wider society, and the fossils he shipped back brought Adam Sedgwick to predict to Robert Darwin that his son would "take a place among the leading scientific men."[29] The prediction was made, of course, not in light of any report of novel speculations on the species question. Most historians agree that during the *Beagle* voyage, though Darwin carefully read the account of Lamarck's theory in Charles Lyell's *Principles of Geology* (the pertinent second volume reaching him in 1832), and though he confronted paleontological vestiges and the Galapagos fauna, he did not seriously entertain the proposition that species had altered over time. By journey's end, however, his biological orthodoxy must have become as weak and insipid as his religious orthodoxy. For it was only a short time after the *Beagle* docked at Falmouth, on 4 October 1836, that Darwin turned his attention not merely to the possibility of species mutation but to the detailing of several, developmentally related theories to make the idea a scientific reality.

Precisely what factors constrained Darwin to adopt the general proposition of species descent remain veiled. The time of his conversion, though, and some of the circumstances that likely led to it can be established. In a note added to his *Journal* for July 1838, he marked the event: "In July opened first notebook on 'Transformation of Species'—Had been greatly struck from about Month of previous March

28. Fitzroy pressed his friend George Peacock, a Trinity mathematican, to find him a naturalist from Cambridge. Peacock first sought Henslow, whose family and university obligations prevented his acceptance, and then asked Leonard Jenyns, an accomplished naturalist and Henslow's brother-in-law, who also turned down the offer. Henslow and Jenyns both thought of their younger friend Darwin, and Henslow wrote him about the position in August 1831. Though Henslow recognized Darwin was not a "finished naturalist," he thought him to be sufficiently equipped. He believed Darwin also had other requisite qualities: "Capt. F. wants a man (I understand) more as a companion than a mere collector & would not take any one however good a Naturalist who was not recommended to him likewise as a *gentleman*." See *Correspondence of Charles Darwin* 1:129.

29. Darwin, *Autobiography*, p. 81.

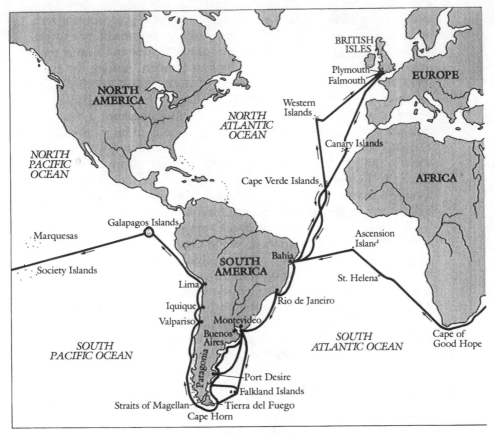

Figure 2.1   Voyage of HMS *Beagle*. Departed Plymouth, December 1831, returned
Falmouth, October 1836.

on character of S. American fossils—& species on Galapagos Archi-
pelago. These facts origin (especially latter) of all my views."[30] Very
probably Darwin's discussions with professional naturalists, such as
Richard Owen and John Gould, stimulated him to entertain the hypothesis of species alteration. Shortly after his return, he gave over his
specimens to Owen, at the Royal College of Surgeons, and Gould, at
the British Museum, to be accurately described and systematically arranged. Apparently he had certain suspicions confirmed that pushed
him toward the new view, when, for instance, Owen declared specimens
from an extant group and from ostensively related fossils yet to be of

30. Charles Darwin, "Journal," ed. Gavin de Beer, in *Bulletin of the British Museum
(Natural History),* historical series 2 (1959): 7.

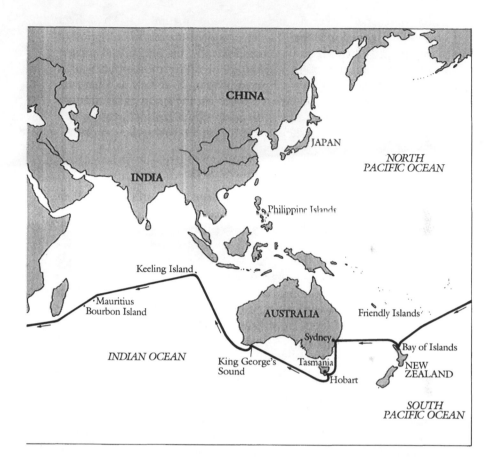

different species. At the same time, he had conventional assumptions cut away when Gould concluded that the Galapagos mockingbirds formed three different species, while Darwin was prepared to explain them as three varieties of one species that had been altered by environmental circumstances.[31]

Though the intellectual pressures began to mount, so that an evolutionary view became a live possibility, Darwin would not likely have

31. Sandra Herbert analyzes the contributions Darwin's collaborators Gould and Owen made to his conversion in the introduction to her edition of *The Red Notebook of Charles Darwin* (Ithaca: Cornell University Press, 1980), pp. 11–12. Frank Sulloway sustains Herbert's basic account, refining some of its details, in his "Darwin and His Finches: the Evolution of a Legend," *Journal of the History of Biology* 15 (1982): 1–53; and in "Darwin's Conversion: The Beagle Voyage and Its Aftermath," *Journal of the History of Biology* 15 (1982): 325–96.

turned the corner except for his own mode of thinking. The salient feature of his mental style that, I believe, contributed to his adoption of the evolutionary hypothesis was, paradoxically, his intellectual conservatism. Unless this trait of his psychology be understood, other aspects of his developmental history will remain opaque, at least as I have reconstructed them below. So before I focus on the strain of his thought that I wish to follow from here—his ideas about instinct, reason, and morality—some brief consideration of his intellectual style is in order.

Darwin's Cognitive Style

At a low level of resolution and with a large perspective taken on an age, Darwin's thought may be rightly perceived as revolutionary and disconnected from that of his predecessors. The trope of the "Darwinian Revolution" does not simply reflect the historian's taste for revolutionary models of science. The intellectual world, as well as the scientific world, at the century's end looked decidedly different from its appearance at the century's beginning, and a major transformation stemmed from the impact of Darwin's ideas. But with finer resolution, when the bloodlines come into focus from Erasmus Darwin, Lamarck, the sensationalist philosophers, the Scottish political economists, and Lyell (from whose brain Darwin believed half of his ideas leaped out), his thought appears more like an emerging new species. Evolutionary connections show his conception as a development, surely novel, but not like Venus riding in on the half shell.

The developmental or evolutionary character of Darwin's thought becomes even more apparent when his views are traced from the *Beagle* years through to their final phase. As we will see, important concepts that Darwin formulated before he read Malthus were retained, though sometimes transmogrified in the descent of his thought. In conservative fashion, he held on to ideas, preferring to change their application or alter their function rather than simply to discard them. No doubt other scientists have displayed a similar cognitive style. One is reminded of Bergson's remark that every great thinker has at most one great idea, which he attempts to develop in a variety of contexts. It could be, of course, that the historiographic model I have chosen makes this developmental conservatism an assumption more than an observation. In Darwin, however, the psychological reality punches through any model employed by historians. A Freudian, for example, would characterize him as anal-retentive: for he not only preserved his original ideas, in some form, but he saved all the scraps of paper on which they were written. The Darwin archives hold huge quantities of odd notes and jottings from his Edinburgh years through his last days, something that has permitted the Darwin industry to grind on. That industry now has begun to refine the history of his thought prior to and during the *Beagle* voyage. And here as well emerges evidence of his conservative style, his persistence in retaining and modifying ideas rather than simply dropping and replacing them.[32]

The hypothesis that has governed these preliminary reflections on Darwin's early development has been, of course, that after the *Beagle* voyage he did not undergo any profound change in cognitive psychology, so that his early style of thinking is adequately reflected in his later mental behavior, about which more is known. But so far, I have really only asserted that conservatism was a salient mode of his thought. In what follows, let me attempt to demonstrate it, albeit indirectly.

Instinct and the Mechanisms of Species Change

After the *Beagle* docked, Darwin stayed for a brief time at his father's house in Shrewsbury. In mid-December 1836, he moved temporarily to Cambridge while his geological specimens were being prepared. Then in early March, he settled in London, taking up lodgings with his brother Erasmus. During his several years in the city (1837–1842), he consulted with Owen and Gould, as well as with many other zoologists and botanists, concerning the *Beagle* collections. He became intimate with leading members of the scientific establishment, dining with the Lyells, renewing friendships with Whewell, Sedgwick, and Grant, and attending meetings of the Geological Society, the Zoological Society, and the Royal Society. This public scientific life stimulated a public literary response. During his London years, Darwin toiled long hours on his *Journal of Researches,* which described the geology and natural history of the countries he had visited on his voyage. The book established his reputation as an accomplished naturalist when it appeared in 1839. He also organized, edited, and contributed to the *Zoology of the Voyage of the Beagle,* and saw its five volumes through the presses between 1840 and 1845. Darwin still found time during this period to compose some twelve scientific papers, which he delivered to various societies and published in their journals.[33] Yet within this public sphere of an extraordinarily active scientific life, he simultaneously inhabited a more private intellectual environment, in which he worked to develop a new theory

32. Sloan's essay "Darwin's Invertebrate Program" provides just such evidence. See note 18 above.

33. All of Darwin's published scientific articles may be found in *The Collected Papers of Charles Darwin,* ed. Paul Barrett (Chicago: The University of Chicago Press, 1977).

of species change. The various features of the public environment—the problems (some suggested by Owen and Gould, as mentioned above), the literature, the strategies of argument (which, when made his own, lacked that Baconian character he later remembered them having),[34] and the zeal for scientific fame (with appropriate caution about besmirching a nascent reputation)—these infiltrated the cognitive space of his emerging species theory, and thus provided important elements of that intellectual ecology against which his theoretical conceptions evolved. It would be a long time, however, before the ideas nurtured in his private notebooks and essays would invade the expansive terrain of Victorian scientific and cultural life.[35]

In summer of 1837, Darwin opened the first of his four (perhaps six) *Transmutation Notebooks.*[36] The theory upon which he labored embraced not only the general supposition that the anatomical structures of species were transformed over time, but the proposition that their behavioral structures and their mental and moral faculties were modi-

34. In his *Autobiography,* Darwin described his method in terms of mid-nineteenth-century scientific orthodoxy: "My first note-book was opened in July 1837. I worked on true Baconian principles, and without any theory collected facts on a wholesale scale . . ." (p. 119). His own recollections notwithstanding, probably no phrase appears more often in his early notebooks than "my theory" In a letter (1859) to the young zoologist John Scott, Darwin revealed more of his actual procedure and his political strategy: "I would suggest to you the advantage at present of being very sparing in introducing theory in your papers (I formerly erred *much* in geology in that way): *let theory guide your observations,* but till your reputation is well established be sparing in publishing theory. It makes persons doubt your observations" (DAR 154).

Jack Morrell and Arnold Thackray discuss the use of Bacon as an icon of science and the various interpretations given his inductive method by the founders of the British Association for the Advancement of Science during the 1830s. See their *Gentlemen of Science: Early Years of the British Association for the Advancement of Science* (Oxford: Oxford University Press, 1981), pp. 267–75.

35. In his "Charles Darwin in London: The Integration of Public and Private Science," *Isis* 73 (1982): 186–206, Martin J. S. Rudwick describes the various complex relations obtaining between Darwin's overt scientific activity and the cloistered development of his species theory. Rudwick's observations are extremely suggestive for understanding the impact of institutional structures on scientific thinking.

36. Gavin de Beer has edited and transcribed *Darwin's Notebooks on Transmutation of Species* and recovered *Pages Excised by Darwin,* in the *Bulletin of the British Museum (Natural History),* historical series 2 and 3 (1960, 1967). The remnants of two further notebooks have been discovered by Sydney Smith and David Kohn. The discovery is reported in David Kohn, Sydney Smith, and Robert Stauffer, "New Light on *The Foundations of the Origin of Species*: A Reconstruction of the Archival Record," *Journal of the History of Biology* 15 (1982): 419–442. The vestiges of these last two notebooks, along with the remains of Darwin's other notebooks, are being readied for publication by Kohn, Smith, Paul Barrett, Peter Gautrey, and Sandra Herbert.

fied as well. In his early notebooks, he advanced the idea that behavior displayed determinate patterns and that these forms, no different than the architecture of the jaguar's foot or the anteater's proboscis, were heritable and changed through generations. But behavior had an added feature in his early speculations: it also functioned as the mechanism of anatomical adaptation.

In the *Transmutation Notebooks,* before hitting squarely on the principle of natural selection (in late September of 1838), Darwin built up the hypothesis that conduct functioned as the chief cause of species alteration. In another set of notebooks, the *M* and *N Notebooks,* and in loose reading notes kept between 1837 and 1839, he focused on the fine aspects of heritable behavior, its neural foundations, and its theoretically important kinds: particularly instinctual, rational, and moral behavior. In the remaining sections of this chapter, I will situate Darwin's early theories of instinct, reason, and moral sense in the context of his initial evolutionary considerations; and then in succeeding chapters, I will describe the ways in which these theories slowly changed in response to separate sets of critical problems he confronted prior to the publication of the *Origin of Species* (1859) and in the decade thereafter.

Darwin's First Two Theories of Species Change

In spring of 1836, about the time the *Beagle* rounded the Cape of Good Hope and headed back to its home port, Darwin began keeping notes in a small, red field notebook, which bore the label *RN*. He continued to jot notes in it through spring of the following year. Sometime in March, when his reflections on South American fossils and fauna of the Galapagos had kindled his belief in the transformation of species, he set down in his *RN Notebook* some tentative hypotheses.[37] He conceived species after the model of individuals and supposed that they gave birth to offspring, to filiated but different kinds.[38] Evidence from distinct though morphologically similar South American forms suggested that the transition would be abrupt, saltative rather than continuous.[39] Moreover, just as individuals reached the end of their years and died, so species, he speculated, grew old and went extinct.[40]

37. Herbert dates the entries in which Darwin first began to discuss species mutability as occurring after January of 1837, probably in March of that year. See her discussion in *Red Notebook,* pp. 10–11.
38. Darwin, *Red Notebook,* MS p. 130 (Herbert, p. 66).
39. Ibid., MS p. 129 (Herbert, p. 66).
40. Ibid.

These sketchy ideas about species alteration gave way in July of 1837 to a more carefully wrought conception, one which attempted not only to characterize the manner of transformation but also to provide mechanisms to account for it. Darwin developed this second theory in his *B Notebook,* his first *Transmutation Notebook,* to which he also gave the same name born by his grandfather's treatise, *Zoonomia.*[41] I call it his "second theory" in that we can perceive a further evolved set of coherently related descriptive and explanatory ideas. In the first nine pages of this notebook, he immediately set the problem of the origin of species. He cited Lamarck's observation that, as collections grew more complete, species tended to disappear; yet like his predecessor, he recognized that species had a functional reality, of which crossbreeding was the test. He then turned promptly to consider the factors that might account for the appearance of new species. First there was the advantage that sexual reproduction had over asexual modes: it produced progeny that varied from the parents and thereby could staunch perpetuation of accidental injuries or maladaptations. Sexual reproduction, moreover, yielded a malleable young, which might become subject to those environmental circumstances that influenced all living things—temperature, soil, and other features of a changing world. Further, since interbreeding kept species fairly uniform, new species could develop only when barriers—geographical as well as instinctual (e.g., repugnance to mating between varieties)—separated some groups from the original stock. Then under the influence of new circumstances, those varietal clusters could become transformed into different species.

In the succeeding pages of his *B Notebook,* Darwin elaborated this theory, the particulars of which resonated of Lamarckian ideas. For instance, he assumed that simple monadic life had spontaneously erupted from the inorganic,[42] and under pressure from the environment gave birth to a continuously progressive series of daughter species.[43] Echoing

41. Darwin, *First Transmutation Notebook,* MS pp. 1–9 (de Beer, pp. 41–42). In the discussion that follows, I have placed Darwin's considerations within the context of his grandfather's, Lamarck's, and Lyell's speculations about species dynamics. In a recent article, Phillip Sloan has broadened and deepened this context to include the German vital-materialistic tradition, which was developed in England by Richard Owen. Darwin frequently spoken with Owen during the spring and early summer of 1837. Sloan convincingly argues that Darwin momentarily imported into his reflections the notion of a vital force that would determine the life-span of species. See Phillip Sloan, "Darwin, Vital Matter, and the Transformism of Species," *Journal of the History of Biology* 19 (1986): 369–445.

42. Darwin, *First Transmutation Notebook,* MS pp. 18, 11 (de Beer, p. 43); and *Pages Excised by Darwin* (from the *Second Transmutation Notebook*), MS p. 102 (de Beer, p. 147).

43. Darwin, *First Transmutation Notebook* MS pp. 18, 26, 204 (de Beer, pp. 43, 44, 65);

Lamarck's similar belief,[44] he presumed that these species would have been uniformly transmuted worldwide if the environment remained constant.[45] But of course it did not. Groups that became isolated on islands, for example, were exposed to the contingent and variable effects of "volcanic activity" (heat?) and "electricity"[46]—environmental agents of distinctly Lamarckian flavor.[47] Darwin, like Lamarck,[48] held that, as a result of local conditions and constant but irregular environmental changes, a heterogeneous branching of species would arise from the trunk and limbs of increasingly more complex biological orders.[49] Nonetheless he affirmed that such branching "offers no (only makes it excessively complicated) Contradiction to constant succession of genera in progress."[50] It would seem, then, that when Darwin claimed to Lyell that he got "not one fact or idea" from Lamarck, he could only have been referring to an alleged debt for his theory of evolution by natural selection, not to his pre-Malthusian conceptions, which bore distinct and manifest hereditary relations to the system of the French zoologist.[51]

On one range of issues, Darwin's connections to Lamarck plunged even more deeply, down to layers of the sensationalist tradition of which they were both a part. Recognition of these connections permits assessment of the frequently made claim that Darwin introduced "population

---

Pages Excised by Darwin (First Transmutation Notebook), MS p. 108 (de Beer, p. 134); and *Third Transmutation Notebook,* MS p. 49 (de Beer, p. 134). Darwin in this early period was a committed progressivist, even though he was sensitive to the relativity of such predicates as "higher" and "lower" (*First Transmutation Notebook,* MS p. 74 [de Beer, p. 50]). He also recognized (*First Transmutation Notebook,* MS p. 204 [de Beer, p. 65]) that his theory of branching transformation allowed the possibility that some creatures might become simpler, if the environment demanded it. In a letter (1859) to Lyell after the publication of the *Origin of Species,* he reiterated this same possibility, but concluded that there would "generally be a tendency to advance in complexity of organization." See Darwin's comprehensive reply to Lyell's review of the *Origin* in *Life and Letters of Charles Darwin,* ed. Francis Darwin (New York: Appleton, 1891), 2:4–10.

44. Jean-Baptiste de Lamarck, *Philosophie zoologique* (Paris: Dentu, 1809), 1:131–33.

45. Darwin, *First Transmutation Notebook,* MS p. 19 (de Beer, p. 43).

46. Darwin *First Transmutation Notebook,* MS p. 18 (de Beer, p. 43).

47. Lamarck, *Philosophie zoologique* 1:369–74; and *Histoire naturelle des animaux sans vertèbres* (Paris: Verdière, 1815–1822), 1:177–85.

48. Lamarck described an irregular and branching tree of species, not unlike Darwin's. See Jean-Baptiste de Lamarck, "Discours d'ouverture" (1800), *Système des animaux sans vertèbres* (Paris: Lamarck et Deterville, 1801), pp. 16–17; *Philosophie zoologique* 1:131–33; and *Histoire naturelle* 1:130–32.

49. Darwin, *First Transmutation Notebook,* MS pp. 21–44 (de Beer, pp. 43–47).

50. Darwin, *First Transmutation Notebook,* MS p. 26 (de Beer, p. 44).

51. Darwin, "Letter to Charles Lyell" (1859), in *Life and Letters* 2:10.

thinking" into biology. Darwin's early training at Edinburgh gave intellectual content to his childhood passion for invertebrates, and formal study of them became his great enthusiasm. Under the tutelage of Robert Grant, he took up Lamarck's *Système des animaux sans vertèbres*; and when he shipped out on the *Beagle,* he carried along the evolutionary volumes of Lamarck's *Histoire naturelle des animaux sans vertèbres.* He came to appreciate, as had his French instructor, the shifting variations and graduated relations of invertebrate species. Because of the manifest variability of individuals within a species, Darwin could not easily think of them as frozen into identical typeforms. Indeed, he never acquired the epistemological framework needed to impress a hidden uniformity on the groups of animals he studied. Later, during the composition of his species notebooks, he did have another sort of framework ready to mind. This derived from Lamarck and his grandfather, as well as from his reading in the sensationalist epistemological literature (e.g., David Hume; see below). This framework permitted him to conceive of universal ideas only as systems of particular varying images. According to the older way of ideas—the Aristotelian way, with which systematists like Linnaeus and Georges Cuvier felt at ease—particular images and the individual organisms they represented had an intelligible structure because of the universals they instantiated (both in the naturalist's mind and in divinely designed nature). The newer way of ideas, sensationalism, denied the existence of universal types. As Erasmus Darwin complained of Locke's residual Aristotelianism, universals had "no existence in nature, not even in the mind of their inventor" (see chapter 1). Charles Darwin's conceptual framework evolved within the sensationalist tradition, binding him ideationally to his grandfather and his French predecessor. Darwin could not, therefore, even begin to conceive of species as anything other than groups or populations of varying individuals. So what is taken as his distinctive contribution to evolutionary theory—population thinking—must be understood, I believe, as a consequence of a particular philosophical-scientific tradition and especially of the ideas of two salient representatives of that tradition.

It would be a mistake, however, to presume that Darwin merely cribbed ideas from Erasmus Darwin or Lamarck, whose works he frequently referred to in his notebooks. The development of his early views was not uniquely determined, but reflected several theoretical influences as well as observations he made during his voyage. Many of his considerations, for example, drew life from Lyell's *Principles of Geology* (1830–1833). Lyell had suggested that periodic creation of preadapted species was necessary to maintain a uniform balance in nature when other species succumbed to a changing environment or competition.[52] Darwin accepted Lyell's assumption of species equability, but reversed his mentor's argument: he urged that when a new species line began, an older, less-adapted one necessarily had to give way. At first he assumed that the death of a species was predetermined; when the end of life's thread was reached, extinction would be the inevitable redress required to make room for better-suited types and to keep species numbers uniform.[53] But later passages in his *B Notebook* indicate that he became suspicious of a fixed duration for a species line. He came to see that only if a line failed to adapt to new circumstances by transmutation would death result.[54] This conclusion brought him farther away from a Lamarckian view of evolution and rather close to the idea of natural selection, though he did not yet grasp it.

Darwin, like Lamarck and others before him, believed that the environment directly affected the heritable structure of animals. Environmental causes appear in the *B Notebook* as the principal forces of species change. Darwin initially supposed that when a few animals became separated from the parent group, the new environment might directly adapt them to its peculiarities. He thought the induced alterations would then be inherited by future generations and so would gradually establish new species.[55] "For instance," as he proposed to himself, "two wrens, found to haunt two islands—one with one kind of herbage and one with other—might change organization of stomach and hence remain distinct."[56] Throughout much of his first transmutation notebook, Darwin continued to work with the hypothesis that the "condition of every animal is partly due to direct adaptation and partly to hereditary taint."[57] But toward the end of that notebook, probably in

52. Charles Lyell, *Principles of Geology* (London: Murray, 1830–1833), 2:126, 141.

53. Darwin, *First Transmutation Notebook,* MS pp. 20, 29 (de Beer, pp. 43, 44).

54. Ibid., MS pp. 38–39 (de Beer, pp. 46, 49).

55. For passages in addition to those cited below that demonstrate Darwin thought of the direct effects of the environment as the chief mechanism of species adaptation, see his *First Transmutation Notebook,* MS pp. 2–4, 7, 10–11, 17–19 (de Beer, pp. 41–43). George Grinnell, "The Rise and Fall of Darwin's First Theory of Transmutation," *Journal of the History of Biology* 7 (1974): 259–73, describes the mechanism of Darwin's first theory only as geographical isolation; he does not consider Darwin's clear presumption that the environment also functioned directly to adapt organisms. Direct adaptation was a device that Lamarck also employed. See my preceding chapter and Lamarck, *Philosophie zoologique* 1: 221–25.

56. Darwin, *First Transmutation Notebook,* MS p. 46 (de Beer, p. 47).

57. Darwin, *First Transmutation Notebook,* MS p. 219 (de Beer, p. 67).

winter of 1837–1838, he seems to have perceived the insufficiency of direct adaptation for finely adjusting an animal to its situation.[58] As Lamarck had earlier understood, a more sensitive and dynamic principle was required; and, indeed, the mechanism Darwin next devised was fairly similar to that of his French predecessor, at least more so that he cared to admit.

The Theory of Habit-Instinct Adaptation

The Mechanism of Use Inheritance

The new theory of adaptation that began to surface in Darwin's reflections, though it owed a good deal to Lamarck, probably had its proximate source in Frédéric Cuvier's essay "On the Domestication of Mammiferous Animals."[59] Darwin seems to have found the essay first mentioned in Lyell's *Principles,* and he himself read it in late 1837.[60] In his *B Notebook,* Darwin copied out this passage from Cuvier's article: "But we could only produce domestic individuals and not races, without the occurrence of one of the most general laws of life—the transmission of a fortuitous modification into a durable form, of a fugitive want into a fundamental propensity, of an accidental habit into an instinct."[61] Frédéric's brother Georges expressly denied this "fundamental law"; it was virtually Lamarck's use inheritance. Frédéric Cuvier, though, was a transformationist only in the limited sense that I have indicated in the previous chapter. He believed that races of animals could be considerably modified, though within certain (but unspecified) bounds, by the exercise of need-provoked behavior. In another part of the same article which Darwin read, Cuvier observed:

58. In his "Essay" of 1842, his first connected sketch of his theory of evolution by natural selection, Darwin retained the idea that direct effects could produce simple heritable adaptations; but he dismissed the notion that they could produce exquisitely articulated adaptations. Darwin's "Essay" of 1842 was first published, along with a longer version written in 1844, by Frances Darwin in *The Foundations of the Origin of Species* (Cambridge: Cambridge University Press, 1909). See p. 4 for his discussion of direct effects.

59. Frédéric Cuvier, "Essay on the Domestication of Mammiferous Animals," translated in *Edinburgh New Philosophy Journal* 3 (1827): 303–18; 4 (1828): 45–60, 292–98.

60. Darwin cited a passage (given in the text, below) from Cuvier's essay in his *First Transmutation Notebook,* MS p. 118 (de Beer, p. 55); he penned this entry in late 1837. His attention was probably drawn to Cuvier by the mention of him Lyell made in *Principles of Geology* 2: 41–45. In his copy of the fifth edition of Lyell's work (1837), Darwin checked the reference to Cuvier's article. He read Lyell's fifth edition early in 1837. Darwin's copy of Lyell, as well as the several other books and articles whose annotations I will cite, is preserved in the Manuscript Room of Cambridge University Library.

61. Cuvier, "Domestication of Mammiferous Animals," p. 297.

Now, the distinctive qualities of animals of the same species, those which have most influence over their particular existence, which constitute their individuality, are those which have been developed by exercise, and whose exercise has been called forth by the circumstances amid which these animals have lived. Hence it follows, that the qualities transmissible by animals to their young, those which give rise to a mutual resemblance in them, are of a nature to arise from fortuitous circumstances; and, consequently, that we are enabled to modify animals and their progeny, or their race, within the limits which bound our power to produce the circumstances calculated to act upon them.[62]

Cuvier was not merely affirming here that domestic animals were more labile than wild breeds, and thus susceptible to the modifying effects of habit; Lyell, who denied species mutability in the *Principles,* admitted this much.[63] Rather the thesis Cuvier defended in the article was that the mechanisms determining both behavior and structure in domestic animals did not differ essentially from those acting in the natural state, and that modifications induced in and passed on to the progeny of domestic animals were of the same type as those which could occur in wild creatures. Lyell, though generally well-disposed to Cuvier's views, could not bring himself to accept this principle of equivalence.[64] For he perceived that the principle smoothed the way to transformism and an easy slide to scientific perdition. Darwin too detected the principle's conceptual direction, but eagerly set out along the path. He immediately penned an addition to the passage he had copied from Cuvier: "I take the higher ground and say life is short for this object and others, that is, not too much change."[65] He thus found in Cuvier the proposal that an accidentally changed environment might indirectly create adaptations by inducing animals to acquire new practices which would produced heritable modifications in their structures. To this hypothesis he added his own uniformitarian notion that these alterations would slowly accumulate over succeeding generations, so producing a gradual transformation of species. In his *C Notebook,* his second transmutation notebook, begun in February of 1838, Darwin explored this new idea. He considered, for example, how exercise might modify the foot of the jaguar: "Fish being excessively abundant & tempting the Jaguar to use its feet much in swimming, & every development giving greater vigour

62. Ibid., p. 59.
63. Lyell, *Principles of Geology* 2 : 39.
64. Ibid., pp. 41–45.
65. Darwin, *First Transmutation Notebook,* MS p. 118 (de Beer, p. 55).

to the parent tending to produce effect on offspring—but whole race of that species must take to that particular habitat.—All structures either direct effect of habit, or hereditary & combined effect of habit."[66] Here then was a way of finely tuning anatomical characters to environmental contingencies, for acquired habits could be as intricate as the structures they introduced. The only requirement was that most members of a group adopt the transforming habits more or less at the same time—a reasonable expectation in a circumscribed environment.

Cuvier's essay was not alone in feeding Darwin's growing conviction that habit might prove the mechanism of adaptation and speciation. Other authors he took up in late 1837 and early 1838 supplied similar notions. He read in his grandfather's *Zoonomia* that an animal could, through habit and association, deliver down "those improvements by generation to its posterity, world without end!"[67] He recognized Thomas Andrew Knight's 1837 paper "On the Hereditary Instinctive Propensities of Animals" to be, as he penned at the bottom of the first page, "important, as showing that instincts probably arise from habit & not from structure."[68] And in his friend John Sebright's *Observations on the Instincts of Animals* (1836), which he esteemed an "admirable essay,"[69] he met the proposition that "acquired habits become hereditary" might explain national character.[70]

The mechanism of use inheritance, of course, was also to be found in Lamarck. Under Grant's supervision at Edinburgh, Darwin had read Lamarck's *Système des animaux sans vertèbres,* which included the transformationist essay "Discours d'ouverture" of 1800.[71] During his voyage, he had many long hours to consider Lamarck's *Histoire naturelle* and the analysis Lyell gave of the *Philosophie zoologique.* Lamarck's treatment of use inheritance certainly provided an important part of the conceptual environment in which Darwin constructed his new mechanism of

66. Darwin, *Second Transmutation Notebook,* MS p. 63 (de Beer, p. 89).

67. Erasmus Darwin, *Zoonomia,* 2d ed. (London: Johnson, 1796), 1:509. See chapter 1, above

68. Darwin's annotation is at the bottom of p. 365 of Thomas Andrew Knight, "On the Hereditary Instinctive Propensities of Animals," *Philosophical Transactions of the Royal Society of London,* part 2 (1837): 365–69.

69. Darwin, *Second Transmutation Notebook,* MS pp. 133–34 (de Beer, p. 96).

70. John Sebright, *Observations upon the Instincts of Animals* (London: Gossling & Egley, 1836), pp. 15–16. Darwin's copy has this passage marked with a vertical line in the margin.

71. Frank Egerton, "Darwin's Early Reading of Lamarck," *Isis* 67 (1976): 452–56, has shown that while at Edinburgh Darwin took notes from Lamarck's *Système des animaux sans vertèbres.* It is likely that he read the prefatory "Discours," though there is no direct evidence that he did so.

adaptation. But his peculiar interpretation of Lamarck's views allowed him to set his own theory at a safe distance from his predecessor's. Lamarck, he supposed, held the fanciful notion that an organism might *will* itself new parts. In his copy of the first volume of Lamarck's *Histoire naturelle,* he curtly penciled his judgment: "Because use improves an organ, wishing for it, or its use, produces it!!!! Oh—"[72] Instead of the foolish idea that willpower could alter species, Darwin proposed that habit functioned as the mechanism of adaptation and speciation. Even plants could acquire habits (e.g., turning their leaves to the sun), which might alter their structures; but as Darwin frequently objected in his notebooks, "Lamarck's willing absurd, [since it is] not applicable to plants."[73] Moreover, organic patterns in a species changed as gradually as the environments which they accommodated; only habits seemed constant and stable enough to introduce them. Will acts were cataclysmic; they fluctuated abruptly. Over generations, however, persistently adopted habits might slowly mold structures to new uses, thus transforming species.

Darwin, of course, misread Lamarck from the beginning, for clearly neither the *Histoire naturelle* nor the *Philosophie zoologique* invoked will to explain the inheritance of acquired characteristics.[74] In Lamarck's scheme, will required consciousness, which appeared only in the vertebrates.[75] Though Darwin's reading had the authority of Georges Cuvier and would be substantiated by Wallace—and by even some modern

72. Darwin's annotation appears at the top of p. 157 of Lamarck, *Histoire naturelle des animaux sans vertèbres,* vol. 1.

73. See, for instance, Darwin's *Second Transmutation Notebook,* MS p. 63 (de Beer, p. 89), and his "Old and Useless Notes," MS p. 35, ed. Paul Barrett, in Gruber's *Darwin on Man,* p. 392.

74. At times Darwin appears to have doubted his interpretation of Lamarck. In his fourth notebook on the transmutation of species, *E Notebook,* he seems to have recognized that Lamarck did not invoke a mysterious power of will to account for species change but only the power of habit. In a passage written early in 1839 (*Fourth Transmutation Notebook,* MS p. 159 [de Beer, p 180]), he cited an appendix to the *Philosophie zoologique*: "'Additions' p. 454—does really attribute metamorphoses to *habits* of animals & takes series of flying mammifers—says lemur volans has skin between its legs." In the text referred to, Lamarck had offered examples of flying squirrels, lemurs, and bats to show how mammals, which had no affinity with birds, might have slowly acquired the habit of leaping from tree branches and through the habit of stretching their legs, have spread the skin and created winglike organs. Lamarck concluded (*Philosophie zoologique* 2:456): "Such, therefore, is the power of *habits* that they have a marked influence on the form of the parts and give to animals which have practiced certain of them faculties not had by animals which have adopted other habits."

75. See, for instance, Lamarck's *Philosophie zoologique* 2:330–45, and my previous chapter.

historians of science[76]—nonetheless its real force flowed from Darwin's desire to distinguish his new conception from the Lamarckian hypothesis, which had already been rejected by Lyell and the rest of the British scientific community.

In his second and third *Transmutation Notebooks,* his *C* and *D Notebooks,* which he kept between February and September of 1838, Darwin worked out his new mechanism of species change. He supposed that habits an animal might adopt to cope with a shifting environment would, during the course of generations, slowly become instincts, that is, innately determined patterns of behavior. Instincts in their turn would gradually modify the anatomy of an organism, adapting the creature to its surroundings. By supposing that habits first became instinctive, Darwin could fit his mechanism to Lyellian uniformitarian principles of gradual change, while at the same time he could reject the presumptive Lamarckian device of conscious will effort, since instincts were automatic and unconscious.

Instinct as Unconscious Memory

Darwin recognized that for instincts to be transmitted through generations, some physical structures had to bear them. His proposal was straightforward, as we see in a penciled annotation, probably made in January of 1840, to Johannes Müller's *Elements of Physiology:* "The inherited structure of brain must cause instincts: this structure might as well be bred as any other adapted structure."[77] The line of thought that led to this conclusion, which Darwin never abandoned, can be traced through his *M Notebook* (July to October 1838). In the first several pages, he recorded clinical observations made by his father and grandfather of the damaging effects of age and cerebral trauma on the operations of thought.[78] A man who suffered a stroke, for instance, could not initiate a conversation but could join one in progress. The implications of such pathologies showed clearly for Darwin: mind must be a cerebral derivative. After all, he asked, "Why is thought being a secretion of brain, more wonderful than gravity a property of matter."[79]

76. See note 99 of chapter 1.

77. Darwin's annotation appears in Johannes Müller, *Elements of Physiology,* trans. W. Baly, vol. 1 (London: Taylor and Walton, 1838), p. 25. Darwin's reading notebook indicates that he examined Müller's *Elements* in January of 1840. See Peter Vorzimmer's transcription in "The Darwin Reading Notebooks (1838–1860)," *Journal of History of Biology* 10 (1977): 123.

78. Darwin, *M Notebook,* MS pp. 1–55, in Gruber, *Darwin on Man,* pp. 266–75.

79. Darwin, *Second Transmutation Notebook,* MS p. 166 (de Beer, p. 101).

Darwin never really plumbed the philosophical depths of the mind-body problem. But he formed a fairly clear and simple idea of the relationship of thought to brain. In annotations to his copy of John Abercrombie's *Inquiries Concerning the Intellectual Powers* (1838), he drew out a bit more the analogy from the relation of matter to its natural powers. For instance, he jotted this proposal to himself: "Elective affinity is a thing not analogous to other quiddities of bodies, yet is supposed property of matter. So would I say thought was—from analogy of organ."[80] And a few pages later, he again conceived thought as comparable to a power naturally arising out of matter organized in a certain way: "From the myriads of animals that have existed we may assume thought as function of matter and then say, to what function of matter shall we compare the phenomenon of attraction?—This assumption is as justifiable as the other. We only know thought as a phenomenon attendant on stimulus, & we only know elective attraction as a function of *matter*."[81] What Darwin required for his emerging theory of species descent was a model of thought that made it a natural power of matter, a power perhaps distinct from its source (i.e., the brain), but one nonetheless completely determined by it. He found his model in the phenomena of chemical affinity and gravitation. For the Newtonian scientist (an ideal toward which Darwin aspired), the occult connections between matter and its powers did not need explanation, only description. So Darwin felt comfortable with the agnosticism expressed by Abercrombie as to the ultimate relation of mind and brain. In a passage Darwin marked, the Scots philosopher declared: "Matter and mind are known to us to be certain properties:—these properties are quite distinct from each other; but in regard to both, it is entirely out of reach of our faculties to advance a single step beyond the facts which are before us. Whether in their substratum or ultimate essence, they are the same, or whether they are different, we know not, and never can know in our present state of being."[82] To this passage, Darwin appended the Newtonian observation: "It is sufficient to point out close relation of kind of thought & structure of brain."[83]

Such determinate correspondences would indeed suffice for Darwin's purpose, which in this instance was twofold. First, the supposed causal

80. Darwin's annotation appears in John Abercrombie, *Inquiries Concerning the Intellectual Powers*, 8th ed. (London: Murray, 1838), p. 29.

81. Darwin's annotation appears in Abercrombie, *Inquiries Concerning the Intellectual Powers*, p. 31.

82. Abercrombie, *Inquiries Concerning the Intellectual Powers*, p. 34.

83. Darwin's annotation appears in Abercrombie, *Inquiries Concerning the Intellectual Powers*, p. 34.

relationships of brain to mind could furnish a perfectly naturalistic (and evolutionary) explanation for apparently nonbiological mental traits. But second, these relationships could also be marshalled in more instrumental ways, as mechanisms of evolutionary change.

Memory, for instance, could be explained as an effect of exercising the brain. "When a muscle is moved," Darwin puzzled out in his notebook, "the motion becomes habitual & involuntary.—when a thought is thought very often it become habitual & involuntary, that is involuntary memory. . . . An intentional recollection of anything is solely by association, & association is probably a physical effect of brain."[84] Senile Miss Cogan, his father's patient, might not recall what was said to her a moment ago, but could sing a childhood song. The muscle of her young brain had been left permanently stretched by constant repetition. The consequence of this view of mental functioning for Darwin's theory of instinct was immediate: "Now if memory of a tune & words thus lied [sic] dormant, during a whole life time, memory from one generation to another also without consciousness, as instincts are."[85] Hence an animal, to adapt itself to its environment, might habitually behave in certain ways. And if its progeny adopted the same habits, to meet the constant necessities of life, then gradually these patterns would inscribe themselves in the heritable structures of the brain. Subsequent generations that inherited these brain structures might thus, in appropriate circumstances, act instinctively, as if from an unconsciously preserved memory.[86] The mysteries of instinct thus yielded to a naturalistic analysis of mind: "These facts showing what a train of thought, action, etc. will arise from physical action on the brain, renders much less wonderful the instincts of animals."[87]

The Complete Theory

Since thought and memory could be interpreted as natural functions of the brain and since modified brain structures could presumably be transmitted by propagation as easily as any other physical traits, Darwin

84. Darwin, *M Notebook,* MS p. 46 (Gruber, p. 274).

85. Darwin, *M Notebook,* MS p. 7 (Gruber, p. 267).

86. In his *Second Transmutation Notebook,* MS p. 173 (de Beer, p. 103), Darwin vividly illustrated his conception of how memory passed into inherited instinct: "Analogy a bird can swim without being web footed yet with much practice and led on by circumstances it becomes web footed. Now man by effort of memory can remember how to swim after having once learnt, & if that was a regular contingency the brain would become web-footed & there would be no act of memory."

87. Darwin, *M Notebook,* MS p. 81 (Gruber, p. 280).

Figure 2.2 Charles Darwin, 1802–1882, wedding portrait done in 1841.

had the elements necessary to make his habit-instinct theory of species adaptation work. He pondered the new theory in his *C Notebook:*

> Reflect much over my view of particular instinct being memory transmitted without consciousness, a most possible thing see man walking in sleep.—an action becomes habitual is probably first stage, & an habitual action implies want of consciousness & will & therefore may be called instinctive.—But why do some actions become hereditary & instinctive & not other.—

We even see they must be done often to be habitual or of great importance to cause long memory,—structure is only gained slowly. Therefore it can only be those actions which many successive generations are impelled to do in same way.[88]

Darwin thus interpreted instinct as a kind of memory, which could be established in earlier generations and biologically transmitted to succeeding generations.

Since my reconstruction of Darwin's early theories of species change differs considerably from Kohn's recent and widely accepted account,[89] let me briefly reiterate what I take as Darwin's most carefully worked out device prior to formulating his mechanism of natural selection. His habit-instinct theory of species change held that if a group of animals entered a new environment (or if the environment slowly altered), then individuals would be compelled to adopt new habits to accommodate themselves to the exigencies of life. These habits, if practiced over several generations, would gradually modify brain structures, producing something like heritable memories. Such memories would be expressed in succeeding generations as instinctive behaviors. And these instincts, unconscious and unwilled, would, by dint of their constant exercise, change other anatomical structures, thus adapting the group and finally generating a new species of animal. In Darwin's summary: "According to my views, habits give structure, therefore habits precede structure, therefore habitual instincts precede structure."[90] Prior to reading Malthus, who turned him to thoughts of population pressure, Darwin had constructed a powerful explanation of species alteration, one which, in his conservative way, he never completely relinquished.

Malthus and After

At the end of September 1838, Darwin picked up Thomas Malthus's *Essay on Population,* and the modern world has reaped the consequences. Darwin remembered the event in his *Autobiography:*

> I soon perceived that selection was the keystone of man's success in making useful races of animals and plants. But how selection could be applied to organisms living in a state of nature remained for some time a mystery to me.
>
> In October 1838, that is, fifteen months after I had begun my systematic enquiry, I happened to read for amusement Malthus on *Population,* and being well prepared to appreciate the

88. Darwin, *Second Transmutation Notebook,* MS p. 171 (de Beer, pp. 102–3).
89. See the appendix to this chapter for a discussion of Kohn's thesis.
90. Darwin, *Second Transmutation Notebook,* MS p. 199 (de Beer, p. 106).

struggle for existence which everywhere goes on from long-continued observation of the habits of animals and plants, it at once struck me that under these circumstances favourable variations would tend to be preserved, and unfavourable ones to be destroyed. The result of this would be the formation of new species. Here, then, I had at last got a theory by which to work.[91]

Darwin's notebooks bear out the general pattern of these recollections, though not the details. The Malthusian revelation came on September 28 rather than in October of 1838, a minor discrepancy.[92] Reading Frédéric Cuvier's essay and several treatises on animal and plant breeding (e.g., those of John Sebright and William Yarrell) undoubtedly did prepare him to liken alterations in domestic species to changes in wild kinds[93]—a continuity of consideration prohibited by Lyell, who thought domestic breeds to be special instances of excessively modifiable species. But Darwin's notebooks suggest that the selection analogy came post hoc. It was a theoretically important insight arrived at subsequent to recognizing selection in nature. More than once in his pre-Malthusian study, he had ruminated on the struggle for existence in nature and the advantage of chance adaptations, even momentarily recognizing their transforming power. One example of this fleeting awareness comes in this entry concerning the struggle for females, which he made in his *C Notebook* in early spring of 1838: "Whether species may not be made by a little more vigour being given to the chance offspring who have any slight peculiarity of structure. ((hence seals take victorious seals, hence deer victorious deer, hence males armed & pugnacious all orders; cocks all war-like))."[94] Darwin recognized that in the wild the fortunately endowed not only had the advantage, but their progeny

91. Darwin, *Autobiography,* pp. 119–20.

92. Darwin's *Third Transmutation Notebook* records his reading of Malthus on 28 September 1838. See *Excised Pages,* MS pp. 134–35 (de Beer, pp. 162–63). His reading notebooks (Vorzimmer, p. 120) do have listed for 3 October 1838 "Malthus on Population." Likely Darwin continued reading Malthus through early October. He returned for "2d time" to Malthus in the first part of April 1847 (Vorzimmer, p. 136).

93. For a subtle analysis of the role the breeder's art played in Darwin's thought, see John Cornell, "Analogy and Technology in Darwin's Vision of Nature," *Journal of the History of Biology* 17 (1984): 303–44. See also L. T. Evans, "Darwin's Use of the Analogy between Artificial and Natural Selection," *Journal of the History of Biology* 17 (1984): 113–40.

94. Darwin, *Second Transmutation Notebook,* MS p. 61 (de Beer, p. 88). Earlier in his *First Transmutation Notebook,* MS p. 90 (de Beer, p. 51), Darwin had also briefly reflected on the survival value of chance adaptations: "Whether every animal produces in course of ages ten thousand varieties (influenced itself perhaps by circumstances) and those alone preserved which are well adapted?"

did as well. He seems to have dimly perceived that this feature of the biological economy might have species-altering consequences. He would draw the analogy with domestic breeding, however, only after he discovered a causal force that could drive nature, much as the desire for novel varieties drove the breeder. He hit upon this causal force only in perusing Malthus.

Malthus brought Darwin to recognize the tremendous fecundity of organisms and their consequent struggle to acquire the necessities of survival. The focusing event of September, cast against the background of his established ideas and reading (especially of Scots moral philosophers—as Schweber has shown in his meticulous reconstruction),[95] led him to formulate the nut of his principle of natural selection.

A few months later, in early December 1838, Darwin (playing the Newton) worked out the axioms of his new theory, from which one could virtually predict species alteration. In his *E Notebook*, he listed these axioms:

> Three principles will account for all
> (1) Grandchildren like grandfathers
> (2) Tendency to small change especially with physical change
> (3) Great fertility in proportion to support of parents[96]

At the present time, these are usually expressed as the principles of inheritance, variation, and fecundity. (In his salad days Richard Lewontin regarded a similar set also as constituting the axiomatic foundation of evolution by natural selection.)[97]

95. Silvan S. Schweber, "The Origin of the *Origin* Revisited," *Journal of the History of Biology* 10 (1977): 231–316.

96. Darwin, *Fourth Transmutation Notebook*, MS p. 58 (de Beer, p. 165).

97. In "Adaptation," *Scientific American* 239 (September 1978): 213–30, Richard Lewontin lists the three principles "necessary and sufficient to account for evolutionary change by natural selection" as: "Different individuals within a species differ from one another in physiology, morphology and behavior (the principle of variation); the variation is in some way heritable, so that on the average offspring resemble their parents more than they resemble other individuals (the principle of heredity); different variations leave different numbers of offspring either immediately or in remote generations (the principle of natural selection)" (p. 220). Darwin's and Lewontin's first two principles agree. Lewontin's third principle would be a consequence of Darwin's three principles together—at least Darwin would have so believed. The largest difference between Darwin's conception and Lewontin's modern—and more agnostic—version is that Darwin furnished by his third principle a causal dynamic. In the context of his emerging theory, another principle was implied or imbedded in his third principle: victory in the struggle for existence went to the fittest organisms. Lewontin's third principle neither specifies nor implies the cause for some variations leaving more offspring. The concepts of fitness and adaptation have become the critical ground in the evolutionary debates of the last quarter of this century. See Lewontin's article for a vetting of the issues.

Most historians of science have assumed that in the period after he read Malthus until the publication of the *Origin*, Darwin was merely mopping up, polishing off the niceties of his theory and gathering ever-larger piles of evidence to support it. If this were so, his delay in publishing the discovery—some twenty years between Malthus and the *Origin*—would be very puzzling. I will attend to that problem at the end of the next chapter, but even now I can begin to suggest a solution. Initially it can be put this way: by concentrating on the Malthus episode and the axiomatic expression of the theory, we untimely rip the emerging conception of natural selection out of its specifying context. Even through the 1840s, Darwin had not, I believe, clearly conceived the mechanism of natural selection (that is to say, the conceptual system at the beginning of the period differed substantially from that in the *Origin*). Certainly he had not worked out the kinds of variation the mechanism processed, nor scaled its scope of operation, nor gauged its role in producing the divergence of species from a common stem, nor precisely determined the units of selection. No light bulb flashed in Darwin's head with the brilliance later illuminating the *Origin*. But even after he had formulated the principle of natural selection with some clarity, he still perceived several deep conceptual difficulties, which required considerable time and effort to overcome.[98] Darwin's hesitation in publishing his theory becomes more comprehensible in view of the gradual evolution of his thought against a shifting problem terrain during the twenty years before the *Origin*.

As we trace out Darwin's ideas on instinct from the late 1830s through the 1850s (here and in the next chapter), we will better understand the kind of conceptual evolution his principle of natural selection underwent. If all were glory for Darwin after reading Malthus, then we would expect him to have brought innately fixed behavior immediately under the aegis of natural selection. This, indeed, is what Swisher, the first historian really to study Darwin's theory of behavior, contends. Shortly after that fateful day in late September of 1838, "Darwin," Swisher presumes, "was sufficiently convinced of the mechanism of evolution by natural selection to use his theory immediately to unravel another prob-

98. Dov Ospovat was an exception among historians of science. He did perceive that Darwin needed to resolve certain crucial problems during the 1840s and early 1850s. In his *The Development of Darwin's Theory* (Cambridge: Cambridge University Press, 1981), Ospovat discussed particularly the problems of whether adaptation was perfect, how species diverged, and whether evolution was progressive. Silvan Schweber also appreciates the conceptual difficulties Darwin confronted prior to the publication of the *Origin*. See his "Darwin and the Political Economists: Divergence of Character," *Journal of the History of Biology* 13 (1980): 195–289.

lem of biology . . . : the origins of behavior in men and animals."[99] I believe that the contrary is true. Evidence derived from Darwin's notebooks, marginalia, and scattered notes indicates that after initially formulating a principle of natural selection in the fall of 1838, he applied it but tentatively to behavior, while continuing to depend on his older mechanism of inherited habit. Only in overcoming certain problems during the early 1840s did he slowly contract the field of application of his more primitive device; he did not, however, completely give it up, as I will try to show in the next chapter. Yet what is the evidence for the assertion that Darwin did not immediately apply the principle of natural selection to behavior?

First consider his *N Notebook* (kept from fall of 1838 to summer of 1839). In this notebook, begun just after he read Malthus, Darwin reflected at length on the whole issue of instinct, but failed to use natural selection as the explanatory key. On 27 November 1838, he did entertain the transient notion that his new mechanism might also account for some instincts:

> An habitual action must some way affect the brain in a manner which can be transmitted—this is analogous to a blacksmith having children with strong arms.—The other principle of those children which *chance* produced with strong arms, outliving the weaker ones, may be applicable to the formation of instincts, independently of habits. The limits of these two actions either on form or brains very hard to define.—Consider the acquirement of instinct by dogs, would show habit.[100]

Only in two other brief passages jotted down in different notebooks, also in the last part of November, did he mention the possibility that natural selection might act with inherited habit to produce instinct.[101]

99. Charles Swisher, "Charles Darwin on the Origins of Behavior," *Bulletin of History of Medicine* 41 (1967): 25.

100. Darwin, *N Notebook*, MS pp. 42–43e (transcribed by Barrett in Gruber, *Darwin on Man*, p. 338).

101. During November 1838, Darwin kept reading notes on John Macculoch's *Proofs and Illustrations of the Attributes of God* (1837); and in those notes he also considered the possible operation of natural selection on behavior: "In the Mollusca /Bees/ the nervous system is endowed with the knowledge of trying a hundred schemes of structure, in the course of ages /step by step/—in man, the nervous system, gains that knowledge, before hand, & can in idea (with consciousness) form these schemes.—I see no reason why structure of brain should not be born, with tendency to make animal perform some action—as well as gain it by habit." (These notes are transcribed by Barrett, in Gruber's *Darwin on Man*, p. 419; the slash marks indicate Darwin's insertions.) About two weeks after he recorded the passages in the *N Notebook*, Darwin again hinted at natural selection as a source of instinct. The relevant remarks occur in his *Fourth Transmutation Notebook*,

But the insights of November faded. By March 16 of the following year, in reflecting on the novel feature of his theory of behavioral evolution, Darwin reiterated his older principle: "Sir J. Sebright has given the phrase "hereditary habits" very clearly, all I must do is generalize it, & see whether applicable at all cases—& analogize it with ordinary *habits* that is new part of the view—let the proof of hereditariness in habits be considered as grand step if it can be generalized."[102] Throughout the entire *N Notebook,* with the exception just quoted, Darwin explained instinct by inherited habit alone.

The second piece of evidence comes from notes Darwin penned in the first volume of Johannes Müller's *Elements of Physiology,* which he read in January 1840. In those sections in which Müller puzzled over the origins of different reflex connections, Darwin offered his own resolution: "established by habit"; "the connection here is hypothetical, Why not custom?" "surely custom"; "it has been said that respiration also subject to the will—Habitual movement shows that any may become reflective [that is, reflexive]."[103] And on the back flyleaf of the volume, he added: "associations may become hereditary, which would account for the alliance of instincts with time, places." These annotations demonstrate that Darwin still supposed innate patterns of behavior had their origins in inherited habit.

Then there is a striking annotation left in John Fleming's *Philosophy of Zoology* (1822),[104] which Darwin appears to have made in winter of 1839–1840.[105] Next to a passage in which Fleming assigned instinct to the "Active powers" of mind, and reason to the "Intellectual powers," Darwin penciled his view of their evolutionary connection: "The individual who by long intellectual study acquires a habit, & can perform action almost instinctively, does that in his life time which succeeding generations do in acquiring true instincts: instinct is a habit of genera-

---

MS pp. 63–64 (de Beer, p. 166): "When two races of men meet, they act precisely like two species of animals—they fight, eat each other, bring diseases to each other &c., but then comes the most deadly struggle, namely which have the best fitted organization, or instincts (i.e., intellect in man) to gain the day."

These passages, written within a few weeks of one another, exhaust, I believe, Darwin's early references to natural selection as a possible source, along with inherited habit, of instinct. After the autumn of 1838, as I will try to show, he continued to appeal exclusively to the habit mechanism of instinct.

102. Darwin, *N Notebook,* MS p. 63 (Gruber, p. 342).

103. The annotations are from Darwin's copy of Johannes Muller, *Elements of Physiology* I : 353, 719, 720, 717.

104. John Fleming, *Philosophy of Zoology* (Edinburgh: Constable, 1822).

105. Darwin's reading notebook indicates that he examined Fleming's *Philosophy of Zoology* in December 1839 (Vorzimmer, p. 126).

tions—each step in each generation being intellectual—for in lower animals [there is?] intellect."[106] Here again Darwin proposed the inheritance of acquired habit as the evolutionary source of instinct.

Several constraints prevented Darwin from applying the principle of natural selection—as we have come to recognize it—to the evolution of behavior. Certainly the ease with which inherited habit could explain instincts produced an inertial force that carried the older theory beyond the fall of 1838. In addition, Darwin seems initially not to have conceived natural selection as a mechanism completely distinct from inherited habit. That he did not immediately distinguish the two should surprise us only if we presume the principle of natural selection to have been created from the beginning with all the sinews that later gave the *Origin* its strength. But if the principle itself underwent an evolution, reacting gradually to different conceptual pressures, then its separation from the mechanism of inherited habit need not have occurred right away. One idea linking natural selection to inherited habit like an umbilical cord was the idea of utility or benefit. In formulating his habit-instinct mechanism of species change, Darwin had supposed that succeeding generations of animals would adopt the same habit only if it were of continuous advantage to them in a given environment. He reasserted this supposition in the spring of 1839, some eight months after reading Malthus: "It is probable," he affirmed, "that becomes instinctive which is repeated under many generations . . . & only that which is beneficial to race, will have reoccurred."[107] Darwin had not yet drawn the distinction between habits preserved because of their usefulness (via use inheritance) and individuals preserved because of their useful habits (via natural selection).

By 1842, in an essay composed in that year, Darwin did engage the principle of natural selection, in recognizable form, to explain behavior—though even then it showed its parentage: "It must I think be admitted that habits whether congenital or acquired by practice <sometimes> often become inherited; instincts influence, equally with structure, the preservation of animals; therefore selection must, with changing conditions tend to modify the inherited habits of animals."[108] In the next chapter I will describe the role certain theologians played in bringing Darwin to apply natural selection to behavior. I will also show

106. This annotation is on pp. 142–43 of Darwin's copy of John Fleming, *Philosophy of Zoology,* vol. 1.

107. Charles Darwin, *Old and Useless Notes,* MS p. 51 (transcribed by Barrett in Gruber's *Darwin on Man,* p. 402).

108. Darwin "Essay" of 1842, *The Foundations of the Origin of Species,* p. 18; the wedge quotes indicate material crossed out in the manuscript.

how they helped him to distinguish natural selection from his earlier mechanism.

The Roots of Rational Thought

One is tempted to credit the English bestowal of generous intellect on animals to their great affection for dogs. The British metaphysician F. H. Bradley once confessed "I never could see any difference at bottom between my dogs & me, though some of our ways were certainly a little different."[109] This explanation will not do for the French, however. They shed their sentiments on other things. The reason lies deeper, in their sensationalist epistemology. French psychological writers enthusiastically adopted the ideas of British sensationalists—Locke, Hume, Hartley, the Mills. Those in the sensationalist tradition argued that ideas were derived from sensations alone, that thought merely reflected darkly the images of sight, hearing, touch, smell, and taste. Intelligence, in this view, was simply the ability to arrange and associate images, something animals—especially dogs—gave ample evidence of doing. On this epistemological basis, Locke concluded that if beasts "have any *ideas* at all and are not bare machines (as some would have them) we cannot deny them to have some reason."[110] Locke did, however, deny them the capacity to formulate abstract ideas of the sort Berkeley found so hard to fathom and Hume interpreted as vague images that readily called up previous associations. Erasmus Darwin's theory of intelligence ran deeply in the grain of the sensationalist tradition; and, as I have argued in the previous chapter, it promoted his own peculiar evolutionary views. His grandson's first introduction to the epistemological-evolutionary question likely came through study of the first part of Erasmus's *Zoonomia*.[111]

Charles Darwin, like his grandfather, assumed that sensory images supplied the content of thought: "thinking consists," he recorded in his *M Notebook*, "of sensation of images before your eyes, or ears (language mere means of exciting association) or of memory of such sensation, & memory is repetition of whatever takes place in brain, when sensation is perceived."[112] Darwin expanded these sensationalist views about cognition in his *N Notebook*, where he proposed that reason in its elemen-

109. F. H. Bradley to C. Lloyd Morgan (16 February 1895), in the Papers of C. Lloyd Morgan, DM 612, Bristol University Library.
110. John Locke, *An Essay Concerning Human Understanding*, ed. John Yolton, 5th ed. (New York: Everyman Library, 1965), 1:127.
111. See chapter 1 and my "Influence of Sensationalist Tradition on Early Theories of the Evolution of Behavior," *Journal of the History of Ideas* 40 (1979): 85–105.
112. Darwin, *M Notebook*, MS p. 61e–62e (Gruber, p. 277).

tary form was merely a comparison of sensations and in its more complex manifestations, only the recollection of several sensory images that
resolved themselves into a lively and pleasant conception. As he put it
in that notebook: "Reason in simplest form probably is single comparison by senses of any two objects—they by VIVID power of conception
between one or two absent things.—reason probably mere consequence
of vividness & multiplicity of things remembered & the associated pleasure as accompanying such memory."[113] Darwin's account of thought
resonates of Hume's psychology, which his grandfather had also endorsed. Darwin began reading Hume's *Inquiry Concerning Human Understanding* in August of 1838, just before penning the first of the preceding remarks. Hume's conception lent itself perfectly to Darwin's
efforts to link human mental abilities with their precursors in animal
mind. But Hume had relatively little to say about the specifically biological connections. Here Darwin drew his inspiration directly from
Erasmus.

Darwin readily acceded, as we have seen, to his grandfather's theory
that cerebral processes determined mental activity. Continuity of physical development of the brain between animals and men thus could be
used to argue for the continuity of mental development. Yet this physi-
calist argument did not quite bridge what was usually perceived as the
gulf separating animal mind from human mind: animals acted instinctively; men reasoned. Darwin realized that he had to define further the
relationship of instinct to reason, since these appeared to be the polar
traits of two different kinds of mind. He worked out his position
against the opposing views of Lamarck and Edward Blyth, the hapless
naturalist who later became his friend.[114]

At the beginning of his evolutionary theorizing, Lamarck had yet to
escape the grasp of Cartesian rationalism (few Frenchmen ever seem to
break entirely clear). In his *Recherches sur l'organisation des corps vivans*
(1802), he granted the higher animals intelligence, yet refused them reason—a putatively more numinous faculty that separated man from
beast. In the *Philosophie zoologique* (1809), however, Lamarck adjusted

113. Darwin, *N Notebook,* MS p. 21e (Gruber, p. 334).

114. Edward Blyth (1810–1873), whose economic circumstances and frail health prevented university training, undertook his own tutelage in natural history. He published
several papers early in his career that caught the eye of fellow naturalists, and in 1841 he
accepted a position as Curator of the Museum of the Royal Asiatic Society of Bengal.
Through considerable effort, he virtually established by himself the study of the zoology
of the subcontinent. But his life in India, as Francis Darwin judged (in *Life and Letters of
Charles Darwin* 2:109), was a constant struggle with poverty and unhappiness, especially
after the death of his wife. Charles Darwin carried on a lengthy correspondence with
Blyth and received him as a guest at Down in 1868.

his theory of reason to allow its presence in animals. Following Cabanis, he came to regard reason as but the guide of experience in the correct use of judgment—something of which all intelligent animals were capable. But the nether end of the chain of being, the lower animals—for Lamarck those wanting cerebral hemispheres—these he still held incapable of reason or intelligence. Darwin recognized that such an apparent break in the linkage of species damaged the case for their gradual transmutation. In this instance again he thought his "theory very much distinct from Lamarck's."[115] Darwin's sensationalist construction of reason as sensory association permitted him no distinction between intelligence and reason. And the tradition of his intellectual forebears counseled empirical evidence, not a priori presumption, be used to decide whether an animal exercised reason. Darwin judged that flexible behavior in contingent circumstances and the ability to learn from experience gave evidence that even insects, on occasion, made rational decisions and that their intellectual faculties differed from man's only in degree of complexity.[116] In counting ants, bees, and other small creatures among the rational animals, he set no precedent with fellow naturalists. Even the natural theologians John Fleming, Algernon Wells, Lord Brougham, and William Kirby—all of whom Darwin read with attention (see the next chapter)—would not deny "some feebler rays of reason" to the lower animals.[117]

Nor did Edward Blyth, whose papers on animal instinct and intelligence Darwin read in early 1838.[118] Blyth, also tinctured with the doctrine of sensationalism, conceded that animals did sometimes reason and reflect; though like Condillac and Le Roy in the previous century, he assumed that they engaged their cognitive powers only when compelled by a rapidly changing environment.[119] Their instinctive knowl-

115. Darwin, *First Notebook on the Transformation of Species,* MS p. 214 (de Beer, p. 66).

116. Darwin, *First Notebook on the Transformation of Species,* MS pp. 207–208 (de Beer, p. 66); *M Notebook,* MS pp. 62e–63e, 72–73 (Gruber, pp. 277, 278).

117. Algernon Wells, *On Animal Instinct* (Colchester: Longman, Rees, Orme, Brown, Green, and Longman, 1834), p. 20.

118. Edward Blyth, "On the Psychological Distinctions between Man and all other Animals," *Magazine of Natural History* 1, n.s. (1837): 1–9, 77–85, 131–41. Darwin first mentioned this essay by Blyth in his *Second Transmutation Notebook,* MS p. 198 (de Beer, p. 106). His own copy of the article is annotated and well marked.

119. Blyth, "On the Psychological Distinctions between Man and all other Animals," pp. 3–4: "I wish not to defend the untenable doctrine, that the higher groups of animals do not individually profit by experience; nor to deny to them the capability of observation and reflection, whereby to modify, to a considerable extent, their instinctive conduct." Blyth understood "reason," as other sensationalists had, to be the name for learning from experience through observation and reflection. "*Reason* . . . in human beings," he remarked (pp. 4–5), "can, of course, be only the result of observation and reflection."

edge—he thought of instinct precisely as a kind of innate knowledge—had been foreordained by a beneficent Providence; and it usually
served their wants.[120] But if the innate repertoire of animals did fail in
alien surroundings, he suspected they might, as a kind of aid to Divinity, restock from natural sources. Alterations in the instincts of domestic animals attested that such changes in the wild might occur. Blyth
consequently ventured that the experiences—and even anatomical
modifications—acquired by an animal in nature might be transmitted
to its progeny through inheritance, thus securing for future generations
a legacy of reformed adaptations appropriate to new circumstances.
Blyth supposed, however, that such transformations only happened occasionally. And like Lyell, he denied that inherited characteristics were
potent enough or animals plastic enough to produce new species from
what were mere varieties.

Blyth discovered in the human animal neither preternatural nor acquired instincts. "The human race," he asserted, "is compelled to derive
the whole of its information through the medium of the senses," while
"the brute is, on the contrary, supplied with an innate knowledge" of
those things affecting its welfare.[121] When a juvenile chimpanzee, for
instance, confronts a python, its natural enemy, "it 'instinctively' recoils
with dread." "But," he asked, "does a human infant evince the like recognition?"[122] He thought not. His reader, however, was not so easily
convinced. In the margin of his copy of Blyth's article, Darwin offered
a refuting observation: "Child fears dark before reason had told it."[123]

To Darwin's mind, Blyth's separation of instinct and reason only preserved the conventional distinction between animals and men. Darwin
rejected what he regarded as Blyth's assumption of a "saltus" between
instinct and reason. He preferred to term the difference a "hiatus," a
description which for him implied a matter of degree rather than of
kind; for he found sufficient warrant to consider reason and instinct as
really continuous powers.[124] In his *C Notebook,* where he responded to
Blyth's article, Darwin considered that human mentality was subject to
acquirable and hereditary determinations: the nervous factors which
enabled men to use language, for instance, must have been gradually
acquired as innate structures. This meant that animal instincts, also the
result of acquirable and hereditary determinations, could not be so radi-

120. See Blyth's remarks about "innate knowledge" quoted in the following paragraph
of the text.
121. Ibid., p. 2.
122. Ibid.
123. Darwin's annotation appears in ibid., at the bottom of p. 2.
124. Darwin, *Second Transformation Notebook,* MS p. 198 (de Beer, p. 106).

cally different from the reasonable acts of men.[125] Intellect might be understood, Darwin proposed a few months later, as "a modification of instinct—an unfolding & generalizing of the means by which an instinct is transmitted."[126] In this conception, intelligence emerged when cerebral structures determining instinct became less fixed, more flexible and responsive to environmental contingencies. Intelligence, then, was not opposed to instinct, but rather grew out of it (a conception Herbert Spencer would also advance). This interpretation of the instinct-intelligence relationship echoed Darwin's then favorite philosophical author, David Hume, who characterized human reason as "a wonderful and unintelligible instinct in our souls."[127]

At the present time, when sociobiologists and ethologists make claims for mental continuity between men and animals, the ensuing debate usually turns on the question of rationality: Can animals reason as we do? While this issue played a role both in Darwin's formulation of his theory and in his opponents' attempts to refute it, in nineteenth-century England the question of animal reason was not so strenuously mooted. The British sensationalist tradition, which carried along many of the century's leading scientists and philosophers, had precluded an effective distinction between human and animal intellectual abilities. The more significant question was: Can animals make moral judgments as we do? Or more pointedly: Is man essentially no more moral than a rutting pig? In our day even many committed evolutionists have abandoned our animal forebears on the other side of this Rubicon: they believe that men have become moral creatures and consequently exempt from an evolutionary rendering of their ethical behavior. Only the benighted sociobiologist, they argue, attempts to generalize from the 'altruism' of the ant soldier to the altruism of the human soldier.[128] Dar-

125. Darwin, *Second Transformation Notebook,* MS p. 198 (de Beer, p. 106).

126. Darwin, *N Notebook,* MS p. 48 (Gruber, p. 339).

127. David Hume, *Treatise of Human Nature,* ed. L. A. Selby-Bigge (Oxford: Clarendon Press, [1739] 1888), p. 179. In his *N Notebook,* MS p. 101 (Gruber, p. 348), Darwin seized on Hume's analysis: "Hume has section (IX) on the Reason of Animals . . . he seems to allow it is an instinct."

128. Robert Trivers, in "The Evolution of Reciprocal Altruism," *Quarterly Review of Biology* 46 (1971): 35–57, and Edward Wilson, in *Sociobiology: The New Synthesis* (Cambridge: Harvard University Press, 1975), pp. 562–64, have renewed the earlier efforts of Julian Huxley, in *Evolution and Ethics, 1893–1943* (London: Pilot, 1947) and C. H. Waddington, in *The Ethical Animal* (New York: Atheneum, 1961), to give an evolutionary account of human moral behavior. These attempts have been severely criticized by evolutionists of a different stripe: see, for example, Stephen Jay Gould's "Biological Potentiality vs. Biological Determinism," in his *Ever Since Darwin* (New York: Norton, 1977), pp. 251–59, and Richard Burian, in "A Methodological Critique of Sociobiology," in *The*

win, however, recognized early on that the question of man's moral nature had to be faced and an evolutionary account constructed. For if the moral sense—that property commonly regarded as essentially distinctive of man—were explained in some other fashion, then such explanation might serve for all other traits, thus emasculating his entire evolutionary conception. Darwin's theory of the moral sense, which underwent continuous development from this earlier period until the composition of the *Descent of Man,* reveals, I believe, his peculiar genius. The theory in its finished form, which will be examined in chapter 5, is biologically and philosophically sophisticated. In the second appendix, I will plump for a refined version of it.

The Evolution of Morality

The Ethical Instinct

Darwin's *Beagle Diary* records countless observations, not only of the geology and zoology of South America—his professional mandate—but also of the customs and behavior of its human inhabitants. His upbringing sensitized him to the moral environment through which he passed. Dr. Robert Darwin gave his son example of an honest, just, yet compassionate physician to rich and poor alike. The tender mercies of Charles's four sisters, who looked to the care of their baby brother, could be armored in a fierce passion about moral issues. Susan in particular, following the tradition of her grandfathers, often became quite incensed over the evils of slavery.[129] None of this, however, prepared the young naturalist for the moral climate of South America. He thus stood affronted by the ignorant, vengeful, and indolent Brazilians, who possessed "but a small share of those qualities which give dignity to mankind."[130] He was outraged at the Spanish troops who slaughtered whole Indian tribes, including all women over twenty because "they breed so!"[131] He was no less shocked by the everyday cruelty of the slavery practiced by plantation owners. Yet he did not romanticize the natives. The South American Indians, he knew, were capable of

---

Sociobiology Debate, ed. Arthur Caplan (New York: Harper, 1978), pp. 376–95. The ethical issues aroused by evolutionary theory receive vetting in the following collections: *Sociobiology and Human Nature,* ed. Michael Gregory, Anita Silvers, and Diane Sutch (San Francisco: Jossey-Bass, 1978); *Morality as a Biological Phenomenon,* ed. Gunther Stent (Berkeley: University of California Press, 1980); and *The Sociobiology Debate.*

129. Brent, *Charles Darwin,* pp. 21, 176–77.

130. *Charles Darwin's Diary of the Voyage of H.M.S. Beagle,* ed. Nora Barlow (Cambridge: Cambridge University Press, 1933), p. 76.

131. Ibid., p. 171.

great "refinements in cruelty."[132] But they also displayed moral courage. Darwin detected a Roman nobility in the suicide leap of a recaptured slave, an old woman who preferred death on her own terms.[133] And he greatly admired the loyalty of a captured Indian, who refused to betray his comrades, crying to his Spanish inquisitors, "Fire, I am a man & can die."[134] Darwin judged that the Spaniard's genocidal war against the Indians would leave the country "in the hands of white Gaucho savages instead of copper coloured Indians. The former being a little superior in civilization, as they are inferior in every moral virtue."[135]

During his five-year journey, Darwin experienced the extremes of moral behavior, from the almost endemic venality of the South American gauchos to the frequent valor of the Indians whom they slaughtered. The different societies he lived among each appeared to have its distinctive code of behavior, which often as not repelled the young Englishman. Yet even within the diversity, Darwin detected common attitudes about right and wrong conduct. As with other aspects of his developing species theory, he began to excavate and organize this sedimented experience only after serious reading. At about the time when he edged toward an initial formulation of natural selection, during the summer and early fall of 1838, he was also busy tracing out a theory of the moral sense. Four books in particular helped him along the way: Harriet Martineau's *How to Observe: Manners and Morals,* William Paley's *Moral and Political Philosphy,* James Mackintosh's *Dissertation on the Progress of Ethical Philosophy,* and John Abercrombie's *Inquiries Concerning the Intellectual Powers.*[136]

Harriet Martineau, essayist, novelist, radical social critic, and Unitarian with ear trumpet, had been a friend of Fanny Wedgwood, daughter of James Mackintosh and wife of Hensleigh Wedgwood, Darwin's cousin and confidant. Martineau was a woman of high intellect and voluble opinion. The editor of the *Edinburgh Review,* Sidney Smith, when asked by a friend how he had spent his evening, wickedly replied: "Oh, horrid, horrid, my dear fellow! I dreamt I was chained to a rock

132. Ibid., p. 98.
133. Ibid., p. 50.
134. Ibid., p. 172.
135. Ibid., pp. 172–73.
136. Harriet Martineau, *How to Observe: Manners and Morals* (New York: Harper, 1838); William Paley, *Moral and Political Philosophy,* in *The Works of William Paley* (Philadelphia: Woodward, n.d.); James Mackintosh, *Dissertation on the Progress of Ethical Philosophy* (Edinburgh: Adam and Charles Black, 1836). Reference to Abercrombie has been given. Edward Manier, in his continuously interesting *The Young Darwin and His Cultural Circle* (Dordrecht: Reidel, 1978), briefly describes the views of Martineau, Mackintosh, and Abercrombie.

and being talked to death by Harriet Martineau and Macaulay."[137] Darwin finally met Miss Martineau in 1838 at a dinner party given by his brother Erasmus. Her intellectual charms bedazzled the young naturalist, but no less the other eminent minds of her widening circle, which included the irascible historian Thomas Carlyle, the great amateur scientist and Chancellor of the Exchequer Henry Lord Brougham, and Britain's leading geologist Charles Lyell. Darwin reported to his granny (his affectionately named sister Susan) that Miss Martineau "has been as frisky lately as the Rhinoceros" at the zoological park. He confided his gratitude that Martineau was "so very plain," lest his brother, who "has been with her noon, morning, and night," succumb to more than her intellect.[138] Probably the family tie and his own encounter seduced Darwin into reading her newly published book, *How to Observe: Manners and Morals*.

In her tract, Martineau emphasized the varieties of morally sanctioned behavior found in other times and places. Men of an earlier age, for instance, estimated their virtue by the number of enemies killed, while a British gentleman recognized the dignity in saving life. Those who acknowledged a universal moral sense for right action must be puzzled, she wryly remarked, since "there are parts of this world where mothers believe it a duty to drown their children and . . . Eastern potentates openly deride the king of England for having only one wife instead of a hundred."[139] Martineau urged that an unbiased survey would convince an observer that virtuous or vicious acts were "the result of particular circumstances amidst which the society exists."[140] She allowed that there were "universal feelings about right and wrong," which led to the formation of codes of behavior, but held that these resulted from the "general influences" dispensed by Providence. Experience provided specific ideas and feelings of right and wrong. Individuals, she concluded, lacked any innate sense of particular virtues or vices.[141]

Darwin reacted to Martineau as a natural historian might. He reckoned the variability of conscience in the many races of man no more

137. Quoted by R. K. Webb, *Harriet Martineau: A Radical Victorian* (New York: Columbia University Press, 1960), p. 11.

138. Charles Darwin, "Letter to Granny (Mrs. Wedgwood [sic]), 1838." The letter is on microfilm at the American Philosophical Society, 496.1. The correspondent was, however, his sister Susan, whom Darwin called "Granny" because of her solicitude in raising him.

139. Martineau, *How to Observe*, pp. 29–30.

140. Ibid., p. 33.

141. Ibid., p. 32–33.

Figure 2.3   Harriet Martineau, 1802–1876, portrait done in 1834.

unusual than the variety of instincts displayed by different breeds of dog. Mere variability, as he observed in his *M Notebook,* did not mean that moral behavior lacked an innate foundation.[142] And even Martineau recognized that in all societies men felt obliged to act kindly toward their fellows and not to injure them without cause. Darwin supposed that the universal features of moral response—for example,

142. Darwin, *M Notebook,* MS pp. 75–76 (Gruber, p. 279).

treating neighbors benevolently—might be attributed to man's instinctively social nature, while peculiar moral attitudes might be due to the special circumstances of a society.[143]

Reading Martineau may well have awakened in Darwin memories of an earlier day, when he had to get up Paley's *Moral and Political Philosophy* for his B.A. exam at Cambridge. For Paley, like Martineau, appealed to the practices of different societies to argue that men possessed no "moral instincts." He believed a child's early experience of beneficial acts became associated with the pleasure they brought, so that the child would subsequently approve of and encourage them, whether he was their recipient or agent.[144] Mackintosh, in his *Dissertation on the Progress of Ethical Philosophy,* described Paley's theory as "selfish," since this very secular divine presumed that pleasure for oneself formed the ultimate motive even in moral acts. Though Paley represented a large class of British moralists in this respect—including Adam Smith and Jeremy Bentham—his ethical analyses bespoke a man of action, not of reflection; his moral theory shared this trait with that of Martineau, herself concerned with the lighter sides of radical politics and Malthusian political economy. Mackintosh might thus have estimated her theorizing as he had that of the influential theologian: "The natural frame of Paley's understanding fitted it more for business and the world than for philosophy."[145]

In his later years, Darwin grew ever more dissatisfied with Paley's morality of selfishness, his distaste sharpened by Mackintosh's animadversions. But in this early period, Paley brought him to a critical stage in his continuing effort to provide a biological skeleton for the moral fabric of human nature. On 8 September 1838, as his *M Notebook* records, Darwin considered "Paley's Rule." In *Moral and Political Philosophy,* Paley stated his expediency rule of morality: "Whatever is expedient is right. But then it must be expedient on the whole, at the long run, in all its effects collateral and remote, as well as in those which are immediate and direct."[146] Darwin gave the rule a biological interpretation: "Sept 8th. I am tempted to say that those actions which have been found necessary for long generation, (as friendship to fellow animals in social animals) are those which are good & consequently give pleasure, & not as Paley's rule is then that in long run *will* do good.—alter *will* in all such cases to *have* & *origin* as well as *rule* will be given."[147] Darwin

143. Darwin, *M Notebook,* MS p. 76 (Gruber, p. 279).
144. Paley, *Moral and Political Philosophy,* pp. 29–30.
145. Mackintosh, *Dissertation on the Progress of Ethical Philosophy,* p. 274.
146. Paley, *Moral and Political Philosophy,* p. 40.
147. Darwin, *M Notebook,* MS p. 132e (Gruber, p. 291).

here suggested that those useful and expedient habits which were necessary to preserve animal groups, allowing them to propagate and protect their young (such as habits of sociality, friendship, nurture, etc.), were what we have come to call morally good. The continued practice of such useful behaviors, as he would make explicit in succeeding pages of his notebooks, produced through use inheritance moral instincts that conformed to a temporally readjusted rule of expediency: what has been good, biologically good, we now experience in our bones as a satisfying moral good. Darwin retained right through to the *Descent of Man* this fundamental identity between the moral good and the biological good. The Paley of his youth was preserved, though transmuted, in his evolutionary theory of morals.

Darwin's response to Martineau and Paley, while expressed in the categories of the naturalist, derived support from a framework of moral philosophy erected by Mackintosh. Darwin read Mackintosh's *Dissertation* on and off from the summer of 1838 to the spring of 1839, as he worked out the implications of his broader theory of evolution for the problems of mind and morals. But his introduction to Mackintosh, as with Martineau, came through personal rather than literary contact. Mackintosh was the brother-in-law of Darwin's uncle Josiah Wedgwood, and like the young nephew he frequently visited the Wedgwood country house at Maer. Darwin first met him in 1827 and fell entranced under the power of the older man's intellect. In his *Autobiography*, Darwin recalled that at the time he "listened with much interest to everything he [Mackintosh] said, for I was as ignorant as a pig about the subject of history, politicks and moral philosophy." Darwin judged this imposing man "the best converser on grave subjects to whom I have ever listened."[148]

Like Shaftesbury, Butler, and Hutchinson—philosophers whose theories he analyzed—Mackintosh believed that human nature came equipped with a moral sense for right conduct. He allowed that external circumstances were probably necessary to educate the moral faculty; thus we could expect variation in moral behavior of different societies as well as a progressive development of higher moral standards over generations. Nonetheless, he denied that mere learned associations could produce the unique connection between particular acts and special feelings of obligation.[149] Moreover, though Mackintosh gave associative learning a role in honing the moral sense, unlike Paley, he did not regard moral action as ultimately selfish, that is, as motivated by anticipations

---

148. Darwin, *Autobiography*, p. 66.
149. Mackintosh, *Dissertation on the Progress of Ethical Philosophy*, pp. 254–61.

of pleasure with which such behavior might have become associated.[150] Men, rather, gave spontaneous approval to virtuous acts; they immediately tried to secure the well-being of their children; they despised cowardice and meanness—they did all this without the kind of hedonistic accounting required to estimate possible rewards to oneself. To get clear on this point and to separate his system from that of both Paley and the great pleasure calculator Jeremy Bentham, Mackintosh made what he believed to be a critical distinction for moral theory: between the *moral sense* for right conduct and the *criterion* of moral behavior.[151] Our immediate repugnance for vicious acts was quite different, he maintained, from subsequent estimations of their consequences. We instinctively perceived murder as vile. And in a cool moment of rational evaluation, we could also weigh the disutility of murder. But these processes were not the same. For Mackintosh, then, utility—or the greatest happiness principle—was indeed a basic criterion for measuring behavior; it simply did not function as an immediate motive for action.[152]

There was, however, a tender spot in Mackintosh's analysis. He could not satisfactorily explain the coincidence of the moral motive and the moral criterion. Why was it that what men did from moral impulse, without reflection, happened to be what subsequent evaluation using the moral criterion might sanction? Why did a nonrational sentiment and a consequent rational judgment always seem to agree? Darwin thought that Mackintosh's hesitating retreat to theology—making faint appeal to divine harmonizing—provided no adequate scientific account for this constant conjunction. He found that his own theory of behavior gained proportional strength as it could explain in this instance—and in several others—what Mackintosh ultimately took for granted.

Darwin's early theory of moral behavior, as well as its mature expression in the *Descent of Man,* can best be understood, I believe, as a biologizing of Mackintosh's ethical system. He had begun the process with Martineau and Paley. But in Mackintosh he discovered a moral system with which he was in fundamental sympathy. Darwin set out to give, in natural-historical and evolutionary terms, an interpretation of the faculties and relationships he found described in Mackintosh's *Dissertation*.

Four principal factors seem to have motivated him in this early period to attempt a theory of the moral sense. The striking and exotic varieties of human moral behavior in South America, of course, had to attract the inquisitive eye of this naturalist. But his real enthusiasm for work in

150. Ibid., pp. 192–93, 385, 400.
151. Ibid., pp. 62–67.
152. Ibid., pp. 355–61.

morals came, I believe, from his casual success in applying ideas about animal instinct to those moral problems suggested initially by Martineau and Paley. At the same time, reading Mackintosh stoked this interest, as indicated by his fourteen-page manuscript on Mackintosh's *Dissertation,* in which he organized his early ideas on morality and constructed the framework for his discussion of morals in the *Descent of Man*.

The third impetus to working out a theory of morality involved its linkage to sex and the species question. In November 1838, shortly after having read Malthus, Darwin turned to the question with which he opened his first *Transmutation Notebook*—the function of sex. Echoing that earlier consideration, he suggested in his fourth *Transmutation Notebook* that sex was required to explain distinct species. Sexual intercrossing allowed animals living in various environments to transmit common traits. Sex gave stability and uniformity to species. But sexual intercrossing required sociality. Hence the moral sense, derived as Darwin thought from social instincts, might also be explained by its function in solidifying species, in molding a population into a unit. In his fourth notebook, he expressed the matter this way:

> My theory gives great final cause . . . of sexes /in separate animals/: for otherwise there would be as many species, as individuals, & though we may not trace out all the ill effects,—we see it is not the order in this perfect world, either at the present, or many anterior epochs.—but we can see if all species, there would not be social animals. hence not social instincts, which as I hope to show is /probably/ the foundation of all that is most beautiful in the moral sentiments of the animated beings.[153]

Darwin thus believed that a solution to the origin and nature of the moral sense would shed light on the orgin of species.

The fourth factor drawing Darwin toward what he later came to regard as the treacherous swamp of moral speculation was the location there of the citadel itself—man and his putatively distinctive character. Darwin realized that if he failed to provide a transformationist explanation of the essential and distinguishing trait of the human animal, his whole project would fall into extreme jeopardy. For had he conceded a divine source for ethical behavior, that wedge would have pushed open the gates to a returning flood of creationism. He needed to work out a naturalistic theory of the moral sense.

153. Darwin, *Fourth Transmutation Notebook,* MS p. E48–49 (de Beer, p. 163). Slashes enclose words inserted by Darwin.

The central nerve of Darwin's theory, which sustained its parts, was the proposal that the moral sense be regarded as an instinct. The bare suggestion itself was hardly original. Paley had considered the likeness of the moral sense to animal instinct, though he had rejected the comparison.[154] The importance of Darwin's proposal lay in the enriched notion he had of instinct and its genesis. The more he followed the analogy between instinct and the moral faculty, the more exact he found the fit. The paradigm of moral sensitivity, as Mackintosh conceived it, was a disinterested action directed toward the welfare of another. Mackintosh allowed that moral behavior might vary in different social groups, but he insisted that its common features sprang from the bowels of man's nature. As a naturalist, Darwin quickly identified the classic instances of moral behavior with the parental, conjugal, and social instincts displayed by animals. The mother bird, for instance, acts instinctively to protect her young and provide for their welfare, certainly without thought of possible reward to herself. The social instincts of animals were indeed unselfish and natural affections. Moreover, both the moral sense and social instincts functioned to bind a society together. In Darwin's judgment, "society could not go on except for the moral sense, any more than a hive of Bees without their instincts."[155] Thus moral acts of men and social instincts of animals had the identical qualities of being innate, disinterested, and socially unifying.

The similarity of moral sense to instinct led Darwin gradually to elaborate a wonderfully ingenious explanation for the faculty of conscience, complete with its reprimanding pangs and insistent calls to duty. He distinguished two kinds of instinct: the immediately impulsive instincts, such as a flare of anger or stab of lust; and the more calm and persistent social instincts, which kept a mother bird patiently tending her brood. Often enough, however, these instincts came into conflict—when, for example, a hen would be momentarily overcome by the sight of a migrating flock and abandon her young. Now, Darwin hypothesized, if the reprobate mother possessed sufficient intellect to recall the situation of her chicks, she would be uncomfortable, uneasy in spirit; she would once again feel the unsatisfied urgings of social instinct. A rational animal in such circumstances would confess a troubled conscience. In Darwin's view, all that was necessary to turn social instinct into the voice of conscience was a mind somewhat greater than that possessed by animals. This hypothesis congealed in his reflections on 3 October 1838, when he recorded in his *N Notebook*:

154. Paley, *Moral and Political Philosophy*, p. 30.
155. Darwin, *Old and Useless Notes*, MS p. 30 (Gruber, p. 390).

> Dog obeying instinct of running hare is stopped by fleas, also
> by greater temptation as bitch. . . . Now if dogs mind were so
> framed that he constantly compared his impressions, & wished
> he had done so & so for his interest, & found he disobeyed a
> wish which was part of his system, & constant, for a wish
> which was only short & might otherwise have been relieved,
> he would be sorry or have troubled conscience.—Therefore I
> say grant reason to any animal with social & sexual instincts &
> yet with passion he *must* have conscience—this is capital view.
> Dogs conscience would not have been same with mans because
> original instincts different.[156]

Conscience so conceived, of course, would become progressively more
sensitive over generations as both greater intellect and advancing civilization acted to sharpen the focus of instinct on more finely graded
objects. This indeed signaled the beginnings of a capital theory of
conscience.

In his 1839 manuscript on Mackintosh, Darwin amplified his account
of conscience.[157] He reiterated that the insistent character of social instincts, which formed the very ribs of man's civilized being, should be
regarded as equivalent to the moralist's *ought*. After all, he observed, we
say a hound ought to point, since that is part of its instinctive nature;
just so, we think a father ought to care for his children because of the
natural bond of parentage. The serene pleasure enjoyed in exercise of
the social instincts and the pain suffered in their breach constituted the
primitive feelings of duty and sin. Yet Darwin did not let moral obligation rest solely on instinctive feeling. In the 1839 manuscript, he adumbrated a kind of rule utilitarianism, which supposed one gradually
learned that the momentary pleasures of passion did not outweigh the
more constant satisfactions of the social instincts. And so finally, "by
association one gains the rule, that the passions & appetite should
/ almost/ always be sacrificed to the instincts."[158] The habit of following
such rules, Darwin apparently believed, brought a distinctive feeling of
the exercise of the moral imperative, a feeling more refined than the
elemental urgings of social instincts alone. It also imbued conscience
with a greater rational character and so helped explain how children
advanced beyond their primitive moral condition.

Perhaps the most impressive feature of Darwin's early ethical theory

156. Darwin, *N Notebook*, MS pp. 1–3 (Gruber, pp. 329–30).

157. The manuscript is included in the bundle of notes labeled by Darwin *Old and Useless Notes*, MS pp. 42–55 (Gruber, pp. 398–405).

158. Darwin, *Old and Useless Notes*, MS p. 45 (Gruber, p. 399). The word enclosed by slashes was inserted by Darwin.

was its solution to Mackintosh's problem of the coincidence of moral motive and moral criterion. The theory resolved the difficulty by biologically uniting the psychology of the moral sense with the normative criterion of utility. Darwin worked out the details in a series of notes begun in late 1838 and culminating in the 1839 manuscript on Mackintosh.[159] He proposed that the social instincts of the individual tended to agree with the criterion of utility—the greatest good for the greatest number—because only those instincts that generally benefited past generations would be inherited. Now Darwin at this time did not argue that such instincts had been acquired and transmitted through natural selection, though he had formulated the principle several months before.[160] Rather, he still found sufficient the mechanism of inherited habit, which, as he noted in May 1839, "fully explains the cementation of habits into instincts."[161] Useful behaviors would be fused into the hereditary substance of animals because, since they were useful, such behaviors would have been practiced frequently, so producing instincts in succeeding generations. Darwin thus advanced the law of utility, Paley's criterion of morality, as also that law governing the acquisition of instinct:

> *On Law of Utility* Nothing but that which has beneficial tendency through many ages could be acquired, & we are certain from our reason, that all which (as we must admit) has been acquired, does possess the beneficial tendency. It is probable that becomes instinctive which is repeated under many generations . . . & only that which is beneficial to race, will have reoccurred.[162]

Those behaviors occurring most constantly would be the conjugal, parental, and social habits—the glue of human and animal societies. Hence Darwin's solution to Mackintosh's problem:

> Two classes of moralists: one says our rule of life is what *will* produce the greatest happiness.—The other says we have a moral sense.—But my view unites both /& shows them to be almost identical/ & what *has* produced the greatest good /or rather what was necessary for good at all/ *is* the /instinctive/

159. Darwin, *Old and Useless Notes*, MS pp. 30–55 (Gruber, pp. 390–405).

160. Manier, in *Young Darwin*, p. 146, assumes that Darwin used the principle of natural selection in his early theory to explain the evolution of the moral sense. Manier's effort to place the young Darwin in his cultural and historical context is skillful and illuminating; although on this issue I believe the evidence, which I have present above in the text, is clearly against him.

161. Darwin, *Old and Useless Notes*, MS p. 48 (Gruber, p. 401).

162. Darwin, *Old and Useless Notes*, MS p. 50 (Gruber, p. 402).

> moral sense. . . . In judging of the rule of happiness we must look *far forward* /& to the *general action/*—certainly because it is the result of what has *generally* been best for our good *far back*.[163]

Darwin's solution, then, to Mackintosh's problem amounted to this. The habits that members of a species would most often adopt over many generations and under different circumstances would be those of general utility—ingrained habits of parental nurture, mutual aid, group defense, and so forth. Habits satisfying peculiar individual desires would wash out over generations, and only those remaining practices conducive to the common good would become deeply entrenched instincts. These latter would constitute the moral motives characteristic of a society. When an individual then came to prescribe actions or judge what ought to be done in a situation, the criterion of general utility—the very measure employed by nature—would recommend itself. It would recommend itself since simple reflection on one's own behavior and on that of others would demonstrate the kind of motive actually at work in moral action. Thus what particular motives and consequent actions nature herself deemed useful and worth fostering would also be approved by a reflective moral agent using the same criterion. Hence the conjunction between moral motives and moral evaluations.

Darwin's solution to Mackintosh's problem gave him confidence that his evolutionary system had the kind of robustness that Whewell required of a good theory: it could render consilient inductive generalizations in a variety of different areas. Yet some difficulties remained for this particular moral theory. The most crucial concerned what we now regard as the very pith of a moral act: its disinterested, altruistic character. The social instincts, as Darwin understood them, were actions of this kind. But how could one explain the original acquisition and maintenance of altruistic habits, since they conferred pleasure and advantage not on their agent but on their recipient? In the sensationalist tradition, habits could be acquired only if they produced some pleasure for or advantage to the individual exercising them. The mother's loving care of her infant thus seemed an improbable habit or instinct, as Darwin recognized: "But the love is instinctive, & how does it apply to mother loving child, from whom, she has never received any benefit."[164] He found the resolution of this difficulty initially in Mackintosh's exposition of David Hartley's views. Hartley (with Adam Smith and Paley following) ar-

163. Darwin, *Old and Useless Notes,* MS p. 30 (Gruber, p. 390). Slashes enclose words Darwin inserted.
164. Darwin, *Old and Useless Notes,* MS p. 49 (Gruber, p. 401).

gued that benevolent regard for another arose out of associations built up during childhood: the child receives a thousand acts of kindness and consequently begins automatically to associate pleasure with behavior of a certain pattern and character. Thus acts of benevolence toward another become habituated because of the pleasurable associations they arouse when a person performs them.[165] Darwin, agreeing with Mackintosh, thought this was really no selfish principle of morality, for such habits would eventually be executed without conscious intention of gaining pleasure.[166] When these habitually repeated behaviors became instinctive, as Darwin believed they would, the original link with pleasure would be even more attenuated, leaving only a warm social feeling arising from the contemplation of the altruistic act. Instincts, in any case, were immediately produced by innate factors bred in the bone, not by expectations of pleasure or advantage. Darwin thus judged "Hartley['s] explanation [to] apply perfectly to origin of these instincts."[167]

When Darwin later came to use the principle of natural selection to account for social instincts, the difficulty just mentioned resurfaced. Natural selection operated for the benefit of the individual expressing certain traits. How could it explain the acquisition of instincts that not only failed to confer direct advantage on their bearers but might even be harmful to them? Darwin was able to handle this problem in moral theory only after he solved a more general and serious difficulty, a difficulty which threatened to overturn completely his entire theory of evolution by natural selection. But more about that in the next chapter.

Moral Freedom and Biological Determinism

Darwin realized that a biological explanation of thought and behavior implied that organisms acted under law, that they were not free. Free will could only be equivalent to chance: "I verily believe," he remarked in his *M Notebook*, "free will & chance are synonymous. Shake ten thousand grains of sand together & one will be uppermost, so in thoughts, one will rise according to law."[168] Now there is a certain tension, as philosophers like to put it, between the belief that man can make unfettered moral decisions and the belief that his thoughts fall out like marbles from a bag. Insofar, however, as Darwin meant his moral theory to be descriptive, explaining how people actually came to ethical

165. Mackintosh, *Dissertation on the Progress of Ethical Philosophy*, pp. 257–58.
166. Ibid., pp. 265–67.
167. Darwin, *Old and Useless Notes*, MS p. 50 (Gruber, p. 402).
168. Darwin, *M Notebook*, MS p. 31 (Gruber, p. 271).

decisions and performed moral acts, there really was no conflict between the theory and the conviction that mind was purely a biological
product. But Darwin also regarded his moral system as leading to ethical imperatives.

He sketched out what would be his persistent view of the relationship
between moral choice and biological determinism in a gloss on a passage from Abercrombie's *Inquiries Concerning the Intellectual Powers,*
which he read in late summer of 1838. The Scots philosopher had argued
that the moral law, when clearly understood by the agent, coerced his
judgment, so that he had to choose according to its dictates. Even in
the moral realm, freedom—except as ignorance—did not exist. The
idea of determined action under law struck a resonant cord in Darwin,
and in the margins of his copy of Abercrombie, he penned in his
reaction:

> If believed—pretty world we should be in—But it could not
> be believed excepting by intellectual people—if I believed
> it—it would make as one difference in my life. for I feel more
> virtue more happiness—Believers would [mate] with only
> good women & pay attention to education & so put their chil-
> dren in way of being happy. It is yet right to punish criminals
> for *public* good. All this delusion of free will would necessarily
> be from men feeling power of action. View no more improb-
> able than there should be sick & therefore unhappy men. What
> humility this view teaches. A man [three words illegible] with
> his state of desire (neither by themselves sufficient) effect of
> birth and other accidents. Yes but what determines his *consid-
> eration?* his own previous conduct—& what has determined
> that? & so on—*Hereditary* character & education & chance.
> According to all this ones disgust at villain ought to be is noth-
> ing more than disgust at one under foul disease.[169]

Darwin understood that the doctrine of determinism, if believed,
would unsettle the mass of men; but he also recognized that only men
of reflection would take it seriously. Most men, because of their feelings
of executive agency, would remain deluded in believing themselves free.
But for those who carefully examined the doctrine, Darwin thought
they would see in it a kind of grandeur. For the doctrine counseled
reasoned treatment of those committing crimes, not blind revenge.
Vice, the theory implied, was a disease, due to hereditary disposition,
environment, and chance—influences to which anyone might fall vic-
tim. Yet the person persuaded of the transmutational view of morals

169. Darwin's annotations appear in Abercrombie, *Inquiries Concerning the Intellectual
Powers,* pp. 202–203.

would take care to select a healthy—morally healthy—mate, so that his children would be given the best chance to escape sin. Moreover, a man would see to the proper education of his children, since habits acquired during early life would become as instincts, indeed for his lineage they might actually become instinctive. This view quickened a man to his responsibilities to future generations.

Darwin did not attempt to resolve the dilemma of the moralist who also believes in determinism. But this does not force us to dismiss his theory. At one level it remains a plausible bit of social anthropology. There are, however, considerations compatible with the Darwinian perspective that mitigate the tension between an evolutionary construction of moral behavior and the acceptance of authentic ethical imperatives. These considerations will be offered in the second appendix to this volume. In the next chapter we will turn to a critical difficulty that Darwin stumbled over during the mid-1840s while pursuing his study of several natural theologians. He deemed it the most theory-threatening problem he faced.

Appendix: Kohn's Analysis of Darwin's Early Theories

David Kohn has analyzed Darwin's early theories of evolution in his comprehensive and meticulous "Theories to Work By: Rejected Theories, Reproduction, and Darwin's Path to Natural Selection."[170] Kohn maintains, correctly I believe, that Darwin's transformationist speculations in the *RN Notebook* did not include a mechanism of species change, but that only in the *Transmutation Notebooks* did he concentrate on the dynamics of evolution. Kohn is also right, I think, in holding that two fairly distinct, pre-Malthusian theories emerge in these latter notebooks. Kohn discriminates two key agents of species alteration operating in Darwin's two theories, the environment and acquired habit, respectively. Kohn contends, however, that Darwin's two early theories incorporated an additional and really more important mechanism— sexual reproduction. He argues that Darwin found asexual reproduction insufficient to account for heritable adaptations and intended sexual reproduction "to be an *explanation* of transmutation."[171]

Darwin certainly considered sexual propagation to be important for any theory of species change. He was convinced, for instance, that sexual crossing functioned to hold species constant. He also thought it

170. David Kohn, "Theories to Work By· Rejected Theories, Reproduction, and Darwin's Path to Natural Selection," *Studies in History of Biology* 4 (1980): 67–170.

171. Ibid., p. 87.

significant that the young of sexually dimorphic species did not exactly duplicate the structures of either parent. As a consequence of this, injuries and maladaptations would not necessarily be inherited. Kohn believes, however, that sexual reproduction also operated in Darwin's two theories as the chief mechanism by which species were adapted to their environments. He reads both of Darwin's early causal theories as holding "that sexual reproduction produces automatically adaptive variation and that that variation is the basis for transmutation."[172] When this interpretation is applied to what I have called Darwin's "habit-instinct" theory, it reduces the role of habit to that of simply securing the adaptations already produced by sexual generation. Kohn puts it this way: "adaptation is the direct result of the tendency in sexually reproducing organisms to produce peculiarities, i.e., to vary. . . . The last steps in adaptation were the continued use of new habits, their gradual support by the accumulation of small morphological and behavioral peculiarities, which did not disrupt constitutional stability, and their ultimate hereditary fixation over a succession of generations."[173]

Kohn's rendition of Darwin's theories—i.e., taking sexual reproduction as the agent of adaptation—is, I believe, unsupported by the evidence, which I have indicated in the text of this chapter, and denied by its general implausibility. Kohn suggests that Darwin assumed some *Deus ex sexuale* to harmonize variations with environmental circumstances, to make sure just those modifications occurred which would adapt an organism to its particular surroundings. But Darwin specifically and repeatedly appealed to the direct effects of the environment (in his first causal theory) and habit become instinct (in his second) as the sources of adaptations. Kohn's insistence on sex as the principal mechanism of adaptation precludes these two agents, which Darwin certainly postulated, from doing their job. Insofar as sexual production would originally provide heritable adaptations, environmental influence and habit could have no real role in initiating adaptations—a consequence contrary to Darwin's explicit intentions.

Kohn rests his peculiar reconstruction mostly on the opening pages of Darwin's *First Transmutation Notebook,* where Darwin jotted these reflections:

> We know world subject to cycle of change, temperature & all circumstances which influence living beings. We see the young of living beings become permanently changed or subject to variety, according to circumstances,—seeds of plants sown in

172. Ibid., p. 120.
173. Ibid., p. 132.

rich soil, many kinds are produced, though new individuals produced by buds are constant; hence we see generation here seems a means to vary or adaptation.—Again we know, in course of generation even mind and instinct becomes influenced. Child of savage not civilized man.[174]

But the next lines are crucial to a correct interpretation: "There may be unknown difficulty with *full grown* individual with fixed organisation thus being modified,—therefore generation to adapt and alter the race to *changing* world." I believe Darwin regarded the sexually propagated young of a species to be more malleable, more susceptible to the influences of the environment or habit than either cuttings and buds of plants, which replicated the mature plant, or the "*full grown* individual with fixed organisation." Darwin believed, quite simply, that you can't teach an old dog new tricks, especially adaptational ones. This interpretation conforms to the hereditary theory Darwin sketched in his notebooks, according to which alterations had to be "long in the blood" to be inherited.[175] Darwin believed that transmutation of species, their adaptation to circumstances, occurred only very gradually, over long periods of time. Thus the agents effecting change had to be "long in the blood." This hypothesis had other dividends for Darwin's theory. It explained, for instance, why hybridization could not be the usual source of new species: since "all mules have their whole form of body gained in one generation."[176] It also made passive modifications, particularly mutilations (e.g., clipping the tails of dogs), unlikely instruments of hereditary change—again because their effects were not long in the blood.[177] Darwin seems to have believed that successive generations of plastic young would have slowly impressed upon them, either through direct effects of the environment or by habit become instinct, incremental alterations that would gradually express themselves as adaptations. These impressions would thus be "long in the blood."[178]

174. Darwin, *First Transmutation Notebook,* MS pp. 2–4 (de Beer, p. 41).

175. Darwin, *Second Transmutation Notebook,* MS p. 136 (de Beer, p. 96).

176. Darwin, *Third Transmutation Notebook,* MS pp. 16–17 (de Beer, pp. 129–30).

177. Darwin, *Second Transmutation Notebook,* MS pp. 65–66, 83, 133 (de Beer, pp. 89, 92, 96).

178. In a more recent article, Kohn seems to have reinterpreted Darwin's theory to accommodate such objections as those just mentioned. See M. J. S. Hodge and David Kohn, "The Immediate Origins of Natural Selection," in *The Darwinian Heritage,* pp. 185–206.

3

Contributions of Natural Theology to Darwin's Theory of the Evolution of Mind and Behavior

It is commonly supposed that British natural theologians defended a position that was profoundly inimical to the theory of evolution, that, for instance, they were united in separating the animal kingdom from the realm of man by denying animals any semblance of "conscious reasoning," while Darwin, by contrast, was intent on "showing the evolutionary continuity between men and other animals."[1] This assumption obscures the deep divisions among natural theologians during the early part of the nineteenth century, especially on the question of continuity of mental faculties between men and animals; and it impedes recognition of Darwin's debt to several natural theologians for contributions to his emerging theory of the evolution of behavior.[2]

In the previous chapter I rejected the belief, common enough among historians, that after Darwin arrived at his principle of natural selection, he forthwith brought mind and behavior under its aegis.[3] This view is distorted. It contracts the conceptual struggles of several years into a pinpoint of insight. As I suggested and will attempt to document here, not until some time after his discovery of natural selection did Darwin begin to apply the principle consistently to behavior (without, however, entirely abandoning his original device of inherited habit). One source

1. Howard Gruber, *Darwin on Man* (New York: Dutton, 1974), p. 233. For other expression of this prevalent assumption, see Morton Beckner, "Darwinism," *The Encyclopedia of Philosophy* (New York: Macmillan, 1967), 2:303; and Ernst Mayr, "The Nature of the Darwinian Revolution," in his *Evolution and the Diversity of Life* (Cambridge: Harvard University Press, 1976), pp. 280–81.

2. I have considered an author to be writing natural theology if he has among his principal aims the revelation of God through rational and scientific examination of natural phenomena. The works characterized here as natural theology clearly display this intention.

3. Historians who have made this assumption quite explicit are mentioned in notes 99 and 160 of chapter 2.

of confusion about this matter arises, I believe, from an oversimplified picture of his designated opponents, the natural theologians. Since, as it is assumed, they uniformly rejected continuity of human and animal mental faculties, Darwin could have had little interest in their analyses of mind and behavior. His theory of species change by natural selection must then have formed (the equally simplified) mirror image of their theological accounts.

In the first section below, I will attempt to correct this perception by describing the variety of positions taken by early nineteenth-century natural theologians on the question of the relationship of animal faculties to human. Then, in subsequent sections, I will try to establish that certain of these writers supplied Darwin with considerations that initially facilitated his comprehension of the power of the Malthusian principle to explain animal and human activities, but that they finally restrained his attempts to extend that principle to the full range of species-specific patterns of conduct. The most formidable obstacle the natural theologians erected for Darwin involved the instincts of neuter insects. The difficulties presented by these "wonderful instincts" so confounded his elaboration of evolution by natural selection that he believed his entire argument would crumble if they were not overcome. Significantly for our understanding of Darwin's delay in publishing his theory, it was only shortly before 1859 that he came to an elegant solution that not only removed the difficulties but proved to be the most effective demonstration of his theory. The convictions of the natural theologians cannot, therefore, be reduced to mere background darkness against which the light of Darwinism suddenly burst forth.

Disputes of Natural Theologians over Instinct and Intelligence

The usual assumption that British natural theologians rejected the continuity of human and animal mind remains plausible when William Paley is taken as representative. In his *Natural Theology* (1802), Paley, a self-educated zoologist, attempted to demonstrate the existence of a Grand Designer from the contrivances of living organisms. Animal instincts were such contrivances; they functioned in the brutish realm as substitutes for rational intelligence, the gift bestowed only on man. Paley developed his conceptions quite pointedly against the opposing views of Erasmus Darwin—something that must have bemused the grandson. The elder Darwin had contended that the instincts of animals were not preordained mechanisms but products of their experience and

Figure 3.1 William Paley, 1743–1805, portrait done in 1789.

intelligence.[4] Were this contention true, it would of course have unrav-
eled the design argument that Paley was weaving. Paley made rejoinder
to Darwin by defining instinct as "a propensity prior to experience and
independent of instruction," and he marshalled evidence to confirm his
definition.[5] Thus in examples of mating preference, parental nurture,

4. See chapter 1.
5. William Paley, *Natural Theology,* in *The Works of William Paley* (Philadelphia: Woodward, n.d.), p. 442.

nest building, food gathering, migration, and so on, he isolated classes
of behavior that appeared in an animal's repertoire prior to any relevant
experience or learning: How could the newly mated sparrow know,
Paley inquired, that the purpose of its nest building was to incubate
eggs which it had never seen?[6] He concluded that all complex animal
behavior was instinctive and that a search for reason in the animal tribe,
unless it be divine reason, was vain.[7]

But not all natural theologians accepted Paley's interpretation of animal behavior. John Fleming, a Scots minister and his country's premier
zoologist, discerned a progression of reason stretching through the animal kingdom up to man. Like Erasmus and Charles Darwin, Fleming
adopted the sensationalist analysis of reason. Consonant with this tradition, he proposed as axioms, in his *Philosophy of Zoology* (1822), that
men as well as animals derived their ideas from sensory experience,
formed abstract ideas by attending only to certain aspects of sensation,
recalled ideas through association, and anticipated future impressions
in imagination.[8] These principles led Fleming to reject Paley's assumption that instincts had to be immutably fixed in the nature of animals.
His epistemological convictions allowed him to perceive that reason
and new habits could alter instincts, especially when the original behavioral patterns ceased to be of advantage.[9] Fleming's Scots temper cautioned him, however, against French excess. He denied Lamarck's trans-
formationism, and ultimately agreed with Paley that God had created
animals much as we see them.[10] And because of an economical epistemology, in which the sensory resources of reason and instinct (which
he described as an impulsive act of intelligence) could be found in the
lowest infusorium and also in man, he contemned Lamarck's classification of animals into apathetic, sensitive, and intelligent. The evidence
fell against both Lamarck and Paley, since reason ran with instinct
through the entire animal kingdom.[11]

Darwin's markings in his copy of the *Philosophy of Zoology*, made in
winter 1839–1840, indicate that he agreed with Fleming's assessment of
the modifying power of animal reason.[12] He also liked Fleming's illus-

6. Ibid., pp. 442–43.
7. Ibid., p. 442.
8. John Fleming, *The Philosophy of Zoology* (Edinburgh: Constable, 1822), pp. 220–22.
9. Ibid., pp. 246–47.
10 Ibid., p. 27.
11. Ibid., pp. 311–12.
12. Darwin underscored passages on p. 246 and elsewhere in vol. 1 of Fleming's *Philosophy of Zoology* that suggest this accord. In his early notebooks and essays, Darwin never
hesitated to predicate of animals an ability to adjust their innate behavior; and in *On the*

trations showing that primitive instincts no longer useful to an animal often yet perdured. He added two further examples of this phenomenon in the margins of his copy and drew from them a suggestive conclusion. Referring to the habits that some domestic dogs have of turning around several times before lying down (as if to beat down high grass) and of covering their dung (as if to escape detection), he scribbled: "turning in mind before sleeping—covering dung show that principle may probably be laid down that every instinct preserved if not changed & some of them may once have been important."[13] Many years later, in *The Expression of the Emotions in Man and Animals* (1872), Darwin would use these very examples to propose that anomalous and useless instincts would be maintained in a species if they had once been important—and therefore deeply impressed into its heritable substance—and if no significant changes in the environment made them harmful.[14] Such cases indicated the kinds of incongruity, the failures of fit between animal instincts and environmental requirements, that caught Fleming's eye—and, of course, Darwin's—but escaped the dogmatically blinkered Paley altogether.

Darwin found Fleming's treatment of animal instinct and intelligence congenial to his own analyses. But another naturalist, a natural theologian by the name of John French, detected in Fleming's descriptions unhappy implications for orthodoxy. In the lead article of the first issue (1824) of the *Zoological Journal* and in subsequent numbers of that periodical, French undertook a critical examination of Fleming's ideas, as well as those of Frédéric Cuvier.[15] French proscribed the theories of Fleming and Cuvier because they attributed too much understanding to animals.[16] Like Paley, French assumed that an animal generally acted

---

Origin of Species (London: Murray, 1859), he referred to that "little dose of judgment or reason" by which even very inferior animals might accommodate their instincts to new circumstances (p. 208).

13. This annotation occurs on p. 247 of Darwin's copy of *Philosophy of Zoology,* vol. 1.

14. Charles Darwin, *The Expression of the Emotions in Man and Animals* (Chicago: University of Chicago Press, [1872] 1965), pp. 39, 42–44.

15. John French, "An Inquiry Respecting the True Nature of Instinct, and of the Mental Distinction between Brute Animals and Man," *Zoological Journal* 1 (1824): 1–32; 153–73; 346–67.

16. Ibid., p. 11. French did not think brutes to be mere automata, though they were not to enjoy reason either. Rather, he declared them to have "a middle conscious nature, between mere irritability, (which is an inferior effect of life, related analogically to consciousness) and the intellectual consciousness of man" (p. 347).

French was precipitous in his criticism of Frédéric Cuvier. Had he investigated the latter's views more carefully, he would have found that the Frenchman also refused animals a human sort of reason, leaving them with only a kind of intelligence which could

from instincts which were determinate, unyielding to experience, and of a purpose beyond its awareness.[17] Yet he was not so obtuse as to deny that brutes often engaged in what seemed to be extraordinarily intelligent behavior and that their instincts were truly wondrous. He ascribed these actions, however, not to the individual consciousness of animals, but to the immediate operation of supernatural powers:

> we must allow them [animals] to possess only a subordinate consciousness and discrimination determinable to natural objects; and overruled and directed by powers or agencies operating in them above the sphere or stream of their own proper consciousness, and which powers or agencies must be of a moral, intellectual, and scientific order: thus that brutes are governed by such agencies, good and evil, but under the control of Providence; and that such agencies act by impressions upon their conscious nature, but unperceived by it in a moral or intellectual sense; effecting such operation by means of connate inclinations implanted in their nature, and disposing them to receive the impressions; and which inclinations appear to constitute the ground or basis upon which is formed that lower species of consciousness, volition, and discrimination, which seems the proper attribute of the brute animal.[18]

The intelligent behavior of one's dog, according to this cabalistic theology, could only be the reflection of angelic powers, while the malicious acts of the marauding wolf showed that Satanic forces indeed roamed the world.

The attitudes expressed by French in his series of essays were fully endorsed by the editors of the *Zoological Journal*[19] as consistent with the ultimate aim of their magazine, which was "the contemplation of the works of creation [which] necessarily leads the mind to the Creator

---

profit from experience but not abstract from it any general principles. See the first chapter and Frédéric Cuvier, "Instinct" (1822), *Dictionnaire des sciences naturelles* (Strasbourg: Levrault, 1816–1845), 23:535–39.

17. French, "An Inquiry Respecting the True Nature of Instinct," pp. 28–31.

18. Ibid., pp. 6–7.

19. Editor's "Introduction," *Zoological Journal* 1 (1824), vi: "A most difficult subject is treated by Mr. French, we think, with singular skill and ingenuity; but even this consideration would not have induced us to give it to the public, had it not appeared to us to be strictly consistent with the plan of our work. Its object is to develop the operations of mind, if we may so express ourselves, in the brute creation, from the habits and actions of several of its members. It necessarily enters into the detail of facts purely Zoological and in the highest degree interesting—and if the main question be metaphysical, it is from natural history alone that its ingenious author derives his arguments in discussing it."

himself—[so that] the more intimate our acquaintance with the former, the deeper and more devoted will be our adoration of the great author of all things." [20] French's own particular methods and conclusions might thus be thought typical of those natural theologians who scoured the animal kingdom for signs of the Creator and who often stumbled around foolishly doing so. But his absolute separation of animal from human consciousness and his dependence on anecdotal evidence and hyperbolic metaphysical explanation were so antithetical to the approach of other religiously minded but careful scientists that the author of the Seventh Bridgewater Treatise, the Reverend William Kirby, ridiculed his opinions. Kirby scoffed at the idea of good angels causing beneficent instincts and demons inciting aggressive ones, and he found delicious contradiction in French's invocation of supernatural agencies to explain the often miscued behavior of animals. [21] Certainly compared to Kirby's own rich explorations of animal behavior, which will be considered momentarily, French's efforts were thin and very flat beer.

Darwin examined the first volume of the *Zoological Journal,* though he left no record of having read French's articles. Doubtless he would have found them of little value, since they represented an approach to zoology that he wished to overcome. In the spring of 1839, however, he picked up an essay by another natural theologian whose rather different analyses of animal instinct and intelligence did capture his attention.

The purpose of the Reverend Algernon Wells's pamphlet *On Animal Instinct* (1834) was to advance a theory of matter and mind that postulated a divine agency to account for the immediate powers of inanimate matter and plants, but that showed animals and men to be directed by their own conscious intelligence. Though Wells regarded instinct as a species of intelligence that guided the voluntary and self-determined acts of animals, he distinguished it from intelligence as commonly understood. Instinctive behavior was innately determined, was performed without conscious purpose, never admitted of improvement, and at first execution was perfectly adapted to both an animal's needs and its anatomical structure. [22] Like all natural theologians who investigated the subject, Wells thought instincts the best "proofs of exquisite, benevolent contrivance." [23]

20. Ibid., p. vii.

21. William Kirby, *On the Power, Wisdom, and Goodness of God as Manifested in the Creation of Animals and in Their History Habits and Instincts,* Seventh Bridgewater Treatise (London: Pickering, 1835), 2 : 238—40.

22. Algernon Wells, *On Animal Instinct* (Colchester: Longman, Rees, Orme, Brown, Green, and Longman, 1834), pp. 15—22.

23. Ibid., p. 23.

Darwin took exception in his *N Notebook* to Wells's assumption that instinct was fixed and always in complete harmony with an animal's bodily organization.[24] For he still held, as I have argued in the previous chapter, that newly acquired habits could augment and refine an animal's stock of instincts. Indeed, his comments on Wells suggest that he had not yet (eight months after Malthus) given full weight to natural selection even as a mechanism of anatomical adaptation. Responding to Wells's example of the perfect articulation of instinct and structure in the woodpecker, Darwin objected: "but this is not so, the instincts may vary before the structure does; & hence we get an apparent anomaly, for if anyone has taken the woodpecker as an example fitted for climbing, his arguments partly fall, when a species is found which does not climb."[25] Though Darwin's intention here may simply have been to highlight the kind of variability which was inexplicable on the creationist hypothesis, he seems also to have had in mind his older mechanism of anatomical adaptation, which required instincts to "vary before the structure does."

Darwin reacted more favorably to Wells's theory of animal reason than to his views on instinct. Wells declared that reason was simply that "power by which its possessor is enabled to propose to himself an end he would wish to accomplish, and then to arrange a course of means, or actions, adapted, and sufficient to bring about the result he designs."[26] Wells felt that if reason were so understood, then beasts could not be denied "some feebler rays of reason; some capabilities of knowledge, besides the skill and guidance of instinct."[27] Men differed from animals, in Wells's estimation, by their greater proportion of reason to instinct—though he believed that both animals and men lost instincts and increased reason as they came into human society; domesticated animals and civilized men were more reasonable and less instinctive than their ruder brothers.[28]

24. Darwin, *N Notebook,* MS pp. 68–73, transcribed by Paul Barrett in Gruber, *Darwin on Man,* pp. 343–44.

25. Darwin, *N Notebook,* MS p. 71 (Gruber, p. 344). In the *Origin of Species,* pp. 183–85, Darwin used the case of the anomalous ground-feeding woodpecker to make two related points: that such oddities were unintelligible on the creationist hypothesis; but that they could be understood as the first stage of a species transformation, after which "it would be easy for natural selection to fit the animal, by some modification of its structure, for its changed habits, or exclusively for one of its several different habits" (p. 183). Darwin's mode of reasoning here seems to fit the pattern of his earlier theory of species transformation via habit and instinct. This is just another instance of the conservative character of Darwin's thought—older ideas were often preserved and put to new uses.

26. Wells, *On Animal Instinct,* p. 15.

27. Ibid., p. 20

28. Ibid., p. 21.

Wells's conception of animal reason was shared by two other natural theologians whose theoretical considerations of this faculty and whose careful observations of instinct not only proved more valuable and challenging to Darwin but also measurably shaped his own developing ideas about the evolution of behavior. These men were the statesman and gifted amateur scientist Henry Lord Brougham and the distinguished entomologist William Kirby.

Contributions to Darwin's Emerging Theory of Behavior

Brougham wrote *Dissertations on Subjects of Science concerned with Natural Theology*[29] to accompany his and Charles Bell's illustrated edition (1839) of Paley's *Natural Theology*.[30] Darwin often paused in reading his copy of Brougham's *Dissertations* to underscore and annotate the main essay "Of Instinct." He estimated the piece quite highly, calling it in his notebooks "very good" and "profound."[31] He undoubtedly appreciated Brougham's effort to distinguish clearly between the facts of animal instinct and intelligence, and the theoretical interpretation that might be given those facts. As factually evident, Brougham determined instinct to be a mental trait impelling an animal to perform acts that were beneficial to the species, that were exhibited without any teaching or experience, and that generally were precise in their first execution.[32] Darwin later incorporated these characteristics—found commonly enough in the descriptions of other theologians—into his own definition. But he especially singled out from Brougham's criteria a note he regarded as important for the selection theory he was slowly constructing: that instinct accomplished a purpose of which the animal was ignorant.[33]

Brougham insisted on this last feature of instinct for both empirical and theoretical reasons. He had carefully investigated many of the won-

29. Henry Lord Brougham, *Dissertations on Subjects of Science concerned with Natural Theology: Being the Concluding Volumes of the New Edition of Paley's Work* (London: Knight, 1839).

30. *Paley's Natural Theology Illustrated,* with introductory and concluding volumes by Henry Lord Brougham, ed. Henry Lord Brougham and Sir Charles Bell (London: Knight, 1835–1839). The volumes mentioned in the preceding note are the concluding volumes of this set.

31. Charles Darwin, *Second Transmutation Notebook,* MS p. 266, "Darwin's Notebooks on Transmutation of Species," ed. Gavin de Beer, *Bulletin of the British Museum (Natural History),* historical series 2 and 3 (1960, 1967): 114; and *N Notebook,* MS p. 62 (Gruber, p. 343).

32. Brougham, *Dissertations on Subjects of Science* 1:6–54.

33. Charles Darwin, "Essays written in 1842 and 1844," in Francis Darwin, ed., *The Foundations of the Origin of Species* (Cambridge: Cambridge University Press, 1909), pp. 17 and 117; Brougham, *Dissertations on Subjects of Science* 1:27–8.

derful instincts of animals. Since these instincts were either extremely complex (such as the hivebee's precise geometrical fabrication of its cells) or performed without relevant prior experience (such as the provisions made by the wasp for feeding young that hatched after its death), they could not possible have been executed with knowledge of their end. "In all these cases," Brougham concluded, "the animal works positively and without knowledge, and in the dark."[34] These were fair reasons, he thought, to overturn Lamarck's hypothesis that instincts resulted from acquired habits.[35] Brougham rather held that at each instant of the exercise of instinct the Deity supplied the animal mind with the necessary "knowledge and design."[36]

Darwin was initially hesitant about admitting that an animal might engage in activities without any conscious intention. Next to a passage in which Brougham argued this point, Darwin asked: "is it not that most instincts happen to have some *end* in view."[37] Later, in his copy of Kirby and Spence's *Introduction to Entomology* (2d ed., 1818), in which the authors also defined instinct by lack of individual purpose, Darwin nevertheless discovered in some of their examples evidence that animals, as he countered, "do know end in view or rather what they wish for."[38] And in the *Origin of Species,* as well as in the "Essays" of 1842 and 1844, he made allowance for animals' intelligently adjusting some of their instincts to fit particular situations.[39] Nonetheless, he was struck by Brougham's examples—for instance, the solitary wasp that supplied grubs for its eggs: "and yet this wasp never saw an egg produce a worm—not ever saw a worm—nay, is to be dead long before the worm can be in existence—and moreover she never has in any way tasted or used these grubs, or used the hole she made, except for the prospective benefit of the unknown worm she is never to see."[40] At the foot of this passage Darwin penciled his reflection: "extremely hard to account by habit." And surely it was. An act performed once in a lifetime, without relevant experience, and having a goal of which the animal must be ignorant—this kind of behavior could not possible have arisen from intelligently acquired habit.

34. Brougham, *Dissertations on Subjects of Science* 1 : 18.

35. Ibid., pp. 161–62. Darwin has this passage marked in his copy.

36. Ibid., p. 95.

37. Darwin's annotation appears on p. 52 of his copy of Brougham's *Dissertations on Subjects of Science.*

38. Darwin's annotation occurs on p. 476 of the second volume of William Kirby and William Spence, *Introduction to Entomology,* 2d ed. (London: Longman, Hurst, Rees, Orme, and Brown, 1818).

39. Darwin, *Origin of Species,* p. 208 ""Essays written in 1842 and 1844," in *Foundations of the Origin of Species,* pp. 17–18, 118.

40. Brougham, *Dissertations on Subjects of Science* 1 : 17–18.

Darwin read Brougham's *Dissertations* early in 1840.[41] Up to this time, I have maintained, he still relied on habit as the principal mechanism of instinct. But Brougham's examples, which the natural theologian precisely set against Lamarck's theory of habit-produced adaptations, seem to have jostled Darwin, so that he began to attend more seriously to natural selection as the likely source of such marvelous instincts. Next to a description of the hivebee's wonderful constructions, which Brougham thought explicable only by intervention of the Divine Teacher, Darwin pondered: "very wonderful—it is as wonderful in the mind as certain adaptations in the body—the eye for instance, if my theory explains one it may explain other."[42] This, of course, was not the first time Darwin had entertained the bare notion that his theory of natural selection might also account for some instincts. I have already mentioned his consideration of the possibility in November 1838. Yet it seems clear that Brougham's exposition prepared him to perceive the advantage his new mechanism might have in explaining certain peculiar instincts, those that were very complex or that could not have been acquired through habitual pursuit of an individual goal. Just after reading Brougham, he came back to this point, as a penned note in John Blackwall's *Researches in Zoology* (1834) reveals: "Lord Brough. says not knowing object—one [of] chief criteria of instinct."[43] Certainly in the "Essays" of 1842 and 1844, Darwin emphasized the intentionless character of complex and unlearnable instincts—an observation he expressly attributed to Brougham—to contend that the principle of natural selection could render their origin intelligible.[44] And in the *Origin of Species,* though granting that some of the simpler instincts might have arisen from habit, he insisted that it was by natural selection "that all the most complex and wonderful instincts have originated."[45] It appears, then, that the views of a natural theologian, Lord Brougham, were quite instrumental in bringing Darwin systematically to apply natural selection to behavior.

Another important feature of Brougham's treatment of behavior obviously struck a responsive cord in Darwin—struck it, in fact, even before he actually picked up Brougham's *Dissertations.* This happened by

41. The first of Darwin's reading notebooks shows that he examined the *Dissertations* in February 1840. See Peter Vorzimmer, ed., "The Darwin Reading Notebooks (1838–1860)," *Journal of the History of Biology* 10 (1977): 107–53; citation on p. 123.

42. Darwin's annotation occurs on p. 77 of the first volume of Brougham's *Dissertations on Subjects of Science*.

43. Darwin's annotation occurs on p. 155 of John Blackwall, *Researches in Zoology* (London: Simplin and Marshall, 1834).

44. Darwin, "Essays written in 1842 and 1844," in *Foundations of the Origin of Species,* pp. 17, 117.

45. Darwin, *Origin of Species,* p. 209.

way of an early notice of the work in the *Athenaeum* that caught Darwin's eye. The reviewer went quickly to the philosophical heart of the first volume: "The general conclusion is, that animal intelligence differs from human, not in kind, but in degree; but this conclusion is rather shadowed forth than distinctly propounded; and the great difficulty to its reception, the fact that animal intelligence is not progressive, is evaded."[46] Darwin perhaps smiled when he considered how his own theory resolved this "great difficulty." His *N Notebook,* in any case, records his sense of the significance of this extension of rationality to animals: "Lrd. Brougham Dissert. on subject of science connected with Nat. Theology.—says animals have abstraction because they understand signs.—very profound.—concludes that difference of intellect between animals & men only in kind [*sic,* degree]. probably very important work."[47]

Strangely, Brougham's belief that God intervened to guide an animal's instinctive behavior did not prevent him from supporting as observationally justified the view that animals, even the most inferior, were capable of thought and rationality. For Brougham, like Fleming and Darwin, was imbued with the sensationalist epistemology, which led him to interpret as rational any action in which "the means are varied, adapted, and adjusted to a varying object."[48] He claimed such behavior abounded in the animal world, but regarded a single instance of a cat's learning to open a door from watching men as perfectly sufficient evidence of rational thought in an animal.[49] Brougham even granted that animals could intelligently adapt their instinctive acts to altered circumstances[50]—a concession also seemingly at odds with his belief in divine superintendence of instinct. Darwin would hardly have concerned himself with this anomaly in Brougham's theory. But he surely welcomed the idea that instincts could be intelligently modified, since, as he easily

46. "Review of Brougham's *Dissertations on Subjects of Science,*" *Athenaeum* (2 February 1839), p. 91.

47. Darwin, *N Notebook,* MS p. 62 (Gruber, p. 343). The passage in Brougham's *Dissertations on Subjects of Science* to which his reviewer took exception, but which expressed Darwin's own conviction, is: "If I am to teach a dog or a pig to do certain things on a given signal, the process I take to be this. I connect his obedience with reward, his disobedience with punishment. But this only gives the motive to obey, the fear of disobeying. It in no way can give him the means of connecting the act with the sign. Now, connecting the two together, whatever be the manner in which the sign is made, is Abstraction; but it is more, it is the very kind of Abstraction in which all language has its origin—the connecting the sign with the thing signified; for the sign is purely arbitrary in this case as much as in human language" (Brougham, *Dissertations on Subjects of Science* 1:196).

48. Ibid., p. 135.

49. Ibid., pp. 167–68.

50. Ibid., pp. 203–205.

came to suppose, such modifications could more finely accommodate behavior to a rapidly changing environment than could natural selection alone.

Thus Brougham, unlike Paley, felt no reluctance in assigning the same kind of rational ability to animals and men. In fact he went so far as to assert that no wider gulf separated a dull reasoner from a "sagacious retriever or a clever ape" than from a Pascal, Newton, or Lagrange[51]—a judgment later echoed by Darwin.[52]

The common assumption that natural theologians were uniformly convinced of the chasm between brute instinctive behavior and human rational action cannot stand against the evidence drawn from Brougham. But that assumption is utterly crushed by the forceful opinions of the respected entomologist and author of the Seventh Bridgewater Treatise, *On the Power Wisdom and Goodness of God as Manifested in the Creation of Animals and in Their History, Habits, and Instincts* (1835).[53] The Reverend William Kirby's investigations of animal instinct and reason were always attentively considered by Darwin, even prompting the young naturalist at the end of his *Beagle* voyage to reflect that "one hand has surely worked through the universe."[54] Though he came to recognize another power operating in the animal world, Darwin nevertheless found in Kirby and William Spence's theologically cast *Introduction to Entomology*[55] "the best discussion of instincts ever published."[56] In the *Origin of Species,* he adapted their description of instinct to his own ends.[57]

51. Ibid., p. 175.

52. In the *Descent of Man,* Darwin urged that we ought not to regard the evolution of mental faculties as impossible, "when we daily see their development in every infant; and when we may trace a perfect gradation from the mind of an utter idiot, lower than that of the lowest animal, to the mind of a Newton." See Charles Darwin, *The Descent of Man and Selection in Relation to Sex* (London: Murray, 1871), 1:106.

53. See note 21 for bibliographical information.

54. *Charles Darwin's Diary of the Voyage of H.M.S. Beagle,* ed. Nora Barlow (Cambridge: Cambridge University Press, 1934), p. 383.

55. Kirby and Spence were disturbed that some philosophers had attempted to use evidence from natural history in arguments against "nature's God." In response, the authors "conceived they might render some service to the most important interests of mankind, by showing how every department of the science they recommend illustrates the great truths of Religion, and proves that the doctrines of the Word of God, instead of being contradicted, are triumphantly confirmed by his works" (*Introduction to Entomology* 1:xi).

56. R. C. Stauffer, ed., *Charles Darwin's Natural Selection: Being the Second Part of His Big Species book Written from 1856 to 1858* (Cambridge: Cambridge University Press, 1975), p. 468. The greater part of Darwin's *Origin of Species* was distilled out of this manuscript for his proposed "Species Book."

57. Darwin did not acknowledge his debt to Kirby and Spence in the *Origin of Species.*

When he read Kirby and Spence in 1843, this was the description Darwin marked and later used:

> Without pretending to give a logical definition of it . . . we may call the instincts of animals those unknown faculties implanted in their constitution by the Creator, by which independent of instruction, observation, or experience, and without knowledge of the end in view, they are impelled to the performance of certain action tending to the well-being of the individual and the preservation of the species: and with this description, which is in fact merely a confession of ignorance, we must, in the present state of metaphysical science content ourselves.[58]

Kirby and Spence confirmed this theoretically modest characterization by their extensive and detailed studies of the behavior of insects; they enumerated, for example, some thirty different instincts in but one caste of hivebee (the nurses).[59] Such precision allowed them to observe a certain variability of instinct within a species[60]—a feature of animal behavior overlooked and often denied by more metaphysically oriented naturalists. Though variation in behavior sometimes appeared to be the result of conscious deliberation, the authors balked at granting animals a prescient awareness of the ends of instinctive action or, in the case of the bee, the mathematical skill required to construct its hexagonal cells. They merely recognized a fact which Darwin also appreciated, as annotations in his copy of the *Introduction to Entomology* show: that behavioral patterns as well as anatomical structures varied within a species.[61]

For reasons of haste, the *Origin* was composed without footnote reference to relevant literature; but in the manuscript of his "Species Book" (*Charles Darwin's Natural Selection*, p. 468), he did cite Kirby and Spence. Compare their description of instinct (given in the text) with Darwin's in the *Origin:* "I will not attempt any definition of instinct. It would be easy to show that several distinct mental actions are commonly embraced by this term; but every one understands what is meant, when it is said that instinct impels the cuckoo to migrate and to lay her eggs in other birds' nests. An action, which we ourselves would require experience to enable us to perform, when performed by an animal, more especially by a very young one, without any experience, and when performed by many individuals in the same way, without their knowing for what purpose it is performed, is usually said to be instinctive" (p. 207).

58. Kirby and Spence, *Introduction to Entomology* 2:471.

59. Ibid., pp. 499–505.

60. Ibid., pp. 473–98.

61. Ibid., pp. 481–98. The authors believed that the strict limits within which most variations of instinct occurred argued against those variations' stemming from rational deliberation. Where they reached this conclusion (p. 496), Darwin jotted in the margin: "isn't this because [when] reason goes so far & no further, it is not reason." What Darwin took exception to, however, was the assumption of Kirby and Spence that the same varia-

Yet Kirby and Spence were not loath to attribute reason even to insects; for "though instinct is the chief guide of insects, they are endowed also with no inconsiderable portion of reason."[62] Here, as with Fleming, Brougham, and Darwin, the ascription of reason to these small creatures was based on their manifest ability to deal with contingent circumstances, to learn from experience, to employ memory, and to communicate with one another.[63] And while Kirby and Spence deemed instinct and reason to be distinct faculties, they perceived that these often worked together in producing behavior. Quoting David Hume, they affirmed that "it is instinct which leads a greyhound to pursue a hare; but it must be reason that directs 'an *old* greyhound to trust the more fatiguing part of the chase to the younger, and to place himself so as to meet the hare in her doubles.'"[64] Their observations suggested to them a continuity of behavior and faculties extending from the lowest animals up to man. Certainly this was a conclusion congenial to Darwin's emerging evolutionary theory and hardly the stuff of a Paley-like compartmentalization of living nature.

Without any apparent doctrinal conflict, Kirby transferred to his Bridgewater Treatise the conception of animal abilities that he first elaborated in the *Introduction to Entomology*. In the Treatise, of course, his main concern was to show how the Creator was revealed in his handiwork; and for this the instincts of animals offered him potent evidence. In tendering his account, Kirby contested Lamarck's theory of instinct as too materialistic and dependent upon an extraordinary hypothetical machinery;[65] yet he refused to succumb to the easy theologizing of John French, whose zoological opinions he regarded as absurd. Kirby of course believed the ultimate disposer of animal behavior to be God, but only as operating through secondary causes. He reached this conclusion in an analogical argument that might have been suggested by Erasmus Darwin, whose *Zoonomia* he had read. Kirby saw a close similarity between plant instincts (for instance, movements of plants toward light) and animal instincts. He reasoned that since only physical conditions determined the vegetable world, it was quite likely that physical organization and stimuli were also the proximate sources of animal instinct.[66]

---

tions of instinct had been constant throughout the ages. He simply inquired (on p. 496): "How do we know this?"

62. Ibid., p. 415.

63. Ibid., pp. 418–530.

64. Ibid., p. 512.

65. Kirby, *On the Power, Wisdom, and Goodness of God*, 1: xx–xlii.

66. Ibid., p. 253. Erasmus Darwin perceived a perfect continuity between plants and animals, and ascribed to the former even "ideas of so many of the properties of the exter-

Darwin read Kirby's Bridgewater Treatise during the last months of his voyage, and after returning home he frequently alluded to it in his notebooks.[67] Though he initially had some doubts concerning plants having instincts, he agreed with Kirby's conclusion about the physical determination of instincts.[68] He had been working toward a similar conception himself during the spring and summer of 1838, when he considered how physical structures of the brain might be the immediate cause of instinctive behavior. He even thought of his theory as a "returning to Kirby's view."[69]

Darwin thus found much in the work of Kirby and Spence that influenced or confirmed his own ideas. But his largest debt to them came in a negative way. In the *Introduction to Entomology,* the authors gave a good deal of space to the analysis of the intricate and amazingly diverse instincts of worker ants and bees. These wonderful instincts created the greatest challenge to Darwin's theory of evolution. But in overcoming the difficulty, Darwin found the single most powerful support for his theory.

The Wonderful Instincts of Neuter Insects

In this and the previous chapter, I have tried to demonstrate that Darwin only gradually and with some hesitation arrived at the selection

---

nal world and of their own existence." See Erasmus Darwin, *Zoonomia; or the Laws of Organic Life,* 2d ed. (London: Johnson, [1794{1796] 1796), 1 : 105–107.

67. Darwin, *First Transmutation Notebook,* MS pp. 141–43 (de Beer, pp. 58–9; *Old and Useless Notes,* MS pp. 34–37, transcribed by Paul Barrett in Gruber, *Darwin on Man,* pp. 392–94.

68. Darwin appears at first to have doubted the propriety of attributing instincts to plants. In a note in *Old and Useless Notes,* MS p. 36 (Gruber, p. 393), he wrote: "Kirby extends instincts to plants, but surely instincts imply willing, therefore word misplaced." This demur was penned probably some time in early 1837. A year or so later, he came around to Kirby's position, as this line from his *N Notebook,* MS p. 48 (Gruber, p. 339), reveals: "Instinct is a modification of bodily structures (connected with locomotion) (no! for plants have instincts))." Elsewhere in this notebook, MS pp. 12–13 (Gruber, p. 332), he, like his grandfather, speculated on the rational abilities of plants. In the "Essay" of 1842 (in *Foundations of the Origin of Species,* p. 17), he fairly clearly adopted Kirby's line of argument: "Habits purely corporeal, breeding season &c., time of going to rest &c., vary and are hereditary, like the analogous habits of plants which are inherited." Even in his last works, Darwin still claimed a continuity of powers between plants and animals, ascribing to plants "mental powers" and comparing their various movements and sensitivities to similar faculties in animals. See Charles Darwin, *The Movements and Habits of Climbing Plants* (London: Murray, 1875), p. 306; and *The Power of Movements in Plants* (London: Murray, 1880), p. 571.

69. From Darwin's notes, titled by Barrett "Essay on Theology and Natural Selection" (Gruber, p. 419).

theory of instinct sketched in the "Essays" of 1842 and 1844, and more confidently asserted in the manuscript of his "Species Book" (1856–1858) and in the *Origin of Species*. There were, I believe, three chief impediments to his developing a systematic account of instinct more rapidly after the initial conception of natural selection. The first was the ease with which habit could explain the origin of instinctive behavior. Not only did Darwin find this true, but so did Lamarck, Frédéric Cuvier, and John Sebright, all of whom he carefully read. The second difficulty, I believe, related to Darwin's lack of a clear distinction between animals' selecting habits because of their usefulness and nature's selecting animals because of their useful habits. (A similar ambiguity clouds the conception of natural selection even in the *Origin of Species*.)[70] From our perspective this confuses what we have come to think of as Lamarckian mechanisms with neo-Darwinian. Needless to say, Darwin was not a neo-Darwinian. In the preceding sections of this chapter, I have indicated the ways certain natural theologians aided him in attaining a greater, if not a complete (for our taste) clarity about the differences between natural selection and inherited habit. In this section, I wish to concentrate on the third difficulty inhibiting Darwin from bringing behavior under the explanatory power of natural selection. It

70. In the *Origin of Species,* useful habit has three roles. First, Darwin believed useful habits were one source of variation upon which natural selection operated (pp. 134–39). Second, inherited habit was another mechanism, along with natural selection, of species evolution. It functioned both positively and negatively in this regard: exercise positively enhanced anatomy and produced new instinctive behaviors (pp. 134, 209–16), while disuse was the principal cause of atrophied and rudimentary organs (pp. 134–39, 454). But Darwin's original device of inherited habit also became the conceptual progenitor to his newer mechanism of natural selection, much in the way I have just suggested above in the text. Consider his definition of natural selection in the third chapter of the *Origin* (p. 61): "Owing to this struggle for life, any variation, however slight and from whatever cause proceeding, if it be in any degree profitable to an individual of any species, in its infinitely complex relations to other organic beings and to external nature, will tend to the preservation of that individual, and will generally be inherited by its offspring. The offspring, also, will thus have a better chance of surviving, for, of the many individuals of any species which are periodically born, but a small number can survive. I have called this principle, by which each slight variation, if useful, is preserved, by the term of Natural Selection." In this definition and elsewhere in the *Origin* (pp. 82, 83–84), Darwin seems to have mixed the notion of 'the evolution of a trait' (i.e., its spread in a population) with that of 'the heritability of a trait' (i.e., the chance that it will be passed on to progeny). While advantageous traits will spread in a population if they are heritable, traits do not become heritable because they are advantageous. In other words, we would not now accept as true that "any variation . . . if it be in any degree profitable . . . will generally be inherited." Darwin's formula that "each slight variation, if useful, is preserved" must remain (for us) ambiguous. A vestige of the older conception of inherited habit was thus retained in Darwin's principle of natural selection.

was the near-fatal problem that Kirby and Spence brought to his attention in their *Introduction to Entomology.*

In his chapter on instinct in the *Origin of Species,* Darwin recognized
that the wonderful instincts of animals appeared to present difficulties
"sufficient to overthrow my whole theory."[71] Similarly, in the corresponding chapter of the "Species Book," he admitted that "it is most
natural to believe that the transcendent perfection & complexity of
many instincts can be accounted for only by the direct interposition of
the Creator."[72] Certainly those natural theologians whom he read with
interest offered the wonderful instincts as the most compelling evidence
for the operations of divine agency. By the early 1840s, however, Darwin had become convinced that instincts, like anatomical structures,
varied and that natural selection would preserve and continually accumulate profitable variations to produce the most complex kinds of innate action patterns. But for instincts of one class, those of the neuter
insects, this explanation seemed precluded. The strange and anthropo-
morphic behavior of slave-making ants, for instance, was exhibited by
the neuter castes of those societies. They could not pass their favorable
variations on to progeny, because they left none. Darwin confessed,
both in the "Species Book" and in the *Origin,* that such cases initially
struck him as "fatal to my whole theory."[73]

This lethal objection apparently occurred to Darwin while he was
reading volume 2 of Kirby and Spence's *Introduction to Entomology.* In
their chapter on instinct, the authors described the marvelous instinct
of sterile worker bees to produce new queens after the death of the old
queen: the neuter workers would quickly select several ordinary grubs,
transfer them to special cells, and feed them a royal substance; from
grubs that otherwise would have matured into sterile workers themselves, fertile queens metamorphosed to ensure the survival of the hive.
This community-preserving behavior of the workers was the instinct the
authors thought showed the divine hand most wonderfully. Darwin as
well must have spent some anxious moments wondering about it. At
one point, when the authors cited the neuter workers' related instinct
to retain fertile females in the hive, Darwin scribbled his frustration:
"Neuters do not breed! How instinct acquired."[74] But he at last came
to a perfectly natural (if ultimately inadequate) explanation of such instincts, the gist of which he sketched next to a description of these

71. Darwin, *Origin of Species,* p. 207.
72. *Charles Darwin's Natural Selection,* p. 466.
73. Ibid., p. 365; *Origin of Species,* p. 236.
74. Darwin's annotation occurs in vol. 2, p. 55 of Kirby and Spence, *Introduction to Entomology.*

wonderful behaviors. "One may suppose," he wrote, "that originally many queens were ordinary thus reared & a few workers & the instinct is thus retained."[75] Darwin spelled out what he meant by this elliptical note more exactly on the back flyleaf, where, referring to a passage on the generation of new queens, he penned: "511–queens & no workers—then few queens with workers & lastly one queen & the instinct in neuters retain trace of old instinct [of original] queen." Here, then, was an explanation of how the wonderful instincts of sterile insects might have been established: in a hive originally without neuter workers, beneficial variations of behavior might have been acquired by the fertile females, who subsequently would pass them to their female offspring; later, when owing to particular circumstances these offspring became sterile, they would still retain the primitive instincts.

Darwin did not remain satisfied with this explanation. It seemed unrealistic to believe that in all the species of insects that displayed neuters, the queens originally performed the same tasks the workers subsequently executed. This was, at least, the major difficulty he mentioned in a manuscript bearing the date June 1848. This manuscript is of considerable interest, for it demonstrates that Darwin continued to worry about the problem of the instincts of neuter insects, and that as late as spring 1848, he had not found a satisfactory answer. The manuscript is also important for three other reasons. First, it shows Darwin coming very close to the solution he would finally offer in the *Origin*, though here he sees only its implausibility. Second, it indicates that he had up to this time retained his older mechanism of hereditary habit as an integral element of his concept of instinct. And third, in it Darwin recognized that the case of the neuter insects presented the most serious obstacle to his general theory of evolution. The manuscript runs to four handwritten pages. The script is in ink, and the pages have a vertical pencil line through them, typical of those notes that Darwin used in the composition of his larger works and then set aside. I will quote the manuscript in its entirety, except for a paragraph that he appears to have added some time later, probably during the writing of the "Species Book"; I will transcribe that passage in a moment.

> June/48/ In wasps & Humble Bees, in which <I believe> females at first work, then is no difficulty in their structure and instincts being varyed & transmitting such to their neuters. Even if the females came to cease to work, the neuters might readily retain such instincts when once acquired, & their in-

75. Darwin's annotation occurs in vol. 2, p. 513 of Kirby and Spence, *Introduction to Entomology*.

stincts might be made to vary a very little by traditionary knowledge <where the society is perpetual> & by force of circumstances—But in case of several species of domestic Bees & Melipomes of America & still more of Ants and Termites <the neuters of which are as soldiers> are we to suppose that the parents of each species had a female which was a worker. Surely all the species of ants were probably derived from a form in which the Queen ant was not a worker & so in Termites. How then have the neuters of the several species of Ants and Termites acquired their different structures and instincts, as they never breed <& even when converted into Queens by special food this takes place in earliest growth>. This shows that experience in the neuters plays no part in the change—it is not hereditary habit, but hereditary instinctive sports. / Yet I lately thought that experience was probably hereditary in insects! rising from their power of varying their instincts = Better leave this point open, state arguments on both sides./ Are we to suppose that by sports the neuters vary in instincts & that those hives or nests whose neuters have some better instinct predominate; but this presupposes that all the neuters thus vary contemporaneously & this is opposed to all analogy. Otherwise we should have variations of differences in instincts of neuters in the same nests, which is not very probable /well we have soldiers & workers in ants [added later in pencil]/—All the neuters expressing a new structure would show the variation in some effect of law, then of change.—

I must get up this subject—it is the greatest *special* difficulty I have met with

More facts are wanted:

(1) Are there many species of those genera in which the females are not workers or better do not perform same office as the *neuters*.

(2) Are all in such cases permanent societies <for traditionary knowledge>

(3) Are there not cases where neuters perform different offices & have different structures—

Read Kirby—Jardine on Bees—Rennie insect architecture[76]

Darwin thought his original explanation of the instincts of neuters

76. Darwin's manuscript runs to 4 pp., in DAR 73. The sentences between slash marks were written to the side of the main text, but arrows indicate where they belong. Wedge brackets enclose material scratched out, and square brackets contain my editorial remarks. In addition to Kirby and Spence's *Introduction to Entomology,* Darwin probably had in mind to consult one of the volumes in the *Naturalist Library,* which William Jardine edited, such as James Rennie's *Insect Architecture* (London: Knight, 1830). I owe Richard Burkhardt thanks for identifying this last bibliographic item for me.

might still serve for species in which there was close resemblance between the fertile females and the sterile workers. But he now recognized that the explanation failed for species in which neuters greatly differed from the queens. A kind of cultural transmission of "traditionary knowledge" could perhaps account for some small variations of instinct, but hardly for the vastly different instincts and great jaws of soldier ants. The alternative explanation that occurred to him here was that queens might produce neuters whose spontaneous variations would give the entire community a competitive advantage. The unit of selection would be the hive or nest and not the individual. Yet this explanation, which he finally came to accept as the most powerful, appeared to him in 1848 as improbable. Since observation (as he presumed) did not disclose any "differences of instincts of neuters in the same nest," the neuters would have had to vary simultaneously in the right direction. But this ran counter to "all analogy" with the production of varying progeny in other animal species. Darwin simply did not see how this explanation could be possible, and would not until he gathered "more facts." Till then he was left without a resolution for a difficulty that threatened his entire theory.

In spring of 1856, Darwin began work on his "Species Book," which some three years later would be published in an abridged and altered form as the *Origin of Species*. By the summer of 1857, he had reached the chapter dealing with the special difficulties for his theory, and there those facts he listed as needed in his 1848 manuscript were collected. But Darwin had not yet found a clear path through the thicket. In that chapter, he advanced a common explanation for sterility in the social Hymenoptera (e.g., wasps, bees, and ants) and three separate explanations, depending on the species, for the instincts and structures of the neuter castes.

In accounting for the sterile insects, he supposed it would be advantageous to a community if some of the females were relieved of the burden of reproduction so that they might more efficiently perform domestic tasks.[77] Such a condition might arise if, for example, members of a hive chanced to eat certain foods that reduced or extinguished fertility. Natural selection would then tend to favor those communities in which a portion of the members had been thus rendered infertile.

This explanation of sterility, as he worked it out in the "Species Book," was transitional. It joined a notion of inheritance of acquired traits (i.e., induced impotency) with an inchoate assumption of selection at a level above the individual. Darwin, however, linked the ac-

77. *Charles Darwin's Natural Selection*, p. 366.

count to his earlier hypothesis concerning the origin of the habits of neuter bees. He noted that female humblebees and wasps, which alone survived in the nests each winter, executed tasks in the spring which would eventually fall to the lot of neuters produced in the summer. Given the morphological resemblance of females and neuters, he conceived that "the neuters of the different species of wasps & Humble bees might be modified by inheriting any selected modifications in the females."[78] That is, the neuters of these species might be generated from female grubs which had been subjected to conditions inducing sterility; their instincts would then have been received from fertile females that had acquired them by way of natural selection operating on the individuals.

But this explanation could serve only for those species in which females and neuters were similar. Another account was needed for hive-bees, whose queens differed considerably from the neuters. To supply this, Darwin adapted the preceding explanation, still preserving, though, the kernel of his original idea. He proposed that in ancestral hives the queens and neuters were similar, and that each summer neuters were developed anew from those fertile females. By reason of circumstances, however, the queens in subsequent generations had their structures and instincts altered "either by disuse or through natural selection," and thus was generated the difference between them and their neuter offspring.[79]

Darwin's recourse to several different explanations for the traits of neuter insects has an *ad hoc* quality, which he himself appears to have recognized. This suggests that even as late as 1857 he did not regard the problem of neuter insects as satisfactorily resolved. His third kind of explanation, the one he felt the morphology and instincts of particular species of ants required, expressly reached back to the discarded hypothesis of 1848. But now he saw its potential in another light.

The problem ostensibly motivating Darwin's third explanation was that some species of ants, notably *eciton*, exhibited neuters which differed not only from the queens, but also from each other. *Eciton* had two distinct classes of neuters, soldiers with enormous jaws and defensive instincts and workers with ordinary jaws and constructive habits. Even more impressive, African driver ants displayed three classes of neuters, each of which diverged markedly from the others in size and behavior. Darwin's previous explanations could not serve here, because

78. Ibid., p. 367.
79. Ibid.

no evidence existed of the variety of queens necessary to yield the distinct types of neuters.

The account Darwin devised for neuter ants was elegant in conception and extremely powerful in explanation. He took his cue from cattle breeders, with whose practices he had become familiar, especially through William Youatt's *Cattle: Their Breeds, Management, and Disease* (1834).[80] When breeders wanted a herd with particular characteristics, they would select animals from several families for slaughter and then, for example, inspect the meat for the desired marbling. When this was found, they would breed, not of course from the dead animal, but from its family. This sort of artificial selection—family or community selection—suggested to Darwin that his original insight about "hereditary instinctive sports" (in the 1848 manuscript) was not far wrong. He first sketched his explanation in a paragraph appended to the earlier manuscript. It may have been added when he was reviewing his notes for the "Species Book." The lines were written in a different ink and with a slightly scrawlier hand. Here we catch Darwin at a moment of insight.

> The best way to put is, that a Breeder would be at [three words illegible] to improve the breed of the neuters by selection: if he found one hive with all the neuters in any respect better he could do it—His selection would be by families & not individuals—It would be like selecting in cattle for a point which could be ascertained only after the death of the individual, as meat streaked with fat, he would then breed from parents of such fat streaked beasts—this must be the case—& I have good argument that experience plays little part in acquiring instincts in insects.[81]

In the "Species Book," the discovery is expressed in this way:

> This principle of selection, namely not of the individual which cannot breed, but of the family which produced such individual, has I believe been followed by nature in regard to the neuters amongst social insects; the selected characters being at-

80. Darwin seems to have first read William Youatt's *Cattle: Their Breeds, Management, and Disease* (London: Library of Useful Knowledge, 1834) in March of 1840. See his reading notebooks (Vorzimmer, p. 123). He skimmed the volume again in preparing the manuscript for his "Species Book." Initially he may have been interested in the problem of the ill effects of inbreeding, in which connection he made his first reference to Youatt in the "Species Book" (*Charles Darwin's Natural Selection*, p. 37). Darwin (p. 369) specifically referred to Youatt's examples when formulating his explanation of the instincts of neuter insects.

81. See note 76 for bibliographic information.

tached exclusively not only to one sex, which is a circumstance
of the commonest occurrences, but to a peculiar & sterile state
of one sex.[82]

The manner is certainly more casual than in his earlier manuscript, but
the principle is clear: if a community of ants, for instance, happened to
produce neuters whose structure and instincts benefited the group as a
whole, the nest would have a competitive advantage over the other nests
and would hence be selected. Further, as he noted in the manuscript
and elsewhere in the "Species Book," the explanation had a dividend: it
excluded any role for Lamarckian mechanisms of inherited habit.[83]

Darwin thus came to accept an explanation of the instincts of neuter
insects that he had before found improbable (in the 1848 manuscript).
Though I have already suggested the reasons for this reversal, let me
state them explicitly. There were, I believe, two principal reasons. First,
in preparing his notes for the "Species Book," he reviewed Youatt's
Cattle, which furnished him with the example of artificial selection of a
family to produce the desired kind of stock. The analogy with artificial
selection, though, came after his initial conception (in the 1848 manuscript) of selection in nature, just as it had in the case of his original
discovery of natural selection. While working on the "Species Book,"
Darwin had cultivated the demonstrative power of artificial selection;
now, in the chapter on difficulties for his theory, the model convinced
even him of the actual operation of the principle of community selection in nature—something he had not appreciated in 1848. The analogy
between man's selection and nature's made clear that selection could act
on units larger than the individual.

The second factor leading to his discovery of community selection
stemmed from Darwin's network of communicants. In the late summer
and through the winter of 1857, he had corresponded with the entomologist Frederick Smith about social insects. As a result of this interchange, he came to understand that not only did certain species of ants
and bees display neuters of two or more varieties, but that transitional
grades linked these more distinctive castes. Smith explained that "if all
the neuters in a nest be carefully examined, a considerable number will
be found graduating from one extreme to the other."[84] Darwin had
thus been wrong in supposing that the instincts and anatomy of neuters

82. *Charles Darwin's Natural Selection,* p. 370.

83. Ibid., pp. 365, 510.

84. Frederick Smith, quoted by Darwin in his "Species Book." See *Charles Darwin's
Natural Selection,* p. 372, and the Darwin-Smith correspondence in DAR 177.

were uniform in a species. The analogy with production of progeny in other species was therefore much closer than he had at first thought. In light of these two considerations, then, the explanation in terms of community selection—or kin selection as we would call it—seemed the right one.[85]

Darwin offered this explanation of the instincts of neuter ants in "Difficulties on the Theory of Natural Selection," chapter 8 of the "Species Book." There it appeared along with the rather different explanations for neuter bees, explanations generated by his reading of Kirby and Spence years before. But even as he composed this chapter, between the beginning of July and the end of September 1857, he seems to have become aware that his original account would not work even in the case of neuter hivebees. For, as he noted in passing, Kirby and Spence had indicated two separate classes of neuter hivebees, nurses and builders, whose instincts differed.[86] During the first months of 1858, when he worked on chapter 10 of his "Species Book," which dealt with the instincts of animals, he was ready to rely solely on community selection as the explanation of the structure and instincts of neuter insects, regardless of their species:

> In the eighth chapter, I have stated that the fact of a neuter insect often having a widely different structure & instinct from both parents, & yet never breeding & so never transmitting its slowly acquired modifications to its offspring, seemed at first to me an actually fatal objection to my whole theory. But after considering what can be done by artificial selection, I concluded that natural selection might act on the parents, & continually preserve those which produced more & more aberrant

85. In this and subsequent chapters, I generally refer to Darwin's new principle as "community selection" rather than "family selection." I do this, first, to short-circuit the too easy identification of Darwin's conception with our understanding of family or kin selection. Darwin had no idea, for instance, of the peculiar genomic structure of bees—i.e., males having half the full complement of chromosomes and sterile females being more closely related to each other than to their mother. Further, Darwin was aware that ant communities harbored several fertile females instead of just one; he thus knew that the community need not be composed of only one family. Finally, Darwin stressed that the advantageous traits of members redounded to the benefit of the entire community. It was this feature of community selection that allowed him later to apply the principle to human communities (see chapter 5). Darwin recognized that potency of the principle depended ultimately on members of a community being related in some fashion; though in the last two editions of the *Origin of Species,* he endorsed a generalized principle of group selection (see chapter 5, note 82).

86. Darwin, *Origin of Species,* pp. 367–68. Kirby and Spence described these two classes in their *Introduction to Entomology* 1:491.

offspring, having any structure or instinct advantageous to the community.[87]

Darwin simply abandoned the other hypotheses he had advanced. In the *Origin of Species,* community selection was the only explanation he gave for the instincts of neuter insects. And there he reiterated, though with greater emphasis, that community selection furnished the best proof of the operations of natural selection.[88] For no other naturalistic explanation, certainly not the Lamarckian—nor, as he now clearly perceived, his own early explanation, which also depended on inherited habit—could so easily account for the wonderful instincts of the social insects.

Conclusion: Mind, Instinct, and Darwin's Delay

In late September 1838, Darwin read Malthus's *Essay on Population,* which left him with "a theory by which to work."[89] Yet he waited some twenty years to publish his discovery in the *Origin of Species.* Those interested in the fine grain of Darwin's development have been curious about this delay.[90] One explanation has his hand stayed by fear of reaction to the materialist implications of linking man with animals. "Darwin sensed," according to Gruber, "that some would object to seeing rudiments of human mentality in animals, while others would recoil at the idea of remnants of animality in man."[91] Darwin, according to this hypothesis, closed the link between humankind and animals, and thus chained himself to the dread doctrine of materialism. Gould, supporting Gruber's argument, finds evidence for this reconstruction in Darwin's *M* and *N Notebooks,* which

> include many statements showing that he espoused but feared to expose something he perceived as far more heretical than evolution itself: philosophical materialism—the postulate that matter is the stuff of all existence and that all mental and spiritual phenomena are its by-products. No notion could be more upsetting to the deepest traditions of Western thought than the

87. *Charles Darwin's Natural Selection,* p. 510.

88. Darwin, *Origin of Species,* p. 242.

89. Charles Darwin, *Autobiography of Charles Darwin,* ed. Nora Barlow (New York: Norton: 1969), p. 120.

90. I have discussed the variety of explanations given for Darwin's delay in "Why Darwin Delayed, or Interesting Problems and Models in the History of Science," *Journal of the History of the Behavioral Sciences* 19 (1983): 45–53.

91. Gruber, *Darwin on Man,* p. 202.

statement that mind—however complex and powerful—is simply a product of brain.[92]

The proffered hypothesis suggests, then, that Darwin was acutely sensible of the social consequences of equating men with animals and therefore mind with brain, and that he thus shied from publicly revealing his views until the intellectual climate became more tolerant.

The history we have examined in these two chapters makes this hypothesis implausible. Even if Darwin warily explored the implications of his emerging theory in his notebooks, his subsequent study of Fleming, Wells, Brougham, and Kirby should have quieted any trepidation. If these natural theologians did not flinch at seeing human reason prefigured in the mind of a worm, should Darwin have? Moreover, he recognized in his *M Notebook* that the thesis of evolutionary continuity between men and animals did not require an explicit avowal of his conviction that brain was the agent of thought.[93] And in any case, his materialism was of a rather benign sort; at least he so expressed it in an annotation in Abercrombie's *Inquiries concerning the Intellectual Powers* (commented on in chapter 2): "By materialism I mean, merely the intimate connection of thought with form of brain—like kind of attraction with nature of element."[94] This belief would have held little terror for British intellectuals, who were quite familiar—some even comfort-

92. Stephen Jay Gould, "Darwin's Delay", in his *Ever Since Darwin* (New York: Norton, 1977), p. 24. Silvan Schweber, "The Origin of the *Origin* Revisited," *Journal of History of Biology* 10 (1977): 310–15, concurs with Gruber and Gould that fear of materialism was a considerable restraining influence on Darwin.

93. Darwin, *M Notebook*, MS p. 57 (Gruber, p. 276): "To avoid stating how far, I believe, in Materialism, say only that emotions, instincts degrees of talent, which are hereditary are so because brain of child resembles parent stock." Darwin's strategy was shrewd. He could advance the idea that phylogenesis of mind paralleled phylogenesis of brain, without implying the stronger thesis that mind was reducible to brain; for the assumption that brain causally affected mind—that alterations of brain produced complementary changes of mind—did not violate orthodox theology. At least Peter Mark Roget seems to have felt no guilt when he attested, in his Bridgewater Treatise, *Animal and Vegetable Physiology Considered with Reference to Natural Theology* (London: Pickering, 1834): "It is certain, from innumerable facts, that in the present state of our existence, the operations of the mind are conducted by the instrumentality of our bodily organs; and that unless the brain be in a healthy condition, these operations become disordered, or altogether cease. As the eye and the ear are the instruments by which we see and hear, so the brain is the material instrument by which we retrace and combine ideas, and by which we remember, we reason, we invent" (2:510). Darwin read Roget's treatise in December 1847. See his reading notebooks (Vorzimmer, p. 139).

94. Darwin's annotation occurs on p. 28 of John Abercrombie, *Inquiries concerning the Intellectual Powers* (London: Murray, 1838). See the previous chapter for an analysis of his use of Abercrombie.

able—with Locke's anti-Cartesian argument that there was nothing contradictory in supposing God could make matter to think.[95] Finally, even if the intellectual atmosphere of early nineteenth-century Britain were inhospitable to Darwin's brand of materialism, there is little reason to believe he breathed a different air at mid-century while preparing his manuscript.

That Darwin should not have feared suspicions of materialism, of course, does not mean that he did not. But I think there were other, more persistent sources of anxiety that kept him from rushing to publish, namely, the several conceptual obstacles he had to overcome if his theory of evolution by natural selection were to be made scientifically acceptable. He had not been able to give more than a phenomenal account of the laws of variation; he had no good theory of heritability to guide him. He worked with much difficulty on a mathematical demonstration of speciation, by which he sought to give his theory the true colors of science, though in the end he kept his calculations to himself.[96]

95. John Locke, *Essay concerning Human Understanding*, 5th ed., ed. John Yolton (London: Everyman, 1964), 2:147: "it being, in respect of our notions, not much more remote from our comprehension to conceive that God can, if he pleases, superadd to matter a faculty of thinking, than that he should superadd to it another substance with a faculty of thinking, since we know not wherein thinking consists, nor to what sort of substances the Almighty has been pleased to give that power, which cannot be in any created being but merely by the good pleasure and bounty of the Creator. For I see no contradiction in it that the first eternal thinking Being should, if he pleased, give to certain systems of created senseless matter, put together as he thinks fit, some degrees of sense, perception, and thought."

Robert Chambers, who also proposed a theory of species transmutation with descent, did make explicit his conviction that brain was responsible for thought in both animals and man. He understood this kind of materialism to be compatible with theism, as he made clear in *The Vestiges of the Natural History of Creation*, 6th ed. (London: Churchill, 1847), p. 414: "There is, in reality, nothing to prevent our regarding man as specially endowed with an immortal spirit, at the same time that his ordinary mental manifestations are looked upon as simple phenomena resulting from organization, those of the lower animals being phenomena absolutely the same in character, though developed within much narrower limits." The reviewer of *Vestiges* for *Blackwood's Magazine* 57 (1845): 448–60, objected to Chamber's theory, but not because of its assumption that life and mind arose from matter: "Is not the world *one*—the creature of one God—dividing itself, with constant interchange of parts, into the sentient and the non-sentient, in order, so to speak, to become conscious of itself? Are we to place a great chasm between the sentient and the non-sentient, so that it shall be derogation to a poor worm to have no higher genealogy than the element which is the lightning of heaven, and too much honour to the subtle chemistry of the earth to be the father of a crawling subject, of some bag, or sack, or imperceptible globule of animal life? No; we have no recoil against this generation of an animalcule by the wonderful chemistry of God; our objection to this doctrine is, that it is not proved."

96. See Karen Parshall, "Varieties as Incipient Species: Darwin's Numerical Analysis,"

Only in the early 1850s did he arrive at an acceptable theory of divergence, which he intended his mathematics to support.[97] He spent many summers during the decade of the 1850s experimenting with hivebees, attempting to puzzle out the origin of their geometrically perfect constructions.[98] Added to all these conceptual difficulties, reasons enough to stay his hand from publishing, were the problems surrounding his changing notions of instinct.

The inertia of his older ideas about instinct at first made it hard for Darwin to gauge how far the theory of natural selection might be applied to behavior. By the early 1840s, he finally felt ready to meet the challenge of theologians by providing a naturalistic explanation for the wonderful instincts of animals. In his "Essays" of 1842 and 1844, one sort of instinct is, however, not considered—that of neuter insects. Yet Darwin seems to have appreciated the difficulties such instincts entailed at least by 1843, when he read Kirby and Spence. He simply required time to work out a solution to a problem he had initially perceived as "fatal to my whole theory." Even while writing the "Species Book" in summer of 1857, he was still juggling several possible solutions compatible with natural selection. It was only a short time before he actually began to work on the *Origin of Species* that he appears to have settled on a single explanation for the difficulties posed by the instincts of worker bees and ants. The force of his theory of community selection snapped the last critical support of the creationist hypothesis and, conveniently enough, also ruptured the generalized Lamarckian account of the evolution of behavior.

Darwin's difficulty with the instincts of neuter insects obviously played a significant role in retarding any rush to publish his theory of evolution. It would be a mistake, however, to give this conceptual problem the full weight in delaying Darwin. There were those several other obstacles just mentioned, in addition to factors to which appeal has been made by historians: for example, Darwin's large agenda of pub-

---

Journal of History of Biology 15 (1982): 191–214; and Janet Browne, "Darwin's Botanical Arithmetic and the 'Principle of Divergence,'" *Journal of History of Biology* 13 (1980): 53–89.

97. In addition to the preceding articles that touch on Darwin's theory of divergence, see Dov Ospovat, *The Development of Darwin's Theory* (Cambridge: Cambridge University Press, 1981), pp. 146–209; Silvan Schweber, "Darwin and the Political Economists: Divergence of Character," *Journal of History of Biology* 13 (1980): 195–289; and David Kohn, "Darwin's Principle of Divergence as Internal Dialogue," in *The Darwinian Heritage*, ed. David Kohn (Princeton: Princeton University Press, 1985), pp. 245–63.

98. Darwin gave some idea of his experiments on beehive construction in the *Origin of Species*, pp. 224–35. His notes on bees stretch from 1840 to 1862, and are contained in DAR 13, 46.2, 47, 48, and 68.

lishing between 1839 and 1856, which established his reputation; his desire to convince Lyell, Hooker, and Henslow; and his anxiety over the scientific reaction to Chamber's evolutionary theory in 1844. The error usually committed has been to regard one factor alone as explaining the phenomenon. Those historians that adopt a traditional growth model have depicted the delay as simply a result of Darwin's need to gather facts to support his theory, while those of a more Scottish taste have deployed a social model, declaring all to be the result of social factors.[99] The evolutionary model I have chosen guards against such assumptions of unitary causality: it requires one recognize that a complex intellectual, cultural, social, and psychological environment focuses an array of pressures on developing conceptual systems. This does not mean, of course, that the historian must give up differential weighting of factors. After all, we explain the hummingbird's long beak by selection in respect to certain kinds of flowers, while safely bracketing the gravitational effects of bill mass—though such effects have consequences. My giving considerable weight to Darwin's frustrations with the instincts of neuter insects thus does not deny the role of other influences, though it does adjust the significance assigned to them.[100]

The evolutionary model also urges another kind of weighting of environmental factors—this in view of later developments. With a change in the surroundings of a biological organism, that organism will find itself more or less prepared to deal with new contingencies. It will be more or less preadapted to novel conditions—if the new circumstances are enough like the old in relevant respects. Thus environmental factors earlier in the evolutionary trajectory of organisms become more important in light of later situations. So also for the evolution of conceptual systems. They often exhibit preadaptations to new problems they encounter. Darwin's solution to the problem of the instincts of neuter insects must, then, be accorded even more importance (though not now as a factor in causing his delay) in light of a set of difficulties that arose in the decade after the publication of the *Origin of Species*. His theory of community selection came preadapted to overcome certain objections to evolutionary theory that arose insidiously from within the Darwinian camp after 1859.

99. See my "Why Darwin Delayed."
100. Richard Burkhardt takes some exception to my analysis of Darwin's delay in publishing. See his "Darwin on Animal Behavior and Evolution," in *The Darwinian Heritage,* especially p. 396.

4

Debates of Evolutionists over Human Reason and Moral Sense, 1859–1871

In October 1846, while ensconced with his family in their new home in the village of Downe (twenty miles from London), Darwin began a study of one species of barnacle, *arthobalanus,* and that species led to another, till some four volumes later he had concluded a systematic study of all the known living and extinct species of Cirripedes. Though he judged the work of great value for constructing principles of classification in the *Origin of Species,* he despaired over the time exhausted. Eight years of tedious research, leavened by sickness and sorrow. He lost a total of some two years, by his own reckoning, to intermittent boils, dizziness, trembling, depression, and extended hydropathic treatments by the famous Dr. Gully at Malvern.[1] In November 1847, Dr. Robert Darwin died at age eighty-three. His death was expected but no less mourned. Charles, however, was too sick even to attend the funeral.[2] Not expected was the death of the Darwin's ten-year old daughter Annie in April 1851. She was her father's favorite child, and he grieved for her deeply.[3] And the while he worked on the seemingly unending species of barnacles. At last in September 1854, he packed up his specimens and distributed copies of his great monographs on barnacles, which remain today standard references.[4]

    1. Darwin left this estimate in his "Journal," ed. Gavin de Beer, *Bulletin of the British Museum (Natural History),* historical series 2 (1959): 13.
    2. Charles Darwin, *The Autobiography of Charles Darwin,* ed. Nora Barlow (New York: Norton, 1969), p. 117.
    3. See Darwin's poignant character sketch of his daughter Annie in *Life and Letters of Charles Darwin,* ed. Francis Darwin (New York: D. Appleton, 1891), 1:109–111.
    4. Charles Darwin, *A Monograph of the Fossil Lepadidae; or, Pedunculated Cirripedes of Great Britain* (London: Ray Society, 1851); *A Monograph of the Sub-Class Cirripedia, with Figures of all the Species. The Lepadidae; or, Pedunculated Cirripedes* (London: Ray Society,

Figure 4.1   Charles Darwin, photograph from 1860.

Darwin then turned to the subject he had always with him, doing what he could on it at odd intervals. His *Journal* records that on "Sept 9 Began sorting notes for Species theory."[5] He steadily accumulated more notes on species through the remaining part of 1854 to the spring of 1856. During this period he also performed experiments and conducted observations related to his theory: he studied the fertilization of

1851); *A Monograph of the Sub-class Cirripedia; The Balanidae (or Sessile Cirripedes); the Verrucidae, &c.* (London: Ray Society, 1854); *A Monograph of the Fossil Balanidae and Verrucidae of Great Britain* (London: Palaeontographical Society, 1854).

5. Darwin, "Journal" (de Beer, p. 13).

flowers by bees; he soaked seeds in salt water, to determine their possible dispersal time via ocean currents; he planted plots with different species of grasses, to compare results of competition; and he performed numerous mathematical calculations on the relations of different kinds of species (e.g., large vs. small, common vs. rare) to their genera. Darwin continued to formulate both the problems and the possible solutions regarding the origin of species. But he needed some prod to write.

Charles Lyell knew the right goad. In mid-April 1856, Lyell urged Darwin to publish a preliminary sketch of his theory, lest he be anticipated by someone else.[6] Darwin hesitated because of the complexities of the project, but admitted "I certainly should be vexed if any one were to publish my doctrines before me."[7] With a little more encouragement, in mid-May he began a sketch of his theory for publication.[8] But that writing faltered; he could not bring his theory into small compass without sacrificing argument and evidence. In October, as his *Journal* records, he began work on a large volume that would satisfy the demands of demonstration.[9] The writing on his big "Species Book" progressed apace during the rest of the year, through 1857, and into spring of 1858. By March 9 he had completed the long chapter "Mental Powers and Instincts of Animals." With eight and a half chapters finished of a projected fourteen, he had composed some quarter of a million words. But he did not finish the "Species Book." In mid-June, he received from Alfred Wallace a letter, which enclosed a sketch for a theory of species development. Wallace asked Darwin to forward the essay to Lyell if it had sufficient merit. Darwin believed it did, since it virtually reproduced his own theory.[10] He wrote Lyell on 18 June in great agitation: "Your words have come true with a vengeance—that I should be forestalled. You said this, when I explained to you here very briefly my views of 'Natural Selection' depending on the struggle for existence. I never saw a more striking coincidence; if Wallace had my MS. sketch written out in 1842, he could not have made a better short abstract!"[11]

6. Leonard Wilson rather convincingly fixes this date in his introduction to *Sir Charles Lyell's Scientific Journals on the Species Question,* ed. Leonard Wilson (New Haven: Yale University Press, 1970), p. xliii–xlv.

7. Darwin, Letter to Charles Lyell (3 May 1856), in *Life and Letters* 1:426–27.

8. Darwin, "Journal" (de Beer, p. 14).

9. Ibid.

10. Malcolm Kottler discusses the several important differences between Darwin's version of evolution by natural selection and Wallace's. See Malcolm Kottler, "Charles Darwin and Alfred Russel Wallace: Two Decades of Debate over Natural Selection," in *The Darwinian Heritage,* ed. David Kohn (Princeton: Princeton University Press, 1985), pp. 367–432.

11. Darwin, Letter to Charles Lyell (18 June 1858), in *Life and Letters* 1:473.

With the help of Lyell and Hooker, Darwin's priority yet held secure. They arranged to have a portion of his 1844 essay and part of a letter to Asa Gray describing his views, along with Wallace's sketch, read before the Linnean Society. The assemblage for the 1 July meeting seemed to find the communication interesting, but not remarkable.[12] Its subsequent publication in the proceedings of the society failed to ignite even the volatile. Darwin's intuition about his theory had been correct. No mere outline would convince or even move to high objection. Only a full argument would seize the attention of the scientific community.

So at the end of July, after overcoming some niggling scruples about continuing without giving Wallace a chance also to make his case, Darwin began to abstract his argument from the monstrously long manuscript and then to fill in the remaining parts. He finished the *Origin of Species* in March of 1859, only eight months after he had started. What the book lost in detail, it gained in economy and force of expression. Examples stacked high in the "Species Book" manuscript were distilled into telling illustrations. Experiments of rigor and variety were recorded. The most severe objections were anticipated. And the argument stood out. The theory of evolution, dismissed when Lamarck first elaborated it, ridiculed when Chambers tried again at mid-century, and largely ignored when presented to the Linnean Society, produced cataclysmic scientific controversy in the wake of the *Origin,* but finally widespread acceptance by the scientific community even during Darwin's lifetime. The general theory of evolution triumphed. But in its application to one animal, consensus faded, as Darwin suspected it might.

In the *Origin of Species,* Darwin scrupulously avoided application of evolutionary considerations to the human animal, detouring, as one reviewer detected, that supposed path of "transition from the instinct of the brute to the noble mind of man."[13] Darwin realized that he had to be discreet in order to get the fairest hearing possible for his thesis.[14] But his readers immediately perceived the implications, for which La-

12. Peter Brent recounts the reaction of the Linnean Society in *Charles Darwin: A Man of Enlarged Curiosity* (New York: Harper & Row, 1981), pp. 415–16.

13. In his review of the *Origin of Species,* William Hopkins, a distinguished mathematician, saw immediately that Darwin must have intentionally circumvented human evolution. See William Hopkins, "Physical Theories of the Phenomena of Life," *Fraser's Magazine* 61 (1860): 739–52; 62 (1860): 74–90.

14. Writing from Malaya in fall of 1857, Alfred Wallace inquired of Darwin whether he planned to discuss man in his book on species. Darwin responded, "I think I shall avoid the whole subject, as so surrounded with prejudices, though I fully admit that it is the highest and most interesting problem for the naturalist." The Darwin-Wallace correspondence is preserved in James Marchant, *Alfred Russel Wallace: Letters and Reminiscences* (London: Cassell, 1916). The quoted letter is in 1:131–33.

Though Darwin did not wish to prejudice readers of the *Origin of Species* against evo-

marck and Chambers had earlier prepared them. One friendly critic wrote Darwin to say he was largely persuaded by the *Origin of Species,* except that it implied that man was "to be considered a modified & no doubt *greatly* improved orang!" The correspondent had flinched, since he could not bring himself "to the idea that man's reasoning faculties & above all his *moral sense* cd. ever have been obtained from irrational progenitors, by mere natural selection—acting however gradually & for whatever length of time that may be required."[15] This was also the stumbling block for Lyell, among the first of Darwin's colleagues to broach the problem of man in print, with his *Geological Evidence of the Antiquity of Man* in 1863.

Darwinian Disputes over Human Nature

Lyell, Huxley, and Wallace on the Evolution of Man

Darwin preferred his friend had remained silent, for Lyell so balanced his evaluation of the Darwinian hypothesis that the scales hardly dipped in its favor.[16] Moreover, he allowed that the doctrine of transmutation did not preclude divinely contrived saltations in human mental and moral development.[17] Darwin rested more satisfied with Huxley's *Man's Place in Nature,* also published in 1863. Huxley forthrightly, though still provisionally, adopted the hypothesis that man was related to the apes through descent from a common ancestor.[18] He illustrated the anatomical relations of man to those higher creatures with characteristic skill and brio. He delighted in demonstrating that Richard Owen, the

---

lution by introducing a discussion of human development, he yet thought honesty required he at least allude to his views (see Darwin, *Autobiography,* p. 130). This he did with one sentence in the last chapter of the *Origin* (p. 488): "Light will be thrown on the origin of man and his history."

15. Reverend Leonard Jenyns's letter to Darwin (4 January 1860) is recorded in Lyell's *Journals.* See Wilson, *Sir Charles Lyell's Scientific Journals,* p. 351. Darwin had informed Jenyns, an old friend, of his species theory in 1845, though apparently without convincing him. See Darwin's letter to Jenyns (12 October 1845), in *Life and Letters* 1 : 392–93. Ralph Colp describes Darwin's early communication of his theory to friends in "'Confessing a Murder,' Darwin's First Revelations about Transmutation," *Isis* 77 (1986): 9–32.

16. Darwin wrote to Huxley (26 February 1863): "I am fearfully disappointed at Lyell's excessive caution in expressing any judgment on Species or origin of Man." Darwin's letter is in *More Letters of Charles Darwin,* ed. Francis Darwin (London: Murray, 1903), 1 : 239.

17. Charles Lyell, *The Geological Evidences of the Antiquity of Man* (London: Murray, 1863), pp. 504–05.

18. Huxley thought the missing link that would inextricably bind the scientific mind to the Darwinian hypothesis was a clear demonstration that interbreeding infertility could be produced by artificial selection from a common stock. See Thomas Huxley, *Evidence as to Man's Place in Nature* (London: Williams & Norgate, 1863), p. 127.

Figure 4.2   Thomas Henry Huxley, portrait from photograph of 1857.

British Cuvier and fierce opponent of Darwinism, either lied or exhibited profound ignorance when he denied the ape a hippocampus minor.[19] This had been a small horse to carry human mental superiority, especially as it was obvious to most naturalists that "even the highest faculties of feeling and of intellect begin to germinate in lower forms of life."[20] Huxley, though, never doubted the vast gulf separating civilized man from the brutes and had little to suggest about how a bridge between the two might have been erected. The cofounder of evolution by natural selection did have some idea.

In an essay in 1864, "On the Origin of Human Races,"[21] Alfred Wal-

19. Adrian Desmond discusses the variety of battles that Huxley waged with Owen. See Adrian Desmond, *Archetypes and Ancestors: Palaeontology in Victorian London, 1850–1875* (Chicago: University of Chicago Press, 1984).

20. Huxley, *Man's Place in Nature,* p. 129.

21. Alfred Russel Wallace, "The Origin of Human Races and the Antiquity of Man deduced from the theory of 'Natural Selection,'" *Anthropological Review* 2 (1864): clviii–clxxxvii.

lace employed the theory of natural selection to resolve the dispute between those who considered the races of man as merely local varieties
of one species and those who assumed man to be one genus and the
races originally different species.[22] Wallace proposed that racial varieties
evolved through natural selection and became thus accommodated to
local conditions, but that this occurred before the appearance of distinctively human mental qualities. He thought that once natural selection began to foster in those protomen reason and sympathetic feelings
(especially moral sentiments which led them to care for their unfit
brethren), selective pressure on their physical structures would cease,
while concomitantly it would increase on their mental abilities, making
mind their principal instrument of survival. Wallace contended that at
some critical period in human evolution (perhaps as early as the Mio-
cene), the physical development of man had been arrested, while "the
power of 'natural selection,' still acting on his mental organisation, must
ever lead to the more perfect adaptation of man's higher faculties to the
conditions of surrounding nature, and to the exigencies of the social
state."[23] He answered the question of the unity of human species by
making it a matter of choice. The fixation of different anatomical types
at the advent of distinctively human mind allowed us to speak of many
species of man. Yet if we preferred to regard that original, undifferentiated progenitor, whose mind was still brutish, as the spokesman for
humanity, then we might also declare the species to be one, though at
present displaying many races. The continued operation of natural selection on the minds of men, however, held the promise that "the
higher—the more intellectual and moral—must displace the lower and
more degraded races—till the world is again inhabited by a single
race."[24]

Wallace originally delivered his paper before the Anthropological Society of London in March 1864. The stimulus for his lecture appears to
have been Lyell's *Antiquity of Man*.[25] In his book, Lyell directly inquired
about the unity of the human races and their classificatory relation to
the higher animals. But his treatment of these questions, from the trans-
mutationist's point of view, was most unsatisfactory. First, he found yet

22. George Stocking has explored the various facets of the monogenism-polygenism
controversy in his insightful history of nineteenth- and early-twentieth-century anthropology, *Race, Culture, and Evolution*, 2d ed. (Chicago: University of Chicago Press, 1981).
23. Wallace, "The Origin of the Human Races," p. clxix.
24. Ibid.
25. Malcolm Kottler, in his "Alfred Russel Wallace, the Origin of Man, and Spiritualism," *Isis* 65 (1974): 147, endorses this opinion, though he does not suggest exactly what
it was about Lyell's book that spurred Wallace.

no objection to his own long-standing conviction that "all the leading varieties of the human family sprang originally from a single pair," a belief he read not from the book of nature, but from the book of Moses.[26] Second, while admitting human beings to be morphologically similar to other primates, he held that "man must form a kingdom by himself if once we permit his moral and intellectual endowments to have their due weight in classification."[27] For "we cannot imagine," he reassuringly intoned, "this world to be a place of trial and moral discipline for any of the inferior animals, nor can any of them derive comfort and happiness from faith in a hereafter. To Man alone is given this belief, so consonant to his reason . . . a doctrine which tends to raise him morally and intellectually in the scale of being."[28] According to Lyell's analysis, then, man's mental constitution indicated no gradual passage up from the intellect of an ape; a gap in intellectual quality separated the two primate species. He admitted that the natural order did evince a similar leap in another area of human development—in the birth of a genius from quite ordinary parents; but he could not conceive the cause of the saltation, in the case either of the individual or of the species, being referred to "the usual course of nature."[29] Man's moral and rational fabric was stitched by God, not natural selection—or so Lyell intimated.

In his paper, Wallace responded to both of Lyell's points, so skillfully that Darwin wished his friend "had written Lyell's chapters on Man."[30] Wallace could concede man a single set of forebears, though only that primordial couple which became morphologically distinct from the apes while remaining their mental equal. But if we thought of man as essentially a moral and reasoning creature, then several races of man independently achieved their humanity—when natural selection began to undertake the creative function Lyell and the Prophets had reserved for God. In May 1864, Lyell wrote Wallace to compliment him on the clear and fair distinctions drawn in his paper, which were "no small assistance towards clearing the way to a true theory."[31] He did not, however, suggest the paper embodied the wanted theory.

26. Lyell, *Antiquity of Man*, p. 385.
27. Ibid., p. 495. Next to this passage in his copy of Lyell's book, Darwin penciled a strong "No."
28. Ibid.
29. Ibid., p. 505. Next to this passage in his copy, Darwin gasped "Oh." He wrote to Lyell, confessing the passage "makes me groan." See Darwin, *Life and Letters* 2:197.
30. Charles Darwin to Joseph Hooker (22 May 1864), in *More Letters* 2:31.
31. Charles Lyell to Alfred Wallace (22 May 1864), in Marchant, *Alfred Russel Wallace* 2:18.

If Lyell stimulated Wallace to write, Herbert Spencer furnished him the philosophical framework. Wallace was a Spencer enthusiast for a good portion of his adult life, even naming his first son after the great thinker. In Wallace's estimation, as he wrote Darwin in January 1864, Spencer was "as far ahead of John Stuart Mill as J. S. M. is of the rest of the world, and, I may add, as Darwin is of Agassiz." He recommended his friend try reading Spencer's *Social Statics* (1851), to regard it, in contrast to his usual pursuits, as "light literature."[32] The book had special significance for Wallace, since, as he acknowledged in a note to his Anthropological Society paper, it had inspired his principal thesis.[33] He later wrote Spencer gratefully to confess that "the illustrative chapters of your 'Social Statics' produced a permanent effect on my ideas and beliefs as to all political and social matters."[34]

Spencer argued that man had gradually emerged from a primitive state by adapting to the exigencies of society, which required he adjust his own needs and wants to those of his fellows. In *Social Statics,* he envisioned a continuous natural development of civilized society, whose goal would be that state in which each man might perfectly fulfill his nature without diminishing that of others, a classless society in which a genuine sympathy with the conditions of others would yield the greatest happiness for the greatest number. In that consummation, which Wallace's essay also foretold, all men would be united in the common bond of perfected humanity.[35] According to both Spencer and Wallace, a natural principle of evolution inexorably led to the moral perfection of man. Wallace, of course, had a different principle in mind than Spencer's device of adaptation through the inherited effects of habit. He nonetheless believed that the principle of natural selection would add further support to Spencer's primary vision, the view that man's moral

32. Alfred Wallace to Charles Darwin (2 January 1864), in Marchant, *Alfred Russel Wallace* 1:150.

33. Wallace, "Origin of the Human Races," p. clxx: "The general idea and argument of this paper I believe to be new. It was, however, the perusal of Mr. Herbert Spencer's works, especially *Social Statics,* that suggested it to me, and at the same time furnished me with some of the applications." Wallace's debt to Spencer likely also went to the "general idea," which was that at a certain period in human evolution, man's bodily frame would cease to change, while his intellect and moral sense would continue to evolve. This was the very conclusion Spencer had earlier reached in his paper "A Theory of Population, deduced from the General Law of Animal Fertility," *Westminster Review* 57 (1852): 468– 501. See chap. 6 for a discussion of Spencer's thesis.

34. Alfred Wallace to Herbert Spencer (12 November 1873), Athenaeum Collection of Spencer's Correspondence, Senate House Library, University of London, MS. 791, no. 89.

35. See the discussion of Spencer's moral conception of evolution in chaps. 6 and 7.

character was not only a goal of evolution, but also a chief means of progress toward the perfection of human nature. Wallace's recommendation of *Social Statics* to Darwin was really unnecessary (and unheeded). Those ideas Darwin would have found attractive in Spencer's book were fully articulated in Wallace's paper.

Darwin admired Wallace's subtle analyses, commending his paper to Hooker as "*most* striking and original and forcible."[36] He wrote to congratulate Wallace and to remark on a particular point of convergence in their views. He acknowledged the novelty of his friend's leading idea, "that during late ages the mind will have been modified more than the body," but indicated that his own studies had already led him "to see with you that the struggle between the races of men depended entirely on intellectual and moral qualities."[37] Darwin seems not to have been preparing for another dispute about priority; on the contrary, he quite generously offered to furnish Wallace his collected notes on human descent, since he then (in 1864) doubted he would publish on man. Darwin simply made known to his friend a long standing conviction, which had found expression as early as his *Transmutation Notebooks*[38] and more recently in a letter (1859) to Lyell. That letter, in which he responded to criticisms of the *Origin of Species,* spelled out his hypothesis about the function of the mental and moral qualities as causes of transmutation:

> I suppose that you do not doubt that the intellectual powers are as important for the welfare of each being as corporeal structure; if so, I can see no difficulty in the most intellectual individuals of a species being continually selected; and the intellect of the new species thus improved, aided probably by effects of inherited mental exercise. I look at this process as

36. Charles Darwin to Joseph Hooker (22 May 1864), in *More Letters* 2:31.
37. Charles Darwin to Alfred Wallace (28 May 1864), in *More Letters* 2:33.
38. Charles Darwin, *Fourth Transmutation Notebook,* MS pp.63–4, transcribed by Gavin de Beer, *Bulletin of the British Museum (Natural History),* historical series 2 (1960): 166: "When two races of men meet, they act precisely like two species of animals.—they fight, eat each other, bring disease to each other &c., but then comes the most deadly struggle, namely which have the best fitted organization, or instincts (i.e. intellect in man) to gain the day.—In man chiefly intellect, in animals chiefly organization, though Cont. of Africa & West Indies shows organization in Black Race there give the preponderance, intellect in Australia to the white." The last remark expresses Darwin's conviction that the aborigines of Australia would finally succumb to the encroachments of the white man. In his *Diary* of the *Beagle* voyage, he observed how the new settler's small peace offerings (e.g., some cow's milk) would be the price of the natives' patrimony: "The thoughtless Aboriginal, blinded by these trifling advantages, is delighted at the approach of the White Man, who seems predestined to inherit the country of his children." See Charles Darwin, *Diary of the Voyage of H.M.S. Beagle,* ed. Nora Barlow (Cambridge: Cambridge University Press, 1933), p. 382.

now going on with the races of man; the less intellectual races being exterminated.[39]

Though Darwin believed that competitive struggle brought civilized reason up from a primitive state, his ideas did not completely converge with those of Wallace, despite his own observation to the contrary. The differences may appear slight, but they grew in significance during the latter part of the 1860s. The first difference concerns the unit of selection in human mental evolution. Wallace argued in his 1864 paper:

> Capacity for acting in concert, for protection and for the acquisition of food and shelter; sympathy, which leads all in turn to assist each other; the sense of right, which checks depredations upon our fellows [etc.] . . . are all qualities that from their earliest appearance must have been for the benefit of each community, and would, therefore, have become the subjects of "natural selection." . . . Tribes in which such mental and moral qualities were dominant, would therefore have an advantage in the struggle for existence over other tribes in which they were less developed, would live and maintain their numbers, while the others would decrease and finally succumb.[40]

Wallace assumed that selection would operate on whole communities and tribes, since the traits selected would confer benefit primarily on the group rather than the individual. He thus appears to have endorsed group selection, though undoubtedly without detecting the miasma of difficulties infecting that concept.[41] Darwin, by contrast, hinted at nothing beyond individual selection in his letters to Wallace and Lyell. Dar-

---

39. Charles Darwin to Charles Lyell (11 October 1859), in *Life and Letters* 2:7. John Greene—in "Darwin as a Social Evolutionist," *Journal of the History of Biology* 10 (1977): 5—has found evidence to suggest that Darwin at one time contemplated describing the development of human races in the *Origin*. In his "Species Book," Darwin left an annotation in the table of contents indicating that chapter 6 was to have included a section entitled "Theory Applied to the Races of Man." Annotations in books from his library mark out material to be used in the planned section. Greene also cites Darwin's letter to Lyell in support of his thesis about Darwin's initial inclinations.

40. Wallace, "The Origin of Human Races," p. clxii.

41. In the contemporary literature, the debate over units of selection rages (in an academic sense). The controversy can be followed in George Williams, *Adaptation and Natural Selection* (Princeton: Princeton University Press, 1966), pp. 92–124; Richard Lewontin, "The Units of Selection," *Annual Review of Ecology and Systematics* 1 (1970): 1–23; Michael Wade, "A Critical Review of the Models of Group Selection," *Quarterly Review of Biology* 53 (1978): 101–44; and William Wimsatt, "Reductionistic Research Strategies and Their Biases in the Units of Selection Controversy," in *Scientific Discovery: Historical and Scientific Case Studies,* ed. Thomas Nickles (Dordrecht: Reidel, 1980). The definitive work on the topic, to date at any rate, is Elliott Sober, *The Nature of Selection: Evolutionary Theory in Philosophical Focus* (Cambridge: M. I. T. Press, 1984).

win's initial failure to recognize this difference may have been due to his own acceptance of community selection in the *Origin*. In Darwin's version of community selection, however, the unit was a group of *related* individuals. Wallace himself, on the other hand, may have merely assumed that his "communities" and "tribes" consisted principally of related individuals; he may, in fact, have taken as his model Darwin's analysis of instinct in the social insects. Another possibility, the most likely, is that the conceptual environment in 1864 simply did not exert sufficient pressure on either Wallace or Darwin to refine their ideas on the question of group selection. But that environment shifted, chiefly as a result of the debates stimulated by Wallace's paper. From these debates arose several theoretically insistent difficulties regarding the units of selection in human evolution. By the end of the decade, while writing *Descent of Man* (and emending the *Origin of Species* to recognize group selection), Darwin squarely faced these problems, which sorely tested his own mental capacities. Wallace's thinking about human evolution took a sharply different turn in the late 1860s, as we will see. For him, the problem of the units of selection, at least in human evolution, never emerged as a live issue.[42]

The other point of divergence concerned Wallace's fervent anticipation that natural selection would continue to improve human mental and moral character. Darwin initially lacked this Spencerian vision. He was oblivious to the hand Wallace saw still adjusting the mental equipment of civilized men, eliminating the less intellectually fit in favor of

42. In Wallace's original paper, "On the Tendency of Varieties to Depart Indefinitely from the Original Type," which he communicated to Darwin on that fateful June day in 1858, he spoke of a struggle for existence among "varieties" rather than individuals. Somewhat unreflectively Wallace continued to think in this mode, till the conflict between his formulation and Darwin's became more apparent. As Kottler ("Charles Darwin and Alfred Russel Wallace," pp. 375–79) shows, Wallace had not really distinguished individual from varietal selection in these early years. However, once he did make the distinction, he became persuaded that group selection was possible (p. 388). Darwin himself felt perplexed. He knew community selection occurred among the social insects. But he apparently was initially unsure about other cases. George Romanes reported a conversation he had with Darwin on the topic, in which his mentor related a discussion with Wallace in which he (Darwin) expressed his doubts about "the possibility of natural selection acting on organic *types* as distinguished from individuals" (see Ethel Romanes, *Life and Letters of George John Romanes,* 4th ed. [London: Longmans, Green, 1897], p. 57). Darwin apparently resolved most of his doubts, however, for he did give clear endorsement to group selection both in the fifth edition of the *Origin of Species* and in the *Descent of Man.* I discuss this in chap. 5 and in n. 82 of that chap. Kottler and Michael Ruse have a different assessment of Darwin's stand on group selection. See Kottler, "Charles Darwin and Alfred Russel Wallace," p. 428, n. 10; and Michael Ruse, "Charles Darwin and Group Selection," *Annals of Science* 37 (1980): 615–630.

their superiors. Darwin rather considered his contemporaries to have approximately equivalent cognitive abilities (despite—or perhaps because of—his own mediocre performance at Cambridge). He presumed that "excepting for fools men did not differ much in intellect, only in zeal and hard work." Thus there were no variations among civilized men for natural selection to seize upon and improve. Recognition of intellectual variability, with its consequences for the continued operation of natural selection on the minds of civilized men, Darwin credited to the work of his ingenious cousin Francis Galton.[43]

Galton's Theory of the Heritability of Intellect

Galton had his own familial debt to acknowledge. He avowed that scales fell from his eyes when he read the *Origin of Species*. At once, he related in his autobiography, he brushed aside the occluding film of religious belief, and began to pursue exact science.[44] His studies of heredity, which date from the mid-1860s, undoubtedly had the stimulus of Darwin's work. But they likely had a more personal motive as well.

Like Darwin, Galton had been chosen by his father for a medical career. After a hospital apprenticeship at age sixteen, Francis's desire to study mathematics persuaded Samuel Tertius Galton to send his son to Cambridge. The boy had shown promise, even if his abilities had been somewhat magnified by his eight older brothers and sisters.[45] At uni-

43. Galton's demonstrations of the variability of intellect among civilized men, in his book *Hereditary Genius,* brought Darwin to declare to his cousin: "You have made a convert of an opponent in one sense, for I have always maintained that, excepting fools, men did not differ much in intellect, only in zeal and hard work; and I still think this is an *eminently* important difference." Darwin's letter (3 December 1869) is quoted in Francis Galton, *Memories of My Life,* 3d ed. (London: Methuen, 1909), p. 290

44. Ibid., p. 287.

45. Galton's early education was undertaken by his sister Adèle. She obviously succeeded in producing a tutored child of knowledge and confidence, as this letter indicates:

"My dear Adèle, I am four years old and can read any English Book. I can say all the Latin Substantives and adjectives and active verbs besides 52 lines of Latin poetry. I can cast up any sum in addition and multiply by 2, 3, 4, 5, 6, 7, 8, 9, 10, 11. I can also say the pence table, I read French a little and I know the Clock. Francis Galton, Feb. 15, 1827."

The letter is quoted in Karl Pearson, *Life, Letters, and Labours of Francis Galton* (Cambridge: Cambridge University Press, 1914–1930), 1:66. Since this letter suggests that Galton could accomplish at four what other children normally achieved at eight or nine years of age, Lewis Terman estimated Galton's IQ at 200. But Terman, who developed the "Stanford-Binet IQ test," did a rough calculation. Since Galton wrote the letter one day before his fifth birthday, a more judicious estimate would have fixed his IQ between 160 and 180. But even in this correction, one might also need to bracket the variance due to the utter devotion of Galton's sister to his early education. See Lewis Terman, "The

Figure 4.3 Francis Galton, 1822–1911, portrait done ca. 1880.

versity, Galton enjoyed the company of other well-financed young men, and stood a bit in awe of those who tested well in mathematics. At the beginning of his third year, he started preparing his father for the possibility that he might not take his exam for honors. Such letters as the following were meant to cushion the blow to his father and to his own ego:

> My head is very uncertain so that I can scarcely read at all, however, I find that I am not at all solitary in that respect. Of the year above me the first 3 men in their College examinations are all going out in the poll [i.e., taking a pass examination],

Intelligence Quotient of Francis Galton in Childhood," *American Journal of Psychology* 28 (1917): 209–15.

the first 2 from bad health and the third, Boulton, from finding
that he could not continue reading as he used to do without
risking it. Foxwell Buxton is quite knocked up and goes out in
the poll, so does Bristed, one of the first classics in our year, in
fact the whole of Trinity is crank.[46]

As Galton's heart palpitations and giddiness grew worse, his father took
him out of university for a long vacation. He finally received an A.B.
(finishing forty-fourth in the medical list and third in the mathematical)
and continued on in medicine. But when his father died leaving him a
nice inheritance, Galton, again in the footsteps of his cousin, gave up
medicine for travel. For one who had failed to achieve eminence at university, the doctrine that genius had biological roots and could not be
earned in schoolboy labor must have had an appeal.

As his wife's diary records, Galton began a statistical analysis of the
hereditary transmission of human mental ability in 1864.[47] The aim of
his essay "Hereditary Talent and Character," published the following
year, was to show that conspicuous mental talent—for science, mathematics, literature, painting, and law—ran in families, that mind and
character were biologically transmitted.[48] To demonstrate this, Galton
searched biographical dictionaries of distinguished people, the roster of
lord chancellors, the list of senior classics at Cambridge, and the roll of
past presidents of the British Association. He discovered that the men
recorded there tended, beyond the average in the population, to have
close relatives who were also of noted intellect. His figures implied "that
when a parent has achieved great eminence, his son will be placed in a
more favourable position for advancement, than if he had been the son
of an ordinary person."[49] Galton pursued this discovery in subsequent
works—*Hereditary Genius* (1869), *English Men of Science* (1874), *Human
Faculty* (1883), and *Natural Inheritance* (1889). But these books, as he
admitted, all merely substantiated and elaborated his original essay.[50]

Galton's study was highly prejudicial to Lyell's impressionistic claims
about the development of intellect and moral faculty. In the *Antiquity
of Man*, Lyell had appealed to the presumed usual occurrence of the
"birth of an individual of transcendent genius, of parents who have

46. Francis Galton to Samuel Galton (2 November 1842), in Pearson, *Life, Letters, and
Labours of Francis Galton* 1:170–71.

47. The pertinent passages of Louisa Galton's diary are transcribed by Pearson in *Life,
Letters and Labours of Francis Galton* 2:70.

48. Francis Galton, "Hereditary Talent and Character," *Macmillan's Magazine* 12 (1865):
157–66; 318–27.

49. Ibid., p. 161.

50. Galton, *Memories of My Life*, p. 289.

never displayed any intellectual capacity above the average standard" to argue that nature might make a similar leap from the "unprogressive intelligence of the inferior animals" to the "improvable reason manifested by Man."[51] The jump, in both cases, would, however, require a supernatural shove. But now Galton provided evidence to the contrary: talent was not visited capriciously on individuals; rather they inherited it from forebears. He had demonstrated that mind functioned like a biological trait: it was heritable and subject to the laws of transmission. When such laws became fully known—a goal toward which Galton toiled—appeal to transcendent causes to explain the appearance of genius, or of ordinary human reason, would lose all attraction.

But even with the laws of heredity still obscure, Galton's analysis suggested that hope for continued progress of human reason and morals was well founded. "What an extraordinary effect might be produced on our race," he mused in Platonic reverie, "if its object was to unite in marriage those who possessed the finest and most suitable nature, mental, moral, and physical!"[52] But such a consummation, Galton feared, would not come about unless men took evolution into their own hands. For the melancholy fact stood plain: not only did men propagate their virtues, but they transmitted their vices as well—"craving for drink, or for gambling, strong sexual passion, a proclivity to pauperism, to crimes of violence, and to crimes of fraud."[53] As Galton surveyed his own time, he perceived that civilization actually had a retarding effect on the natural selection of the best. The poverty of men of good character is, he declared, "more adverse to early marriages than is natural bad temper, or inferiority of intellect."[54] The best hope for the unabated advance of human mind was enlightened social policy that encouraged early marriage of men and women of talent and set obstacles to the egregious propagation of intellectual and moral paupers.[55] But what Galton had viewed as a *possible* restraint on mental progress, another enthusiast for evolution detected as a fundamental and perhaps insuperable barrier to the progress of mind and morals.

The Failure of Natural Selection in the Case of Man

William Rathbone Greg, Scots moralist and political writer, worked out the implications of a problem first touched on by Galton. The immediate source of his reflections, however, was not Galton's work,

51. Lyell, *Antiquity of Man,* pp. 504–505.
52. Galton, "Hereditary Talent and Character," p. 165.
53. Ibid., p. 320.
54. Ibid., p. 326.
55. Ibid., pp. 165–66, 319–20.

which only subsequently became known to him, but Wallace's paper "On the Origin of Human Races." Greg, in what turned out to be a highly provocative essay, "On the Failure of 'Natural Selection' in the Case of Man" (1868), initially agreed with Wallace that the struggle *among* races and nations had promoted those groups having superior mental abilities.[56] He demurred, however, at Wallace's assumption that natural selection, after ceasing to mold human anatomy, would continue to operate on the minds of individuals *within* larger social groups. For, Greg proposed, the moral sympathies of the more advanced societies would defend not only the physically unfit from the hand of natural selection (as Wallace himself had recognized), but the intellectually and morally inferior as well. So protected, these retrograde types would procreate at a faster rate than their betters. Hence, the moral and intellectual pillars of a society would be washed over in the high tide of prolific dullards and degenerates. With the finely honed sensitivity of the Scots gentleman, Greg offered the case of the Irish as cautionary: "The careless, squalid, unaspiring Irishman, fed on potatoes, living in a pig-sty, doting on a superstition, multiplies like rabbits or ephemera:—the frugal, foreseeing, self-respecting, ambitious Scot, stern in his morality, spiritual in his faith, sagacious and disciplined in his intelligence, passes his best years in struggle and in celibacy, marries late, and leaves few behind him. . . . In the eternal 'struggle for existence,' it would be the inferior and less favored race that had prevailed—and prevailed by virtue not of its good qualities but of its faults"[57]

Greg's analysis was perfectly consonant with that of other evolutionists, Galton and Spencer, for instance. But some who were friendly to the doctrine of development took exception. They generally responded in two ways. Some concurred with his diagnosis of the problem, but not with his prognosis that only calculated intervention in the social process could reinstate that "righteous and salutary law [of natural selection] which God ordained for the preservation of a worthy and improving humanity."[58] The other class of respondents simply dismissed the entire problem, since they believed that principles other than natural selection operated in civilized society.

The anthropologist Lawson Tait, in the former group, argued in response to Greg that though civilization might momentarily preserve the weak and profligate, eventually more savage eruptions of disease and the accumulated effects of mental incompetency would break their

56. [William R. Greg], "On the Failure of 'Natural Selection' in the Case of Man," *Fraser's Magazine* 78 (1868): 353–62

57. Ibid., p. 361. Darwin, with some relish, quoted this example of the disengagement of natural selection in the *Descent of Man* (London: Murray, 1871), 1:174.

58. Greg, "On the Failure of 'Natural Selection' in the Case of Man," p. 358.

numbers.[59] Darwin himself inclined to this opinion. In the *Descent of Man,* he did tremble with the fear of Greg and Galton that natural selection might be disengaged in advanced societies, since our moral sympathies inhibited us from allowing the physically and mentally unfit to meet their natural end. Darwin ventured, however, that there were already inherent checks on the possible increase among the inferior classes: the poor crowded into towns would die at a faster rate; the debauched would also suffer higher mortality; and jailed criminals would not bear children—fortune would finally, it seemed, snip the thread of the unworthy.[60] And one could hope, as Darwin, Galton, and Greg did, that enlightened social legislation and the impact of moral education would return the propagatory advantage to the more favorably endowed.[61] Nonetheless, it could be that civilized nations faced, after

59. Lawson Tait, "Has the Law of Natural Selection by Survival of the Fittest Failed in the Case of Man?" *Dublin Quarterly Journal of Medical Science* n.s. 47 (1869): 102–13. Darwin indicated that he borrowed ideas from Tait in formulating his own response to Greg. See *Descent of Man* 1 : 168, n. 10.

60. Ibid., pp. 174–80.

61. Galton, Greg, and Darwin did not clamor for immediate changes in law to remove the impediments to natural selection. They, rather, preserved the nineteenth-century faith in the efficacy of moral education. They presumed that once the simple truths of human evolution were appreciated, appropriate social policy would follow. In an article entitled "Hereditary Improvement," *Fraser's Magazine* n.s 7 (1873): 116–30, Galton avowed: "I believe when the truth of heredity as respects man shall have become firmly established and be clearly understood, that instead of a sluggish regard being shown towards a practical application of this knowledge, it is much more likely that a perfect enthusiasm for improving the race might develop itself among the educated classes" (p. 120). Greg was more inclined to advocate legislative changes, if they would be effective. In his original article, he suspected that legislative attempts to conform to natural principles would, in fact, be ineffective, and that continued education was the best hope: "Obviously, no artificial prohibitions or restraints, no laws imposed from above and from without, can restore the principle of 'natural selection' to its due supremacy among the human race. No people in our days would endure the necessary interference and control; and perhaps a result so acquired might not be worth the cost of acquisition. We can only trust to the slow influences of enlightenment and moral susceptibility, percolating downwards and in time permeating all ranks" (Greg, "On the Failure of 'Natural Selection' in the Case of Man," p. 362).

Darwin did not wish to prevent men from bestowing sympathy on their less fortunate fellows, since that would stunt "the noblest part of our nature" (*Descent of Man* 1 : 169). Yet he recommended that social policy not check the salutary work of natural selection: "Hence our natural rate of increase, though leading to many and obvious evils, must not be greatly diminished by any means. There should be open competition for all men; and the most able should not be prevented by laws or customs from succeeding best and rearing the largest number of offspring" (2: 403). John Greene assesses Darwin's social views and their development in "Darwin as a Social Evolutionist," *Journal of the History of Biology* 10 (1977): 11–27.

reaching a peak, an inevitable decline. After all, as Darwin gloomily observed, "progress is no invariable rule."[62]

Another respondent to Greg's analysis, however, rejected the idea that natural selection had—or should have—aegis over civilized society. A writer for *The Spectator* made this rejoinder:

> The real answer to him [Greg] is this,—that directly you reach in the ascending stages of animal life, you reach a point where the competitive principle of "natural selection" is more or less superseded by a higher principle, of which the key-note is not "Let the strong trample out the weak," but "Let the strong sacrifice themselves for the weak." This is really the law of supernatural selection, as distinguished from the law which governs the selection of races in the lower animal world.[63]

The author admitted with Greg that man's altruism, his moral sympathy for his fellows, "prevents the true Darwinian consummation in the case of man."[64] But it was precisely these acquisitions of moral sentiment—as distinguished from the inheritance of physical traits—that made men fit to live a noble and worthwhile life.

The debate over the application of natural selection to human mental and moral faculties spun around what some would now regard as a confusion. Both sides assumed that the most fit of a species were to be defined in terms other than success in propagation. Even Darwin thought that those favored by nature—the intellectually superior, the morally upright (and generally those of the appropriate social class)—were those fit to survive. If they did not, it was an anomaly, which men should correct, or nature herself eventually would. None of the scientists who considered the problem of human evolution attempted to separate nature from the ghost of the since-departed Deity—and Darwin no more than the rest.[65] The traits they valued most—intellect and moral rectitude—had been ascribed by hallowed tradition to a beneficent God. Hence if nature (or nature's God) granted these favors, they could not fail to be valuable properties. The unfit gained an advantage, it was supposed, only because highly "artificial" schemes (e.g., the Poor Laws) had fouled the well-designed machinery of nature.

62. Darwin, *Descent of Man*, 1: 177.

63. "Natural and Supernatural Selection," *Spectator* 41 (1868): 1155.

64. "The Darwinian Jeremiad," *The Spectator* 41 (1868): 1215. This is another editorial on Greg's essay by the same author as in the previous note.

65. Darwin still counted himself a theist when he wrote the *Origin* (*Autobiography*, p. 91). So there is little reason to doubt his sincerity when he spoke of the laws of development being divinely established (*Origin of Species*, p. 488).

By our contemporary lights, of course, if the lower-class vandals of the population were trampling the tender buds of the Victorian best and brightest, it might be because they were actually superior in what mattered most in classic 'Darwinian' terms—survival. This, at least, is how modern sociobiologists might analyze the problem.

But a word must be said for the intuitions of Darwin and his colleagues. While they perhaps did not appreciate the difficulties of determining fitness components—something that occupies the modern evolutionist—yet they operated in light of a distinction that moderns have tended to ignore or confuse. "Survival" is not the *definition* of fitness. Survival is rather caused by fit traits which organisms possess. Survival, therefore, is only a *criterion* of fitness. In order to prevent the principle of natural selection from devolving into a tautology, this distinction must be preserved. Consequently, to apply natural selection as an empirical principle, it must be possible to make assessments of fitness independent of knowledge of survival. We might take exception to the evaluations actually made by Darwin and his followers—when they judged intellect and moral sense to be components of fitness in the particular environment of Victorian society—yet their effort at independent assignment was logically impeccable. If natural selection is really to do empirical work in evolutionary theory, then it must be logically possible to have a situation in which the fit perish, while the unfit inherit the earth.

The controversy generated by Greg's analysis of the "failure of natural selection in the case of man" also gave birth to a kind of problem for evolutionary theory different from that of the decline of morals and intellect in Victorian society. If moral sympathy arose in the initial stages of human evolution, so that tribes of protomen formed cohesive bonds secured by altruistic habits, then human evolution should have gone no farther. In those primitive societies, sympathetic responses would preserve the mentally as well as the physically feeble. Hence, the high intellect and acute moral sense of civilized man could never have been achieved by natural selection alone. It began to appear that a completely naturalistic account of human evolution foredoomed itself. At least this is how the cofounder of evolution by natural selection came to perceive the matter.

Wallace and the Challenge of Spiritualism

Alfred Wallace had hardly uttered his last words about human evolution in his essay of 1864. Five years later he returned to the question, but in the interim had completely thrown over his former opinion. His

Figure 4.4   Alfred Russel Wallace, 1823–1913, portrait done ca. 1864.

new belief left Darwin reeling: "But I groan over Man—you write like a metamorphosed (in retrograde direction) naturalist, and you the author of the best paper that ever appeared in the *Anthropological Review!* Eheu! Eheu! Eheu!"[66]

What occasioned Darwin's exasperation was his friend's review in

66. Charles Darwin to Alfred Wallace (26 January 1870), in Marchant, *Alfred Russel Wallace* 1 : 251.

1869 of some new editions of Lyell's geological works.[67] At the end of his article, Wallace revealed the pitch his ideas had quietly taken in the previous few years:

> Neither natural selection or the more general theory of evolution can give any account whatever of the origin of sensational or conscious life . . . But the moral and higher intellectual nature of man is as unique a phenomenon as was conscious life on its first appearance in the world, and the one is almost as difficult to conceive as originating by any law of evolution as the other.[68]

Evolutionary theory, as Darwin himself admitted in the *Origin*, remained mute concerning how life and consciousness first arose in the universe; it could only account for subsequent transformations.[69] Just so, Wallace now proclaimed, natural selection brought no clearer perception of the origins of specifically human intellect and moral feeling. He was persuaded that these distinctive capacities must have originated under the influence of higher powers, intelligences who shepherded the progressive development of mind through the ages.

Wallace's metamorphosis had an unusual precipitating cause. He had undergone a conversion to spiritualism and as a result saw man in a new light.[70] He had been a materialist, baptized in that creed as an adolescent when he read the socialist and anti-free-will tracts of Robert Owen.[71] And as a young man, he became hardened in his naturalistic views by a study of mesmerism and phrenology. Even in his old age he prized the delineations of his cranium done by Edwin Hicks and James Rumball, the latter having also read Herbert Spencer's head.[72] But in July of 1865, only a short while after seeing the publication of his famous paper, he attended a seance at the home of an acquaintance. The hand-holding ceremony, done without medium, produced a few of the usual

67. Alfred Russel Wallace, "Review of *Principles of Geology* by Charles Lyell, 10th ed., 2 vols. (London, 1867, 1868); *Elements of Geology* by Charles Lyell, 6th ed. (London, 1865)," *Quarterly Review* 126 (1869): 359–94.

68. Ibid., p. 391.

69. In the chapter on instinct (chap. 7) of the *Origin of Species*, Darwin cautioned (p. 207): "I have nothing to do with the origin of the primary mental powers, any more than I have with that of life itself. We are concerned only with the diversities of instinct and of other mental qualities of animals within the same class."

70. Malcolm Kottler provides the most detailed and persuasive account of Wallace's conversion to spiritualism and of its effect on his theory of human evolution. See his "Alfred Russel Wallace, the Origin of Man, and Spiritualism."

71. Alfred Russel Wallace, *My Life* (New York: Dodd, Mead, 1905), 1:88–104, Marchant, *Alfred Russel Wallace* 2:182.

72. Wallace, *My Life* 1:257–62.

occurrences—table rappings, vibrations, movements. These sparked the curiosity, but not yet the faith of the empiricist. Beginning in November of 1866, Wallace held a series of seances in his home in order carefully to test the reality of spiritualistic phenomena. He engaged the services of a Miss Nichol, a medium whose divine performances quickly convinced him. At one sitting, with the lights extinguished and hands held tight around the table, the very rotund Miss Nichol suddenly disappeared. When the lights were struck, she was found in her chair centered on the top of the table. To all assembled, she appeared to have floated up like a hot-air balloon. She also possessed the knack of producing, during seances held in the dead of winter, fresh flowers damp with morning dew. This gift particularly impressed Wallace. In his pamphlet *A Defense of Modern Spiritualism* (1874), he relates that he dried and preserved his bouquet.[73]

In the latter part of the nineteenth century, many men of prominent intellect walked upright into the fold of spiritualism. The conjurer's tricks that convinced Wallace also persuaded the likes of Francis Galton, the mathematician Augustus De Morgan, the Catholic evolutionist St. George Mivart, the physiologist Charles Richet, the psychologist Frederic Myers, the physicist Oliver Lodge, and the chemist and editor of the *Quarterly Journal of Science* William Crookes.[74] In the United States, William James published his informal experiments with a medium, whom he "believed to be in possession of a power as yet unexplained."[75] The wave of spiritualism also rushed onto German shores. The medium Henry Slade, traveling in Germany after his prosecution in England for fraud,[76] converted the astrophysicist J. F. Zöllner, the physicist W. Weber, the physicist-psychologist G. T. Fechner, and the mathematician W. Scheibner.[77] Wallace was not a lone crank isolated by his delusion.

73. Alfred Russel Wallace, *A Defense of Modern Spiritualism* (Boston: Colby and Rich, 1874). This originally appeared in the *London Fortnightly Review* in 1874.

74. Kottler gives a full account of the spiritualist movement among Wallace's contemporaries. See his "Alfred Russel Wallace, the Origin of Man, and Spiritualism." For more general surveys, see Brian Inglis, *Natural and Supernatural: A History of the Paranormal from Earliest Times to 1914* (London: Hodder & Stoughton, 1977); and Janet Oppenheim, *The Other World: Spiritualism and Psychical Research in England, 1850–1914* (Cambridge: Cambridge University Press, 1985).

75. In 1886 William James wrote an account of seances with the medium Mrs. Piper, in which he, his wife, and his in-laws participated. The report appeared in the first volume of *Proceedings of the American Society for Psychical Research*. The passage quoted is from Gay W. Allen, *William James* (New York: Viking, 1967), p. 284.

76. Wallace testified for Slade at his trial. See Inglis, *Natural and Supernatural,* pp. 277–81; and Oppenheim, *The Other World,* pp. 22–23.

77. For a description of the disputes among German scientists over spiritualism, see

Yet most of his close scientific associates were not taken in. Huxley rejected Wallace's entreaties to attend a seance—he thought it would be a waste of time;[78] John Tyndall accepted the invitation, but laughed at the episode;[79] and William B. Carpenter, in his *Mental Physiology,* gave a thorough analysis of spiritualism, specifying the kinds of errors to which men like Wallace fell prey.[80] At the one seance Darwin attended, he became bored and left before the furniture started flying.[81] He remained an unbeliever.

Wallace interpreted his experiences at seances as empirical evidence demonstrating the reality of the spirit domain and its action in the physical world. It was this new conviction, he confided to the much dismayed Darwin, that led him to alter his ideas about man:

> My opinions on the subject [of man] have been modified solely by the consideration of a series of remarkable phenomena, physical and mental, which I have now had every opportunity of fully testing, and which demonstrate the existence of forces and influences not yet recognised by science. This will, I know, seem to you like some mental hallucination, but as I can assure you from personal communication with them, that Robert Chambers, Dr. Norris of Birmingham, the well-known physiologist, and C. F. Varley, the well-known electrician, who have all investigated the subject for years, agree with me both as to the facts and as to the main inferences to be drawn from them, I am in hopes that you will suspend your judgment for a time till we exhibit some corroborative symptoms of insanity.[82]

Among the several important inferences Wallace derived from his commerce with the spirit world were: "Man is a duality, consisting of

Marilyn Marshall and Russel Wendt, "Wilhelm Wundt, Spiritism, and the Assumptions of Science," in *Wundt Studies,* ed. W. Bringmann and R. Tweney (Toronto: Hogrefe, 1980), pp. 158–75.

78. See the correspondence between Wallace and Huxley (November and December 1866), in Marchant *Alfred Russel Wallace* 2:187–88.

79. See Wallace's letter to Barrett (9 December 1877), in Marchant, *Alfred Russel Wallace* 2:198; and Kottler, "Alfred Russel Wallace, Origins of Man, and Spiritualism," pp. 171–72.

80. William B. Carpenter, *The Principles of Mental Physiology* (New York: D. Appleton, 1874), chap. 16. Carpenter mentions Wallace as one who deluded himself about spiritualism (p. 627).

81. Charles Darwin, letter (18 January 1874), in *Life and Letters,* 2:364–65. The excerpted letter concludes: "The lord have mercy on us all, if we have to believe in such rubbish. F. Galton was there, and says it was a good seance."

82. Alfred Wallace to Charles Darwin (18 April 1869), in Marchant, *Alfred Russel Wallace* 1:244.

an organized spiritual form, evolved coincidentally with and permeating the physical body, and having corresponding organs and developments"; and "Progressive evolution of the intellectual and moral nature is the destiny of individuals."[83] It was precisely this destiny that appeared to outrun the power of natural selection. Natives of the Americas and Malaya revealed capacities for greater intellectual attainment and moral sensitivity than they required for survival in the wild. Wallace's early training in phrenology led him to regard these capacities, which he thought a future destiny would actualize, as real reservoirs well marked in the savage brain. Contemporary primitives and our ancestors thus had latent mental qualities that could not be explained by natural selection, which demanded that selected traits confer immediate advantage, not simply promise it. Wallace's contacts with the spirit world convinced him that higher intelligences rather than natural selection controlled human evolution.

Wallace forthrightly claimed that a conversion to spiritualism proximately caused his rejection of natural selection as an adequate principle to explain human evolution; and virtually all historians have taken him at his word.[84] But we need not. For after all, Wallace might well have chosen to regard natural selection as the disposing instrument of higher spiritual powers and to have held survival of the fittest as a secondary cause. Darwin himself tactfully hinted at this idea, and the American Darwinist Asa Gray happily proclaimed it.[85] Thus Wallace's new belief was not itself a sufficient reason for him to deny that natural selection produced man's high intellect and refined moral sense. A fuller explanation of his decision must mention four factors that exerted pressure during the mid-1860s on the shape of his ideas about human evolution. First was his abiding faith in the inherent progressiveness of human

83. Wallace, *Defense of Modern Spiritualism*, p. 56.

84. Kottler, in his "Alfred Russel Wallace, Origins of Man, and Spiritualism," and Frank Turner, in his *Between Science and Religion* (New Haven: Yale University Press, 1974), chap. 5, portray Wallace's adoption of spiritualism as leading immediately to his new theory of human evolution. But this view, as suggested below in the text, foreshortens the transformation in Wallace's opinions.

85. In the last chapter of the *Origin of Species* (p. 488), Darwin observed that it accorded better with sound theology to believe that the Creator worked through secondary causes in "the production and extinction of the past and present inhabitants of the world." And Lyell (*Antiquity of Man*, pp. 502–506) cited Asa Gray's article "Natural Selection not Inconsistent with Natural Theology" to argue that it was compatible with Darwinism to believe that selection was the instrument by which the Creator accomplished his designs. Darwin arranged to have Gray's essay published in England in 1861. Thus Wallace had several models which might have sanctioned his postulation of higher powers and his retention of natural selection as the proximate cause of human evolution.

nature, a central creed of Spencer's *Social Statics,* which Wallace so much admired. Second was the argument directed against his 1864 essay by Greg, that several natural barriers inhibited progressive evolution. Third was the awareness, gained during the many years he spent in South America and Malaya, that members of native tribes often exhibited great cleverness and tender moral sentiments, traits that could be further developed. In his article "How to Civilize Savages," published in June 1865—a month prior to his first spiritualistic encounter—Wallace expressed full confidence that missionary activity, particularly the example of good men, would produce natives who had "improved in morality and advanced in civilization."[86] The final factor was indeed his conversion to spiritualism. But I suspect that its principal effect was to make Wallace sensitive to an objection that Spencer had raised with great force against the "all-sufficiency" of natural selection: that natural selection could not produce attributes that were of little or no use to the survival of the organism.[87] This principle fueled the various specific difficulties Wallace enumerated in his review of Lyell's books and carefully specified in a subsequent article in 1870, entitled "The Limits of Natural Selection as Applied to Man."[88] Just after Wallace's review appeared, he made explicit to Lyell the argument that led him logically to reject natural selection in the case of man. The argument hinged on the Spencerian objection:

> I am glad the article gives satisfaction to yourself & your friends. Darwin tells me he likes it greatly, all except the conclusion *which he would have thought written by someone else.* He says however that he will consider the points I mention in what

86. [Alfred] W[allace], "How to Civilize Savages," *Reader* 5 (17 June 1865): 671.

87. In his *Principles of Biology* (1864–1867 in serial form), Spencer acknowledged that natural selection could account for simple modifications of organisms but not complex coadaptations. He precluded natural selection in this latter case, since elements of a co-adaptive system must arise before they can be integrated into a useful organic system (see chapter 6, below). Spencer specifically mentioned the case of higher mental adaptations (*The Principles of Biology* [New York: D. Appleton, (1867) 1884], pp. 454–55): "but it appears to me that as fast as the number of bodily and mental faculties increases, and as fast as the maintenance of life comes to depend less on the amount of any one, and more on the combined action of all; so fast does the production of specialties of character by natural selection alone, become difficult. Particularly does this seem to be so with a species so multitudinous in its powers as mankind; and above all does it seem to be so with such of the human powers as have but minor shares in aiding the struggle for life—the aesthetic faculties, for example."

88. Wallace's "The Limits of Natural Selection as Applied to Man" was the concluding chapter of his *Contributions to the Theory of Natural Selection* (1870). I have used the edition which combines this book with his *Tropical Nature* (1878): *Natural Selection and Tropical Nature* (London: Macmillan, 1891).

he is now writing about man. I fear that being in the "Quarterly" it will not attract the attention of scientific men.

It seems to me that if we once admit the necessity of any action beyond "natural selection," in developing man we have no reason whatever for confining that action to his brain. On the mere doctrine of chances it seems to me in the highest degree improbable, that so many points of structure should concur in man, and in man alone of all animals. If the *erect posture, the freedom of the anterior limbs for purposes of locomotion,* the *powerful* and *opposable thumb,* the *naked skin,* the *great symmetry of force,* the *perfect organs of speech;* and in his mental faculties—*calculation of numbers, ideas of symmetry, of justice,* of *abstract reasoning,* of the *infinite,* of a *future state,*—& many others can not be shewn to be each and all *useful* to man in the very lowest state of civilization, how are we to explain their coexistence in him alone of the whole series of organized beings? Years ago I saw a *Bushman* boy & girl in London, & the girl played very nicely on the piano. Blind Love the idiot *negro* had a "musical ear" or *brain,* superior perhaps to that of any living man. Unless Darwin can shew me how this rudimentary or latent musical faculty in the lowest races can have been developed by survival of the fittest—can have been of *use* to the individual or the race, so as to cause those who possessed it in a fractionally greater degree than others to win in the *struggle for life,* I must believe that some other power caused that development,—and so on with every other especially human characteristic. It seems to me that the *onus probandi* will lie with those who maintain that man, *body & mind,* could have been developed from a quadrumanous *animal* by "natural selection."[89]

Wallace's spiritualism may have added the critical pressure that finally overturned his early ideas about human evolution. But his particular objections to the application of natural selection to man could not be dismissed because of an outre motive. They were based on firm empirical evidence and compelling Spencerian principles. They proved quite formidable. In his review of Lyell, Wallace drew on his own experience of native cultures and on recent anthropological findings. These, he thought, indicated that the brain and attendant intellectual and moral capacities of savages were little inferior to those of "the average members of our learned societies."[90] But, as he argued, the "higher moral

89. Alfred Wallace to Charles Lyell (28 April 1869), in the Lyell Correspondence, BD 25.L., American Philosophical Society, Philadelphia.

90. Wallace, "Review of Lyell," p. 392.

faculties and those of pure intellect and refined emotion are useless to them."[91] Hence they could not have been produced by natural selection, which only worked on useful traits. In his essay "The Limits of Natural Selection as Applied to Man," Wallace did allow that benevolence and honesty might have aided the tribe whose members practiced these virtues; yet, he cautioned, natural selection could not explain "the peculiar sanctity attached to actions which each tribe considers right and moral."[92] The conspicuous mental attributes of modern man—his "capacity to form ideal conceptions of space and time, of eternity and infinity"—which must, upon the evolutionary hypothesis, have initially arisen in the primitive state, these could hardly have made a difference in producing more progeny.[93] Darwin would have to show him, as he related to Lyell in the above-quoted letter, how rudimentary or latent mental attributes "in the lowest races can have been developed by survival of the fittest,—can have been of *use* to the individual or the race." The scientific evidence appeared to Wallace to march toward an obvious consequence, that after natural selection had produced a creature of anthropoidal anatomy, another hand must have intervened to shape the large brain of emerging man and endow it with faculties requiring only the right cultural setting for their manifestation. It must be, Wallace concluded, that "a superior intelligence has guided the development of man in a definite direction, and for a special purpose, just as man guides the development of many animal and vegetable forms."[94]

91. Ibid.

92. Wallace, "The Limits of Natural Selection as Applied to Man," p. 199.

93. Ibid.

94. Ibid., p. 204. Robert M. Young—in "Malthus and the Evolutionists: the Common Context of Biological and Social Theory," in his *Darwin's Metaphor* (Cambridge: Cambridge University Press, 1985), pp. 23–55—suggests that it was Wallace's reading of Henry George's *Progress and Poverty,* an anti-Malthusian tract, which convinced him that "Malthus's law did not apply at all to human evolution" (p. 48). While Wallace expressed to Darwin (letter of 1881, in Marchant, *Alfred Russel Wallace* 1 : 317–18) admiration for George's work, it most assuredly had nothing to do with his abandonment of natural selection in human evolution. George's book was first published in 1879 and Wallace read it in 1881—over a decade after he had argued that natural selection failed in the case of man.

5

Darwin and the Descent of Human Rational and Moral Faculties

In biological evolution, the strongest selective forces on individuals arise from competition with conspecifics and members of related species. So too with the evolution of intellectual systems. The greatest inducements to development come from encroachments by closely allied conceptions. If the only intellectual pressures on Darwin to render an explicit theory of human evolution were attacks like that of Bishop Samuel Wilberforce, who challenged Huxley to declare whether it was "through his grandfather or his grandmother that he claimed his descent from a monkey," he might well have left it to Huxley and other supporters to make appropriate rejoinder.[1] Certainly against an intellectual and social environment formed entirely by the declared enemies of evolution, Darwin's *Descent of Man* would have evolved quite differently—if at all. But Darwin experienced mounting pressure from men of like mind, evolutionists who perceived the selectionist mechanism as straining to a halt when applied to man's rational and moral faculties. Charles Lyell and Asa Gray regarded the Darwinian device as deficient. Both suggested that human evolution required a supernatural impetus. Francis Galton and William Greg seemed to show that even if selection were able to produce the beginnings of human reason and moral sentiment, intensified social and sympathetic feelings would prevent beneficial culling of the mentally and morally inferior—natural selection would slowly be disengaged. And Wallace, who Darwin now feared "murdered too completely your own and my child,"[2] set out a series of persuasive objections to the idea that natural selection had designed human nature. Wallace estimated that, after all, for sheer survival man

1. See Leonard Huxley, *Life and Letters of Thomas Huxley* (New York: D. Appleton, 1900), 1:197–98. See also my conclusion to this volume.
2. Charles Darwin to Alfred Wallace (27 March 1869), in James Marchant, *Alfred Russel Wallace: Letters and Reminiscences* (London: Cassell, 1916), 1:241.

required a brain "little superior to that of an ape."[3] Finally, if the super-natural powers to which Lyell, Gray, and Wallace appealed failed the needs of science, Spencer stood ready with the mechanism of inherited habit. Darwin felt the heat of that engine as Spencer cranked it up to explain what he believed natural selection could not—complex human mental and moral faculties.[4]

Pressure from several sides thus increased on Darwin to formulate in detail a theory of human evolution and to reveal it to an awaiting pub-lic. Recognition of these forces makes more comprehensible Darwin's shifting attitude toward publishing. Shortly after the appearance of the *Origin of Species,* he wrote Wallace of his intention to bring out an "Es-say on Man."[5] In January 1860, he had begun work on what was to become a two-volume study, *On the Variation of Animals and Plants under Domestication,* which he completed in 1867. Darwin planned to include his essay on man in the *Variation,* since, as he whimsically but pointedly suggested to Wallace, man seemed "an eminently *domesticated* animal."[6] Several events intervened, however, to prevent an attack forthwith on human evolution. First, from 1860 to 1862, even while stockpiling curious and pedestrian information for the *Variation,* he put out two new editions of the *Origin*[7] and a book on orchids.[8] Then

3. Alfred Wallace, "Review of *Principles of Geology* by Charles Lyell, 10th ed., 2 vols. (London, 1867, 1868); *Elements of Geology* by Charles Lyell, 6th ed. (London, 1865)," *Quar-terly Review* 126 (1869): 392.

4. See chapter 6. Darwin first examined Spencer's *Principles of Psychology* (London: Longman, Brown, Green and Longmans, 1855) in 1860, as he mentioned in a letter to the philosopher dated 2 February 1860 (Athenaeum Collection of Spencer's Papers, Senate House Library, University of London, MS. 791, No. 47). Darwin's own copy of Spencer's volume (in Cambridge University Library) is well marked, particularly part 4, which deals with the phylogenetic development of reason from animal instinct. Darwin penciled his judgment of one aspect of Spencer's theory on the back flyleaf of the book: "good discus-sion on necessity of evolution hypoth. to unite experience and transcendental hypothe-sis." This referred to Spencer's use of inherited habit to explain the relatively a priori forms of human thought (*Principles of Psychology,* pp. 577–83). In a letter to J. S. Mill, Spencer applied the principle of inherited habit to account for the human faculty of making a priori moral judgments. The letter was published in Alexander Bain, *Mental and Moral Science* (London: Longmans, Green, 1868), pp. 722. Spencer's moral theory is discussed in chapter 7, below.

5. Charles Darwin to Alfred Wallace (7 March 1860), in Marchant, *Alfred Russel Wallace* 1:140.

6. Charles Darwin to Alfred Wallace (March 1867), in Marchant, *Alfred Russel Wallace* 1:181.

7. The second and third editions of the *Origin of Species* were published in January 1860 and March 1861 respectively.

8. Charles Darwin, *On the Various Contrivances by which Orchids are Fertilised by Insects* (London: Murray, 1862).

during the first several months of 1864, he suffered a severe aggravation of his chronic maladies.[9] Later, in May, Wallace sent Darwin a copy of his article "Origin of the Human Races." Darwin replied by offering his friend a few collected notes on man, since he now despaired of ever publishing on the topic.[10] Given an illness that tied his digestive tract in knots and left him palpitating in miserable weakness, while still facing mountains of tedious work on the *Variation,* it is quite understandable that he palled at tackling so large a question as human origins, especially since Wallace had already made an admirable start. But in the early part of 1867, feeling more vigorous than he had for some time and with the end of the *Variation* in sight, Darwin renewed his intention to publish "a little essay on the Origin of Mankind."[11] He would, as he related to Lyell, treat the subject using his theory of sexual selection.[12] To Wallace he confessed that his "sole reason" for taking up human evolution was "that I am pretty well convinced that sexual selection has played an important part in the formation of races, and sexual selection has always been a subject which has interested me much."[13]

When Darwin actually sat down in February 1868 to begin work on what would appear as *The Descent of Man and Selection in Relation to Sex,* his consuming efforts were first directed to the general theory of sexual selection and to heaping up mountains of evidence on the operations of that mechanism in the animal kingdom. His appetite for this part of the work had been stimulated by his differences with Wallace, with whom he had engaged in a protracted correspondence on the subject during the previous year.[14] Both initially agreed that the conspicuous coloration of male birds stemmed from female choice—females of a given species preferred, for contingent reasons, to mate with males exhibiting certain patterns of color. The dispute between the two friends arose over the dull plumage of the less conspicuous sex, usually the female. Wallace believed that the distinctive coloration of a species had been inherited equally by both sexes, but that the sex most often in

9. Darwin's "Journal"—transcribed by Gavin de Beer, *Bulletin of the British Museum (Natural History),* historical series 2, no. 1 (1959): 4–21—for 1864 reads: "Ill all Jan. Feb. March, Last sickness April 13th" (de Beer, p. 16).

10. Charles Darwin to Alfred Wallace (28 May 1864), in Marchant, *Alfred Russel Wallace* 1 : 155.

11. Charles Darwin to Alfred Wallace (26 February 1867), in Marchant, *Alfred Russel Wallace* 1 : 180.

12. Charles Darwin to Charles Lyell (19 July 1867), in *Life and Letters,* ed. Francis Darwin (New York: D. Appleton, 1891), 2 : 254.

13. Charles Darwin to Alfred Wallace (March 1867), in Marchant, *Alfred Russel Wallace* 1 : 182.

14. The bulk of the correspondence is in Marchant, *Alfred Russel Wallace* 1 : 177–230.

danger would give up bright colors, due to the operations of natural selection. While Darwin at first agreed that selection for protective camouflage likely caused the drab feathers of most female birds, he also recognized another possibility: that the inheritance of coloration was sex-linked from the start. During the course of the debate, Darwin came to reject Wallace's explanation altogether. He argued rather that both males and females had begun with equally muted hues, and while sexual selection would dress out the male in brighter colors, the female's original pattern would simply be handed down to her daughters. Darwin had worked up considerable evidence in his *Variation* to demonstrate that sex-linked inheritance often operated from the first appearance of a variation.[15] So, for instance, in some fish, the male tended the nest, but displayed more brilliant colors than the female—protective selection thus played no role. The female's colors, Darwin believed, had to have arisen initially in her female ancestors and been passed on only to daughters. If sex-linked inheritance were a common mechanism of transmission, it could serve from the beginning to keep females drably feathered.[16]

Darwin filled most of part 2 (i.e., the last half of volume 1 and the whole of volume 2) of the *Descent of Man* with reams of evidence for his theory of sexual selection, concentrating his efforts on the animal kingdom. Only in the last two substantive chapters of the book, under the rubric of "Secondary Sexual Characters of Man," did he use his theory of sexual selection to explain: the love of ornament found in primitive tribes, the racial traits of different groups, and the human male's superiority in strength and intelligence over the female. His discussions of this last 'fact' show him to be a representative man.

Darwin assessed the qualities of women as would most Victorian gentlemen, save Wallace. Women abounded in the tender virtues, but lacked the distinctively male intellect. Were they to be schooled over many generations as were the sons of the gentry, they might—via the inheritance of acquired traits—achieve intellectual muscle, but at the cost of the role for which evolution naturally fitted them. Darwin's letter to a curious American correspondent, Caroline Kennard, makes explicit his views on the intellectual and moral character of women.

> The question to which you refer is a very important one. I have it briefly in my "Descent of Man." I certainly think that

15. Charles Darwin, *Variation of Animals and Plants under Domestication,* 2d ed. (New York: D. Appleton, 1899), 2:47–51.

16. Malcolm Kottler has fully detailed the debate between Darwin and Wallace. See his "Darwin, Wallace, and the Origin of Sexual Dimorphism," *Proceedings of the American Philosophical Society* 124 (1980): 203–26.

women, though generally superior to men in moral qualities, are inferior intellectually: and this seems to me to be a grant [guard?] from the laws of inheritance (*if I understand the laws rightly*) in their becoming the intellectual equals of men. On the other hand, there is some reason to believe that aboriginally (& to the present day in the case of savages) men & women were equal in this respect, & this would greatly favour their recovering this equality. But to do this, as I believe, women must become as regular "bread-winners" as are men; & we may suspect that the easy education of our children, not to mention the happiness of our homes, would in this case greatly suffer.[17]

Darwin's journal indicates that he finished the two chapters on man by the end of spring 1869.[18] There is no evidence that he planned an extensive treatment of human intelligence and morality. Up to this time, he seems to have been satisfied with explaining sexual dimorphism and the evolution of the different human races. In April 1869, however, he obtained Wallace's review of Lyell, in which his friend's heretical views about man were openly declared. A year later, Wallace sent him a copy of *Contributions to the Theory of Natural Selection*, which contained the essay "Limits of Natural Selection as Applied to Man" and a 'spiritualized' revision of his 1864 paper "Origin of the Human Races." By January 1870 (and probably before), he had read Greg's article on "The Failure of Natural Selection in the Case of Man" and the several responses to it.[19] And between December 1869 and March 1870, he received the first three installments of the second and greatly expanded version of Spencer's *Principles of Psychology*. Thus through 1869 and into early 1870, the pressure grew extreme. Darwin could not easily avoid giving a detailed evolutionary interpretation of human mind and morals.

In spring of 1870, a few months before completing the *Descent* (in August), Darwin wrote his daughter Henrietta, asking her criticism of two chapters of his manuscript. He feared parts were too like a sermon. "Who would ever have thought," he wondered, "that I sh'd turn parson?"[20] The chapters (2 and 3 of the first edition) dealt with the evolution of man's rational and moral faculties. His request of his daughter

17. Charles Darwin to Caroline Kennard (9 January 1882), in DAR 185, the Darwin Papers, Cambridge University Library.

18. Darwin's "Journal" entry for 1869 reads: "Feb. 11th Sexual Selection of Mammals & Man & Preliminary Chapter on Sexual Selection (with 10 days for notes on Orchids) to June 10th when I went to North Wales. On Augt 4 recommenced going over all chapters on Sexual Selection" (de Beer, p. 18).

19. On 28 January 1870, Francis Galton wrote to inform his cousin that Greg was the anonymous author of the article. The letter is in DAR 80.

20. Charles Darwin to Henrietta Darwin (spring 1870), held at British Library, MS. Add. 58373.

and the events of 1869 and 1870 argue that the chief problems of part 1 of the book, the evolution of intelligence and moral sense, were taken up last, only after (or likely because) Greg and Wallace had made their challenges. Darwin, of course, may have intended all along to explore the genesis of man's distinctive faculties. He did have early on the stimulus of Lyell's and Gray's theological versions of human reason and morality. And there was the persistent protest of Spencer that only inherited habit could explain the evolution of complex traits, such as the higher mental powers. It seems nonetheless likely that Darwin formulated his ideas with special attention to the troubling observations of Greg and the infanticidal claims of Wallace. This conclusion will be further supported by results from an analysis of the first part of the *Descent,* particularly Darwin's theories of the evolution of human intelligence and morality.

Darwin's *Descent of Man*

Structure of Part 1 of the *Descent*

The first chapter of the *Descent of Man* advances homologous traits, similar stages in embryonic development, and rudimentary organs as strong evidence for the descent of man and the higher apes from the same primitive mammalian stock. Man shares a vertebral structure with the apes, often suffers from the same diseases (e.g., "hydrophobia, variola, the glanders, syphilis, cholera, herpes"), which implies kindred blood and tissue types, and even engages in similar courtship, mating, and child-rearing rituals. Darwin maintained, as he had in the *Origin,*[21] such homologous structures implied that the respective lineages diverged from a common ancestor. Turning to the evidence of embryology, he drew on studies by Thomas Huxley, Richard Owen, Jeffries Wyman (the teacher of William James), and Ernst Haeckel, which showed that man and the higher animals went through virtually identical stages of early fetal development. Huxley suggested[22]—and Haeckel proposed it as a biological principle (the "biogenetic law")[23]—that

21. Charles Darwin, *On the Origin of Species* (London: Murray, 1859), chap. 13. This chapter presents evidence from morphology, embryology, and rudimentary organs for the common descent of creatures.

22. Thomas Huxley, *Evidence as to Man's Place in Nature* (London: Williams & Norgate, 1863), pp. 79–83.

23. Haeckel discussed what he later termed the biogenetic law in his *Generelle Morphologie* (Berlin: Reimer, 1866), 2:300. He explicitly promulgated the law in 1874, in his *Anthropogenie oder Entwicklungsgeschichte des Menschen:* "The fundamental law, upon which we must ever return and upon whose recognition our complete understanding of evolutionary history depends, can be expressed briefly in the proposition: The history of

these similar developmental stages could be explained if we supposed that variations supervened rather late in development, pre- and postnatally, so that common ancestral forms would be preserved in early stages. Examples of rudimentary organs in man held the greatest fascination for Darwin, and he spent the bulk of the first chapter of the *Descent* discussing them. The simian features of human ears, the ability of some people to wiggle theirs, instances of hairy-shouldered men, reduced molars, the veriform appendix, human tailbones, rudimentary male mammae—all became *lusus naturae,* except under the evolutionary hypothesis. These traits could be easily explained, however, by supposing that the ancestors of men had them in more perfect form, but that when early men changed their habits, the traits became atrophied through generations of disuse or through natural selection. Only our "natural prejudice," Darwin thought, led us to avoid the obvious conclusion of the common descent of man with the other animals.

Chapters 2 and 3 turn from man's physical frame to his mental and moral character. I will focus on these chapters and chapter 5, which continues their discussion, in a moment. Chapter 4, which in the second edition of the *Origin* is relocated more reasonably as chapter 2, brings the causal armamentarium of the *Origin*—that is, the laws of variability and heredity, the effects of the environment and habit, and natural selection under various conditions—to account for the modified descent of man's physical structure. Under the rubric of the laws of variability, Darwin rehearsed those possible sources of variation in organisms, finding that they most likely affected man as well as other animals. So variegated human traits might well be due to the direct effects of the environment, habits of use and disuse (thus the children of watchmakers tend to be shortsighted), arrested development, reversions, and correlation of parts. These sources, Darwin surmised, furnished the variations upon which natural selection operated to bring man up from his apelike progenitors.

Chapter 6 attacks the problem of human origins from the perspective

---

the embryo is an expression of the history of its lineage [Stamm]; or in other words: Ontogeny is a recapitulation of phylogeny. Stated more fully, it may be put: The series of forms which the individual organism passes through during its development from a single cell to its completed condition is a short, compressed repetition of the longer series of forms through which the animal predecessors of the organism, or the lineage forms of the species, have passed from the earliest times of organic creation to the present." See Ernst Haeckel, *Anthropogenie oder Entwicklungsgeschichte des Menschen,* 4th ed. (Leipzig: Engelmann, 1891), 1:7. Stephen Jay Gould discusses Haeckel's biogenetic law and its fate in *Ontogeny and Phylogeny* (Cambridge: Harvard University Press, 1977). A different slant on Haeckel's law is provided by Ruth Rinard, who connects it with the ideal morphology tradition. See her "The Problem of the Organic Individual: Ernst Haeckel and the Development of the Biogenetic Law," *Journal of the History of Biology* 14 (1981): 249–276.

of zoological systematics. The unbiased mind, Darwin felt, had to acknowledge that the differences between man and the higher apes were considerably less—in both anatomical and mental characters—than between, say, the scale insect *coccus* and the ant. Yet while systematists placed *coccus* and ants in the same class, they have made of man a separate order. For this, Darwin insisted, there was no justification. In chapter 6 he also traced the systematic relations of mammals and their predecessors, and ventured the educated guess that man had achieved his human form in Africa—an insight further illustrating the hypertrophied instincts of a fact-sniffing genius.

In the final chapter of part 1, Darwin addressed the question posed by Wallace in his 1864 paper: Whether the human races formed one species of many varieties or many distinct species? He weighed the evidence on both sides of the question, noting on the one hand the strikingly different characters of the several races and on the other the smooth gradations linking the races and the complete hybrid fertility among them. Darwin found especially significant the studies of such anthropologists as E. B. Tylor (*Early History of Mankind*) and his Downe neighbor Sir John Lubbock (*Prehistoric Times*).[24] Anthropological evidence of common social practices, psychological attitudes, and mental habits of widely divergent races bespoke a specific kinship. But since the concept of species itself remained vague and drained of importance by evolutionary theory, "it is almost a matter of indifference whether the so-called races of man are thus designated, or are ranked as species or sub-species." Though if required to select, Darwin preferred the category of subspecies.[25]

Darwin concluded his considerations in chapter 7 with the problem that furnished the pretext for the expansive treatment of sexual selection in part 2 of his book: How are we to account for the several races of men and the distinctive features of the two sexes? The differences in stature, structure, skin, hair, and habits that distinguish human groups could not be due, Darwin judged, to natural selection, since such traits appeared to have little bearing on survival. Nor, it seemed, could the male's superior strength, intelligence, and artistic abilities be explained simply by invoking survival of the fittest. But Darwin thought both could be explained by the mechanism of sexual selection, through either male and female choice or male combat. To reach these conclusions about human racial traits and sexual dimorphism in chapters 19 and 20,

24. George Stocking supplies a historically acute account of Tylor and Lubbock in his *Race, Culture, and Evolution,* 2d (Chicago: University of Chicago Press, 1981).
25. Darwin, *Descent of Man* 1: 235.

he amassed eleven chapters bulging with studies of sexual selection in insects, fish, reptiles, birds, and mammals. In sheer page numbers, man was swamped in butterflies, lizards, and hummingbirds. As might be expected, though, problems of human evolution, especially the question of evolution of mind and ethical behavior, captured the attention of most of Darwin's readers. But before turning directly to Darwin's account of human mental and moral evolution—and the critical reaction it evoked—a moment should be spent on a theoretical issue, certainly germane to human evolution, which surfaces in the fourth chapter of the *Descent*. It involves the relative explanatory power to be accorded natural selection.

The Role of Natural Selection in Evolution

Though Darwin gave natural selection and sexual selection the principal shares in his causal account of human evolution, he also recognized other factors. The direct effects of the environment and inherited habit, he believed, not only provided variations for selection to operate on but themselves functioned as transforming forces. The assignment of explanatory weight to these elements in the study of human evolution offered him occasion to reflect more generally on the question of appropriate balance. In an often cited passage from the fourth chapter of the *Descent*, Darwin confessed:

> in the earlier editions of my 'Origin of Species' I probably attributed too much to the action of natural selection or the survival of the fittest. . . . I may be permitted to say as some excuse, that I had two distinct objects in view, firstly, to shew that species had not been separately created, and secondly, that natural selection had been the chief agent of change, though largely aided by the inherited effects of habit, and slightly by the direct action of the surrounding conditions. Nevertheless I was not able to annul the influence of my former belief, then widely prevalent, that each species had been purposely created; and this led to my tacitly assuming that every detail of structure, excepting rudiments, was of some special, though unrecognized, service.[26]

In his own time, Darwin may well have been the shrewdest man alive in conceptually dissecting the problems of animal structure and behavior, but when anatomizing his own theoretical practice, he wielded a very blunt scalpel. In the early editions of the *Origin*, he did attempt to explain most adaptations by use of natural selection, but he certainly admit-

26. Ibid., pp. 152–53.

ted the significance of other factors. For example, he allowed that "habit, use, and disuse, have, in some cases, played a considerable part in the modification of the constitution, and of the structure of various organs."[27] And in explaining the instincts of domestic animals, he forthrightly acknowledged the role of inherited habit: "In some cases compulsory habit alone has sufficed to produce such inherited mental changes; in other cases compulsory habit has done nothing, and all has been the result of selection, pursued both methodically and unconsciously; but in most cases, probably, habit and selection have acted together."[28]

When Darwin composed the above-quoted passage of the *Descent,* recent struggles with critics burned freshly in his mind. Some of the objectors were themselves evolutionists of sorts, but certainly not ultra-selectionists. For example, St. George Mivart, the Catholic lawyer-zoologist whose often disingenuous attacks caused Darwin much grief, lodged an objection that would continue to vex Darwinians through the turn of the century. He marshalled examples of complex adaptations whose first stages provided no benefit without the later stages. These initial traits, being of no advantage, could not, therefore, be selected; and so the system of coadaptations could never be introduced—at least not by natural selection. In the fifth and sixth editions of the *Origin,* Darwin faced off against such cases from Mivart as that of eye location in flatfish. The early stages in the transition of an eye from one side of the head to the other, Mivart argued, could be of no earthly use to the fish. He attributed the migration to an "inherent tendency" of the organism, whose source science could not fathom, though religion could. Darwin responded to examples of this kind by admitting the explanatory inadequacies of natural selection. But, as made plain in several places in the *Origin,* he never claimed natural selection to be the sole cause of evolutionary change, only the most pervasive. The flatfish, in Darwin's view, did not suffer from a theological "tendency," but simply from the habit of trying to look upward with both eyes while resting on its side at the bottom. The evolved modification could be explained through inherited habit. As objections of the flatfish kind mounted, Darwin more often employed mechanisms that had always been a part of his theory, but never before so often required.

Darwin's own estimation that his early explanatory efforts had displayed a residual tincture of natural theology may have had some justice. But I think more potent forces led him to 'exaggerate' the importance of natural selection. Surely the very nature of the explanatory enterprise

27. Darwin, *Origin of Species,* pp. 142–43.
28. Ibid., p. 216.

compels the scientist to attempt an account of all relevant data. Moreover, the demonstrative requirements for fielding a new theory quite naturally would lead him to apply the chief principle of that theory in all quarters. Finally, Darwin's success in giving ingenious natural selection accounts (e.g., the instincts of neuter insects) would encourage him in an unstinting effort to reduce initially recalcitrant data. When he later called up the more latent resources of his theory of evolution (e.g., direct effects of the environment and inherited habit), he certainly did not play false to his discovery. Some historians, however, have calculated that under the brunt of attack and with the special difficulties he faced in the *Descent,* Darwin allowed Lamarckian mechanisms to usurp the role of natural selection. The quotation from the *Descent* appears to them to admit as much. They see the *Descent* as marking Darwin's decline into his dotage.[29] But a more careful evaluation of Darwin's self-judgment and a sustained analysis of his finely elaborated theory of human mental and moral evolution dispose, I believe, of that historiographic conceit.

The Argument from Common Mental Abilities

Logic of the Argument

In chapter 2 of the *Descent,* Darwin continued the strategy of his first chapter, where he had argued from homologous anatomical structures in men and animals to a common ancestor. He now focused on homologous mental structures, detailing the intellectual heritage men shared with the lower animals. Therewith he set out on the final stretch of an evolutionary path, along which not even some of his strongest

---

29. Commenting on Darwin's controversy with Mivart, Gertrude Himmelfarb, in *Darwin and the Darwinian Revolution* (New York: Norton, 1968), thinks that "more and more, the Lamarckian principle of the inherited effects of use and disuse came to replace natural selection" (p. 367). This view is shared by Loren Eiseley, who observes in *Darwin's Century* (New York: Doubleday, 1961): "Darwin was essentially a transitional figure standing between the eighteenth century and the modern world. He had never entirely escaped certain of the Lamarckian ideas of his youth . . . As a consequence it is not surprising that in time of stress he grew doubtful that natural selection contained the full answer to the sallies of his critics" (p. 245). Peter Vorzimmer, in *Charles Darwin: The Years of Controversy* (Philadelphia: Temple University Press, 1970), concurs in the sentiments of Himmelfarb and Eiseley. He imagines that Mivart's criticisms brought Darwin to a "state of frustrating confusion" (p. 251), and that as a result of the attack, the old biologist had to rely on Lamarckian mechanisms and "never succeeded in reestablishing either the necessity or the sufficiency of the thesis of natural selection" (p. 249). The pleasure that comes from smashing the carbuncles of great men is, however, fleeting.

supporters—Lyell and Wallace in particular—could travel. It led to a thoroughly naturalistic account of the very defining qualities of human beings. Some modern readers have also been wary of what awaits at the end of this evolutionary road, but for them to object straight on might betray an unenlightened metaphysics. So they usually scruple for a reason other than that of Darwin's contemporaries. They complain of Darwin's use of anecdotal evidence in support of his genetic explanation of the animal origins of human mind and morals. Himmelfarb, for instance, ever sensitive to "Darwin's failures of logic and crudities of imagination," found his argument in this chapter to reveal the naked defects of the theory itself.[30] For the modern critic, it is easy to become bemused and then to cluck at Darwin's stories of the humanlike traits of animals, while missing the logical force that lies behind them.

Darwin assembled his evidence to show that the elemental emotions—fear, courage, affection, shame, and the like—and the basic mental faculties of imitation, attention, imagination, and reason were possessed by animals as well as by man. He drew from scattered literature tales of the grief of female monkeys for the loss of their offspring, the curiosity of young apes, the jealousy and shame of dogs, the keen memories of a host of creatures, and the reasoning abilities of higher animals, which "few persons any longer dispute."[31] His account was persuasive—at least for many of his Victorian readers.[32] For Darwin compiled his examples from recognized, authoritative sources—from such naturalists as John Blackwall, Christian Ludwig Brehm, Pierre Huber, J. C. Jerdon, Charles-Georges Le Roy, John Lubbock, Charles Lyell, L. H. Morgan, Georges Pouchet, Johann Rengger, E. B. Tylor, Carl Vogt, Alfred Wallace, and men of somewhat lesser fame. And generally, the individual cases recounted were unexceptional—at least most Englishmen of a natural-history bent would not have demurred. Once

30. Himmelfarb, *Darwin and the Darwinian Revolution*, pp. 370–75. Even Peter Bowler, an historian in complete sympathy with the basic features of evolutionary theory, unlike Himmelfarb, yet concludes that in the *Descent* Darwin "fell into the all-too-obvious trap of anthropomorphism in his anxiety to make the case for evolution." See Peter Bowler, *Evolution: The History of an Idea* (Berkeley: University of California Press, 1984), p. 219.

31. Darwin, *Descent of Man* 1:46.

32. Even Darwin's harshest critics did not object to his attribution of emotional and elementary intellectual states to animals. The writer for the *Edinburgh Review*, who believed Mr. Darwin's thesis "has broken down at every point where it has been tried," yet happily added his own stories of animal character and talent, concluding that "it is universally allowed that in all these particulars the mental constitution of man strongly resembles that of the higher animals." See "Darwin on the Descent of Man," *Edinburgh Review* 134 (1871): 195–235. Quotations are from pp. 235 and 210.

Darwin had gathered and designedly arranged his evidence, however, the consequence slipped out: no insuperable barrier, on this level at any rate, prevented the descent of man from the lower animals. In constructing his argument, Darwin applied a tactic—surely unconsciously—that is stitched into virtually every scientific explanatory attempt, namely, beginning with evidence that competent professionals could accept. Darwin's peers would not have quibbled with the kind of evidence he produced, though we might. But standards of evidentiary acceptability evolve over time; it bespeaks a failed imagination and crude logic to demand that all earlier standards meet the tests of future science.

A closer examination of some of Darwin's arguments, however, reveals assumptions that most modern biologists of behavior would themselves not hesitate to make. First is Darwin's general theory of psychological attribution, that is, the principles he employed to justify attributing different mental states to other creatures. Today few would deny that we predicate psychological traits of other people, as well as of animals, on the basis of manifest behavioral responses and similarity of nervous systems. Thus when I ask my friend to close the window, and he gets up and shuts it, I unblinkingly presume he has understood my request. The same kind of evidence would appear to warrant a similar presumption when my dog, upon command, retrieves the paper—at least this is the conviction of a considerable number of contemporary behavioral biologists.[33] Of course, we may be as mistaken in the one case as in the other. And we cannot count on the same subjective states occurring in a human being and in a member of a different species. But then we cannot with perfect confidence count on the same subjective states occurring in any two people, even under like circumstances. The opposing parties in the current debate over animal language, whether chimps (or Skinner's pigeons) can use signs as we do, agree that understanding and linguistic ability will be demonstrated if the animal's behavior meets certain standards. Most modern psychologists employ such behavioral criteria, regardless of their finer-grained theories of mentality—that is, whether they are behaviorists or some brand of mentalist.[34] Similarly Darwin, working in the British empiricist tradi-

33. See, for example, Donald Griffin, *The Question of Animal Awareness: Evolutionary Continuity of Mental Experience* (New York: Rockefeller University Press, 1976), especially pp. 68–71. Thomas Nagel, in his instructive article "What Is It Like to be a Bat?" *Philosophical Review* 83 (1974): 435–50, makes the further point that simply because we cannot well imagine what the subjective states of animals are like—nor frequently those of other human beings—is no reason to reject the idea that they have subjective states.

34. The Skinnerian behaviorist may deny that "linguistic understanding" depends on

tion, assumed that a psychic trait could legitimately be attributed to other creatures if their behavior met certain standards. The modern reader of Darwin's chapter may, of course, dispute his claim that animals do in fact reliably meet the criteria or that the criteria are adequate. But that is a matter quite other than poverty of imagination or faulty logic.[35]

A second assumption that would find a number of contemporary advocates concerns the theory of intelligence that underlay Darwin's examples. British psychologists working in the sensationalist tradition—Locke, Hume, Erasmus Darwin, James and John Mill, Spencer, and their philosophic descendants—have conceived rational intelligence as a matter of association of faded sensory images. While modern psychologists and philosophers may have enhanced this basic theory, the modifications do not preclude animals' enjoying a modicum of reason. Thus Darwin's recitation of cases of animal intelligence, for the bulk of his fellows—and for the less captious of his modern readers—really served more to illustrate than to demonstrate; for the fact of animal reason, as Darwin confidently pointed out, "few persons any longer dispute."[36]

Tool Use and Sense of Beauty

But Darwin recognized that it would not be enough, at least for some (including, perhaps, Lyell and Wallace), to show that animals and men shared the same basic mental abilities. Human beings, after all, were capable of apparently distinctive mental acts of a higher sort: they used tools, had language, appreciated beauty, and displayed religious feeling. At the end of the second chapter, Darwin considered each of these four classes of mental activities and attempted to show that their rudiments could be found in animals below man. Concerning tool use, he had

---

any nonphysical specter hovering about cortical neurons, but his theoretical parsimony extends to men as well as animals.

35. A wonderful example of Darwin's efforts to go beyond description came in his support for a tale he drew from Brehm's *Thierleben*. The great nineteenth-century German naturalist had recounted the story of an affectionate baboon that had adopted a kitten. The kitten scratched the baboon, who, as Darwin tells it, "certainly had a fine intellect, for she was much astonished at being scratched, and immediately examined the kitten's feet, and without more ado bit off the claws" (*Descent* 1:41). One of Darwin's critics (St. George Mivart in an anonymous review), denied that such display of intelligence could have taken place, since it would have been impossible for a baboon to have clipped a kitten's claws with its teeth. In the second edition of the *Descent* (in the *Origin of Species and the Descent of Man* [New York: Modern Library, 1936], p. 449) Darwin provided the crucial experiment, as he related in a note: "I tried, and found that I could readily seize with my own teeth the sharp little claws of a kitten nearly five weeks old."

36. See chapter 2 for further discussion of this issue.

little difficulty culling stories of apes manipulating sticks and pebbles for a variety of instrumental purposes—some that recall Wolfgang Köhler's experiments demonstrating the insightful use of tools by chimps.[37] Nor would the aesthetic sense appear to distinguish man. For if a sense of beauty meant—as it inevitably did for the British empiricist—a nonrational delight in particular colors, sounds, and shapes, then animals quite obviously possessed this sense; indeed, as Darwin delightfully suggested, a woman who decorated her head with brightly colored plumes might find both men and birds admiring her display.[38] When he took up the questions of religious feeling and language, however, his technique ascended from gathering persuasive cases to very shrewd analysis.

Religious Sentiments

Darwin trod carefully, but, I think, with some gleeful malice in approaching the topic of religion. His strategy was to corrode by insinuation and barely visible logic the usual barriers surrounding the discussion of religious belief. He quickly and defensively admitted that the question of aboriginal religious sentiment and belief appeared "wholly distinct from that higher one, whether there exists a Creator and Ruler of the universe." That latter question, he declared, "has been answered in the affirmative by the highest intellects that have ever lived."[39] Darwin wished to deal explicitly only with the problem of the religious sentiments of savages, though implicitly to suggest an answer to the question of God's existence quite different from that given by the high intellects.

The theologically disposed Victorian could dismiss as rank superstition the native's beliefs in spirits and ghostly beings, yet, as Darwin knew, most would recognize that the primitive's feelings were of a genuinely religious cast. He proposed that as soon as the faculties of imagination and wonder had developed, along with some reasoning ability, early man, like the contemporary savage, would have begun to speculate on terrifying natural phenomena, as well as on his own existence. Yet the tendency of primitives to imagine that natural objects were animated by spirits seemed, Darwin observed, little different from that exhibited by his own dog, which barked and growled at a parasol blown by the wind, apparently believing that its flight indicated the presence of some invisible agent. Moreover, the feeling of religious devotion,

37. See Wolfgang Köhler, *The Mentality of Apes* (New York: Harcourt Brace, 1925).
38. Darwin, *Descent of Man* 1 : 63.
39. Ibid., p. 65.

which he analyzed as consisting of "love, complete submission to an exalted and mysterious superior, a strong sense of dependence, fear, reverence, gratitude, hope for the future, and perhaps other elements"—such feeling bore strong resemblance to that simpler emotional complex displayed by his dog's worshipful devotion to its master.[40] Darwin's comparison surely struck a resonant cord in the heart of every English huntsman.

Most of the *Descent*'s more pious readers would not likely have taken offense at the suggestion that savage man's sentiments and beliefs hardly differed from those of an English hound.[41] Darwin would not, therefore, have excited the zealous by explicitly drawing the conclusion that primitive religious feeling and credulity were not unique to man. But that conclusion, coupled with Lubbock's, McLennan's, Spencer's, and Tylor's anthropological studies of the evolution of religious creeds—to which Darwin constantly referred his reader—brought the further, implicit inference that the Christian's disposition to believe in higher powers might have decidedly ancient roots.[42] Religious sentiments, therefore, proved no obstacle to the evolutionary hypothesis as applied to man.

Language

Darwin had initially connected language and the evolutionary hypothesis in an oblique way. In the *Origin of Species*, he used the genealogical relationships of languages to illustrate how the goal of the early systematists—that of achieving a "natural" classification of species—might be achieved by his theory. In the *Origin*, descent and modification of languages supplied a model for the descent and modification of species.[43] He also compared the retention of unpronounced

40. Ibid., pp. 67–68.

41. The reviewer of the *Descent* for the *Athenaeum* did not object to the comparison between the sentiments of dogs and primitive men; rather he argued that "what is said about savages is beside the mark as to true religion and true belief in God." See "The Descent of Man," *Athenaeum* 4 (March 1871), p. 275.

42. Darwin cited E. B. Tylor, *Researches into the Early History of Mankind*, 2d ed. (London: Murray, 1869); John McLennan, "The Worship of Animals and Plants," *Fortnightly Review* 12 (1869): 407–27, 562–82; 13 (1870): 194–216; Herbert Spencer, "The Origin of Animal Worship," *Fortnightly Review* 13 (1870): 535–50; and John Lubbock, *Prehistoric Times* (London: Williams and Norgate, 1865) and *The Origin of Civilization* (London: Williams and Norgate, 1870). In this last, Lubbock explicitly supplied Darwin's suppressed premise: "Yet, while savages show us a melancholy spectacle of gross superstitions and ferocious forms of worship, the religious mind cannot but feel a peculiar satisfaction in tracing up the gradual evolution of more correct ideas and of nobler creeds" (p. 114).

43. Darwin, *Origin of Species*, p. 422–23.

letters in a word, which gave witness to its derivation—to rudimentary organs in animals, which also bespoke descent from a no-longer-extant ancestor.[44]

Charles Lyell, in his *Antiquity of Man*, developed Darwin's passing suggestions into a more explicit, if highly qualified, argument for his friend's theory. Lyell's chapter "Origin and Development of Languages and Species Compared" displayed both the force and the hedging subtlety of a man trained in the craft of the Old Bailey.[45] He showed that those features of the evolutionary hypothesis that provoked major objections also characterized historical linguistics: ancient linguistic forms appeared to be distinct from modern forms; there were wide gaps between dead and living languages, with no transitional dialects preserved; and the forces altering contemporary languages seemed impotent to produce new languages from old. But competent linguists, he maintained, did not doubt the descent of modern languages from more primitive ones. Therefore, an objection overruled in the one case should not be sustained in the other.

But thus far, Lyell had merely brought to the bar the precedent of Darwin's earlier analysis, though now considerably elaborated and ably defended. The real originality of Lyell's case rested in his arguing a common explanation for the two kinds of descent, linguistic and biological. The development of languages and their proliferation were due, he proposed, to "fixed laws in action, by which, in the general struggle for existence, some terms and dialects gain the victory over others."[46] Lyell contended that the processes of biological evolution were thus comparable to those of linguistic evolution—in both the more fit types were selected. Yet in working out this novel support for Darwin's theory, Lyell hesitated: the principle of natural selection alone appeared unable to account completely for the intricately designed fabric of language, even that of savages, or for the similarly refined intellectual and moral faculties characterizing the human species. Lyell judged (as Darwin groaned his great frustration) that the natural selection of language and life forms could only be a secondary cause, operating under the guidance of higher powers. "If we confound 'Variation' or 'Natural Selection' with such creational laws," he cautioned, "we deify secondary causes or immeasurably exaggerate their influence."[47]

During the same year that Lyell published his *Antiquity of Man*, 1863,

44. Ibid., p. 455.
45. Charles Lyell *The Geological Evidences of the Antiquity of Man* (London: Murray, 1863), chap. 23.
46. Ibid., p. 463.
47. Ibid., p. 469.

Figure 5.1   Charles Lyell, 1797–1875, sketch done in 1853.

the eminent philologist August Schleicher addressed to his colleague at Jena, Ernst Haeckel, a tract that undertook to apply Darwinian theory to account for language descent and diversity. In *Die Darwinsche Theorie und die Sprachwissenschaft,* Schleicher, like Lyell, compared the geneal-ogy of species to that of languages and proposed a natural struggle for existence as the engine propelling both.[48] "Languages," he declared, "are natural organisms which, without any determination from the hu-man will, have emerged and, in accord with determinate laws, have

48. August Schleicher, *Die Darwinsche Theorie und die Sprachwissenschaft* (Weimar: Bö-lau, 1863).

grown, evolved, and finally have become old and died."[49] Even the great philologist Friedrich Max Müller, who denied that the mental gap between animals and man could be bridged, found Darwin's mechanism an attractive explanation for the linguistic evolution that *followed the advent* of human speech.[50] He differed from Schleicher, though, in holding that the struggle went on among words and grammatical forms within languages, rather than directly between distinct languages. But the developmental consequences, he thought, would be the same.

Darwin had planted the seed of the comparison between biology and language, and now in the *Descent* he reaped the fruit. Citing Lyell, Schleicher, and Max Müller, he endorsed the arguments for common explanatory principles. On the one hand, these arguments further supported his general theory; and on the other, they suggested that language development could be given a perfectly naturalistic account. But to sustain this latter conclusion fully, he had to deal with the reservations of Lyell and Max Müller. Both men regarded language as the exclusive possession of man. Max Müller, striking the most combative stance of the two, pushed forward the human accomplishment as the decisive challenge to the theory of evolution:

> Where, then, is the difference between brute and man? What is it that man can do, and of which we find no signs, no rudiments, in the whole brute world? I answer without hesitation: the one great barrier between the brute and man is *Language*. Man speaks, and no brute has ever utter a word. Language is our Rubicon, and no brute will dare to cross it. This is our matter-of-fact answer to those who speak of development, who think they discover the rudiments at least of all human faculties in apes, and who would fain keep open the possibility that man is only a more favoured beast, the triumphant conqueror in the primeval struggle for life.[51]

49. Ibid., pp. 6–7.

50. Friedrich Max Muller, "The Science of Language," *Nature* 1 (1870): 256–59. Max Muller had no objection to explaining the development and differentiation of language through natural processes. But he believed that a divine source had to be sought for the origin of the human species, whose reason and language ability set it distinctly apart from lower kinds. See Friedrich Max Muller, *The Science of Thought* (New York: Charles Scribner's Sons, 1887), 1:149–75. See also the review article by P. Giles, "Evolution and the Science of Language," in *Darwin and Modern Science*, ed. A. C. Seward (Cambridge: Cambridge University Press, 1909), pp. 512–28.

51. Friedrich Max Müller, *Lectures on the Science of Language delivered at the Royal Institution of Great Britain in April, May, & June, 1861*, 4th ed. (London: Longman, Green, Longman, Roberts, & Green, 1864), pp. 367–68. Elizabeth Knoll dexterously discusses Max Muller's reaction to evolutionary theory in her "The Science of Language and the

This neo-Kantian philologist declared, turning up a great metaphysical nose at British empiricism, that animals, though capable of intelligent manipulation of sensory images, were yet devoid of the higher faculty of reason.[52] Language marked the great divide, since, "it is as impossible to use words without thought, as to think without words."[53] Language and thought had to make their appearance simultaneously and instantaneously, so no gradual evolution of either was possible. Moreover, as Lyell added, the perfection of language precluded a simple, mechanistic explanation, of the kind his friend had proposed. The tongues of men had to be loosed by higher powers.

Darwin responded to these philological objections—that is, those based on the necessity of saltation and perfection—in a style typical of his handling of others of this kind. He constructed a plausible account of the gradual origin for the trait in question and showed that perfection could be generated from imperfect antecedents. In the case of the intricacy of even the primitive's language, Darwin pointed out that complexity itself was no sure sign of perfection—a Crinoid consisted of up to 150,000 pieces of symmetrically tessellated shells, yet it was no advanced type. Rather a division of labor, the result of irregular terms, borrowings, bastardizations, and abbreviations—the emblems of speech in civilized nations—would better indicate the perfections of language, he thought. In any case, the savage's speech had already undergone generations of development.[54] So the real difficulty had to be resolved in an account of the beginnings of language.

Darwin advanced what Max Müller maliciously termed the "bow-wow" and "pooh-pooh" theories of language origins. Darwin suggested that our apelike ancestors might have begun to communicate by imitating the sounds of objects and by using the spontaneously emitted vocalizations induced by strong feelings. Might not these onomatopoeic expressions and interjections form the basic roots of language, which further linguistic evolution might modify? Hensleigh Wedg-

Evolution of Mind: Max Muller's Quarrel with Darwinism," *Journal of the History of Behavioral Sciences* 22 (1986): 3–22

52. Max Muller, ever the romantic, thought that had Darwin become acquainted with Kant's *Critique of Pure Reason* he would likely have given up his theory of human descent. See Max Müller, *Science of Thought* 1:150.

53. Friedrich Max Müller, *Lectures on the Science of Language delivered at the Royal Institution of Great Britain in February, March, April, & May, 1863*, second series (London: Longman, Green, Longman, Roberts, & Green, 1864), p. 72. Darwin quoted Max Müller's epigram in the second edition of the *Descent* (p. 464), adding: "What a strange definition must here be given to the word thought!"

54. Darwin, *Descent of Man* 1:61–2.

wood believed so, and Darwin cited his cousin's detailed defense of these hypotheses against Max Müller.[55] Darwin admitted that our progenitors must have evolved considerable intellectual powers before employing anything like human speech; but once the threshold was passed over, the naturally evolving structure of language would rebound on mind, leading to greater development of the rational faculty:

> The mental powers in some early progenitor of man must have been more highly developed than in any existing ape, before even the most imperfect form of speech could have come into use; but we may confidently believe that the continued use and advancement of this power would have reacted on the mind by enabling and encouraging it to carry on long trains of thought. A long and complex train of thought can no more be carried on without the aid of words, whether spoken or silent, than a long calculation without the use of figures or algebra.[56]

In this passage Darwin laid down an assumption that would ground his extremely important accounts of the natural selection of high intelligence and the moral sense in chapters 3 and 5 of the *Descent*. The assumption was simply that social and cultural relationships constituted the proximate environment for the evolution of those traits we think of as distinctively human. In this specific case, however, he suggested that the environment acted not as the agent of selection, but as the stimulus for use inheritance: continued efforts at communication, using a naturally evolved linguistic structure, must reciprocally influence the phylogenetic growth of brain, the organ of mind.[57]

In the *Descent of Man*, Darwin often enough fell back on the doctrine of use inheritance to tidy up his account of the higher mental and moral

55. Hensleigh Wedgwood, *On the Origin of Language* (London: Trubner, 1866). Darwin, of course, adopted his cousin's theory, since it was the most congenial to his view of human evolution. He later admitted to Max Müller that his competency on the question was negligible; but if one were "fully convinced, as I am, that man is descended from some lower animal, [he] is almost forced to believe *a prior* that articulate language has been developed from inarticulate cries." See Charles Darwin to Max Müller (3 July 1873), *More Letters of Charles Darwin*, ed. Francis Darwin (New York: D. Appleton, 1903), 2:45.

56. Darwin, *Descent of Man* 1:57. While preparing this chapter of the *Descent*, Darwin peppered Hensleigh Wedgwood with inquiries about language. Here he may have tacitly adopted a suggestion of his cousin, who thought that the complexity of the natives' speech did not signify an advanced intellect, rather that highly developed language made civilized mind possible. Wedgwood wrote to Darwin: "But if you suppose speech to have actually grown from the beginning, the supposition of an originally civilized condition seems wholly cut away as it is impossible to conceive a civilized state antecedent to the acquisition of speech" (Hensleigh Wedgwood to Charles Darwin [1870], in DAR 80).

57. Darwin, *Descent of Man* 1:58.

faculties, just as he did in the *Origin* to explain certain morphological traits. But he did not retreat to the doctrine in the way commonly supposed.[58] Rather, he pushed forth natural selection as the main force in his explanation of those crowning human faculties of intellect and moral sense. But he had to deploy his account so as to overcome the dangerous objections raised chiefly by his friend Alfred Wallace.

Evolution of the Moral Sense and Intelligence

Objections from the Friends of Evolution

In the previous chapter we examined several objections to the Darwinian hypothesis as it might be applied to man. The most pungent of these difficulties arose from Darwin's own colleagues and sympathizers. Lyell could not conceive that man's intellect and moral sensibility naturally grew by slow degrees from animal stock. Galton and Greg isolated another crucial problem for the Darwinian approach to man: as soon as protomen formed social bonds and through sympathy became solicitous for their mutual welfare, natural selection ought to be disengaged; for sympathy would prevent the salutary elimination of mentally and morally inferior individuals. Wallace, after his profound change of heart, pressed these difficulties home. He urged that man's great intellect and refined moral sense far exceeded what was required for mere survival in the wild; hence, natural selection could not have produced them.

These unpleasantries revealed by the friends of evolution flowed from the very bowels of the concept of natural selection. Darwin had to face the objections squarely. In meeting the challenge, he devised an ingenious explanation of the rise of human intelligence and morality. His account, though, was not always understood even by those seeking to adopt aspects of it, and it thus suffered distortion when incorporated into subsequent theories of human evolution—several of which we will examine in later chapters. Because of these distortions, or at least odd refractions, produced by later Darwinians, it became easier to reject efforts at a thoroughgoing evolutionary explanation of human mental and moral faculties, even while unflinchingly accepting evolutionary constructions of the physical man. Yet, Darwin's original formulation of his theory, in chapters 2, 3, and 5 of the *Descent,* has redoubtable strength, enough, I think, to overcome the objections usually raised against it. This is especially true of his conception of moral evolution. In the second appendix, I will try to show how it might function at the

58. See note 29 for mention of representatives of the craven school of Darwinian interpretation.

Figure 5.2   Alfred Russel Wallace, photograph from 1902.

heart of a philosophically acceptable evolutionary ethics. Now, at last, we should turn to his moral theory, upon which he spent most of his time, and then, in briefer compass, to his proposals about the evolution of intelligence.

Darwin's Theory of Conscience

Darwin expended considerable effort on a theory of moral evolution, because he judged the moral sense, or conscience, to be by far the most important distinguishing feature of human nature. It was a judgment he found endorsed by James Mackintosh, in whose philosophical shadow he stood, and by Kant, who provided the inspirational motif of his study. Darwin quoted the German philosopher's hymn to duty, which asked "whence thy original?" Darwin, in one of those rhetorically modest recommendations that have at times introduced the grand-

est of ideas, allowed as how his perspective furnished a unique answer
to Kant's question:

> This great question has been discussed by many writers of con-
> summate ability; and my sole excuse for touching on it is the
> impossibility of here passing it over, and because, as far as I
> know, no one has approached it exclusively from the side of
> natural history. The investigation possesses, also, some inde-
> pendent interest, as an attempt to see how far the study of the
> lower animals can throw light on one of the highest psychical
> faculties of man.[59]

Darwin's method of approach to the moral sense had already been
established during the period of his great creative effort, from late summer of 1838 through spring of the next year. As we have seen in chapter
2, he had interpreted Mackintosh's ethical theory in biological terms,
providing explanations of faculties and relationships left dangling in the
philosopher's account. He now resurrected those early ideas, but altered, reformulated, and greatly refined them. Their conceptual environment had markedly changed; not only were they now faced with
dangerous objections, but they had available a new resource—the concept of community selection.

Darwin began his reconstruction by postulating four stages in the
evolution of conscience. In the first, animals (our ancestors) would develop social instincts, which would initially bind together closely related
and associated individuals into a society. The second stage would arrive
when the members of this society had evolved sufficient intellect to recall instances when social instincts went unsatisfied because of the intru-
sion of momentarily stronger urges. The third stage would be marked
by the acquisition of language, which would enable these early men to
become sensitized to mutual needs and to be able to codify principles
of their behavior. Finally, habit would come to mold the conduct of
individuals, so that acting in light of the wishes of the community, even
in matters of small moment, would form a second nature. As is evident
from Darwin's descriptions, he conceived these stages as sequential, but
largely overlapping, with faculties in continuous interaction. He gave
most of his attention to the first two stages.

Sympathy as an Instinct

As a good ecologist, and more as a patient reader of natural history
literature, Darwin reproduced many instances of the social behavior of

59. Darwin, *Descent of Man* 1:70–71.

diverse animal types, from the foot stomping of rabbits signaling danger to an old male baboon's heroic rescue of an infant attacked by dogs. Most of these behaviors, particularly those of the lower animals, he thought to be instincts, social instincts. In classifying the social and sympathetic behaviors of animals as instinctive, Darwin thereby rejected the assumption of the sensationalist school that such conduct had been produced in each individual through associations of pleasure and pain.

The principal British psychologists and moral theorists of the period generally subscribed to the doctrines of sensationalism, at least in the case of human behavior. Adam Smith, whose views Darwin had initially adapted to his own early theory (see chapter 2), had explained our sympathetic response to another's distress as a consequence of our own similar experience: the evoked painful association would lead us to remove its source. Alexander Bain preserved this hypothesis in his account of social behavior.[60] And it was to Bain that Darwin usually turned for the orthodox psychological and moral opinions he wished to overthrow.[61] His objections to the theory of Smith and Bain were of two sorts, one simple and straightforward and one enmeshed in his own theory. The simple objection was that Smith and Bain could not explain why we should have a much stronger sympathetic response to a relative or friend than to a stranger.[62] On their account, the response ought to be the same, since it stemmed from our own memories and associations. The second rejoinder really amounted to a demonstration of the superiority of Darwin's own theory of conscience; briefly put, it was that social and sympathetic reactions were instinctive in character, not learned associations.[63]

60. Darwin read very carefully Bain's account of sympathy in *Mental and Moral Science* (London: Longmans, Green, 1868), pp. 276–83. Darwin's reading notes are in DAR 80.

61. Darwin thought he also detected the associationist hypothesis in John Stuart Mill's *Utilitarianism*. In a long note to p. 71 of the *Descent,* he cited Mill's belief that "'the moral feelings are not innate, but acquired.'" Darwin was not quite certain about the position of Mill, since the philosopher had also spoken of the social feelings as "'a powerful natural sentiment.'" In preparation of the second edition of the *Descent,* he asked his son William to attempt to puzzle out the theory. But William Darwin also found interpretation difficult, as wrote to his father: "It seems to me that he considers the social feelings in man the result of association and depends upon intellect to a great extent. Which is very extraordinary that he should recognize the social instincts to be natural to animals, which he can hardly put down to intellect, and should consider them almost entirely the result of intellect & association in man." After discovering passages where Mill suggested that the moral faculty might also have some roots in human nature, William concluded that Mill was "rather in a muddle on the whole subject." The correspondence of father and son is in DAR 88.

62. Darwin, *Descent of Man* 1:81.

63. Ibid., pp. 81 and 87.

As he had in his early theory of morals, Darwin based his argument for the instinctive nature of human sympathy largely on analogy: social and altruistic responses in animals were instincts, so why should they not also be in that most social of animals? It would be no objection that, though man was instinctively social, he yet made war on others. Darwin simply pointed out that among lower animals, "the social instincts never extend to all the individuals of the same species."[64] Nor was it a potent objection that men did not naturally display fixed patterns of social or sympathetic response. In the human species, Darwin held, the instinct amounted to a plastic urge, which required the guidance of reason.

The social instincts were not themselves sufficient to form what we recognized as conscience or the moral sense, else we would be forced to regard animals as moral beings. Darwin believed something more was required, and this was adequate intelligence. "Any animal whatever," he proposed, "endowed with well-marked social instincts, would inevitably acquire a moral sense or conscience, as soon as its intellectual powers had become as well developed, or nearly as well developed, as in man."[65]

Reason in Moral Theory

An evolved intellect played two critical roles in Darwin's moral theory. First, reason and experience would guide conduct that had been stimulated by social instinct: "Although man, as just remarked, has no special instincts to tell him how to aid his fellow-men, he still has the impulse, and with his improved intellectual faculties would naturally be much guided in this respect by reason and experience."[66] Darwin thought this guidance would become more routinized after speech had developed in early human groups, since the language of praise and blame would help channel conduct into stable though culturally diverse patterns.[67]

The second role Darwin assigned to developed intellect formed the kernel of his theory of conscience. An evolved intelligence, he argued, allowed the individual to compare an unsatisfied social instinct with a more powerful urge, such as hunger, fear, or the sexual itch, to which it had been sacrificed. This was the hypothesis that Darwin had developed very early in his reflections on the evolution of behavior, as will

64. Ibid., p. 85.
65. Ibid., pp. 71–2.
66. Ibid., p. 86.
67. Ibid., pp. 86, 164–66.

be recalled from chapter 2. He advanced it now in the *Descent* with little modification. He maintained that conscience ultimately derived from a persistent social impulse: when an intelligent creature contemplated giving in to a momentarily stronger urge—or actually did so—but then recollected and so rekindled the more persistent social instinct, the voice of duty, about which Kant rhapsodized, would be heard. Hence moral obligation ultimately stemmed, in Darwin's biological ethics, from the demands of social instincts: "The imperious word *ought* seems merely to imply the consciousness of the existence of a persistent instinct, either innate or partly acquired, serving him as guide, though liable to be disobeyed. We hardly use the word *ought* in a metaphorical sense when we say hounds ought to hunt, pointers to point, and retrievers to retrieve their game. If they fail thus to act, they fail in their duty and act wrongly."[68] What, to Darwin's mind, prevented us from according the hound moral stature was that it lacked the essential ability to reflect on its behavior, and so to reestablish a suppressed social instinct. In this respect, then, a moral being is "one who is capable of comparing his past and future actions or motives, and of approving or disapproving of them."[69] Had Darwin not insisted on this distinction, his theory would have been liable to the reductio ad absurdum that it would make ants and termites into ethical beings.

In formulating a dual role for intelligence in his theory of the moral sense, Darwin by no means meant to suggest that all moral conduct required rational deliberation. Indeed, he took this to be another consideration that distinguished his theory from that of the utilitarians. He believed (with some, though not entire, justice) the Benthamites supposed that a hedonistic calculation went on or should go on in advance of a moral act. But we would not hesitate, as Darwin pointed out, to consider moral a reflexive act of courage. When a man jumps into the river to save a drowning child, literally without giving thought to his own safety, we surely and correctly judge him virtuous. Darwin thought the utilitarians could not easily explain such immediate moral responses; though his theory could, since it interpreted moral acts as ultimately caused by impulse and habit.

The Role of Habit

After men had evolved social instincts, a considerable intelligence, and a faculty of language, they could be classified as authentically moral beings. The final distinguishable stage in evolution involved the refine-

68. Ibid., p. 92.
69. Ibid., p. 88.

ment of habit. Darwin believed that in the development of individuals, the frequent conflicts of strong passion and persistent social feeling would be resolved now in favor of the one, now of the other. But as an individual matured, especially as he became indoctrinated into the special mores of his tribe or class, he would then learn that it was better to forego even the pleasures of the strongest desires in order to enjoy the more lasting satisfactions of fulfilled social instincts. It was even likely, or so Darwin believed, that the habit of self-command—the automatic rejection of powerful desires when they competed with the quieter pleasures of sociability—would, with due repetition, itself become innate, ground into the heritable structure. In this way a Darwinian might explain the inbred restraint that governed the behavior of the Victorian gentleman.

Though Darwin gave some explanatory force to inherited habit, he denied that it formed the foundations of the moral sense. That was left to natural selection, in the guise of community selection. He thought he had a truly powerful account of the moral sense in community selection (which will be discussed in a moment), so he could yield to inherited habit the refined points of moral sensitivity, the delicacies and idiosyncrasies of taste that could have no conceivable use in struggles amongst men. But on this matter, he danced a diverting pas de deux with Herbert Spencer. In a letter to John Stuart Mill, which Bain had reprinted in his *Mental and Moral Science,* Spencer outlined a moral theory very similar to Darwin's.[70] They differed principally on the source of the moral sense: for Spencer it was inherited habit, for Darwin community selection. Darwin, citing Spencer's letter, approved of the general claim that the moral sense evolved, but disputed the mechanism of evolution. His major objection was simply that some bizarre customs and superstitions—for example, the Hindu's aversion to unclean food—were not inherited.[71] But Darwin, who certainly believed in the inheritance of acquired habits, gave no clue as to why such behavior as the Hindu's should not become an instinct. His major concern seems to have been to undercut Spencer and thus protect what he regarded as his most original contribution to ethical theory—the derivation of the moral sense from community selection.

Community Selection

Darwin's theory of conscience displays his particular genius. He reinterpreted a general philosophical system—the ethics of Mackin-

70. Spencer's letter and moral theory are examined in chapter 7.
71. Darwin, *Descent of Man* 1 : 101–3.

tosh—in biological terms; he sifted numerous volumes in natural history for pertinent evidence; he cautiously integrated into his conception the work of leading philosophers and social thinkers; he constantly balanced his theory against sober English intuitions about human nature; he persevered in the patient assembly and reassembly of his ideas over many years; and he created a novel moral theory of great power, some parts of which have been rediscovered by sociobiologists in our day.[72] Darwin also dealt squarely with the most serious objections to his hypothesis, the ones suggested by Lyell, by Galton and Greg, and particularly by Wallace. These went to the heart of his general theory of evolution and its particular application to man. They remain prototypes of objections most often brought against contemporary efforts to biologize ethics.

The objections were of three kinds. Lyell complained that the chasm between animal and human mind was too wide; nothing like moral behavior could be discovered in the animal kingdom. Darwin countered by tracing out the gradual transformation of low intelligence and social instinct into rational mind and moral sense. The second two objections proved more formidable, and their resolution consequently more dramatic. Greg and Galton suggested that the evolution of a moral faculty and high intelligence would not get beyond the first establishment of the social instincts, since the unfit would tend to be preserved. And Wallace judged that an exquisite moral sense or great intellectual capacity—such as civilized men did exhibit and natives held in reserve—brought no advantage for survival; indeed, altruistic behavior often harmed those persisting in it. St. George Mivart, for whom Darwin developed a deep loathing, pressed this last objection in his *Genesis of Species*. He argued that "on strict utilitarian principles," of which natural selection was only the biological expression, acts of altruism, which had to be useless to the individual, could not evolve. So we had to conclude, he insisted, "that admiration which all feel for acts of self-denial done for the good of others, and tending even toward the destruction of the actor, could hardly be accounted for on Darwinian principles alone."[73] The power of these objections, however, bled away under the force of Darwin's conception of community selection.

Darwin rooted the moral sense in the social instincts, as he had in his earlier constructions of moral theory. But between that time and the

72. Edward Wilson's theory of morality bears strong resemblance to Darwin's, though he does not seem to recognize any debt on this score to his predecessor. See Wilson's *On Human Nature* (Cambridge: Harvard University Press, 1978), pp. 149–67, for a brief discussion of Wilson's views; see also the second appendix.

73. St. George Mivart, *The Genesis of Species* (New York: D. Appleton, 1871), pp. 207–8.

period when he worked on the *Descent,* he had resolved a problem that allowed him to bring a powerful device to explain the origin of the social instincts and thus the foundations of the moral sense. The difficulty was the evolution of the instincts and anatomy of social insects, and his solution was community selection (see chapter 3). The challenge to an evolutionary construction of social-insect morphology exactly mirrored the most formidable objection to a natural selection account of the moral sense: in both cases natural selection working on the individual could not produce the traits in question. But now the explanation of the one, opened the way to the explanation of the other, as Darwin perceived:

> With strictly social animals, natural selection sometimes acts indirectly on the individual, through the preservation of variations which are beneficial only to the community. A community including a large number of well-endowed individuals increases in number and is victorious over other and less well-endowed communities; although each separate member may gain no advantage over the other members of the same community. With associated insects many remarkable structures, which are of little or no service to the individual or its own offspring, such as the pollen-collecting apparatus, or the sting of the worker-bee, or the great jaws of soldier-ants, have been thus acquired.[74]

Darwin actually formulated three complementary explanations for the origin of the moral sense, each designed to overcome the problem of a trait that offered the individual no direct benefit. The first is now known as reciprocal altruism. Darwin put it this way: "as the reasoning powers and foresight of the members [of a tribe] became improved, each man would soon learn from experience that if he aided his fellowmen, he would commonly receive aid in return."[75] Darwin thought this would be a lesson learned each generation, but in time the acquired associations might become hereditary.[76] The second boost to our moral nature would derive from the social forces of praise and blame.[77] The first two explanations were hardly novel, being commonplaces of the literature. But the most powerful explanation of the origin of the altruistic instincts, at least in Darwin's judgment (not surprisingly), was community selection:

74. Ibid., p. 155.
75. Ibid., p. 163.
76. Ibid., p. 164.
77. Ibid., pp. 164–65.

It must not be forgotten that although a high standard of morality gives but a slight or no advantage to each individual man
and his children over the other men of the same tribe, yet that
an advancement in the standard of morality and an increase in
the number of well-endowed men will certainly give an immense advantage to one tribe over another. There can be no
doubt that a tribe including many members who, from possessing in a high degree the spirit of patriotism, fidelity, obedience, courage, and sympathy, were always ready to give aid
to each other and to sacrifice themselves for the common good,
would be victorious over most other tribes; and this would be
natural selection.[78]

Community selection not only showed how a trait harmful to the
individual (though of benefit to the community) might evolve, it also
dissolved two other objections to an evolutionary theory of morals.
Community selection might, first off, overcome the difficulty recognized by Greg and Galton, that the social instincts would work to preserve the unfit. While that might occur *within* a tribe, yet such entropic
force might well be overcome by selection *between* tribes, so that the
evolution of morality would continue. "At all times throughout the
world," Darwin explained, "tribes have supplanted other tribes; and as
morality is one element in their success, the standard of morality and
the number of well-endowed men will thus everywhere tend to rise and
increase."[79]

In his 1864 paper "The Origin of Human Races," Wallace had proposed that tribal communities having certain moral qualities would
have an advantage in their struggle with other tribes—a conception
quite congenial to Darwin's own (see chapter 4). In his reconsideration
of human evolution, however, Wallace appears either to have forgotten or to have suppressed his original proposal. His subsequent
objection to a Darwinian account of human evolution supposed that
selection pressures would remain static, so that only a low level of an
intellectual or moral trait would be needed for sheer survival. But in a
dynamic human environment, in which the most proximate selection
pressures on one tribe would be exerted by its changing relations with
others, a constant refinement of the intellectual and moral faculties
would result. This development would of course be augmented by
those other agencies of language, social structure, and habit, all of
which, in Darwin's estimation, might alter the heritable characters
of men. From Darwin's point of view, Wallace had initially started off

78. Ibid., p. 166.
79. Ibid.

along the right path in his 1864 paper, but had later detoured onto a
slippery slope that endangered the entire theory of evolution by natural
selection.

There is one significant problem that Darwin's contemporaries
missed, though modern biologists have not. This has to do with an
apparent disparity between the model for community selection and its
application to human groups. Community selection, as Darwin defined
it in the *Origin of Species,* works because the community members are
related: neuter workers in a hive of bees are siblings, so any communally
beneficial traits can be passed on to future generations through the
queen. But this is not usually the case in human communities. Darwin
seems to have been aware of the difficulty, since he obliquely offered
several suggestions that mitigated the problem, if they did not quite
eliminate it.

Darwin believed that the social instincts themselves were extensions
of the "parental and filial affections," which could be explained by natural selection.[80] So altruistic behavior might initially gain a foothold in a
group whose members were indeed closely related. He knew also, of
course, that small tribes would consist of only a few clans. He alluded
to this fact when he considered that the inventiveness and rational acumen of some members of a tribe could benefit the whole group in competition with other tribes. In so fortunate a tribe, the new inventions
and clever plans would spread by means of imitation. This would then
give the tribe a selective advantage. Now, Darwin argued, if these
primitive Newtons

> left children to inherit their mental superiority, the chance of
> the birth of still more ingenious members would be somewhat
> better, and *in a very small tribe decidedly better.* Even if they left
> no children, the tribe would still include their blood-rela-
> tions; and it has been ascertained by agriculturists that by pre-
> serving and breeding from the family of an animal, which when
> slaughtered was found to be valuable, the desired character has
> been obtained.[81]

Darwin did, at least dimly, recognize that small tribes would have
greater biological cohesiveness, and thus possess group traits with
greater heritability. But he certainly did not insist that a high degree of
relatedness would be necessary to get group selection going or to sustain it, especially when the traits—for example, altruistic ones—might
be harmful to the individuals who possessed them. He had already sug-

80. Ibid , pp. 80–81.
81. Ibid., p. 161, emphasis added.

gested in the fifth edition of the *Origin of Species* (1869) that natural selection would generally alter individuals if that benefited the group, though he did not stipulate that members of the group need be related.[82] He seems not to have realized that in a tribe without sufficient relatedness, the next generation would not have, in the case of his example, any greater representation of intelligent members (though on average no less). Traits a tribe copies from its few geniuses might give it a selective advantage, but not necessarily an effectively heritable one. Only under certain conditions might group selection prove a force in evolution—for example, small groups that remain stable, large numbers of them in competition, little migration between groups, sufficient biological relatedness of members, small disadvantage to members initiating the group-enhancing trait, and so forth. Darwin seems not to have been completely oblivious to these problematic requirements, though not focally aware either—hardly surprising, except to hagiographers.

Once, however, tribes had been established with highly developed social instincts and an enlarged mental capacity, then several nonselection mechanisms, Darwin believed, would continue to hone the moral sense. And as the evolution of mind reached a pitch where individuals of different societies and nations would begin to recognize each other as brothers, members of the same global tribe, then quite automatically social sympathy would also be mutually extended. So in this way, what began in our ancestors as primitive instincts, might evolve, Darwin supposed, into a respect for the moral nature of all men.[83]

Not the Greatest Happiness, but the Greatest Good

Meager scholarly attention has been paid to Darwin's ethical theory. One reason few historians have closely examined it, I suspect, is that it seems too redolent of utilitarian selfishness—not the sort of thing a

82. In the fifth edition of the *Origin of Species* (1869), Darwin invoked group selection in a passage that in the first four editions spoke only of individual selection. In those early editions, the passage in question, which dealt with the power of natural selection to affect social groups, read: "In social animals it [natural selection] will adapt the structure of each individual for the benefit of the community; if each in consequence profits by the selected change" (*Origin of Species*, p. 87). The fifth edition amended the last phrase to read "if this in consequence profits by the selected change"; and the sixth edition more clearly put it: "if the community profits by the selected change." Darwin thus seems to have thought of community selection as operating on social groups without respect to the relatedness of the members. For a comparison of the texts of the several editions, see *The Origin of Species by Charles Darwin: A Variorum Text*, ed. Morse Peckham (Philadelphia: University of Pennsylvania Press, 1959), p. 172. I am grateful to Jane Masterson for calling this emendation to my attention.

83. Darwin, *Descent of Man* 1: 100–101.

friend would care to mention, and most historians of evolution embrace Darwin as a friend. Michael Ghiselin, however, has turned this seeming indelicacy into cold scientific virtue. Darwin, he declares, recognized that "since it furthers the competitive ability of the individual and his family, an 'altruistic' act is really a form of ultimate self-interest."[84] It would appear, then, that in advance of the sociobiologists, Darwin had already discovered the selfish gemmule spoiling the core of every good intention. But this judgment, while it keeps Darwin right up to date, entirely misses what he regarded as the most important consequence of his theory: that it overturned utilitarianism.

Darwin pointed out that the utilitarians had initially claimed that "the foundation of morality lay in a form of Selfishness; but more recently in the 'Greatest Happiness principle.'"[85] His own theory, by contrast, proposed that moral action was not motivated by self-interest nor calculated to achieve the greatest amount of pleasure. Altruistic behavior stemmed from an immediate instinct and was guided by social habits. Its goal was not the general happiness but "the general good of the community." He interpreted "general good" according to the criteria endorsed by natural selection. "The term, general good," he offered, "may be defined as the means by which the greatest possible number of individuals can be reared in full vigour and health, with all their faculties perfect, under the conditions to which they are exposed."[86] Hence, Darwin concluded, "the reproach of laying the foundation of the most noble part of our nature in the base principle of selfishness is removed."[87]

In terms of his own set of biological assumptions, Darwin gave an accurate assessment of his ethical theory. Pleasure for oneself or even

84. Michael Ghiselin, "Darwin and Evolutionary Psychology," *Science* 179 (1973): 964–68; quotation from p. 967.

85. Darwin, *Descent of Man* 1:97.

86. Ibid., p. 98.

87. Ibid. Darwin attempted to excise the stigma of selfishness from his general theory, especially its application to man. Historians, either unaware of Darwin's effort or perhaps not believing him to be the best judge in the matter, have depicted his theory as deeply tinctured with utilitarianism. Robert M. Young expresses a common view: "The debate which we summarize by the idea of evolution also embraced the associationist, utilitarian philosophy of the Philosophic Radicals, Bentham, the Mills, and their followers, who would apply natural laws to men and morality and apply sanctions to induce men to act for the greatest good of the greatest number. The pleasures and pains of utilitarian psychological theory became the rewards and punishments of radical reform movements. In effect, Darwin extended this point of view to the ultimate natural sanctions of survival or extinction. One can also say that Darwinism was an extension of *laissez-faire* economic theory from social science to biology." See Robert M. Young, *Darwin's Metaphor* (Cambridge: Cambridge University Press, 1985), pp. 2–3.

for another would not, as a rule, form the overt motive for virtuous acts, since such behavior often sprang from spontaneous choice, which precluded toting up pleasures and pains. Moreover, natural selection would ignore emotional states that had no use. (The utilitarian standard depicted actions as beneficial if they produced pleasure; natural selection produced pleasure only if that pleasurable state induced beneficial actions.) Nor would an individual be guilty of false consciousness when he intended to act for the welfare of others. Since the kind of selection that produced the moral sense did not directly operate on the individual but on the whole community, community welfare stood as its ultimate motive and object; hence, altruistic behavior might be pure in the biological depths as well as at the surface of intention. Some sociobiologists today would deny the existence of group selection, claiming all evolved behavior serves the needs of the selfish gene. Darwin concluded, by contrast, that group selection gave rise to the biologically unselfish moral sense. The reality of group selection remains a controverted topic in contemporary evolutionary biology,[88] but Darwin's historical position should not.

Response of the Critics

John Murray published twenty-five hundred copies of *The Descent of Man and Selection in Relation to Sex* on 24 February 1871; and before the year ended, he printed five thousand more. Almost immediately the book dominated fashionable discussion, as the *Edinburgh Review* observed: "In the drawing room it is competing with the last new novel, and in the study it is troubling alike the man of science, the moralist, and the theologian. On every side it is raising a storm of mingled wrath, wonder, and admiration."[89] The initial response from friends and sympathizers was generous, yet usually hesitant about the theory of morals. In the cruel spring, though, the critical breezes grew harsh, climaxing with three large reviews, which Darwin regarded as damaging and one particularly malicious. He feared that the rejection of his ideas about man, especially by friendly naturalists, might endanger his general theory. He wondered whether, perhaps, "it was a mistake on my part to publish it."[90] Nonetheless, in the revised second edition, appearing

88. The controversy may be traced in the literature cited in chapter 4, note 41. Though Edward Wilson admits group and kin selection may produce altruistic attitudes and behavior, he yet argues that reciprocal altruism constitutes the highest kind of morality. Wilson's position is discussed in the second appendix.

89. "Darwin on the Descent of Man," p. 195.

90. Darwin, *Life and Letters* 2 : 313.

in 1874, he reinforced the essential arguments of the theory, while also making important accommodations to telling objections. The controversy stirred by the book in England, Germany, and America refused to die down during the remainder of the century, since the subject of mental and moral evolution continued to be stoked by the likes of Herbert Spencer, Ernst Haeckel, and William James.

The Reaction of Friends

Darwin received the first extended review of his ideas from his cousin Hensleigh Wedgwood. In a long manuscript sent a few days after the *Descent* came out, Wedgwood touched on a problem that others—for example, Huxley in his debates with Spencer—would seize upon.[91] He wondered why an intelligent creature, who compared the satisfactions of a brief but stronger instinct with the gentler pleasures of a social instinct, should prefer the latter. No metric of pleasure would obviously tip the scale for virtue. An intelligent creature who had initially followed the stronger instinct might subsequently regret the choice, but the regret would be over a mistake; it would not produce "shame," which Wedgwood took to be the "true essence of conscience."[92] In two follow-up letters, he pressed this difficulty about the special character of the moral sentiment.[93] A mere recollection of unsatisfied social instinct could not, he thought, be a "vera causa"; it could never evoke the pain we recognized as the prick of conscience. He insisted that the shame felt over transgressions resulted from the disapprobation of fellow creatures.

Darwin responded to his cousin in letters of 3 and 9 March 1871.[94] He first pointed out that his theory never supposed that in the heat of action the moral agent would take time for a balancing of pleasures. Quite the contrary. The individual would be immediately impelled either to virtuous or to selfish behavior. Further, the pain of conscience would not consist in a recollection of unsatisfied instinct, but would well up from the actual renewal of that instinct during the time of reflection. Darwin did not wish to deny the role of social approval or disapproval in forming the moral outlook. Natural selection might provide the in-

91. Hensleigh Wedgwood's manuscript analysis is preserved in DAR 88. Hereafter I will refer to it as the "Wedgwood manuscript." The manuscript bears no date, but subsequent letters indicate it was composed at the very end of February or first few days of March 1871.

92. Wedgwood manuscript, MS p. 5.

93. The two letters of Hensleigh Wedgwood to Charles Darwin are undated, but Darwin's replies came on 3 and 9 March 1871. The letters are in DAR 88.

94. Darwin's letters to Hensleigh Wedgwood are in DAR 88.

stinct to aid one's fellows, but social approbation or disapprobation would suggest the means by which this instinct could be satisfied.[95] While Darwin rejected Wedgwood's principal complaint, that he had no vera causa, he did emphasize the role of social approval and disapproval in the second edition of the *Descent*. He also added a long passage on shame, agreeing that it had its chief source in the judgment of our fellows, but also pointing out that our sensitivity to such judgment ultimately stemmed from instinctive sympathy.[96]

Among the very first published reviews was Alfred Wallace's in the *Academy* (15 March 1871). Wallace presented a judicious account of the book, sprinkled with the sort of tributes that always marked his discussions of "Mr. Darwin's theory." Rather gently he sketched his differences with his friend on the subject of sexual selection. He admired the novelty of Darwin's theory of conscience, but reiterated that natural selection seemed inadequate to account for the moral sense. He understood, it seems, how Darwin's appeal to the dynamics of tribal interactions had been used to counter his own original objection; for in his review he suggested that the struggle of family with family and tribe with tribe assumed a "large population inhabiting an extensive area." This essential condition, he contended, was not given in his friend's formulation of the problem, and so community selection would not likely occur.[97] Darwin made no response in his second edition to this last charge by Wallace; for already in several passages of the first edition he had indeed suggested that intertribal competition would take place in large, densely inhabited areas.[98]

The Popular Press

In the popular press, Darwin's book received decidedly mixed reviews. The *Spectator*, to Darwin's astonishment, gave it quite a decent

95. Darwin, *Descent of Man* 1 : 86; additions in *Descent of Man*, 2d ed., p. 483.
96. Ibid., pp. 485–86.
97. Alfred Wallace, "The Descent of Man," *Academy* 2 (1871): 177–83.
98. Darwin, *Descent of Man* 1 : 157, 160. As his species theory developed from 1838 through the several editions of the *Origin*, Darwin gave more prominence to the processes of sympatric speciation, as opposed to allopatric speciation. He believed that species formation would occur more rapidly when incipient groups were competitively differentiated in large planes and open areas (p. 101–9). He tacitly assumed that this mode of development would also characterize human evolution. For discussions of Darwin's views about speciation processes, see Ernst Mayr, "Darwin and Isolation," in his *Evolution and the Diversity of Life* (Cambridge, Mass.: Harvard University Press, 1976), pp. 120–28; and Malcolm Kottler, "Charles Darwin's Biological Species Concept and Theory of Geographic Speciation," *Annals of Science* 35 (1978): 275–97.

notice.[99] The reviewer discovered no incompatibility between evolutionary theory and religion; on the contrary, he regarded "Mr. Darwin's investigation of the origin of man a far more wonderful vindication of Theism than Paley's *Natural Theology.*" Darwin bested Paley in this respect, since he implied that a "Creative Force" produced variations that divinely promulgated law molded into human nature; the processes of evolution, in short, gave evidence of an infinitely creative power and designing intelligence at work in the universe. The sticking point, however, was the theory of conscience. While the reviewer believed Darwin came "nearer the kernel of the psychological problem, than many of his eminent predecessors," the estimable naturalist yet failed to account for the peculiar authority of conscience, which only a more far reaching teleological analysis might provide. The *Athenaeum* reacted more predictably. Its critic judged the chief merit of the book to be the accumulation of a vast array of facts, but found Darwin at his "feeblest" in attempting to give an evolutionary account of the moral and intellectual faculties.[100] The *Times* also complimented Darwin for the drudge's work of gathering observations, and likewise regarded him as "quite out of his element" in treating mind and morals. The reviewer settled on Darwin's inattention to the peculiar authority of moral sentiment as the most egregious deficiency of the theory.[101] The account in the *Pall Mall Gazette,* while critical, yet struck a more conciliatory note.[102] It peaked Darwin's interest in discovering the identity of its anonymous author.

The writer had several insightful objections to Darwin's theory of conscience. First, he attempted to correct a misconception Darwin had about John Stuart Mill's theory of morals. Mill, the reviewer observed, proposed the greatest happiness principle not as a "foundation" for the moral sense, but as a "standard" by which to judge actions, whether those actions were acquired or innate. Second, since Mill located the natural source of the moral sense in social feelings, "between Mr. Darwin and utilitarians, as utilitarians, there is no such quarrel as he would appear to suppose." Third, intellectual activity should be considered, he ventured, the primary requisite of conscience, not the secondary as Darwin maintained. This last objection followed from Darwin's own recognition that as social conditions changed, rational discrimination had to take them into account. Even for an evolutionist, the reviewer

99. "Mr. Darwin's Descent of Man," *The Spectator* 44 (11 and 18 March 1871): 288–89, 319–20.

100. "The Descent of Man," *Athenaeum* (4 March 1871): 275–77.

101. "Mr. Darwin on the Descent of Man," *The Times of London* (7 April 1871).

102. [John Morley] "The Descent of Man," *Pall Mall Gazette* (21 March 1871), pp. 11–12; and "Mr. Darwin on Conscience," *Pall Mall Gazette* (12 April 1871), pp. 10–11.

pointed out, "the foundations of morality, the distinctions of right and wrong, are deeply laid in the very conditions of social existence," which had to be understood and reflectively negotiated.[103]

Darwin appreciated these careful remarks, and wrote to thank the author, who revealed himself to be John Morley, the editor of the *Pall Mall Gazette*.[104] But Darwin also wanted to join issue. He admitted his blunder in ascribing to Mill the notion that greatest happiness acted as foundation of the moral sense; yet he held fast to his criticism of Mill, that the philosopher did regard moral feeling as acquired, not innate. Darwin also acknowledged that his reduction of intelligence to a secondary place was misleading, since he really meant secondary only in respect to development, not function or significance. Finally, he did not wish to deny the social character of morality, but insisted, as against the idea that such character came with high intelligence, that it must have evolved under natural selection prior to the acquisition of a large cognitive capacity.

As a result of his correspondence with Morley, Darwin confessed in the second edition of the *Descent* his incomplete reading of the utilitarians; he remained resolute, however, in his objection to their moral views. He allowed that these philosophers did regard the greatest happiness "as the standard, and not as the motive of conduct." They nonetheless supposed men to be moved by the amount of pleasure associated with different kinds of behavior. This ignored, to Darwin's mind, the often impulsive and instinctive character of moral acts. Further, an evolutionary analysis implied that the standard of morality ought to be regard as "the general good or welfare of the community, rather than the general happiness." In amending his criticisms of utilitarianism, Darwin reintroduced a distinction he had made in his early speculations (1838–1839), but only hinted at in the first edition of the *Descent*—the distinction between the motive of moral behavior and the standard for judging it. Concerning the character of both the motive and standard, Darwin had not been persuaded by Morley. Community selection fixed in our natures the motive of acting for the general good, while reflection on our native impulses produced the standard, shaded and particularized though it might be by changing social circumstances. He therefore maintained that his theory differed considerably from that of Bentham, Bain, Mill, and others of the "derivative school of morals."[105]

103. Ibid.

104. Darwin's correspondence with Morley is DAR 88 and DAR 146. See also portions published in Darwin, *More Letters* 1 : 324–29.

105. Darwin, *Descent of Man*, 2d ed., pp. 489–90.

The Learned Journals

As might be expected, Darwin received the most searching and extensive criticism from the learned journals. He took notice of three especially: Miss Frances Cobb's article in the *Theological Review,* and the more damaging analyses of writers for the *Edinburgh Review* and the *Quarterly Review.*

Miss Cobb had Darwin "crowning the edifice of Utilitarian ethics" with his moral theory, and thus lending scientific stature to "the most dangerous [doctrines] which have ever been set forth since the days of Mandeville."[106] Because Darwinian moral theory devilishly identified the "Right" and the "Useful," it offered no stability or authority for conscience; the moral sense would shift aimlessly in the winds of expediency.[107] In the second edition of the *Descent,* Darwin calmly pointed out that feeling for the community good provided the anchor of conscience, though for those persons blown to licentiousness by his argument, perhaps no sentiment could hold them steady.[108]

The *Edinburgh Review* sounded the same alarm, but with more authority.[109] Its critic also spied the stone ax poised to bludgeon the social organism:

> If our humanity be merely the natural product of the modified faculties of brutes, most earnest-minded men will be compelled to give up those motives by which they have attempted to live noble and virtuous lives, as founded on a mistake. . . . If these views be true, a revolution in thought is imminent, which will shake society to its very foundations, by destroying the sanctity of the conscience and the religious sense; for sooner or later they must find expression in men's lives.[110]

The reviewer, despite his express fear of a possible truth, wished to protect against what he believed certain error in Darwin's moral theory. Borrowing his shot from Wallace, he unloaded with the argument that natural selection could not have produced the excessive intelligence and moral feeling in primitive men, since these had no immediate use. By this time (July 1871), Darwin must have grown weary of hearing this retort. He seems to have thought no specific response necessary, for he had already tried to defend against this objection, and moreover the

106. Frances Cobb, "Darwinism in Morals," *Theological Review* (1872): 167–192; quotations are from pp. 170 and 175.

107. Ibid., pp. 176–77.

108. Darwin, *Descent of Man,* 2d ed., p. 473n.

109. "Darwin on the Descent of Man," *Edinburgh Review* 134 (1871): 195–235.

110. Ibid., p. 195–96.

author had treated him with the respect due a worthy opponent. The reviewer for the *Quarterly*, however, descended from that high critical plane, not only to bury the *Descent*, but also to shovel dirt in Darwin's face.[111]

Mivart's Attack and Huxley's Defense

The Quarterly's reviewer confessed acute chagrin in pointing out the "grave defects and serious short comings" of Darwin's arguments, but honesty compelled him.[112] Certainly he could not overlook Darwin's "singular dogmatism," which led the eminent naturalist to assert what required proof, to beg all the important questions, and to fawn over his own supporters.[113] The reviewer allowed the *Descent* bulged with examples, but had to report that "Mr. Darwin's power of reasoning seems to be in an inverse ratio to his power of observation."[114] In the first part of his analysis, he pitted Darwin's theory of sexual selection against Wallace's, concluding that the neutralizing result pointed to "the existence of some unknown innate and internal law which determines at the same time both coloration and its transmission to either or to both sexes."[115] The same law, he believed, governed the development of the human animal, about which Mr. Darwin could offer only "hasty and inconclusive speculation."[116] The second part of the review took up the more important questions of man's mental and moral faculties. The reviewer did not deny that human beings shared with animals fundamental powers of sensation, which would be subject to laws of association; but he insisted that man possessed "a superior nature and faculties of which no brute has any rudiment or vestige."[117] Human reason transcended animal cognition entirely, though if Darwin were right, one ought to find high intelligence, that most useful of traits, spread throughout the animal kingdom.[118] Concerning the moral faculty, the reviewer dismissed Darwin's attempt to found ethics on an instinct, since "men have a consciousness of an absolute and immutable rule *legitimately* claiming obedience with an authority necessarily supreme

111. [St. George Mivart], "Darwin's *Descent of Man*," *Quarterly Review* 131 (1871): 47–90.
112. Ibid., p. 47.
113. Ibid., p. 85–86.
114. Ibid., p. 87.
115. Ibid., p. 60.
116. Ibid., p. 66.
117. Ibid., p. 71.
118. Ibid., p. 77.

and absolute—in other words, intellectual judgments are formed which imply the existence of an ethical ideal in the judging mind."[119]

Darwin responded to this last-mentioned objection in the second edition of the *Descent;* it had been voiced by many of his other critics as well. In the first edition, he had directly connected the moral imperative with instinct: "Thus at last man comes to feel, through acquired and perhaps inherited habit, that it is best for him to obey his more persistent instincts. The imperious word *ought* seems merely to imply the consciousness of the existence of a persistent instinct, either innate or partly acquired, serving as a guide, though liable to be disobeyed."[120] He emended this passage in the second edition to read: "Thus at last man comes to feel, through acquired and perhaps inherited habit, that it is best for him to obey his more persistent impulses. The imperious word *ought* seems merely to imply the consciousness of the existence of a rule of conduct, however it may have originated."[121] Darwin's several critics made their point, though it was one he had already appreciated in his earliest constructions of a moral theory.[122] In the second edition, he made explicit that the ethical imperative involved a rational consideration, one, however, formulated in light of the moral instinct.

When Darwin first read the review in the *Quarterly,* he became extremely agitated. He thought it painted him "the most despicable of men."[123] He also detected a familiar sneering obsequiousness. He wrote Huxley that "the skill and style make me think of Mivart."[124] Huxley agreed, and came to his friend's defense in a counter-review—"Mr. Darwin's Critics"[125]—that absolutely delighted the older man. Huxley took the opportunity to consider also Wallace's objections as well as the theologically structured evolutionary ideas of Mivart's *The Genesis of*

119. Ibid., p. 79.

120. Darwin, *Descent of Man* 1:92.

121. Darwin, *Descent of Man,* 2d ed., p. 486.

122. See chapter 2.

123. Darwin, *Life and Letters* 2:326.

124. Ibid. Mivart was indeed the author of the review in the *Quarterly.* He reprinted the article in his *Collected Essays and Criticism* (London: Osgood, McIlvaine & Co., 1892), 2:1–59.

125. Thomas Huxley, "Mr. Darwin's Critics," in his *Collected Essays* (New York: D. Appleton, 1896–1902), 2:120–86. The article originally appeared in *Contemporary Review* 18 (1871): 443–76. Huxley composed the review while vacationing at St. Andrews, Scotland, in September of 1871. During the same three-week period, he also penned "Administrative Nihilism," a piercing attack against another old friend, Herbert Spencer. It was on this occasion that Huxley remarked "I am Darwin's bull-dog." See L. Huxley, *Life and Letters of Thomas Huxley* 1:390–91.

Species, which had been published in the same year as Darwin's *Descent*. [126]

Huxley tarried only a short while over Wallace's demur about natural selection in the case of man. He derived from Wallace's own writings about savage life descriptions of the extraordinary mental feats such life actually required—knowledge of a vast territory, reading signs of game or enemies, discovery of properties of plants and habits of animals, and so forth. "In complexity and difficulty," Huxley estimated, "the intellectual labour of a 'good hunter or warrior' considerably exceeds that of an ordinary Englishman." [127] Wallace had simply miscalculated the brain power the savage actually needed for survival; thus neither primitive man nor modern native likely had in excess what could be delivered by natural selection or augmented by entering into civilized life. On the question of the moral sense, Huxley could "find nothing in Mr. Wallace's reasonings which has not already been met by Mr. Mill, Mr. Spencer, or Mr. Darwin." [128] Huxley treated Wallace rather gently. But with Mivart, his onetime protégé and friend, he turned Darwin's bulldog, playing a bit first, then crushing his prey.

With much solicitude, Huxley advised the lawyer Mivart that he might have an action against the *Quarterly* reviewer, who had plagiarized his ideas without giving direct quotations. [129] Since Mivart and the reviewer agreed on so many essentials, Huxley decided to treat their positions as one. He first noted that both the reviewer and Mivart reproached Darwin with the charge that his theory was mired in a radically false metaphysical system, which suffocated true philosophy and religion. Huxley, to counterbalance this objection, decided to look at the metaphysics which supported Mivart's own version of evolution. The Catholic lawyer had protested that though Darwinian theory failed because of its materialism, the basic notion of species descent could be sustained if framed by the right philosophy. Indeed, Mivart maintained that the theory of "derivative creation," or evolution, had actually been taught by the doctors of the Church—Augustine, Thomas, Suarez. As luck would have it, Huxley had available a set of the ponderous tomes of Suarez, and could check the citations Mivart had made in his *Genesis of Species*. Huxley produced several long passages from Suarez's Latin text and accompanied each with a profound commentary done in the scholastic style. He discovered that despite Mivart's representations to

126. St. George Mivart, *The Genesis of Species* (New York: D. Appleton, 1871).
127. Huxley, "Mr. Darwin's Critics," p. 176.
128. Ibid., p. 179.
129. Ibid., p. 122.

the contrary, the learned Jesuit held to the literal interpretation of *Genesis*. This would seem to imply that Mivart's own metaphysics and theology were highly suspect, for "if Suarez has rightly stated Catholic doctrine, then is evolution utter heresy."[130]

Huxley obviously enjoyed chewing at the heels of Mivart's metaphysics, but he did not neglect the more pointed objections. First, he argued that Mivart's and the *Quarterly* reviewer's psychology, which demanded a complete separation of sensation and feeling from reason, must ignore the plain lessons of physiology, which showed that the same nervous structures underlay cognitive processes in man and the higher animals, so that if we could not suppose that animals reasoned, then neither could we suppose they had sensations or feelings. Analogy, moreover, was the only evidence we had that other men could feel and think as we do, so if we could not analogically postulate thought of animals, we should be debarred from doing so of other men. Huxley felt confident, however, that most would recognize that the greyhound did judge the distance of the rabbit he watched, did recall it to be like others he had chased, and did decide to go after it.[131] Concerning moral theory, Mivart had decried Darwin's lack of understanding that an act could be formally moral only if the agent at the time rationally intended to do his duty—spontaneous actions, he contended, were devoid of moral worth. Here Huxley remonstrated that Darwin did recognize the difference between action done spontaneously and action done with a self-reflective moral intention; and he sided with his friend, as he thought most men would, in holding that an individual who without a moment's reflection jumped into the river to save a drowning child nonetheless performed a precious moral act.[132]

Darwin's delight with Huxley's essay approached ecstasy. "How you do smash Mivart's theology," he wrote his younger colleague. His gratitude waxed:

> I must tell you what Hooker said to me a few years ago. "When I read Huxley, I feel quite infantile in intellect." By Jove I have felt the truth of this throughout your review. What a man you are. There are scores of splendid passages, and vivid flashes of wit. I have been a good deal more than merely pleased by the concluding part of your review; and all the more, as I own I felt mortified by the accusation of bigotry, arrogance, &c., in

130. Ibid., p. 147.
131. Ibid., pp. 152–67.
132. Ibid., pp. 167–73.

Figure 5.3 Charles Darwin, photograph from 1881.

the 'Quarterly Review.' But I assure you, he may write his worst, and he will never mortify me again.[133]

As far as Darwin was concerned, his theory of human mental and moral evolution had received the imprimatur of the man whose judgment he

133. Charles Darwin to Thomas Huxley (30 September 1871), in Darwin, *Life and Letters* 2:328–29.

most prized in these matters. He might then, perhaps, have been rather disappointed at Huxley's reevaluation of evolutionary ethics at the end of the century (the story of which is told in chapter 7). But that occurred a decade after Darwin had died, so he was saved from his friend's later doubts about evolution in the case of man. After 1874, when the second edition of the *Descent* appeared, Darwin never publicly expressed hesitation over his theory. He did, however, continue to study the less philosophical aspects of the biology of behavior right to his last years.

Expression of the Emotions

In the introduction to the *Descent,* Darwin mentioned that he had intended to consider the expression of various emotions in men and animals. He was prompted to do so, he explained, because Sir Charles Bell's work on expression deeply interested him and issued a challenge. Bell had maintained that man was endowed with certain muscles for the purpose of expressing emotions to his fellows, a view Darwin regarded as "opposed to the belief that man is descended from some other and lower form."[134] Darwin decided, however, to hold his study of the emotions for another volume, since, as he suggested, even a gentleman's patience might be tested were the *Descent* further swollen. He finished correcting page proofs for the *Descent* on 15 January 1871. On 17 January he began work on *The Expression of the Emotions in Man and Animals,* which he saw published on 8 November 1872.

The *Expression of the Emotions* is an original and disconcerting book. In his introduction to a recent reprint, Konrad Lorenz credits Darwin with foreseeing "in a truly visionary manner the main problems which confront ethologists to this day and [with mapping] out a strategy of research which they still use."[135] Lorenz routinely dismissed as benighted any biologist who accepted the inheritance of acquired characters; yet he failed to notice that Darwin relied exclusively on this suspect principle. Moreover, while modern ethologists attribute vital communicative functions to expressions in animals, Darwin denied that emotional responses had any use at all, which is why he did not invoke natural selection to explain them. To understand how the biological significance of expression escaped Darwin's notice requires some consideration of the background against which he developed his theory.

134. Darwin, *Descent of Man* 1:5.
135. Konrad Lorenz, "Introduction," in Charles Darwin, *The Expression of the Emotions in Man and Animals* (Chicago: University of Chicago Press, [1872] 1965), pp. xi–xii.

In his account, Darwin took his departure from the work of Sir Charles Bell on the emotions. Bell's *Expression: Its Anatomy and Philosophy* (3d ed., 1844) had the advantage of his precise knowledge of the muscles and innervations of the human face, as well as his thorough acquaintance with emotional representation in art and literature, all of which Darwin admired and made use of in his own analysis. What Darwin objected to, however, was Bell's explanation of the meaning of emotional expression. Bell claimed that grimaces and smiles, frowns and blushes had been inscribed in human physiognomy as a kind of natural language which allowed immediate communication of one soul with another. This language bespoke its originating source: "in every intelligent being," Bell avowed, "[the Creator] has laid the foundation of emotions that point to Him, affections by which we are drawn to Him, and which rest in Him as their object."[136]

Darwin, of course, refused to admit that instinctive expressions had such origin and transcendental use. But he went further. He denied they had any intrinsic use at all.[137] When assessing the work of natural theologians, Darwin was always ready to acknowledge the functional, though not the theological, significance they assigned to animal traits. The famous anatomist and natural theologian Bell did not recognize any singularly biological function of the emotions, only their transcendental purpose. When Darwin rejected the latter, he had no better idea than Bell about the former. Neither Bell nor he could imagine any strictly biological use human emotions might have. Both men were simply unaware of the kind of important communicative functions the emotions serve in the animal and human economy. But Darwin did insist on the theoretical implications of the emotions. The universality of many expressions among human varieties and the presence of allied instincts in man and higher animals (e.g., laughter of men and monkeys) indicated common evolutionary origins. Conversely, the commonality of expressions and instincts were "rendered somewhat more intelligible, if we believe in their descent from a common progenitor."[138] But evolved instincts which had no use could not be explained by natural selection. The problem for Darwin, then, was how instincts expressing emotion might have evolved.

Darwin had already spent considerable effort on a theory of the emo-

136. Charles Bell, *Expression: Its Anatomy & Philosophy*, 3d ed. (New York: Wells, [1844] 1873), p. 78.

137. Darwin (*Expression of the Emotions*, p. 354) did grant that vocal expression in some species might have been sexually selected.

138. Darwin, *Expression of the Emotions*, p. 12.

tions in his *M* and *N Notebooks*, when he still lingered with the habit-instinct mechanism of evolutionary change.[139] Those deeply ingrained considerations prepared him for a simple explanation of the emotions: intentional behavior long practiced would become inherited and instinctive, even though it had no biological function. This was his general principle of explanation, but he specified it differently in his account of three classes of expression.

The principle of "serviceable associated habits," the first specification, recognized that actions intentionally undertaken, especially to remove an unpleasant sensation, might become habitually associated with a certain emotional state and might afterwards be called up by that emotional state alone.[140] For example, the "vulgar man" who scratched his head to relieve an itch might thereafter continue to do so whenever he was perplexed in mind, or the man who closed his eyes before a ghastly sight would likely shut them again when telling of his experience.[141] Connections of this kind, Darwin proposed, could become inherited and passed on to succeeding generations. Thus his own child, when sixteen weeks old, would violently blink when a box was rattled before it, though it could not have known that a rattling sound might signal danger. "But such experience will have been slowly gained at a later age during a long series of generations," or so Darwin believed.[142]

His second and third principles were aimed at explaining instinctive expressions not originally acquired as habits. The "principle of antithesis" held that when certain actions were linked with a particular state of

139. William Montgomery discusses the contribution Erasmus Darwin made to his grandson's early views about the emotions in his "Charles Darwin's Thought on Expressive Mechanisms in Evolution," in *The Development of Expressive Behavior: Biology-Environment Interactions* (New York: Academic Press, 1985), pp. 27–50. See also Janet Browne, "Darwin and the Expression of the Emotions," in *The Darwinian Heritage,* ed. David Kohn (Princeton: Princeton University Press, 1985), pp. 307–26; and Richard Burkhardt, "The Development of an Evolutionary Ethology," in D. Bendall, ed., *Evolution from Molecules to Men* (Cambridge: Cambridge University Press, 1983), pp. 431–44.

140. This principle was suggested to Darwin by Alexander Bain, whose *The Senses and the Intellect* (2d ed., 1864) and *The Emotions and the Will* (2d ed., 1865), he had read in preparation for the composition of the *Descent.*

141. Darwin, *Expression of the Emotions,* p. 42: "my object is to show that certain movements were originally performed for a definite end, and that, under nearly the same circumstances, they are still pertinaciously performed through habit when not of the least use. That the tendency in most of the following cases is inherited we may infer from such actions being performed in the same manner by all individuals, young and old, of the same species."

142. Ibid., p. 39. This and other observations of his child were later reported in his "A Biographical Sketch of an Infant," *Mind* 2 (1877): 285–94.

mind, the appearance of an opposite state would tend to elicit behavior of an opposite kind. For example, a dog, which stands rigid with hair erect and tail stiff when hostile, will crouch low with back bent and tail curled when affectionately disposed. Darwin assumed that the link between the opposite disposition and the corresponding behavior would gradually "become hereditary through long practice."[143] Why he felt it necessary to add this last provision, since he thought contrary emotional states naturally evoked contrary behavior, is hard to say. It seems but another attempt to secure the evolutionary foundations of all patterned behavior.

Darwin derived his third principle, at least in part, from Herbert Spencer. The principle stipulated that certain expressions resulted from an excess of nervous energy spilling over into other pathways. An example of such nervous overload would be the trembling produced by fear. In Darwin's view, such emotional expression was "of no service, often of much disservice, and cannot have been at first acquired through the will, and then rendered habitual in association with any emotion."[144] Spencer had argued a similar thesis and added "that an overflow of nerve-force, undirected by any motive, will manifestly take the most habitual routes; and if these do not suffice, will next overflow into the less habitual ones."[145] Citing Spencer, Darwin happily adopted this latter presumption as a corollary to his third principle.[146]

The assumption grounding Darwin's three principles was, of course, that habits could become instincts. In the *Variation of Animals and Plants under Domestication,* he elaborated a genetic theory (his pet, pangenesis) that he thought could explain the heritability of acquired structures.[147] But it was only in the *Expression of the Emotions* that he considered the details of how habit might modify structures in the first place. This development is important for understanding his general theory of the evolution of mind. Its history thus deserves at least brief mention.

In his early notebooks, Darwin supposed that habit could alter cerebral anatomy and that such changes could be inherited. This conviction, as we have seen, sprang from his grandfather's proposals in *Zoonomia*

143. Darwin, *Expression of the Emotions,* p. 65.

144. Ibid., p. 67.

145. Herbert Spencer, "The Physiology of Laughter," in his *Essays, Scientific, Political and Speculative* (New York: D. Appleton, 1892), 2:458–59. Darwin referred to this passage in *Expression of the Emotions,* p. 71.

146. Ibid.

147. Darwin's discussion of pangenesis comes in his *The Variation of Animals and Plants under Domestication* (London: Murray, 1868), chap. 24.

and from his father's studies of how brain pathology affected mental processes.[148] But Darwin had not developed in his early speculations a specific theory with which to buttress the belief that habit could affect heritable brain structures. This deficiency probably caused him little worry, though, since after giving up the instinct mechanism of species change, he felt no pressing need for an elaborate theory. But the need arose again when he worked on the genesis of emotional expression. He turned for help to the neural theories of Johannes Müller's *Elements of Physiology,* which he had originally studied in the early 1840s. There he discovered the explanation he required in Müller's hypothesis that "the conducting power of the nervous fibers increases with the frequency of their excitement."[149] This simple hypothesis confirmed for him the easy notion that habitual use of nervous pathways physiologically altered them. For more specific details of nerve physiology, he seemed to find consolation, if not perfect enlightenment, in the elucubrations of Spencer on the colloidal chemistry of neural action.[150] His reading of Müller and Spencer satisfied him that the technical problems of habit modifying cerebral structures were soluble, if not already solved.

Conclusion: Utilitarian and Darwinian Moral Theories

Darwin's moral ideas met considerable resistance, even from friends who accepted the general theory of evolution by natural selection. Critics, from Darwin's day to the present, have lodged three related objections to an evolutionary ethics: that it is merely a biologized version of utilitarianism, that it robs conscience of a distinctive authority, and that it commits the naturalistic fallacy of sliding from an "is" to an "ought." Let us focus on the first of these objections, and then consider the other two more briefly, in an effort to determine whether and to what degree Darwin's theory is liable to the charges.

Utilitarianism and Darwinian Theory

We have seen that Darwin took great pains to distinguish his conception of moral behavior from that of the utilitarians. He believed three

148. See chapter 2.

149. Johannes Müller, *Elements of Physiology,* trans. W. Baly (London: Taylor and Walton, 1838–1842), 2:939. Darwin cited this passage in the *Expression of the Emotions,* p. 29.

150. Spencer discussed the chemistry of nerve action with precisely the same aim as Darwin—to provide physiological justification for belief in the heritability of habits. See Herbert Spencer, *The Principles of Biology,* 2 vols. (New York: D. Appleton, [1866] 1884), 1:432–42.

traits in particular saved his theory from being classified with theirs. His theory located the moral motive essentially in an innate, instinctive response to the needs of the community, while the utilitarians construed the moral sense as an acquired response. The evolutionary interpretation detached the moral sense from expectations of pleasure, whereas the utilitarians, Darwin claimed, derived the moral sense from pleasurable associations. Finally, his theory took the community good rather than the greatest pleasure as the objective of moral behavior. These marks did distinguish Darwin's ethical conception from Bentham's. Bentham distrusted any appeal to an intuitive or innate moral sense. While he recognized sympathy as a native feeling, he regarded it as an unreliable guide in achieving the moral end of man, the greatest happiness.[151] The only natural sentiment the legislator and moralist could count on was an individual's desire for personal pleasure, which Bentham installed as the mainspring of human action.[152] This motive for conduct, however, needed restraint. It had to be focused on the general happiness. Accordingly, he required the application of external sanctions of law, religion, and public opinion to secure the greatest happiness for those whose interests were at stake.[153] Bentham's construction of morals thus differed considerably from Darwin's. But what of the enlightened utilitarianism of John Stuart Mill? A comparison of Mill's theory with Darwin's can serve to bring into further relief the logical structure of the evolutionary conception of morals and to gauge the direction of its flow with respect to a major project of ethical thought in the nineteenth century.

Ethical Theories of Mill and Darwin

Mill endorsed Bentham's general happiness principle, but interpreted happiness more subtly than his mentor. Mill did not conceive the plea-

151. Jeremy Bentham, *Introduction to the Principles of Morals and Legislation,* corrected ed. (New York: Hafner,[1823] 1973), pp. 13–23.

152. Bentham begins his *Principles of Morals* (p. 1) with the famous lines: "Nature has placed mankind under the governance of two sovereign masters, *pain* and *pleasure*. It is for them alone to point out what we ought to do, as well as to determine what we shall do."

153. Mill regarded Bentham's perception of human nature to be so jejune, so narrowly directed to the individual's sensible pleasures, that the great moralist could not conceive anyone acting for the common good without the external sanctions of law, religion, and public opinion. As Mill put it in his essay "Bentham" (1838, reprinted in Mill's *Dissertations and Discussions* [New York: Holt, 1882], 1: 387): "Bentham's idea of the world is that of a collection of persons pursuing each his separate interest or pleasure, and the prevention of whom from jostling one another more than is unavoidable, may be attempted by hopes and fears derived from three sources—the law, religion, and public opinion."

sures of life to be of equal value; pushpin was definitely not as good as poetry. "Better to be Socrates dissatisfied, than a fool satisfied," he maintained in his tract *Utilitarianism*. [154] But for an ethical system based on a hierarchy of pleasures, the question of scaling became acute: Who would settle the dispute of Socrates and the fool over which pleasures were to be pursued? Mill suggested that the wise men of their time should be consulted, though he offered no clue as to how to select these men. [155] One is left with the impression that Mill judged the wise to be those who would make the right choices about which pleasures to cultivate; though, of course, that was exactly the question at issue. Darwin avoided the main thrust of this difficulty, if not the entire problem. For him the purpose of moral action was the community good, defined precisely and objectively in evolutionary terms—the lives, health, and welfare of members of the community, not their diverse pleasures. Community selection, the main principle of Darwin's theory, calibrated even more finely the ethical response: it fixed moral obligation most strongly toward one's family, with the moral bonds becoming weaker as they stretched to remote kin, neighbors, other community members, and men in general. Darwin conceived this hierarchy as established not by wise men but by a wiser nature, which, as he put it in the *Origin*, "is daily and hourly scrutinising, throughout the world, every variation, even the slightest; rejecting that which is bad, preserving and adding up all that is good." [156] It must be recognized, however, that Darwin's theory does not escape Mill's problem altogether, since it is indeed left to human wisdom to decide the means for achieving the various moral ends falling under the general rubric of community good.

Mill attempted to demonstrate, in two egregious arguments, that happiness (by which he meant "pleasure and the absence of pain") [157] was the general aim of moral action and that the happiness of all formed the specific aim. His argument for the first ran: if men can only desire happiness or pleasure (all else being means to this), then "happiness is the sole end of human action, and the promotion of it the test by which to judge of all human conduct; from whence it necessarily follows that it must be the criterion of morality, since a part is included in the whole." [158] This argument by itself, however, could easily be redirected to conclude that pleasure was the criterion of immorality, for immoral acts also formed a part of the whole of human action. Darwin, by con-

154. John Stuart Mill, *Utilitarianism* (1861), in his *Dissertations and Discussions* 3 : 312.
155. Ibid., p. 314.
156. Darwin, *Origin of Species*, p. 84.
157. Mill, *Utilitarianism*, p. 308.
158. Ibid., p. 354.

trast, denied that pleasure served as the sole object of human conduct. Men, he believed, did act without intending their own pleasure: when they acted instinctually or from habit, for instance, or when they disinterestedly sought the welfare of another. According to Darwin's evolutionary ethics, acting from pleasure never indicated that an action was moral, but it often marked immoral behavior.

Mill's demonstration that the happiness of all was the specific measure of morality unblushingly revealed its own naked fault. The argument simply went: since "each person's happiness is a good to that person . . . the general happiness, therefore, [is] a good to the aggregate of all persons." [159] Mill outrageously linked the plausible premise with the unwarranted conclusion by an illegitimate dictum de omni: he assumed that individual desires by themselves would coalesce into the single desire of the collective, which, since it was a desire predicated of the whole, could then be repredicated of each of the parts. It is obvious, though, that each individual might desire his own good without a flicker of desire for the good of others or the common good. Mill's own benevolence outran his logic. His hasty and inelegant move came as he labored with a difficulty that utilitarians of his generation understood but could not resolve, namely, how to transfuse impartiality into the heart of the moral standard. As Bentham expressed this requirement of the moral principle but could not defend it: each to count as one and none more than one. Yet, as the critic might point out, it is entirely conceivable, indeed rather likely, that each individual could anticipate greater pleasure if he counted himself for more than one. Only external sanctions could harmonize the self-aggrandizing atoms of society. But then, one would have to ask what motives the harmonizing rulers of society could possess, if they themselves lacked constraints on personal interests. Darwin's theory escaped this difficulty, since it postulated that nature both selected impartially and had impartiality as the trait selected for: those communities whose members utterly failed to regard the common good would have perished from the earth.

Mill, in his essay on Bentham, found his mentor's reforming merit, the way by which he introduced scientific procedure into morals, to

159. Ibid., p. 349. Mill followed very closely the argument of Bentham's *Principles of Morals* (p. 31): "Sum up all the values of all the *pleasures* on the one side, and those of all the pains on the other. The balance, if it be on the side of pleasure, will give the *good* tendency of the act upon the whole, with respect to the interest of that *individual* person. . . . Take an account of the *number* of persons whose interests appear to be concerned; and repeat the above process with respect to each. . . . Take the *balance;* which, if on the side of *pleasure,* will give the general *good tendency* of the act, with respect to the total number or community of individuals concerned."

consist in attending to the measurable consequences of action: "the morality of action depends on the consequences which they tend to produce."[160] (Though Bentham often spoke as if actual behavior were to be measured and judged, Mill more perceptively recognized that only the "tendency" of a particular act should be the object of moral evaluation, since a good act might, unforeseeably, have bad consequences. What Mill failed to recognize, or what his Benthamite disposition shielded from view, was that the same logic required the "intention of the agent" rather than the "tendency of the act" to be the ultimate object of moral evaluation. After all, many acts tend to produce pain and unhappiness, although the intention that guides them is benign: should we judge physicians of centuries past morally evil because their leeching tended to produce more harm than health? The logic that protects the virtue of an act from contingent circumstances must locate the source of moral value in the motive and intention of the agent.) The utilitarian thus applies the moral criterion to the results of a calculation of the probable consequences of action: those actions are deemed good which, as calculation reveals, tend to produce the greatest happiness. The utilitarian stipulation would, however, seem to require that spontaneous, instinctive behavior be regarded as morally worthless—the hero who without calculation jumped into the river to save the drowning child could not be judged virtuous by the utilitarian standard. He could not be judged virtuous, either because he would not have made a felicity estimate, or because no observer (of the utilitarian persuasion) could regard spontaneous, nonrationally motivated action to "tend" to produce the greatest happiness. Bentham and Mill contemned moral sense theory precisely because it postulated a nonrational, and thus unreliable, motive for human moral behavior. Darwin's theory, on the other hand, would judge the impetuous hero virtuous, since in such a case nature would have already made the calculation: only those actions which on balance did contribute to community good would be retained as a moral inheritance, as an instinct prompting spontaneous action. Evolutionary ethics, like utilitarianism, postulates a calculation as the determiner of moral value, but leaves the weighing in the hands of nature. Darwin did not, of course, deny that the individual's own rational estimate formed a part of the moral intention. The moral motive is blind. It demands a rational assessment of the situation in order to become engaged; the means to achieve nature's end must be calculated. But in Darwin's view, the moral worth of the action does not reside in the agent's calculation, nor in the actual consequences of his behavior, but

160. Mill, "Bentham," p. 411.

rather in acting from the moral motive, which nature has already determined. So the evolutionary ethician would not characterize the primitive who sacrificed virgins for the community good as morally evil, only as mistaken. Mill's utilitarianism was a morality of consequences; Darwin's evolutionary ethics was a morality of intentions.

These, then, are the palpable and profound differences separating the utilitarian and the evolutionary constructions of moral theory. Yet below these differences, often in the logic if not in the overt expression, lie several common features that indicate a deep harmony of moral outlook sounding through the ethical ideas of the nineteenth century.

Darwin and Mill (and Bentham as well) shared a common view on the general character of moral theory: they attempted to build their ethics from the lower ground of nature; they made no appeal to the Divinity for the source of moral feeling, or for the sanctions of ethical behavior, or for the object of virtuous action. The utilitarians, like the evolutionists, advanced a naturalistic ethics, and Mill, among the utilitarians, also an ethical naturalism. Though Bentham thought little about nature, in the large or in man, Mill did. The logic of Mill's argument required appeal to nature as a quasi-moral resource—something Darwin detected in his own reading of Mill's *Utilitarianism*.[161] In *Utilitarianism,* Mill distinguished between the moral standard and the moral motive, a distinction that John Morley brought to Darwin's attention. The standard was general happiness, but the motive or "internal sanction" (as opposed to Bentham's external sanctions) was a feeling, a pain attendant on violation of duty or an urge to follow duty. Mill expressed two opposite opinions about the status of the moral feeling, one stemming from the overt demands of his doctrine, the other from the logic of his argument. On the one hand he expressed the belief that "the moral feelings are not innate, but acquired . . . [though] they are not for that reason the less natural."[162] Yet on the other hand, he described these insistent feelings for the general happiness as if they were innate. He called them a "powerful natural sentiment," which in association with an understanding of the moral standard would "make us feel it congenial, and incline us not only to foster it in others (for which we have abundant interested motives), but also to cherish it in ourselves." If this natural sentiment to promote the general happiness were not operative, he supposed, "it might well happen that this association [be-

161. In preparing his chapters on the moral sense for the *Descent of Man,* Darwin examined Mill's *Utilitarianism,* in which he discovered the philosopher's reference to the social feelings as "a powerful, natural sentiment." Darwin also recognized Mill's vacillation on the question of the innateness of social feelings. See the *Descent of Man* 1:71n.

162. Mill, *Utilitarianism,* p. 342.

tween the urgent feeling and the general happiness], even after it had
been implanted by education, might be analyzed away."[163] Darwin, in
preparing the second edition of the *Descent,* asked his son William to
read over Mill's *Utilitarianism* to get straight on the philosopher's opinion about the moral feeling. William concluded that Mill "must have
been very close to allowing the moral faculty to be inheritable, but [is]
rather in a muddle on the whole subject"[164]—a view most readers
might share. Mill seems to have implicitly recognized that were it not
for a "natural feeling" for the general happiness, there could be no guarantee, there could not even be a rational expectation that people would
act in accord with the moral standard. The Darwinian hypothesis of
community selection supported such expectation; Mill could only postulate it.

Both Darwin and Mill understood the logical requirement for forging a noncontingent link between the moral motive and the moral standard. Darwin, it will be recalled, came to appreciate the necessary connection in his study of James Mackintosh's ethics. Mackintosh proposed
that God constructed human nature such that the moral feeling, an ara-
tional sentiment that warmed to thoughts of producing happiness in
others, agreed with a rational evaluation and application of the moral
standard of utility. Darwin replaced God with nature, holding that community selection linked the moral motive with the moral standard: the
general and more particularized social instincts that promoted the vigor
and welfare of the community would be selected; and as individuals
came to reflect on these native motives, they would set them as standards by which to assess behavior, their own and that of others in their
society. Mill also replaced God with nature, but failed to account for
the natural coincidence between motive and standard.

Bentham and Mill, along with Darwin and Spencer, wished to introduce scientific procedure into the discipline of morals. They differed, of
course, in the methods and resources they brought to the study of morals, but each intended to illuminate every corner of human concern with
the light of science. Since the Enlightenment, and certainly through the
nineteenth century, emancipated thinkers believed the techniques and
rational methods of science held the promise of true progress in human
welfare and of the establishment of a progressive social structure to realize that end. A principal assumption of both the utilitarians and the
evolutionists that grounded the possibility of a scientific analysis of
moral behavior was that men were not free, that the causes determining

163. Ibid., p. 343.
164. William Darwin to Charles Darwin (n.d.), DAR 88. See also note 61.

their actions could be isolated, understood, and controlled. This presumption of determinism appears inimical to any moral theory that pretends to be more than social anthropology, to one that not only accounts for the way men behave, but recommends how they should behave. To the degree that both utilitarian and evolutionary ethics indulge in this latter, to that degree they confront perhaps the major ethical conundrum: the reconciling of determinism and moral responsibility. Both Darwin and Mill attempted a resolution, though Darwin kept his to private notes.[165] The biologist believed that acceptance of the doctrine of determinism actually produced moral obligations: to treat criminals more humanely, as if they were ill, and to mold the characters of children, so as to set them in the path of virtue. The utilitarian concluded, after more extensive argument, much as the biologist, that determinism harmonized with a refined conception of moral responsibility, that it enforced the idea that one's character could be strengthened or restrained through benevolent social controls.[166] Neither Darwin nor Mill may have satisfactorily solved all the problems of the causal interpretation of ethical behavior, but in this respect they cannot be distinguished from moral philosophers of other doctrinal persuasions that have been working since the mid-Victorian period.

The Authority of Conscience and the Naturalist Fallacy

Mill and Darwin thus concurred that the authority of conscience resided in the bowels of human nature, not in a transcendent power. Their critics, of course, immediately perceived that neither the utilitarian nor the evolutionary conscience possessed the traditional authority. Even more than a half century after Kant, fear still gripped many eminent Victorians when they confronted an ethical perspective that did not rely on God to enforce sanctions for moral action. This fear, I believe, elicited the charge that Darwin's theory voided conscience of its authority. The Victorian critics heard the sea of faith's melancholy, long, withdrawing roar.

Modern critics are not so much concerned with the question of the authority of conscience, at least not in quite those terms. They are much concerned, however, with what might be regarded as the equivalent—the naturalistic fallacy, that is, the founding of the authority of

165. See chapter 2 for a discussion of Darwin's early ideas about freedom and determinism.

166. Mill attempted the reconciliation of moral responsibility and determinism in his *An Examination of Sir William Hamilton's Philosophy,* 4th ed. (London: Longmans, Green, Reader, and Dyer, 1872), pp. 586–606.

the moral "ought" on the apparently slack and unsupporting "is." This
form of the objection, though, could not be precisely expressed until
later in the century, when the appropriate ethical concepts became available. It was lodged specifically against that other great promoter of evolutionary ethics in the nineteenth century, Herbert Spencer. Its potency
will be assayed in the following chapters on Spencer and in the second
appendix.

6

Spencer's Conception of Evolution as a Moral Force

Herbert Spencer's intellectual stock has steadily declined since his death in 1903, and there appears no prospect of an immediate upturn. Indeed, estimates by historians of science, who might be expected to render a more contextualizing judgment, make us wonder what even his contemporaries found of value in his thought. Gertrude Himmelfarb calls his work "a parody of philosophy."[1] Charles Singer adds that one point of agreement among modern philosophers is "that the philosophical system of Spencer is an object of derision."[2] Derek Freeman, comparing Darwin's theory of evolution with Spencer's, judges the former to be scientific and the latter to rest only on a "metaphysical supposition."[3] And Ernst Mayr will not allow the "confusion" of Spencer's theories to cloud his monumental *Growth of Biological Thought*. "It would be quite justifiable," he maintains, "to ignore Spencer totally in a history of biological ideas because his positive contributions were nil." Mayr grudges three paragraphs only because Spencer's views inspired "various popular misconceptions about evolution."[4]

Reading Spencer today usually evokes spontaneous reactions that

1. Gertrude Himmelfarb, *Darwin and the Darwinian Revolution* (New York: Norton, 1968), p. 222.

2. Charles Singer, *A Short History of Scientific Thought* (Oxford: Clarendon, 1962), p. 512

3. Derek Freeman, "The Evolutionary Theories of Charles Darwin and Herbert Spencer," *Current Anthropology* 15 (1974): 215.

4. Ernst Mayr, *The Growth of Biological Thought* (Cambridge: Harvard University Press, 1982), p. 386. Historical judgments such as those of Mayr and the other historians quoted are common enough. But in a recent and certainly uncharacteristic study, C. U. M. Smith finds in Spencer a seedbed of modern thought. Smith cultivates the image of Spencer as an anticipator of Piaget, Prigogine, Craik, Sartre, and many developers of contemporary neurophysiology. See C. U. M. Smith, "Evolution and the Problem of Mind: Part I: Herbert Spencer," *Journal of the History of Biology* 15 (1982): 55–88. More balanced treatments of Spencer's accomplishments are provided in the following important studies:

support these historical judgments. He seems to give us cause. The leaden sentences of his later books gradually sink all interest. Not even his habit of dictating manuscripts between sets of tennis and quoits could make his ideas spring to the imagination.[5] Many of his biological theories went under, at the turn of the century, with the Lamarckian craft on which they were carried. And that odor of "Social Darwinism" that seems to permeate his thought stifles any admissions of historical relationship by those sociobiologists whose ideas bear strong resemblance to his.

The evaluations of Himmelfarb and the others mentioned derive from models of the development of science that give free play to our immediate and unhistorical responses to Spencer's works. In the introduction to this book and in the first appendix, I have referred to these as growth and revolutionary models of science. They isolate certain standards of contemporary science and use them like Geiger counters to detect among older conceptual systems those that give off the familiar glow, and Spencer's do not glow. The model that I have adopted, the natural selection model, distinguishes two different ways of historically evaluating older conceptual systems: first by use of those standards actually employed by contemporaries in the scientific community of the time, and then by estimating the role a system played in generating what we regard, from our perspective, as modern science. The historians just cited fail really to do even the latter for Spencer, since they judge progenitors of modern science by the sole criterion of similarity of ideas—as if Triassic reptiles could not have given rise to modern mammals because they look so different. Though many evolutionary conceptions today bear little resemblance to Spencer's, this does not preclude historical parentage.

Spencer did not always receive accolades from his scientific community, but he was recognized as a leading and imposing member of it. Alexander Bain thought Spencer's *Psychology* constituted him "the philosopher of the doctrine of Development, notwithstanding that Darwin has supplied a most important link in the chain."[6] Conwy Lloyd Mor-

---

J. W. Burrow, *Evolution and Society* (Cambridge: Cambridge University Press, 1966), chap. 6; John M. Greene, "Biology and Social Theory in the Nineteenth Century: Auguste Comte and Herbert Spencer," in *Critical Problems in the History of Science,* ed. M. Claggett (Madison: University of Wisconsin Press, 1959), pp. 419–46; J. D. Y. Peel, *Herbert Spencer: the Evolution of a Sociologist* (New York: Basic Books, 1971); and Robert M. Young, *Mind, Brain and Adaptation in the Nineteenth Century* (Oxford: Oxford University Press, 1971), chap. 5.

5. Spencer's books after 1859 were dictated and scarcely revised.

6. Alexander Bain to Herbert Spencer (17 November 1863), Athenaeum Collection of Spencer's Correspondence, MS. 791, no. 67, University of London Library.

gan confessed to Spencer that "to none of my intellectual masters do I owe a larger debt of gratitude than to you."[7] Alfred Wallace, who named his first son in honor of Spencer, avowed after his friend's death: "The only metaphysics I could ever really understand & that was any good to me was that in H. Spencer's 'First Principles.'"[8] And Darwin himself, upon finishing Spencer's *Principles of Biology,* felt abashed, since the book indicated that Spencer was "a dozen time my superior."[9] In his later years, though, Darwin tempered his feelings because of Spencer's "deductive manner of treating every subject."[10] It was this inveterate mode of thought that led Huxley to quip that "Spencer's idea of a tragedy is a deduction killed by a fact."[11] Huxley nonetheless held Spencer in high regard as both a thinker and friend. Indeed he, along with John Stuart Mill, Charles Darwin, Charles Babbage, Charles Lyell, Joseph Hooker, John Tyndall, Henry Buckle, Alexander Bain, John Herschel, and a host of scientific stars of somewhat lesser magnitude subscribed in 1860 to Spencer's program of "Synthetic Philosophy," which would issue volumes in biology, psychology, sociology, and morality. These men allowed their names to be used in soliciting subcriptions to Spencer's project—certainly testimony to their high estimate of his thought. Finally, if a scientist's ability to enchant the muse is a sign of eminence, then along with Newton (of whom Pope sang), Spencer merits the palm. Grant Allen, an Oxford graduate who became Spencer's champion, rhapsodized:

> Deepest and mightiest of our later seers,
> Spencer, whose piercing glance descried afar
> Down fathomless abysses of dead years
> The formless waste drift into sun or star,
> And through vast wilds of elemental strife
> Tracked out the first faint steps of unconscious life.[12]

Some might perhaps judge that Spencer got the poet he deserved.

When Spencer is placed within his intellectual environment, so that

7. Conwy Lloyd Morgan to Herbert Spencer (27 July 1898), Athenaeum Collection of Spencer's Correspondence, MS. 791, no. 269, University of London Library.

8. Alfred Wallace to C. Lloyd Morgan (23 September 1905), in the C. Lloyd Morgan Papers, 128/143, Bristol University Library.

9. Charles Darwin, *The Life and Letters of Charles Darwin,* ed. Francis Darwin (New York: D. Appleton, 1891), 2:239.

10. Charles Darwin, *The Autobiography of Charles Darwin,* ed. Nora Barlow (New York: Norton, 1969), p. 109.

11. In a rare display of humor, Spencer tells this story on himself. See his *Autobiography* (New York: D. Appleton, 1904), 1:109.

12. Grant Allen to Herbert Spencer (10 November 1874), Athenaeum Collection of Spencer's Correspondence, MS. 791, no. 102, University of London Library.

the problems with which he dealt can be uncovered and the selective pressures that shaped his ideas delineated, then not only will the obtuse and constricted nature of his thought stand out, but, I think, its peculiar greatness as well. We can then better appreciate in what measure Spencer's ideas played the sort of role in the development of theories about the evolution of mind and behavior that can be neither ignored nor denigrated.

The conclusion a natural selection model helps deliver is that the central problem with which Spencer struggled and in terms of which his scientific efforts must be understood is that of moral behavior: more precisely, how the natural processes of evolution could produce a moral society. Spencer, like many other Victorian scientists who serenely shed their belief in the Christian God, passionately struggled to retain the moral interpretation of nature that the ancient creed had supported. Spencer's "ultimate purpose, lying behind all proximate purposes," as he judged his own intellectual trajectory, "has been that of finding for the principles of right and wrong in conduct at large, a scientific basis." [13] It was in pursuit of this purpose that he undertook the project of "Synthetic Philosophy." I will, however, argue that he was able to construe evolution as having a moral goal because he conceived of the mechanisms and processes of evolution according to the logical pattern of his ethical principles. This runs counter to the usual assumption that his moral ideas derived from a theory of evolutionary survival of the fittest. Two further aspects of his conception of evolution have also passed without due historical consideration. First, his utopian socialism. The young Spencer envisioned an ideal society in which government would have withered away, classes would have dissolved, and wealth in land would be held in common, a society in which individual freedom would operate harmoniously with social dependence. Spencer, I believe, attempted to give scientific substance to this dream by projecting it as the inevitable outcome of the evolutionary process. But to cast evolution in that role required he formulate his evolutionary theory so that it could have such consummation.

Spencer's discovery of a via media between Lockean and Kantian epistemologies furthered his efforts to construct a scientifically sanctioned morality. This second feature of his evolutionary theory has, however, elicited little interest from historians of science, despite the importance Spencer himself placed upon it: he thought his discovery of

13. Herbert Spencer, "The Data of Ethics," part 1 of *The Principles of Ethics* (Indianapolis: Liberty Classics, [1893] 1978), 1:31. "The Data of Ethics" appeared originally as a separate book in 1879.

an evolutionary compromise would put his *Principles of Psychology* on the shelf next to Newton's *Principia Mathematica*.

Darwin, as we have seen, elevated evolutionary processes to produce moral consequences; Spencer reduced his moral principles to the developmental mechanisms of nature. But in the end, their evolutionary perspectives merged: each came to regard nature, especially human nature, as intrinsically moral.

Early Life as a Nonconformist

Education in Dissent

Spencer descended on both sides of his family from religious nonconformists who settled in Derbyshire. The Spencer family and the family of his mother, the Bretells, were among John Wesley's earliest followers. His paternal grandmother, Catherine Spencer, who died at age eighty-four in 1833, when Spencer was thirteen, knew John Wesley personally, and "was among the few who attached themselves to him in the days when he was pelted by the populace."[14]

Spencer recalled his mother as a sweet, practical-minded woman, who "always criticized discouragingly" his enlarged plans and "urged the adoption of some commonplace career."[15] Like many other eminent Victorian thinkers, Spencer was the child only of his father, whom, despite a lack of warming sympathy, he revered. Of William George Spencer (called George), his son judged that "save in certain faculties specially adapting me to my work, inherited from him with increase, I consider myself as in many ways falling short of him, both intellectually and emotionally as well as physically."[16] George Spencer was a frustrated inventor, who, for instance, devised a new kind of shorthand, which his son tried to promote, with no success. His keen interest in mechanical contrivances undoubtedly recommended him for the post of honorary secretary to the Derby Philosophical Society (founded by Erasmus Darwin and associated business men of the area). The son shared this penchant for curious invention, adding as a young man to his father's list of gadgets a binding pin to hold newspapers together, an instrument for calculating train velocities, an engineer's level, a scheme for a universal language, and other such devices. (In his middle years, Spencer's inventions, as well as his habits, became more domestic: he designed a collapsible fishing rod and an invalid's bed with elevator.)

14. Spencer, *Autobiography* 1:21.
15. Ibid., p. 66.
16. Ibid., p. 48.

But more importantly, the father contributed to the intellectual character of the son an intractable independence and an extreme distaste for authority. Herbert, though, was not the only one to have been infected by his father's intellectual dispositions. The Reverend Thomas Mozley, who achieved some small literary fame in mid-century, wrote of his former teacher: "I had derived straight from the elder Spencer a constant repugnance to all living authority and a suspicion of all ordinary means of acquiring knowledge." [17]

George Spencer, raised in a dissenting religion, dissented even from that. He frequently disputed with the Methodist ministers about books for the parish library: he wanted materials that were really instructive, while they simply wanted more tracts on religion. George, finally exasperated with Methodist governance, fell in with the Quakers, going to their meetinghouse on Sundays. Herbert later judged the system of the Friends, "in respect of its complete individualism and absence of ecclesiastical government," to be congenial to his father's nature. [18] Because of illness, George Spencer had to give up teaching in grammar school. He acquired lace-manufacturing equipment and moved the family to Nottingham when Herbert was about five years old. His business declined over three years, and after losing a considerable sum, he was forced to return to Derby and teaching, though only taking on private pupils.

Herbert Spencer was born in 1820 and received an indifferent education up to age thirteen. His first schoolmaster complained that Spencer "was exceedingly unwilling to learn the Latin grammar, and with some trouble we found the objection to consist in its want of system." [19] When he was ten, his education was undertaken by his uncle William, who inherited the school of Herbert's grandfather. In 1833, the Spencers decided to entrust the further instruction of their son to Thomas Spencer, George's younger brother.

Thomas Spencer, born in 1796, was the only one of the older generation to achieve any fame. He received honors at St. John's, Cambridge, and had a small living as curate at Charterhouse Hinton, near Bath. At college he was a leader of the evangelical movement in religion and the radical movement in politics. Spencer remembered his uncle as a personally generous man, tending to the needs of his parishioners. Yet Thomas Spencer argued against the state dole, since he thought it tended to make the poor dependent and encouraged mischief. The

17. Ibid., p. 52
18. Ibid., p. 94.
19. Ibid., p. 95.

theme of his many pamphlets against the Poor Laws (which mandated a tax of the parish to feed the poor), the Corn Laws (which put tariffs on imported grain), and a centralized ecclesiastical authority was that government had no right to interfere in the private affairs of its people. This became the pulse beat of his nephew's social philosophy.[20]

Herbert spent two and a half years at the school for boys run by his uncle. There he received instruction in mechanical arts, physics, chemistry, and mathematics, which last his uncle had read at Cambridge. He was also tutored in Latin, Greek, and French, but retained nothing of the ancient languages and merely a smattering of the modern. Of poetry, literature, and history, his uncle offered only the dried ends of translation exercises. If Spencer initially had any feeling for the humanities, it must have withered at Hinton. As an adult he amply demonstrated a magnificent indifference to poetry, history, and the fine arts.[21] Shelly's "Prometheus Unbound," he admitted, was "the only poem over which I have ever become enthusiastic."[22] When in his fifties he took up the *Iliad* as part of his study of ancient superstition, he could get through only six books. He said "I would rather give a large sum than read to the end."[23] And after his first trip to an opera, his judgment reflected those aesthetic principles of his early trade, mechanical engineering: "The inconsistencies of recitative dialogue, the singing words of wholly opposite meanings to the same harmony, &c., &c., so continually annoyed me as to destroy all the pleasure due to the music or the story."[24]

During his youth and early adulthood, Spencer read very little. His curiosity was piqued by neither the famous nor the pedestrian. His father's library had on its shelves Locke's *Essay,* but Spencer never took it down. He did peruse several of Harriet Martineau's novellas in the series *Illustrations of Political Economy,* in which the prolific social critic attempted to inculcate through interesting stories the lessons of Adam Smith and Thomas Malthus; but Spencer claimed that he read only for the tale and not the moral.[25] He glanced at the recent translation of

20. J. D. Y. Peel describes in rich detail the political and philosophic currents sweeping through the radical movement of the 1840s. See his *Herbert Spencer,* pp. 56–81.

21. For Spencer's own evaluation of his education, see his *Autobiography* 1:31–47, 113–42.

22. Ibid., p. 299.

23. Ibid., p. 300.

24. Ibid., p. 314.

25. Some of the lessons Miss Martineau propounded may, nonetheless, have disposed Spencer for their self-conscious adoption later. For instance, in the first of her nine volumes of *Illustrations of Political Economy* (1832–1834), entitled *Life in the Wilds* (Boston:

Kant's *Critique of Pure Reason,* but upon learning of the Kantian theory of space and time, and rejecting it, "went no further" in the book. He confessed in his *Autobiography* that "it has always been out of the question for me to go on reading a book the fundamental principles of which I entirely dissent from."[26]

Spencer's attitudes about the arts and literature indicate not simply a lack of negative capacity, as Keats called it, the inability momentarily to suspend disbelief, but a shriveled curiosity. George Eliot met Spencer after he had completed his first book, *Social Statics,* and was surprised that the great expenditure of thought had not left any lines on his broad forehead. He replied: "I suppose it is because I am never puzzled."[27] Darwin, by contrast, always felt puzzled, and in attempting to reach solutions, climbed across great mountains of books, constantly stumbling over new facts and contrary opinions. Spencer met few such obstacles. A letter from a friend captured the young Spencer's habit of mind: "You talk of your power of writing a long letter with very little material; but that is a mere trifle to your facility for building up a formidable theory on precious slight foundations."[28]

Dalliance with Phrenology

Spencer thought his scant literary and historical education kept him "free from the bias given by the plexus of traditional ideas and sentiments."[29] More likely the lack of intellectual ballast left him free, during his youth at least, to tilt toward any theory, no matter how queer, that chanced to attract his attention. His early involvement in phrenology suggests this.

When Spencer was eleven, he heard lectures by Gall's itinerant disciple Spurzheim. He converted to phrenology, and remained a believer for many years. When in January 1842, Mr. J. Q. Rumball, a phrenolo-

---

Bowles, 1833), she tells a tale designed to bring home the virtues of the "Division of Labor" (p. xvii); in volume 6, *Weal and Woe in Garveloch* (Boston: Bowles, 1833), she inspired her readers with the great principle of social behavior: "the greatest happiness of the greatest number" (p. 200); and in volume 7, *Cousin Marshall* (Boston: Bowles, 1833), she taught that "all arbitrary distribution of the necessities of life is injurious to society, whether in the form of private almsgiving, public charitable institutions, or a legal pauper-system" (p. 186). These several doctrines would become important elements in Spencer's social theory.

26. Spencer, *Autobiography* 1:289.

27. Ibid., p. 462.

28. Letter from E. A. B., 1842, quoted in David Duncan, *Life and Letters of Herbert Spencer* (New York: D. Appleton, 1909), 1:45.

29. Ibid. 2:523.

gist of some repute, came to Darby, Spencer had his head read. Rumball thought "such a head as this ought to be in the Church"—a diagnosis Spencer later prized.[30] Perhaps this early experience indicated to him the less than scientific character of the discipline, and encouraged him in 1846 to design a machine that would aid in the more accurate interpretation of cranial topography.[31] Typically, Spencer went his own way as a phrenologist, writing several articles for a phrenological quarterly called *The Zoist,* in which he propounded variants of mainstream phrenology.[32] A dissenter always, he finally abjured in his late twenties the specific text of phrenology, but not the context. There were several broad features of phrenological doctrine that seem to have remained with Spencer, impelling him along the particular path he took in his evolutionary biopsychology.[33]

One can pull from George Combe's immensely influential phrenological tract *Essay on the Constitution of Man* (1827) representative ideas that, if they were not precisely responsible for aspects of Spencer's later views, at least nurtured them.[34] The leading thesis of phrenology was that character and intellect were functions of the brain and subject to natural causation. Though Spencer came to reject the idea that traits and abilities could be read directly from the mountains and valleys of an individual's head, he retained belief in a naturalistic and deterministic psychology. The phrenologists derived certain moral and social doctrines from their science. Combe held that the harmonious exercise of the mental and physical faculties, in conformity to physical, organic, and moral laws, produced happiness, while excessive activity of any of the faculties, such that it would abridge the free action of others, led to pain, and was what we called vice.[35] The ideal of a harmonious exercise

30. Spencer, *Autobiography* 1:229.

31. Spencer describes this machine in ibid., pp. 634–38.

32. Ibid., pp. 227–31.

33. George Denton was the first to call attention to the debt of Spencer's psychology to phrenology. See Denton's "Early Psychological Theories of Herbert Spencer," *American Journal of Psychology* 32 (1921): 5–15. Robert Young has insightfully explored the relationship. See his *Mind, Brain, and Adaptation,* pp. 150–62.

34. Spencer read little and cited less. He never mentioned having read Combe's *Essay on the Constitution of Man.* But the circumstantial evidence is strong that he did. Combe's book sold over seventy thousand copies by 1837 (Young, *Mind, Brain and Adaptation,* p. 158), and it would have been the sort of book the library of the Darby Philosophical Society would have had on its shelves. Firmer evidence, however, is provided by Spencer's letters to the journal the *Nonconformist,* "The Proper Sphere of Government." Certain passages from these letters bear the strong stamp of specific ideas found in Combe's *Essay.* I will note these passages in the text.

35. George Combe, *Essay on the Constitution of Man and Its Relation to External Objects,*

of the faculties, where none impinged on the proper sphere of the others, functioned for Spencer as a fundamental principle uniting his psychological, moral, and social views: the well-being of the individual and of the society were, he came to argue, in direct proportion to the free but harmonious activity of the parts. The phrenologists maintained that the moral sense, like other sentiments, was a divinely deposited instinct, not, as the utilitarians held, a learned acquisition.[36] Spencer agreed that the moral sense was an innate faculty, though he later proposed a less than divine origin for it. Combe maintained, borrowing from Erasmus Darwin (among others), that parents, through the exercise of their moral and intellectual faculties, could biologically transmit increased capacities to their offspring;[37] and the theory of inheritance of acquired characteristics became a linchpin of Spencer's biopsychological theory. Though not an evolutionist, Combe did accept Buckland's conviction that the earth had gone through many upheavals, transformations by which God had successively adapted the natural environment as a fit abode for man,[38] a progressivist conception of nature that Spencer shared. Finally, Combe's Enlightenment sentiments about man as a progressive being resonated with the Methodism that took deep root in Spencer: each conception represented the system of nature as basically good, despite appearances of cruelty, and urged that if we could but understand the design of the Creator, we would perceive that all pain and momentary suffering were necessary accommodations that would prepare the way for a progressively better future.[39] Of course, there were important issues separating Spencer from Combe: the phrenologist sought a theological explanation of nature's origin, while Spencer came

---

3d ed. (Boston: Phillips & Sampson, [1835] 1845), pp. 40–51: "the due exercise of the osseous, muscular, and nervous systems, under the guidance of intellect and moral sentiment, and in accordance with the physical laws, contribute to human enjoyment. . . . Thus there is, 1st, a wide sphere of action provided for the propensities, in which each may seek its gratification in its own way, without exceeding limits of morality, and this is a good and proper action: 2dly, There is ample scope for the exercise of each of the moral and intellectual faculties, without infringing the dictates of any of the other faculties belonging to the same classes; and this action also is good. But on the other hand, the propensities in opposition to the dictates of the whole moral sentiments and intellectual powers enlightened by knowledge and acting in combination; and all such actions are wrong. Hence right conduct is *that which is approved of by the whole moral and intellectual faculties, fully enlightened, and acting in harmonious combination*."

36. Ibid., p. 50.

37. Ibid., pp. 151–52, 183.

38. Ibid., pp. 3–5.

39. Ibid., p. 35. Peel (*Herbert Spencer*, pp. 102–11) describes the religious ideology of deterministic optimism that saturated the intellectual atomosphere in which early-nineteenth-century dissenters lived.

increasingly to rely on physical law, claiming that a further account could not penetrate to what was in principle unknowable. And even as a youth, Spencer would not have countenanced Combe's suggestion that a legislature should "accelerate improvements, by adding the constraining authority of human laws to enactments already proclaimed by the Creator."[40] Spencer's early enculturation at the hands of father and uncle persuaded him that government had no right to interfere in the relations among men, save as protector from physical violence and adjudicator of contracts.

Along with the training in dissent Spencer received from his family, and with few other correcting intellectual constraints, phrenological doctrine would seem to have provided a conceptual ambience in which his own scientific and philosophical ideas could be selectively refined and further differentiated. The influence of this early environment is reflected in his first important literary effort. His published letters, "The Proper Sphere of Government," contained seeds that would develop during the rest of his career.

Social and Moral Science as Foundation of Evolutionary Theory

"The Proper Sphere of Government"

Through the intercession of his uncle William, Spencer received a job offer from Mr. Charles Fox (a former pupil of his father), who was the resident civil engineer for the London and Birmingham Railway.[41] From ages seventeen to twenty-one, Spencer worked as an engineer, designing bridges, planning excavations, and so forth, his longest tenure being with the Birmingham and Gloucester Railroad. He rather enjoyed the companionship of his fellows and the mechanical nature of the work. But in 1841, just before his twenty-first birthday, he refused an advancement in order to return to Derby and collaborate with his father on the design of an electromagnetic engine.[42] The project quickly derailed when he realized that a battery-powered device could not compete economically with a steam engine. In January 1842, he again left to board with his uncle Thomas, whose own social interests encouraged the young man to exercise his literary and polemical talents. So stimulated, he composed a series of letters to the newspaper the *Nonconformist,* the editor of which, Edward Miall, was a friend of his uncle.

40. Ibid., p. 222.
41. Spencer, *Autobiography* 1:141–46.
42. Ibid., pp. 215–16.

In his twelve letters, "The Proper Sphere of Government," Spencer erected the basic framework that his book *Social Statics* would fill out some nine years later. The letters successively addressed the Poor Law, commercial restrictions, a national church, war, education, sanitation administration, and colonization. The considerations governing his discussions derived mainly from his family's dissenting philosophy, the doctrines of phrenology, an engineer's sense of balance in natural forces, and adolescent righteousness.

The first letter propounded the general principle, which remained for Spencer a touchstone of political sanity, that government was not ordained to educate, administer charity, teach religion, or make roads, but only to protect person and property. His justification for this dissenting policy advanced upon Combe's principle that the natural laws of the three realms must be obeyed for man's happiness.[43] He argued that inexorable law governed not only the physical, organic, and mental realms, in which transgression brought punishment, but the social realm as well. Government interference, then, could only deform the self-correcting natural forces controlling social development:

> In short, do they want a government because they see that the Almighty has been so negligent in his arrangements of social laws that everything will go wrong unless they are continually interfering? No; they know, or they ought to know, that the laws of society are of such a nature that minor evils will rectify themselves; that there is in society, as in every other part of creation, that beautiful self-adjusting principle which will keep everything in equilibrium; and, moreover, that as the interference of man in external nature destroys that equilibrium, and produces greater evils than those to be remedied, so the attempt to regulate all the actions of a people by legislation will entail little else but misery and confusion.[44]

Spencer drew out the implications of these considerations in subsequent letters, touching on the several governmental policies the nonconformists diagnosed as sources of the national complaint. But most of the "political pills," as he called his tracts, were concocted as a specific for the Poor Law, the malady that seemed to most aggrieve uncle and

43. Combe, p. 19: "The natural laws may be divided into three great and intellectual classes,—Physical, Organic, and Moral; and the peculiarity of the new doctrines is, its inculcating that these operate independently of each other; that each requires obedience to itself; that each, in its own specific way, rewards obedience and punishes disobedience; and that human beings are happy in proportion to the extent to which they place themselves in accordance with all of these divine institutions."

44. Herbert Spencer, "Letter I," *Nonconformist,* 15 June 1842.

nephew.[45] First, Spencer urged that the Poor Law redressed no natural injustice. The improvident reaped their reward. Even those who came to poverty by accident could make no claim in justice on their neighbors.[46] The Poor Law, in Spencer's estimation, either disrupted the salutary consequence of transgressions against natural law or else it deprived the industrious of opportunity to exercise voluntary charity.[47] In most cases the distresses of the poor resulted from heritable character deficiencies of parents; and "the moral disease requires a cure—under a healthy social condition that cure will be found in the poverty which has followed in its train."[48]

Combe had suggested that exercise of the faculties prevented their atrophy.[49] Spencer concurred that instincts and organs diminished when not exercised against external circumstances, a principle with important implications for human moral development:

> One generation after another will the faculty, or instinct, or whatever it may be, become weaker, and a continued degeneracy of the race will inevitably ensue. All this is true of man. He has wants, many and varied—he is provided with moral and intellectual faculties commensurate with the complexity of his relation to the external world—his happiness essentially depends upon the activity of those faculties; and with him, as with all the rest of creation, that activity is chiefly influenced by the requirements of his condition.[50]

When government established a national education, medical rules, and poor laws, it inhibited the exercise of the faculties of those who, as a consequence, no longer needed to secure requisite goods for themselves; and this must retard the progress of the race.

In "The Proper Sphere of Government," Spencer also insisted on the converse of the principle of disuse, and thereby introduced, in a unconsidered way, the mechanism that would become his chief engine of evolutionary advance. He maintained that exercise of the faculties against a variety of external (usually social) circumstances increased their capacities, and that this too could be transmitted to progeny, thus producing continued social progress. He harbored few doubts about the kind of social circumstances that would form the hard flinty edge against which social development might be honed:

45. Duncan, *Life and Letters of Herbert Spencer* 1:52.
46. Spencer, "Letter II," *Nonconformist,* 22 June 1842.
47. Spencer, "Letter III," *Nonconformist,* 13 July 1842.
48. Spencer, "Letter IV," *Nonconformist,* 27 July 1842.
49. Combe, p. 83.
50. Spencer, "Letter IX," *Nonconformist,* 23 November 1842.

the Almighty has given laws to the general mind of mankind which are working together for its ultimate advancement. It may be fairly assumed that, in this case as in the more tangible ones, the apparently untoward circumstances are in reality eminently conducive to the attainment of the object sought after. That all the prejudices, the mental idiosyncrasies, the spirit of opposition, the tendencies to peculiar views, and a host of other qualities, in their infinitely varied proportions and combinations, are all conspiring to bring about the mental, and moral, and, through them, the social perfection of the human race. If it be granted that man was created a progressive being, it must be granted also that the constitution given to him by his Creator was the best adapted to secure his progression.[51]

The Lord, it seems, was good to have provided dissenters, whose "mental idiosyncracies" and "spirit of opposition" would ensure the social development of the race.

The Ripening of Spencer's Social Theory

Spencer spent the next two years, from 1842 to 1844, in desultory reading and at efforts to pursue a literary career. In 1843 he wrote to a friend that "I have been reading Bentham's works, and mean to attack his principles shortly, if I can get any review to publish what will appear to most of them so presumptuous."[52] Shortly thereafter his interest in a phrenological treatment of sympathy brought him to pick up Adam Smith's *Theory of the Moral Sentiments*. And a year later, in the library of the Derby Philosophical Society, he dipped into Mill's newly published *System of Logic*. But he spent most of his time in writing: he composed articles for the soon-to-be-defunct journals the *Eclectic* and *Tait*, neither of which published his work; he wrote an unorthodox piece for the *Phrenological Journal*, though only the more heretical *Zoist* brought it out; he revised his letters on "The Proper Sphere of Government" for issuing in pamphlet form, the sales of which failed to recover its printing cost; and he produced an account of his father's system of shorthand, which went unprinted. In 1844 Spencer secured a subeditorship of the new radical newspaper *Pilot*, for which he wrote many of the lead articles during his first month; but he cut short his journalistic career to return to engineering at the urgent behest of a former employer. It was a time of the great railway mania in England, when both commerce

51. Spencer, "Letter VII," *Nonconformist,* 19 October 1842.
52. Spencer, *Autobiography* 1: 260.

and stock greatly increased in value. Finally, in 1848, when he was twenty-eight, Spencer found an ideal situation for prosecution of a profession in literature and philosophy. He became a managing editor of the *Economist,* a classically liberal newspaper founded (1843) and edited by James Wilson.

The working conditions at the *Economist* were extremely congenial. Not only did Spencer have ample time to engage in private study and writing, but the laissez-faire philosophy of Wilson and the radical socialism of his coeditor Thomas Hodgskin nurtured his own social evolutionary thought. Wilson's view of society as an organism with each part efficiently performing its particular functions, a conception that formed the bloodline of the *Economist's* pages,[53] likely predisposed Spencer to be receptive to the idea when he rediscovered it in the anatomical work of Milne-Edwards in 1851. Spencer became particularly close to Hodgskin, often spending evenings in talk with him.[54] Hodgskin undoubtedly encouraged Spencer in his particular brand of socialism and anti-utilitarianism.

Hodgskin, like Combe whom he frequently cited, believed that social development was governed by natural laws and that legislation could only protect, not confer, human rights, especially those of property.[55] "The master principle of all modern production," he asserted "is division of labour, or mutual co-operation," whence flowed social improvement.[56] He opposed the Benthamites, who regarded governmental legislation as the source of property rights. Indeed, Hodgskin rejected virtually all legislation, because it interfered with the natural operation of social laws. These ideas and their further elaboration in Hodgskin's *Labour Defended against the Claims of Capital* (1822) and *Natural and Artificial Rights in Land Contrasted* (1832) found warm endorsement by Karl Marx. Spencer also adopted and further refined his friend's ideas in his first book, *Social Statics,* begun in summer of 1848.

53. Scott Gordon furnishes an excellent account of the *Economist's* social theory. See his "The London *Economist* and the High Tide of Laissez Faire," *Journal of Political Economy* 43 (1955): 461–88.

54. Spencer, *Autobiography* 1 : 398.

55. See for example, Hodgskin's quotations from Combe's *Constitution of Man* in his *The Natural and Artificial Right of Property Contrasted* (London: Steil, 1832), pp. 49–50, 153–54.

56. Ibid., p. 138. While Spencer's introduction to the master principle of the division of labor likely came from Hodgskin, he may yet have encountered it in the first chapter of Adam Smith's *Wealth of Nations.* Had Smith discussed the principle in a later chapter, he might be more easily dismissed as a possible source for Spencer. See note 25 for another possibility.

Social Statics

In *Social Statics* (1851), Spencer pursued ideas first laid down in the *Nonconformist* and which had matured during the intervening period. He envisioned a progressive development of society according to the moral laws of nature. The two chief laws were (his version of) the greatest happiness principle and the principle of equal freedom. An individual's happiness consisted in the proportionate exercise of all his faculties, without any one faculty unduly restricting the freedom of others; and a society's happiness depended upon the proportionate functioning of all its different parts, with no one part attempting to regulate the rest. If that were so, and Spencer's early study of phrenology convinced him it was, then a man "must have liberty to do all that his faculties naturally impel him to do."[57] Since God wills human happiness, a man has a natural right to liberty. And since all men have this right, it can only be limited by others exercising their freedom.[58] The only legitimate function of the state, Spencer concluded, would be the protection of the equal rights of all; to do more would necessarily infringe on the rights of some in order to give unfair advantage to others.[59]

Like his fellow workers at the *Economist,* Spencer rejected utilitarian social theory. He did so on three grounds. First, the utilitarians assumed that human nature remained stable, so that happiness meant the same for all men at all times. Spencer's study of phrenology suggested to him the great variability of human constitutions;[60] and his own experience indicated that his pleasures were not apt to be those of others—Bentham liked pushpin as well as poetry, but Spencer enjoyed neither. Second, the expediency philosophers confused, as Darwin also pointed out, the general end of society with the motive for immediate moral judgments. The greatest happiness may be the "creative purpose" of a perfect society, but men instinctively judged actions right or wrong without intellectual calculations of general felicity.[61] Finally, the utilitar-

57. Herbert Spencer, *Social Statics: or, the Conditions Essential to Human Happiness Specified, and the First of them Developed* (London: Chapman, 1851), p. 77.

58. Ibid., pp. 76–77.

59. Ibid., p. 278: "Let it [government] undertake many additional duties, and there will be scarcely a man who does not object to being taxed on account of one or more of them—scarcely a man, therefore, to whom the state does not in some degree do the very opposite of what it is appointed to do. Now this thing which the state is appointed to do is the essential thing—the thing by which society is made possible; and these other things proposed to be done are nonessential, for society is possible without them. And as the essential ought not to be sacrificed to the nonessential, the state ought not to do anything but protect."

60. Ibid., pp. 4–5.

61. Ibid., p. 66.

ians charged government with determining the means, through legislation, by which society would be brought to the greatest happiness;
but this presumption both violated the law of equal freedom and ignored the natural developmental forces that impelled society toward a
perfect end. These last two considerations, as expressed in *Social Statics*,
already contained those features that formed the backbone of Spencer's
theory of mental and moral evolution, so they deserve some further
exploration.

As a legacy both of his ideas about natural rights and his phrenological faculty psychology, Spencer argued that individuals came endowed
with a moral sense. "This instinct or sentiment," he assured his readers,
"being gratified by a just action and distressed by an unjust action, produces in us an approbation of the one and a disgust toward the other;
and these readily beget beliefs that the one is virtuous and the other
vicious."[62] Spencer analyzed the moral sense into two principal components: the instinct for personal rights—"a purely selfish instinct,
leading each man to assert and defend his own liberty of action"; and
the sentiment of sympathy.[63] Our native feeling of pleasure in the free
exercise of our faculties reached altruistically to our fellow man because
of sympathetic response. Spencer's explanation of sympathy generally
followed the account in Adam Smith's *Theory of the Moral Sentiments*,
which he had read in preparation for reanalysis of certain phrenological
hypotheses. Smith held that sympathy was a natural faculty of imagination by which we put ourselves in the circumstances of another:
"Sympathy, therefore, does not arise so much from the view of the passion, as from that of the situation which excites it."[64] Thus we do not
take pleasure in another's happiness, but imaginatively situate ourselves
in the position of another and experience an original joy. The motive,
then, of beneficent action must be individual happiness, a proposition
Spencer fully endorsed: we "are led to relieve the miseries of others,"
he concluded, "from a desire to rid ourselves of the pain given by the
sight of misery, and to make others happy because we participate in
their happiness."[65] Though we may be morally outraged at an injustice
done to another by the state, or mercifully disposed to offer a hand to
someone in need, it is ourselves we mourn for. Altruistic behavior must
at root be selfishly motivated. Yet only at root. The innate determinants

62. Ibid., p. 26.
63. Ibid., p. 96.
64. Adam Smith, *The Theory of the Moral Sentiments*, ed. A. Macfie and D. Raphael,
Glasgow edition of the Works of Adam Smith (Oxford: Oxford University Press, 1976),
p. 12.
65. Spencer, *Social Statics*, p. 97.

of such behavior, for Spencer, lay below the reflective plane of charitable intention. We might be Benthamites under the skin; but it was
the evolutionary processes of a wiser nature that would perform the
hedonistic calculations. (When Spencer later introduced natural selection into the moral equation, even nature would be seen to perform her
calculations more unselfishly.)

Spencer's Utopian Socialism

Spencerian ethics, even in its initial formulations, seems to our sensibilities rather mean spirited. Spencer, after all, refused the ameliora-
tions of the state to those in distress and contended that we helped
others only to help ourselves. But these easily formed sentiments need,
I think, the restraint that the original environment of Spencer's thought
provides. He grew up during the enforcement of the old Elizabethan
Poor Laws, which mandated a tax on the parish to give meager aid to
widows, orphans, the sick, and the momentarily deprived.[66] The law's
institution varied from parish to parish, but it generally produced a
debilitating and degrading effect everywhere. Parish vestries had the
financial obligation to care for their poor, but they were constrained by
London authority in administration. Poor women often felt compelled
to bear as many children as possible, frequently illegitimately, since al-
locations thereby increased. Spencer undoubtedly heard his uncle
Thomas rail against the abuses of the laws, especially the Speenhamland
plan. Named for the village of its origin, the plan was formulated to
quell the food riots that erupted in 1795, after a series of bad harvests.
It originally allowed use of the poor tax to supplement the wages of
workers. Its calming effect forestalled the kind of social upheaval recently experienced across the Channel. This insidious use of the poor
laws caught on among the large landowners and manufacturers, since it
allowed them to keep wages below subsistence. Thus the wealthy had
the self-sufficient working poor help subsidize the wages of the insufficient working poor. It was a practice that Spencer found particularly
reprehensible, vying in probity with horse-trading. Even after the re-
forms of 1834, these sorts of abuses often continued, preserving what
Edmund Burke had earlier assumed as the natural lot of humanity—a
large and permanent underclass of the wretchedly destitute.[67] Govern-

66. The Elizabethan Poor Law and subsequent legislation are discussed by R. K.
Webb, in his *Modern England: from the 18th Century to the Present* (New York: Dodd and
Mead, 1971), pp. 30–31, 54–56, 135, 193, 242–47.

67. See Gertrude Himmelfarb's discussion of Burke's view of poverty in her *The Idea
of Poverty* (New York: Knopf, 1984), pp. 66–73.

ment, to Spencer's mind, fostered this assumption, because its own bungling and corruption made it a living reality. But another dispensation, he believed, was a possibility—rather, if his evolutionary calculations proved correct, an inevitability. One could realistically hope for the advent of a state of constitutionally perfect individuals, who had so mentally and physically adapted to living in society that each could exercise the full freedom of his or her faculties without infringing on the free activity of others. In this consummation, social distinctions of class and caste would be absorbed in a respect for the dignity of each person's share in the division of labor; and privately held land would be restored as the common patrimony of all citizens. Government, the largest impediment to this development, would wither away. It was a socialist's utopia achieved through an evolutionary process.

Spencer's evolutionary ideas, to which I will specifically turn in a moment, began to smolder during the 1840s, initially ignited by his reading of phrenology and Lyell's account of Lamarck's views. The evolutionary theory adumbrated in *Social Statics* served as a vehicle to give some scientific substance to his utopian socialism. Spencer believed that the progressive development of man toward a perfect society was a logical necessity.[68] He argued that individuals were ever adapting to different social circumstances through inherited use and disuse. Increasing adaptation, on its side, improved social relations and circumstances, thus creating a new environment against which individuals would, perforce, continue to adapt. This developmental dynamic would, in Spencer's estimation, have two consequences, one social and one moral. Progressive adaptation would, first, produce greater specialization within society, each part of the social organism becoming more articulated and adapted to particular functions. However, not only was the principle of the division of labor a mark of advanced society, as every reader of the *Economist* already knew, but it found confirmation in the science of zoology. Spencer cited many examples from T. R. Jones's *General Outline of the Animal Kingdom* to demonstrate that: "By greater individuality of parts, by greater distinctness in the nature and functions of these, are all creatures possessing high vitality distinguished from inferior ones."[69] With articulation of parts came a corresponding ability to withstand the ravages of external circumstances, since the functions of one part could make good the deficiencies of another. In the animal organism, so in the social organism, greater individuation brought a more perfect kind of life. In drawing this conclusion, Spencer attempted to infuse a social

68. Spencer, *Social Statics*, pp. 63–5.
69. Ibid., p. 438.

principle[70] with the logical force of an accepted scientific axiom of his day.[71]

Adaptation to the social state had, in Spencer's estimation, a further consequence, this a moral one. He argued that behavior injurious to our neighbors would diminish, since their disagreeable reactions would tend to repress the desires in us that produced such behavior. Accordingly, social and moral evil, which he interpreted as "non-adaptation of constitution to conditions,"[72] would gradually die out: "Equally true is it," he wrote, "that evil perpetually tends to disappear. In virtue of an essential principle of life, this non-adaptation of an organism to its conditions is ever being rectified; and modification of one or both continues until the adaptation is complete. Whatever possesses vitality, from the elementary cell up to man himself, inclusive, obeys this law."[73]

Adaptation to the social state thus produced an evanescence of evil and a social organization with the most elaborate "subdivision of labor"—that is, an extreme mutual dependence, because of adaptation, but with each individual exercising his faculties in complete freedom, since all pernicious desires will have been purged from society.[74] The vision differed little in essence from that of the two Germans exiled in London at the time, who had a similar dream: "The old bourgeois society, with its classes and its conflicts of classes gives way," they foretold, "to an association where the free development of each individual is the condition of the free development of all."[75]

Spencer believed that such legislation as the Poor Law stood as a barrier, temporary to be sure, to the realization of the perfect society. The Poor Law had three characteristics that suggested it really belonged to a more primitive stage of social development. First, it prevented adaptation to the requirements of social advance by allowing the propagation of the manifestly unfit. In an anticipation of Darwin's principle

70. Ibid., pp. 85–86.

71. Young (*Mind, Brain, and Adaptation*, p. 162) assumes that Spencer got the principle of division of labor from Jones's *Outline of the Animal Kingdom*, a strange mistake for a Marxist historian.

72. Spencer, *Social Statics*, p. 59: "All evil results from the non-adaptation of constitution to conditions. This is true of everything that lives. Does a shrub dwindle in poor soil, or become sickly when deprived of light, or die outright if removed to a cold climate? It is because the harmony between its organization and its circumstances has been destroyed."

73. Ibid., pp. 59–60.

74. Ibid., pp. 441–42.

75. Karl Marx and Friedrich Engels, *The Manifesto of the Communist Party*, in *Karl Marx, The Revolutions of 1848: Political Writings, Volume 1*, ed. David Fernbach (New York: Vintage, 1974), p. 86.

of natural selection, to which anticipation he would later point with wounded pride, Spencer invoked the rule of the animal kingdom in order to demonstrate the manner of nature's cleansing ministrations:

> their carnivorous enemies not only remove from herbivorous herds individuals past their prime, but also weed out the sickly, the malformed, and the least fleet or powerful. By the aid of which purifying processes, as well as by the fighting, so universal in the pairing season, all vitiation of the race through the multiplication of its inferior samples is prevented; and the maintenance of a constitution completely adapted to surrounding conditions, and therefore most productive of happiness, is insured.[76]

A Poor Law disrupted the natural force that would swiftly slough off the reprobate from the social organism. That miscreants would eventually be removed, Spencer had no doubt, since progress was inevitable. Misguided legislation only painfully prolonged the social sickness.[77]

Yet, this engineer's heart was not so unfeeling that he would allow the unfortunate to be ground in the dust. He advocated private charity, of which his uncle Thomas gave good example, in place of state almsgiving. He truly believed that private charity adapted the instincts of the benefactor to the needs of others, and thus, through personal exercise, altruistic behavior would become a heritable legacy. But a poor law inhibited the sort of character change in individuals that in the long run would produce a more humane society. The practice of voluntary charity, he judged, would be worth the price of preserving some of those who by natural right perhaps deserved a less kind fate.[78]

Finally, Spencer detected in the Poor Law an evil that would also inflame the young Marx. Poor Law legislation served to repress a potentially rebellious population. The people cry for land and an equitable

76. Spencer, *Social Statics*, p. 322.

77. Ibid., pp. 323–24: "we must call those spurious philanthropists, who, to prevent present misery, would entail greater misery upon future generations. All defenders of a Poor Law must, however, be classed among such. That rigorous necessity which, when allowed to act on them, becomes so sharp a spur to the lazy and so strong a bridle to the random, these pauper's friends would repeal, because of the wailing it here and there produces. Blind to the fact that under the natural order of things, society is constantly excreting its unhealthy, imbecile, slow, vacillating, faithless members, these unthinking, though well-meaning, men advocate an interference which not only stops the purifying process but even increases the vitiation—absolutely encourages the multiplication of the reckless and incompetent by offering them an unfailing provision, and *discourages* the multiplication of the competent and provident by heightening the prospective difficulty of maintaining a family."

78. Ibid., pp. 318–21.

share of its wealth, but the moneyed classes keep them groveling at the almshouse. In a passage that might be mistaken for an excerpt from the *Communist Manifesto,* Spencer declared:

> All arrangements, however, which disguise the evils entailed by the present inequitable relationship of mankind to the soil postpone the day of rectification. "A generous Poor Law" is openly advocated as the best means of pacifying an irritated people. Workhouses are used to mitigate the more acute symptoms of social unhealthiness. Parish pay is hush money. Whoever, then, desires the radical cure of national maladies, but especially of this atrophy of one class and hypertrophy of another, consequent upon unjust land tenure, cannot consistently advocate any kind of compromise.[79]

Here and in many other passages of *Social Statics,* Spencer's indignation reached levels of eloquence never again attained. He described with great feeling the wretched conditions of the working poor and with great scorn the advantages of those whose mercantile practices reeked of adulterated products, dishonest weights, and viciously low wages. In a voice raised to a martial tempo, he declaimed of the inevitability of violent revolution in those cases in which a repressive government attempted to constrain the progressive development of a society: "The existence of a government which does not bend to the popular will—a despotic government—presupposes several circumstances which make any change but a violent one impossible."[80] Whether the British monarchy and its parliament held an iron lid on the bubbling injustices of his society so that violent explosion had to result, Spencer did not say. But he did foresee the final social settlement. There would be a revolution to restore "equilibrium" between character and institutions, perhaps peaceful, perhaps violent.[81] Out of this would eventually come a society in which land would be held in common, class structures would have crumbled, and legislative government would have withered away.

Is this socialism or even communism?[82] The critic of the *North British Review* thought so. He declared that in *Social Statics,* "Mr. Spencer but repeats the well-known cardinal speculation of Proudhon, whose notion also it is that history is an evolution of the doctrine of equal rights, and that the goal to which the human race tends is that of *anarchy,* or the

79. Ibid., p. 316.
80. Ibid., p. 432.
81. Ibid., pp. 431–34.
82. The terms "socialism" and "communism" were often used interchangeably in the middle of the nineteenth century. See Spencer's own usage in *Social Statics,* p. 131.

absence of all forms of government."[83] But Spencer even went further
than Proudhon, since he included both women and children in this
social evolution, arguing that they should achieve equal rights with
men. For the reviewer this latter proposal conjured up nightmares of
domestic anarchy. He was, however, relieved that Spencer's socialism
extended only to land and not to all private property.

Spencer's restriction of common dominion to land indeed provided
him that criterion by which he distinguished his theory from Proudhon's. He rebuked the communists for failing to recognize that the faculty of acquisition formed an intrinsic part of our nature and, indeed,
was the reason for demanding all men have opportunity to enjoy their
original patrimony of land.[84] Yet Spencer undoubtedly recognized
among the radicals men of kindred vision, for he conceded basic consilience, admitting that his own principles did "not at all militate against
joint-stock systems of production and living, which are in all probability
what Socialism prophesies."[85]

Spencer was not the only philosopher of the period to bend laissez-
faire far enough to have it join with socialism. John Stuart Mill, fed on
Benthamism as a child, raised on his father's extreme individualism, yet
came to call himself a socialist in his old age. Mill's changing attitudes
toward communism and socialism can be followed through the several
editions of his *Principles of Political Economy,* from the first (1848), where
he could detect only impracticalities and unsound psychology, to the
seventh (1871), in which he declared that if the choice were between
"Communism with all its chances" and "the present state of society with
all its sufferings and injustices," the difficulties of Communism "would
be as dust in the balance."[86] He found Fourierism particularly attractive, since that brand of socialism did not contemplate the abolition of
private property but distributed ownership rights through stock trans-
fers. In his *Autobiography,* Mill classified himself and his wife as socialists, and considered that the social problem of the future would be "how
to unite the greatest individual liberty of action with a common ownership in the raw material of the globe, and an equal participation of all

83. "Reviews of Michelet, Comte, Mill, Fichte, Spencer, and Newman," *North British Review* 15 (1851): 324. An anonymous critic for the *Leader* (probably the editor and Spencer's friend George Lewes) also classed Spencer with Proudhon. See "Proudhon on Gov-
ernment," *Leader* 2 (1851): 997.

84. Spencer, *Social Statics,* pp. 132–33.

85. Ibid., p. 132n. In the much abridged version of *Social Statics,* which Spencer issued
in 1892, no quarter was given to communism, and this conciliatory note dropped from
sight.

86. John Stuart Mill, *Principles of Political Economy,* vols. 2 and 3 of *The Collected Works
of John Stuart Mill,* ed. J. M. Robson (Toronto: University of Toronto Press, 1965), 2:207.

in the benefits of combined labor."[87] As Mill grew older his political philosophy became more utopian and socialistic. But age produced an opposite effect on Spencer.

Spencer's social vision dimmed in his later years. In the greatly abridged version of *Social Statics*, which he issued in 1892 (along with *Man versus the State*—a work more in keeping with the Spencer who dwells in memory as the *bête noire* of modern social idealism), the notion that individuals could become perfectly adapted to life in the state vanished, for good biological reason: "The rate of progress towards any adapted form must diminish," he estimated, "with the approach to complete adaptation, since the force producing it must diminish; so that, other causes apart, perfect adaptation can be reached only in infinite time."[88] Spencer also judged his early proposal that land ownership revert to the community to have neglected certain considerations that indicated individual ownership was more equitable.[89] With these alterations and the growth of Marxism during the later part of the century, Spencer erased all hint of an earlier shared cause with socialism.[90]

Several of the elements of Spencer's later, more explicitly expressed theories of mental and organic evolution appear in his letters to the *Nonconformist* and in *Social Statics*. Some of them have a decidedly social-moral source, for example, his notions that the social organism progresses through ever-more-complex stages of the division of labor and that greater individuality within a context of social dependence characterizes those stages. Some of the roots of his later theories, however, bespeak a mixed parentage, reflecting also phrenological and Lamarck-

87. John Stuart Mill, *Autobiography of John Stuart Mill* (New York: New American Library, [1873] 1964), pp. 167–68.

88. Herbert Spencer, *Social Statics, Abridged and Revised; together with The Man versus the State* (New York: D. Appleton, [1892] 1904), p. 32

89. See Spencer's reconsiderations in *Principles of Ethics*, 2:appendix B.

90. Bruce Wiltshire, in his *The Social and Political Thought of Herbert Spencer* (Oxford: Oxford University Press, 1978), compresses the end of Spencer's thought into its beginning when he asserts of the young Spencer: "Evolution, again, is endless. The processes of modification and adaptation cannot reach a final, static resolution. Spencer disowned Utopianism" (p. 193). Compare this observation with Spencer's own description of the process of social evolution in the first edition of *Social Statics* (p. 65): "Progress, therefore, is not an accident, but a necessity. Instead of civilization being artificial, it is a part of nature; all of a piece with the development of the embryo or the unfolding of a flower. The modifications mankind have undergone, and are still undergoing, result from a law underlying the whole organic creation; and provided the human race continues, and the constitution of things remains the same, those modifications must end in completeness. . . . So surely must the human faculties be molded into complete fitness for the social state; so surely must the things we call evil and immorality disappear; so surely must man become perfect."

ian doctrine (as well as engineering principles)—thus his conception of adaptation to a complex social environment and of an equilibration of forces powering ever more articulated development. And some of his later evolutionary notions were simply taken over from the phrenologists and Lamarck, for example, the idea of the inheritance of individual characteristics through use and disuse. The later changes in Spencer's social doctrine cover over that original conceptual environment in which his evolutionary theory grew. But it is his youthful utopian vision of a morally perfect terminus to society that makes intelligible to the historian the selective incorporation of these just-mentioned elements into his developing theory of evolution.

The Development of Spencer's Theory of Evolution

Spencer's ideas about evolution developed against certain moral and social conceptions. The impact of these conceptions penetrated to the very root of his scientific considerations, leading him to identify physiological law with moral principles: "moral truth, as now interpreted," he concluded in the last chapter of *Social Statics*, "proves to be a development of physiological truth; for the so-called moral law is in reality the law of complete life."[91] But if moral ideals provided Spencer the mold for his evolutionary conceptions, it is also true that his concentrated effort to work out a theory of evolution produced a reciprocal effect on his moral notions. For example, as he came to have more confidence in the validity of his evolutionary perspective, he depended more on natural law to serve as sanction for his moral principles. Even in *Social Statics,* divine authority for moral principles had begun to give way by the end of the book to the inexorable rule of nature. Science had replaced God as guarantor of the validity of moral laws. We need, then, to examine Spencer's original elaborations of evolutionary theory in order to gauge its rebounding impact on his ethics.

Influence of Lamarck and Chambers

During his work as a civil engineer, from ages seventeen to twenty-one, Spencer became interested in collecting the fossils he sometimes stumbled across in excavations. To better acquaint himself with the subject—and geology in general—he purchased in 1840 Lyell's three-volume *Principles of Geology.* In the second volume, he read the detailed description of Lamarck's evolutionary theory and Lyell's lengthy rebuttal. Expressing an inveterate recalcitrance, he found the ideas of the

91. Ibid., p. 461.

French zoologist more attractive than Lyell's biological orthodoxy.[92] Spencer's "Letters on the Proper Sphere of Government" gives evidence that the evolutionary notions he met in Lyell's account, particularly ideas of progressive advance and adaptation to external circumstances, had seeped quite early into the fabric of his social theory.

Spencer's interest in the developmental hypothesis warmed considerably in the conflagrational reaction to Robert Chambers's *Vestiges of the Natural History of Creation,* when it appeared in 1843. Chambers, who had the good sense to publish anonymously, advanced an evolutionary theory that drew from Lamarck, but added enough of the outrageous—for example, a member of one species giving birth to a very different one—that it could be easily dismissed by the learned. Even Darwin grew more cautious when leading naturalists savaged the *Vestiges* in review. Aside from Spencer's own perversity of taste, there was enough in the book to sustain his growing interest in the idea of biological progress. Chambers, after all, endorsed Gall's phrenology, which linked animal mental operations with human, and stretched the developmental hypothesis to cover social progress: "So is human society," Chambers observed, "in its earliest stages, sanguinary, aggressive, and deceitful, but in time becomes just, faithful, and benevolent. To such improvements there is a natural tendency which will operate in all fair circumstances."[93] By the time Spencer began *Social Statics* in 1848, the evolutionary hypothesis had taken firm hold.

What the argument of *Social Statics* lacked, however, was hard scientific confirmation. Spencer sought this in physiology, which he proposed, as he expressed it in a letter at the time, to make "my special study."[94] To this end, he attended Richard Owen's lectures on comparative osteology at the British Museum, and he took to reading on physiology. The two books having the greatest impact on him were by Milne-Edwards and William Carpenter. In Henri Milne-Edwards's *Outlines of Anatomy and Physiology,* which he read in 1851, Spencer discovered that his leading principle of political economy was also recognized as a principle in physiology. The great French physician and student of Cuvier declared that the "principle which seems to have guided nature in the perfection of beings, is, as we see, precisely the one which has had the greatest influence upon the progress of human industry: *the division of*

92. Spencer, *Autobiography* 1:201.

93. [Robert Chambers], *Vestiges of the Natural History of Creation,* 6th ed. (London: Churchill, 1847), p. 168.

94. Spencer, *Autobiography* 1:426.

labor."[95] Spencer had already tentatively advanced the principle in *Social Statics* as governing not only social economy but also biological economy; now he had an unexpected endorsement. The scientific status of the principle of the division of labor became further enhanced when he read in Carpenter's *Principles of General and Comparative Physiology* of Karl Ernst von Baer's analysis of embryonic development. Von Baer held that the embryo passed from a more general, homogeneous condition to a specifically organized and more heterogeneous one.[96] Carpenter, however, thought von Baer's law—that "a heterogeneous structure arises out of one more homogeneous"—could be applied with greater generality. Peering from just this side of orthodoxy, he urged that "if we watch the progress of evolution [i.e., embryonic development], we may trace a correspondence between that of the germ in its advance towards maturity, and that exhibited by the permanent conditions of the races occupying different parts of the ascending scale of creation."[97] Though Carpenter held to the fixity of species, his generalization of von Baer's law strongly suggested—to those with ears to hear—a dynamic interpretation of species relations. Spencer heard the message. He now understood the relevant social and physiological principles, and he had at hand an array of examples from political economy and zoology. The conclusion had only to be clearly drawn, and in fall of 1851 he began constructing the argument.

In an October issue of the socialist newspaper *The Leader,* edited by his friend George Lewes, Spencer published in discreet anonymity a broadside against the anti-progressivist, anti-developmental geology of Charles Lyell. He thought that Richard Owen had completely exploded the geology of stasis and had shown that development could be clearly observed in the paleontological deposits. Undoubtedly the conservative Owen squirmed a bit when he read that his argument supported the

95. H. Milne-Edwards, *Outlines of Anatomy and Physiology,* trans. J. Lane (Boston: Little and Brown, 1841), p. 12.

96. For a discussion of von Bear's theory of development and its impact on Owen and Darwin, see Dov Ospovat, "The Influence of Karl Ernst von Baer's Embryology, 1829–1859," *Journal of the History of Biology* 9 (1976): 1–28.

97. William Carpenter, *Principles of General and Comparative Physiology,* 2d ed. (London: John Churchill, 1841), pp. 194–95. Among embryologists "evolution" simply meant the unfolding of the germinal structures. Carpenter applied the term also to the hierarchy of species, without, however, claiming any actual unfolding of one species out of another. Yet Carpenter, like many of his contemporaries, did believe that varieties of a species could be produced by the inheritance of acquired characters, as well as by the occurrence of accidental variations which men attempted to perpetuate through restricted breeding. See pp. 501–12.

hypothesis of the author of the *Vestiges of the Natural History of Creation*; for that was the polemical—though not unreasonable—twist Spencer gave his teacher's ideas. In March 1852, Spencer published a more general defense of the evolutionary proposal, again in *The Leader*. The essay, entitled "The Developmental Hypothesis," pitted a naturalistic account of species genesis against the creationist account. With acute, if devilish, logic, he showed that, contrary to the usual assumption, the mythical events of special creation strained the imagination infinitely more than the natural idea of a gradual transformation of species, which was, he suggested, similar to the metamorphosis of the simple, undifferentiated embryo into the heterogeneously articulated adult.[98]

The declarations in *The Leader* were bold, even if the author shielded himself in anonymity. Yet they lacked the conceptual power to slice through the Gordian tangle of religious and scientific orthodoxies. But shortly thereafter he found a cutting argument, which he revealed in the *Westminster Review* that April. Sensing the genuine power of his essay, he unabashedly circulated it to several eminent scientists and critics.

In his "Theory of Population," Spencer turned up the withered hams of the Reverend Thomas Malthus to the sun.[99] The political economy of Malthus's own "Essay on Population" was indeed a dreary science, since it forecast neutralizing regressions in the efforts at human progress. When food supplies were ample and life promising, population surges would wash out any gains by creating famine, which in turn would decimate peoples. Malthus's iron calculations seemed to crush Enlightenment hopes of continued human progress. But Spencer, like Darwin and Wallace, discovered in population pressure a dynamo driving the progressive evolution of life.

Spencer's new theory of population had four elements: (1) a population dynamics; (2) a physiological division of labor both in organisms and society; (3) a mechanism of functional inheritance of accommodating modifications; and (4) a physiological principle of inverse relationship between fertility and reason. As populations grow, individuals—animals and men—must perforce accommodate themselves to increasingly difficult circumstances; they must adapt to more complex and varied relationships with their environments and with each other.

98. [Herbert Spencer], "Lyell and Owen on Development," *Leader* 2 (October 18, 1851): 996–97; and "The Developmental Hypothesis," *Leader* 3 (March 20, 1852): 280–81. The latter essay is reprinted in his *Essays: Scientific, Political, and Speculative* (New York: D. Appleton, 1896), 1:1–7.

99. [Herbert Spencer], "A Theory of Population, deduced from the General Law of Animal Fertility," *Westminster and Foreign Quarterly Review* 57 (1852), 457–501.

Such accommodations must consequently further articulate the individual and social organism, creating more complex but efficient divisions of labor. In his essay, Spencer argued the necessity of physiological division of labor from some bits of inductive evidence, but mostly from analogy with those human social processes that operated to secure survival:

> this "physiological division of labour," as it has been termed, has the same effect as the division of labour amongst men. As the preservation of a number of persons is better secured when, uniting into a society, they severally undertake different kinds of work, then when they are separate and each performs for himself every kind of work: so the preservation of a congeries of parts, which combining into one organism, respectively assume nutrition, respiration, circulation, locomotion, as separate functions, is better secured than when those parts are independent, and each fulfills for itself all these functions."[100]

Several years later, in his essay of 1857, "Progress: Its Law and Cause," Spencer would furnish considerable inductive evidence to support his thesis that organic and social life developed from a more homogeneous to a more heterogeneous state.[101] But in the present essay, he remained satisfied with a few examples appended to the analogical deduction.

The third element of his theory of population remained muted in the published essay, and it is really only from our favored position that we can detect its presence. This is the principle of the inheritance of functionally acquired traits: new habits produced by the division of labor, he suggested, would induce heritable structural changes. Though this idea had already surfaced in the evolution of Spencer's thought, it would begin to play a designedly major role only in his *Psychology*. In the present essay Spencer did, however, give it expression, but linked it with another hazy, inchoate notion. The passage rings in our ears of Darwinian triumph and Spencerian tragedy. In referring to the "dan-

100. Ibid., p. 486.
101. [Herbert Spencer], "Progress: Its Law and Cause," *Westminster and Foreign Quarterly Review* 67 (1857): 445–485. Though Spencer marshalled evidence from cosmology, embryology, human racial development, growth of civilizations, changes in languages, and progress in the arts to support his law of homogeneity to heterogeneity by a consilience of inductions, he nonetheless delivered the coup de grace with an a priori argument. Every active force he maintained, must produce more than one change, since minimally there will always be a reaction and in a varied medium causal influence will be transmitted. Therefore, "starting with the ultimate fact that every cause produces more than one effect, we may readily see that throughout creation there must necessarily have gone on, and must still go on, a never-ceasing transformation of the homogeneous into the heterogeneous" (p. 467).

gers of excessive fertility" and the difficulties of adjusting to complex social life, Spencer wrote:

> it is clear, that by the ceaseless exercise of the faculties needed to contend with them [i.e., the complexities of society], and by the death of all men who fail to contend with them successfully, there is ensured a constant progress towards a higher degree of skill, intelligence, and self-regulation—a better co-ordination of actions—a more complete life.[102]

The principle of natural selection arose and then subsided in Spencer's early thought. It was not without some cause that he groused in his old age that he had not received credit for his contribution to the full range of evolutionary conceptions.

The final part of Spencer's new theory consisted of some very hoary but yet presumptively sound physiology, which supported the no less venerable notion that greater intellectual activity produced proportionately less biological fertility. The roots of the idea can be traced to the Hippocratic tract *De aere, aquis et locis* and its ramifications through Leonardo Da Vinci right up to William Carpenter and Richard Owen. The older versions of the theory held that pangenes, bearing the impress of the external environment, traveled from all parts of the body to the brain and medullary core, and then finally passed to the organs of generation. Leonardo even depicted a connection between the spermatic cord and the base of the spine. As the theory evolved within the scientific and broader folk traditions, it came to suggest that large generation of thought produced small generation of children (as well as the popular idea that masturbation caused insanity). Spencer filled out the theory with some rejuvenated physiology and with inductions indicating that animals with more primitive nervous organization produced larger numbers of offspring, while those with greater cerebral complexity left few behind them. This principle of inverse relation between intellectual and biological conception had, Spencer, proposed, great implications for human development.

Further evolution would not likely produce a swifter, stronger, or more agile race, since the inventiveness of man replaced the need for physical alterations.[103] Rather, in order to accommodate to new situations resulting from population growth and the division of labor, men would increase in intelligence and moral capacity. Such mental and moral advance (especially in matters of sexual behavior) would have the consequence of reducing fertility, according to the principle of the in-

102. Spencer, "Theory of Population," p. 500.
103. Wallace seems to have borrowed this idea from Spencer. See chap. 4.

verse ratio. Hence, population dynamics would cause greater adaptation to the social state, including damping down population growth. The result would be, Spencer concluded, a perfected society and the ultimate happiness of man:

> it is manifest that, in the end, pressure of population and its accompanying evils will entirely disappear; and will leave a state of things which will require from each individual no more than a normal and pleasurable activity. [Cessation of development] implies that the nervous system has become fully equal to all that is demanded of it—has not to do more than is natural to it. But that exercise of faculties which does not exceed what is natural constitutes gratification. Consequently, in the end, the obtainment of subsistence will require just that kind and that amount of action needful to perfect health and happiness.[104]

Spencer's dizzying reversal of Malthus fitted his evolutionary theory with a dynamics sufficient, he thought, to carry the human race to perfection.

The essay "On Population" had important social consequences for Spencer: it initiated him into the regnant scientific community. Robert Chambers liked the essay, while William Greg objected to it, but certainly noticed it. Owen declared it a good article, though, as Spencer told his father in 1852, the great anatomist would not give an opinion of the theory expressed.[105] Perhaps the most socially significant result, however, was that the essay served to introduce him to Thomas Huxley. Spencer sent the article to the young biologist, and thus began a friendship lasting, with only momentary breech, until Huxley's death. From that time onward, the two spoke often of the developmental hypothesis, usually with Huxley demonstrating that modern biology would not allow it. Spencer recalled that in their wrangling over the topic Huxley "habitually demolished now this and now that argument which I used. But though continually knocked down, I continually got up again."[106] Spencer showed the resilience of a young thinker in his bouts with Huxley. He prized their close friendship that grew over the years. The depth and strength of the relation, surely not often remarked on by those two Victorian gentleman, was given warm description by Spencer in a letter (1 January 1872) responding to an invitation to the Huxleys' new home:

104. Spencer, "Theory of Population," p. 500–501.
105. Duncan, *Life and Letters of Herbert Spencer* 1:83–84.
106. Spencer, *Autobiography* 1:519.

My dear Huxley:

I have often wished to express in some way more emphatic than by words, my sense of the many kindnesses I have received at your hands during the twenty years of our friendship. Remembrance of the things you have done in furtherance of my aims, and of the invaluable critical aid you have given me, with so much patience and at so much cost of time, has often made me feel how much I owe you.

The invitation to your new house suggests to me a fit occasion for expressing my feelings. Let me do this by asking your acceptance of the drawing-room clock sent herewith.

I am afraid I have often too much disguised my real attitudes of mind by those impertinences of speech into which my natural irritability, made worn by my nervous disorder, often betrays me. I fear too, that the like effect may have resulted from inadvertencies of writing, due to a mental trait which my father frequently reproached me as a boy—saying "as usual, Herbert thinking only of one thing at a time." But inadequately as I may ordinarily show it, you will (knowing that I am tolerably candid) believe me when I say that there is no one whose judgment on all subjects I so much respect, or whose friendship I so highly value,

ever yours sincerely,
[signed] Herbert Spencer[107]

Given the vintage of their friendship, Spencer became especially troubled in the late 1880s, when Huxley misinterpreted and then took considerable offense at a remark he had made. The cooling of their friendship reached extremely painful levels in 1893, when Huxley delivered his Romanes Lecture, which was directed against Spencer's evolutionary ethics. But the rupture was healed in time to escape the final breach.

Spencer's other great friendship initiated at this time (in the early 1850s), though one that did not as often descend below the high intellectual plain, was with John Stuart Mill. Three times Mill furnished occasion for Spencer to make important advances on his theories of mental and moral evolution. The first occasion came when Spencer attempted to arbitrate the raging (in a scholarly sense) epistemological dispute between Mill and William Whewell.

107. Herbert Spencer to Thomas Huxley (1 January 1872), Huxley Papers, VII, 129, Library of Imperial College, London.

Figure 6.1   Herbert Spencer, 1820–1903, photograph from 1866.

Mental Evolution: Empiricized Kantianism

The Mill-Whewell Debate

During his tenure, first as fellow and tutor (1818 and 1823) and then master (1841–1866) of Trinity College, William Whewell made virtually all scientific knowledge his personal domain: he wrote books on mathematics, mechanics, astronomy, and mineralogy, as well as on moral philosophy and education. His vast erudition and gravely worn authority prompted Sidney Smith, editor of the *Edinburgh Review,* to crack that "science was his forte, and omniscience his foible." In his own time, as in ours, he was best known as the author of two monuments of scientific, historical, and philosophical scholarship: *The History of the Inductive Sciences* (1837) and *The Philosophy of the Inductive Sciences* (1840). Both bore the mark of Kant.

In his *Philosophy of the Inductive Sciences,* Whewell distinguished between the two kinds of truth with which science dealt, contingent truths and necessary truths. The task of the scientist, he contended, was to reduce the contingent to the necessary, to show that objective facts could be explained by universal law.[108] To accomplish his task, however, the scientist had to appreciate the original distinction between necessary and contingent truths, which Whewell cast in hard Kantian terms. The former, he declared, "are true universally and necessarily, the latter, only learnt from experience, and limited by experience." In his treatise, he would demonstrate of the first kind of truths that

> the universality and necessity which distinguish them can by no means derive from experience; that these characters do in reality flow from the ideas which these truths involve; and that when their necessity is exhibited in the way of logical demonstration, it is found to depend upon certain fundamental principles, (Definitions and Axioms,) which may thus be considered as expressing, in some measure, the essential characters of our ideas.[109]

Whewell, like most philosophers writing after Hume, recognized that limited experience could not logically demonstrate laws of universal and necessary validity. For sensationalist epistemologists such as Hume, Condillac, Erasmus Darwin, and James Mill, this simply meant that we could never achieve universal and necessary knowledge. But Whewell, like Kant, felt the foundations of science quake at the very thought. Mathematics, logic, mechanics, astronomy—all provided examples of sciences governed by universal and necessary laws. We were assured that the principles of these sciences really did have those traits of universality and necessity, since they conformed to the appropriate definition of necessary truth, which Whewell formulated in this way: "Necessary truths are those in which we not only learn that the proposition is true, but see that it must be true; in which the negation of the truth is not only false, but impossible; in which we cannot, even by an effort of imagination, or in a supposition, conceive the reverse of that which is asserted."[110] Necessary truths, according to Whewell, were such that their denial could not even be consistently conceived.

If the universality and necessity of scientific laws could not be assured through experience, then, according to Whewell, they had to arise independently of experience. Following Kant, he assigned them to the

---

108. For a discussion of Whewell's epistemology, see Robert Butts, *William Whewell's Theory of Scientific Method* (Pittsburgh: University of Pittsburgh Press, 1968).

109. William Whewell, *The Philosophy of the Inductive Sciences, Founded upon their History* (London: Parker, 1840), 1:54.

110. Ibid., p. 54.

organizational structure of the human mind. We can be absolutely sure, for example, that any spatial relations logically derived from accurate descriptions of some domain, say in mechanics or astronomy, will themselves be necessarily and universally true in the system. We can be sure because we must conceptually and perceptually organize spatial experience according to invariant principles (i.e., those of Euclidean geometry).[111] Kant attempted to demonstrate this sort of necessary application of mental structures to experience with what he termed a transcendental deduction: an argument showing that the very possibility of having any experience at all implied the imposition of such pure intuitional structures as time and Euclidean space, and of the categories of substance, causality, and the rest. Whewell did not develop this more generalized kind of deduction, but rather constructed what might be called a transcendental deduction from science: "scientific" experience could not be accounted for (e.g., Kepler's discovery of the laws of the motion of Mars) or even exist, except by the imposition of spatial, temporal, and causal "Ideas."[112] From Whewell's Kantianism John Stuart Mill vigorously dissented.

In 1843 Mill published his magnificent *System of Logic*, which attempted to set out the laws of reasoning, both in their general mode (i.e., the rules of induction and deduction) and in their special applications (i.e., the methods of experimental inquiry). He devoted special attention to the "logic of the moral sciences," arguing that social patterns were reducible to the aggregate actions of individuals, and that the behavior of individuals could be comprehended in a science of character (i.e., "ethology"), into which sociology itself could be reduced. Mill's treatment of the laws of logic, of mathematics, and of the physical and social sciences lay deep in the tradition of Hume and Mill's own father. He maintained that all laws came out of inductions from experience. "The proposition, straight lines which have once met, do not

111. Ibid., p. 84

112. William Whewell, *On Induction* (London: Parker, 1849), pp. 28–33: "I conceive that Kepler, in discovering the law of Mars's motion, and in asserting that the planet moved in an ellipse, did this;—he bound together particular observations of separate places of Mars by the notion, or, as I have called it, the *conception*, of an *ellipse*, which was supplied by his own mind. . . . Before [the imposition of an idea], the facts are seen as detached, separate, lawless; afterwards, they are seen as connected, simple, regular; as parts of one general fact, and thereby possessing innumerable new relations before unseen. Kepler, then, I say, bound together the facts by superinducing upon them the conception of an ellipse, and this was an essential element in his Induction. . . . That the fact of the elliptical motion was not merely the sum of the different observations, is plain from this, that other persons, and Kepler himself before this discovery, did not find it by adding together the observations."

meet again, but continue to diverge," even this kind of mathematical proposition, he claimed, "is an induction from the evidence of our senses."[113] What set off logical and mathematical generalizations from others, such as "The sun rarely shines in Glasgow," was simply that our experience remained more constant in cases of the former than of the latter. But if the laws of the various sciences, as well as those of logic and mathematics, flowed from experience, we could never be sure of their truth. Moreover, Mill retorted to Whewell, in some instances we commonly denied that the laws of mathematics were true: no mathematician, after all, thought the straight lines geometry described could be found in the real world.

In the *System of Logic*, Mill directly confronted Whewell's epistemology with examples from the history of science—a sweet rhetorical turn. Whewell claimed that we had evidence of necessary truths because their contrary was inconceivable. But, Mill pointed out, the history of science contained remarkable instances "in which the wisest men rejected as impossible, because inconceivable, things which their posterity, by earlier practice and longer perseverance in the attempt, found it quite easy to conceive, and which everybody now knows to be true."[114] The battle between Whewell and Mill was joined, and into the fray jumped Spencer.

The Universal Postulate

In the *Westminster* for spring 1853, Spencer used the occasion of a review of the classics of Reid, Berkeley, Hume, and Kant, as well as of Mill's *System of Logic* and Whewell's *Philosophy of the Inductive Sciences*, in order to settle the issue between the two antagonists.[115] The epistemological problem that served as prolegomenon, prompted by Kant's theory of space and time, Berkeley's immaterialism, and Hume's skepticism, was that metaphysics often stood in opposition to common sense, that is, to that set of beliefs which utterly resisted efforts at disbelief, which our intelligence delivered to us as ineluctable (e.g., that physical objects and other people exist). Any metaphysics, Spencer argued, that required us to abandon common sense must be self-contradictory. For to deny common sense would be to overthrow the basic

---

113. John Stuart Mill, *A System of Logic, Ratiocinative and Inductive; being a connected view of the Principles of Evidence and the Methods of Scientific Investigation* (New York: Harper & Brothers, [1843] 1848), p. 152 (bk. 2, chap. 5, no. 4).

114. Ibid., p. 157 (no. 6).

115. [Herbert Spencer], "The Universal Postulate," *Westminster Review* n.s. 3 (1853): 513–50.

principles by which our intelligence operated. Yet these constituted the very principles that metaphysicians used to establish their own peculiar congeries of beliefs. "How then," Spencer inquired, "can we legitimately end in proving something at variance with our primary beliefs, and so proving our intelligence fundamentally untrustworthy? Intelligence cannot prove its own invalidity because it must postulate its own validity in doing this."[116] Our commonsense realism, Spencer thought, formed the bedrock of all our beliefs, but especially our scientific beliefs.

What convinced us of the truth of our commonsense beliefs, Spencer contended, was simply that we could not believe otherwise. Indeed, adopting a strategy that Whewell might have admired (and William James later did), he argued that immutability of belief had to be our ultimate criterion of truth. When expressed in propositional form, it became, as we might say, virtually an analytic principle. "For an invariably existent belief is, by virtue of its being one, incapable of being replaced by another." With the triumph of the young logician catching up his master, Spencer concluded that it was "not that we ought to adopt that belief [i.e., an invariable one], but that we can do nothing else."[117] On the bare question, then, of what we must accept as the criterion of truth, Mr. Mill simply had to bow. Whewell was right. It could only be invariability of belief, that is, the inability to conceive otherwise.

Nonetheless, as Spencer evenhandedly acknowledged, our criterion of truth was not infallible. Presumably, as Mill insisted, our invariably existent beliefs arose from invariable experience. But at any time that experience could change, producing new beliefs.[118] Like Lutheran epistemologists, we had to stand on our invariable beliefs; we could do no other. But this "universal postulate" left us naked to the world, which at any time might undo us.

Spencer's review in the *Westminister,* "The Universal Postulate," had raised up a conception of considerable epistemological subtlety, which Mill treated accordingly in a chapter added to the fourth edition (1856) of his *Logic.* Spencer's taste of success impelled him beyond his review to work out supporting ideas in a new theory of evolutionary psychology. Through Whewell he came to appreciate the need for preestablished mental structures to organize the data of experience. Such structures or ideas formed the invariable—and thus a priori—set of beliefs that served both as preconditions to further knowledge and as pillars of

116. Ibid., p. 514.
117. Ibid., p. 520.
118. Ibid., p. 523.

a realistic metaphysics. Yet these beliefs could only be relatively a priori; they stemmed—and here he agreed with Mill—from constant and uniform experience. To these propositions Spencer would join his already fixed conviction that a uniform experience could alter heritable physiological structures, including mental structures. Finally his "universal postulate" provided him an analytic method by which to conduct psychological science: through intuitions of invariable relations, he would attempt to reduce complex mental phenomena to their essential elements. Spencer's arbitration of the Mill-Whewell dispute thus led him to the brink of a thoroughgoing evolutionary psychology and epistemology. He crossed over shortly thereafter with his *Principles of Psychology* (1855), a book that he predicted would "ultimately stand beside Newton's *Principia*."[119]

The Principles of Psychology

Composition and Structure

While vacationing on the French coast in August 1854, Spencer began his *Principles of Psychology*. The first part of the writing went well. He modified slightly his article "The Universal Postulate," and it became part 1, "General Analysis." He next turned to what would become part 3, "General Synthesis," which he finished by 1 November 1854. Part 2, "Special Analysis," came next, and then the last part, "Special Synthesis."[120] For publication he arranged the parts into symmetrical form, expressing nothing so much as an engineer's aesthetic: "General Analysis," "Special Analysis," "General Synthesis," "Special Synthesis."

The "General Analysis" establishes the "universal postulate" of invariability of belief as the methodological instrument, realism as the grounding philosophy the postulate yields, and psychology as the master science that adjudicates the epistemological claims of the other sciences. The "Special Analysis" investigates the nature of reasoning and its various forms (i.e., quantitative, qualitative, inductive, and deductive). The analysis produces two major conclusions: that all forms of reasoning are reducible to intuitions of likeness or difference, and that the "universal process of intelligence is the *assimilation* of impressions."[121] This latter conclusion embodies two results of Spencer's examination: first, that even perception is a form of judgment of similarity

---

119. Duncan, *Life and Letters of Herbert Spencer* 1:98.
120. Spencer, *Autobiography* 1:536.
121. Herbert Spencer, *Principles of Psychology* (London: Longman, Brown, Green, and Longmans, 1855), p. 332.

or difference in respect to already established categories;[122] and second, that "differences displayed in the ascending grades of intelligence are consequent solely upon the increasing complexity of the impressions assimilated."[123] The "General Synthesis," where Spencer actually began composing his *Psychology,* erects the evolutionary framework structuring the whole. This part advances the principle of adaptation—the organic adjustment of internal relations to external circumstances—as the definition of both life and mind. Dynamic correspondence with the environment moves organic life and its specialized adaptation, mind, to higher orders of complexity. The final part, the "Special Synthesis," brings the work to a climax. Using the "law of intelligence," which reiterates the principle of adaptation, Spencer shows how the dispute between Mill and Whewell must be solved: through an evolutionary Kantianism.

Spencer finished the *Psychology,* which runs over six hundred printed pages, at the end of July 1855. He then suffered a nervous breakdown for a year and a half. The signs of his collapse manifested themselves about two weeks before he completed the manuscript. He remembered the event clearly:

> One morning soon after beginning work, there commenced a sensation in my head—not pain, nor heat, nor fullness, nor tension, but simply a sensation, bearable enough but abnormal. The seriousness of the symptom was at once manifest to me. I put away my manuscript and sallied out, fishing-rod in hand, to a mountain tarn in the hills behind the hotel, in pursuance of a resolve to give myself a week's rest; thinking that would suffice.[124]

But it did not, and Spencer became a mental invalid for the next eighteen months. His condition reminds one of Bertrand Russell's. Russell, after completing with Whitehead that great three-volume work on logic, *Principia mathematica,* also suffered severe mental exhaustion and

122. Like Kant, Spencer conceived perception as an implicit form of judgment: "A perception of it [an object] can arise only when the group of sensations is consciously coordinated and their meaning understood. And as their meaning can be understood only in virtue of those past experiences in which similar groups have been found to imply such and such facts; it is clear that the understanding of them—the act of perception, involves the assimilation of them to those similar groups—involves the thinking of them as like those groups, and as having like accompaniments. The perception of any object, therefore, is impossible save under the form either of Recognition or Classification" (pp. 185–86).

123. Ibid., p. 332.

124. Spencer, *Autobiography* 1:544.

collapse, which were aggravated by complications Spencer would never experience, the frustrations of the end of a marriage and the entanglements of a new liaison. The highly abstract nature of both Spencer's and Russell's work and the great effort they invested in rapid construction provide some understanding of their travail. A careful reading of either produces an absolutely sympathetic response.

The Evolutionary Unity of Life and Mind

The "General Synthesis," with which Spencer began, displays clearly the evolutionary framework in terms of which he conceived his project. It traces the development of life from simple organisms immersed in their environment through the gradual differentiation of individuals into more complex species by reason of, what he called, the law of intelligence. Differentiation produces a division of labor, which eventually bifurcates life into its two main branches—the physiological and the mental. The development of mental life leads, as Spencer traced it, from simple reflex reaction to instinct, memory, reason, and will. Three related principles guided him through this synthetic development. The first was that life and mind were essentially the same—each was an adjustment of internal organic relations to external relations in the environment.[125] The second principle was that all mental operations were the same, differing only in complexity—each was an ever more complex adjustment of inner to outer relations.[126] And the final principle was that adaptations to a continuously changing environment (especially a social environment) forced organisms to respond with progressively more complex adjustments of internal to external relations—"the evolution of life," he urged, "is an advance in the Speciality of the correspondence between internal and external relations."[127]

Since mind was not visited on creatures as a special gift of the creator, it must have had a natural origin; and for Spencer this meant it must have arisen from the organic. Darwin shared this same conviction, and, as we have seen, it was a staple of the sensationalist tradition, from which Spencer also drew sustenance. Though life and mind evolved together and in their primordial forms did not differ, yet a progressive

---

125. Spencer, *The Principles of Psychology*, pp. 374, 435–6.
126. Ibid., p. 486: "Every form of Intelligence being, in essence, an adjustment of inner to outer relations; it results that as, in the advance of this adjustment, the outer relations increase in number, in complexity, in heterogeneity, by degrees that cannot be marked: there can be no valid demarcations between the successive phases of Intelligence."
127. Ibid., p. 423.

division of labor produced a specialized nervous system and a consequent branching into a distinctive mode of life, intelligence.

Reflex activity constituted the most primitive stage of intelligent life in Spencer's scheme. Reflexes, he held, grew out of simple irritative responses characteristic of elementary organisms. Reflexes differed from irritative responses principally in that reflexes traveled through nervous fibers, while simple irritative responses remained localized. Following Carpenter's discussion of reflexes in the *Principles of Human Physiology*, Spencer conceived them as chained, serial reactions of organisms to external stimuli. As individuals became adjusted to more complex environments, their reflexes became more complicated, and instincts, which were but compound reflexes, gradually emerged. Memory, the next level of mental development, arose early in the evolution of mind; it too was a reflex process, though one that remained nascent, a partial response to environmental situations. As animals developed more complex reflexes, they advanced beyond simple instincts to intelligent and rational reactions to their surroundings. But this, according to Spencer's theory, did not mean that they passed to an emergent mode of behavior; it was still reflexes all the way up. In an individual's response to novel situations, several opposing reflexive and instinctive behaviors, while yet nascent, would compete for expression, till the most forceful flowed into action. Since, according to Spencer's theory, "this strongest of them must, in the average of cases, be the one that has been the most uniformly and frequently repeated in experience, the action must, in the average of cases, be the one best adapted to the circumstances. But an action thus produced, is nothing else than a rational action."[128]

Though rational behavior grew out of stereotyped instinctive acts, Spencer contended the reverse also occurred: rational acts constantly repeated would become instinctive.[129] This theory of instinct from "lapsed intelligence," which had its antecedents in the sensationalist tradition, became an integral part of evolutionary analyses of mind a bit later in the century, especially in works of a neo-Lamarckian cast. For Spencer it formed an important element of his solution to the Whewell-Mill controversy, as will be indicated in a moment.

Spencer received from Carpenter an introduction to von Baer's theory of developmental recapitulation. But his debt to Carpenter amounted to much more. Spencer's theory of mental and cerebral activity derived in large measure from the great British physiologist. Car-

128. Ibid., p. 567.
129. Ibid., pp. 568, 571.

penter, for instance, maintained that the cerebrum engaged in reflex action, "so as to evolve *intellectual products*."[130] This developmental platform for conscious reasoning Carpenter called "Unconscious Cerebration."[131] Spencer appropriated this term to describe both the competition of nascent reflexes during "rational deliberation" and the implicit judgments proper to perception, that is, "the unconscious classing of . . . impressions, relations, and conditions, with the like before-known ones."[132] Carpenter also developed a conception that, with due modification, provided the bloodline for Spencer's own system of mental evolution (as well as for that of James and Baldwin). This was the "Ideo-Motor" hypothesis. According to Carpenter, when volition remained in abeyance, the strongest or most dominant idea resident in the mind would automatically lead to a corresponding action. "Thus, then, we see," Carpenter proposed in *Principles of Human Physiology*, "that in all those states in which the directing power of the Will over the current of thinking is suspended, the course of action is determined by some dominant idea, which for the time has full possession of the mind, and from which the individual has no power of withdrawing his consciousness."[133] Spencer merely came to regard any presumed "willful" activation or inhibition itself as the competitive outcome of a previously nascent tendency.

The Dynamic Principle of Mental Evolution

The genetic relation between mind and body strongly implied for Spencer that they were cut from the same cloth. He found, however, another guarantee of their essential unity: they both obeyed the same principle, one which he had implicitly assumed in his earlier book and essays but only now fully articulated. This was the principle of the persistent correspondence between inner organizational relations and external environmental relations. In the case of the evolution of intelligence, it took special form in "The Law of Intelligence," which Spencer promulgated as:

> the *persistency* of the connection between . . . two states of consciousness, is proportionate to the *persistency* of the connection between the phenomena to which they answer. The relations

130. William Carpenter, *Principles of Human Physiology*, 5th ed. (Philadelphia: Blanchard and Lea, [1852] 1853), p. 791.

131. Ibid.

132. Spencer, *Principles of Psychology*, p. 229.

133. Carpenter, *Principles of Human Physiology*, p. 811.

between external phenomena are of all grades, from the absolutely necessary to the purely fortuitous. The relations between the answering states of consciousness must similarly be of all grades, from the absolutely necessary to the purely fortuitous.[134]

The law of intelligence served as a dynamic principle of evolution, since the external conditions encompassed by an individual's experience continually changed. In the case of advanced organisms, social relations formed their proximate environment. Consequently, continuous adjustment—a moving equilibration, as he later termed it—drove organisms to more complex mental states.

Evolutionary Kantianism

In the first part of his "Special Synthesis," Spencer reiteratively applied the law of intelligence to demonstrate how consciousness first emerged in primitive animals and successively but gradually progressed from stages characterized by simple reflex responses and instinct to those of memory, reason, and will. But it was in dealing with the problem of instinct that Spencer cultivated the theoretical ground for his major discovery. He believed that instincts could be best explained as complex integrations of reflexes, which themselves became implanted in the organism through adaptive habits responding to a changing environment.[135] Yet it was not simply that new habits produced a new organization. More was involved. "[T]he modified nervous tendencies produced by such new habits of life," he claimed, "are also bequeathed: and if the new habits of life become permanent, the tendencies become permanent."[136] Inherited nervous tendencies might account not only for the instincts peculiar to each variety of animal species, but even to those complexes of human instincts characteristic of different nationalities. Undoubtedly with some feeling of triumph, Spencer made good

134. Spencer, *Principles of Psychology,* p. 509.

135. Spencer's conception of instinct as a complex of reflexes was a traditional enough notion, but one forcefully endorsed by Carpenter, upon whom he greatly depended for his anatomy and physiology. Carpenter and other anatomist had, by mid-century, utterly abandoned the attempt to explain instincts as quickly learned behavior: caterpillars and bees had taught naturalists differently. Instincts were innate. Carpenter explained them in the orthodox way as providential designs of the Creator. He did not, however, confine instincts to animals, since man too manifested "mental instincts." Spencer, of course, rejected the "Hebrew myth" as an explanation. Instead he turned to Lamarck, Chambers, and his own devices.

136. Ibid., p. 526.

on Mill's promised "political ethology or the science of national character."[137]

The key doctrine of Spencer's *Principles of Psychology*, which he thought would secure his work a place beside Newton's, can be appropriately termed 'evolutionary Kantianism.' Spencer contended that the "so-called 'forms of thought,'" as well as "all psychological phenomena," could be explained by two laws.[138] The first was the one whose explanatory power he had demonstrated throughout his general and special analyses: the law of intelligence—that the frequency of encountered external environmental relations produced a proportionate frequency of internal psychological relations. The other law expressed an assumption underlying virtually all of Spencer's writing in the early 1850s, but now for the first time pronounced clearly and self-consciously: the law that "habitual psychical successions entail some hereditary tendency to such successions, which, under persistent conditions, will become cumulative in generation after generation."[139] This was the law of the inheritance of characteristics persistently maintained. Spencer was sensible of the more tenuous nature of this law. But when yoked to the law of intelligence, it offered the only naturalistic explanation of animal and human mental instincts. The unacceptable alternative was the Hebrew myth. In Spencer's eyes, however, its real power derived from its use in solving the deep epistemological problem dividing Whewell and Mill, that of necessary laws in mind and nature. He expressed this rationale when he declared that "I adopt the hypothesis [of the inheritance of acquired characters] until better instructed: and I see the more reason for doing this, in the facts, that it appears to be the unavoidable conclusion pointed to by the foregoing investigations, and that it furnishes a solution of the controversy between the disciples of Locke and those of Kant."[140]

Mill was correct in claiming that all laws governing mental operations ultimately derived from experience, but wrong in supposing that each individual had to acquire such modes of activity anew. According to Spencer's novel theory, the laws of mind resulted from the adaptational experiences of the race. Hence, Whewell was also correct: the structures of thought and perception in each individual were a priori and necessary, but as the evolved consequences of the inheritance of acquired mental habits. We now regard this conclusion as essentially correct (except for the Lamarckian theory of inheritance), even perhaps a tru-

137. Ibid., p. 527.
138. Ibid., pp. 578–79.
139. Ibid., p. 579.
140. Ibid., p. 578.

ism—certainly a sign that Spencer deserves at least honorable mention for the Newton prize. But historians of science have ignored the importance of Spencer's resolution. Like the biological community, they too have been washed in the high tide of Darwinism, but are ever wary of the undertow of social Darwinism, or anything that reminds them of that despised doctrine.

I have suggested that Spencer gradually constructed his evolutionary theory with certain moral ends in view. He attempted to demonstrate scientifically that nature, particularly human nature, inexorably moved toward perfection, which in the case of man he interpreted as complete adaptation to the social state, a consummation in which evil and stupidity, both inadaptations, would be finally replaced by freedom from social coercion and by expansion of individual rational capacities. This state of adaptation to society would yield happiness. Spencer gained confidence in his evolutionary theorizing in the measure that these moral ends—the evanescence of evil, the growth of freedom and rational mind, and the achievement of happiness—could be derived from it. Now, in working out his psychology, he had added support for his theory: it was also capable of resolving a profound epistemological controversy.

After Spencer had demonstrated, in his evolutionary-Kantian fashion, how the forms of space, time, motion, and force had been derived from experience and had then solidified into heritable categories of thought, he turned to emotions. Here too the explanatory concept of an empirically acquired a priori relation served to account for characteristic human sentiments. Spencer gave special consideration to the passion for liberty. That emotion stemmed from our animal fear of restricted action; but in adaptation to the social relations of an evolving society, it became honed into the sentiment for individual rights. The desire for individual freedom, which incorporated a desire for the freedom of others, arose organically out of man's intercourse with his natural and social environment. This imprinted sentiment, then, was not due to a rational calculation, as the Benthamites believed. And any extrinsic interference with these natural relationships, as, for example, government attempted, could only arrest, he concluded in the last sentence of his book, "that grand progression which is now bearing Humanity onwards to perfection."[141]

141. Ibid., p. 620.

Further Developments of Spencer's Kantianism

Spencer never read Kant's *First Critique* beyond the "Aesthetic," whose doctrines of space and time he found unacceptable (and in the translation he read, undoubtedly unintelligible).[142] He got his Kant through the philosopher's British disciples, particularly Whewell, William Hamilton, and Henry Mansel. Spencer's reading of Hamilton's essay "Philosophy of the Unconditioned" and Mansel's books *Prolegomena Logica* and *Limits of Religious Thought*[143] moved him to develop further certain epistemological views only implicit in his *Psychology*. This he did with extraordinary philosophical acumen in the cornerstone of his Synthetic Philosophy, *First Principles* (1862), and in the second edition of *Principles of Psychology* (1872).

Both Hamilton and Mansel argued that the conditions of human knowledge precluded any penetration to noumenal reality. The categories of thought—unity, substance, cause, and so on—legitimately subtend only appearances; when we attempted to push beyond phenomena, the antinomies of reason would call us up short and demonstrate the impossibility of even thinking that which was beyond human mind. Hamilton, with Mansel, was thus led to hold that our knowledge of unconditioned reality could only be negative:

> In our opinion, the mind can conceive, and, consequently, can know, only the *limited, and the conditionally limited*. The unconditionally unlimited, or the *Infinite*, the unconditionally limited, or the *Absolute*, can not positively be construed to the mind; they can be conceived, only by a thinking away from, or abstraction of, those very conditions under which thought itself is realized; consequently, the notion of the Unconditioned is only negative—negative of the conceivable itself.[144]

142. Spencer first looked into Kant while visiting the home of James Wilson, the editor of the *London Economist*, in 1844. See Spencer's account in the *Autobiography* 1:289. He ostensibly reviewed the second edition of Haywood's translation—*Critick of Pure Reason*, trans. Francis Haywood, 2d ed. (London: Pickering, 1848)—in his essay "The Universal Postulate."

143. Spencer probably read Sir William Hamilton's essay "Philosophy of the Unconditioned" (originally published in 1829) in the collection of Hamilton's essays *Discussions on Philosophy and Literature* (New York: Harper, 1853). Mansel's *Prolegomena Logica: an Inquiry into the Psychological Character of Logical Processes* (Oxford: Graham, 1851) was among the books reviewed in "The Universal Postulate." Mansel's *Limits of Religious Thought* is quoted by Spencer in *First Principles* (London: Williams & Norgate, 1862), pp. 39–43 especially. Spencer was probably put on to Mansel by reading Carpenter's *Principles of Human Physiology*, where Mansel is cited frequently.

144. Hamilton, "Philosophy of the Unconditioned," p. 20.

In *First Principles,* Spencer acknowledged the power of these considerations. He also elaborated two further arguments, sanctioned by his psychology, that gave them even greater force. First, if all perception and conception involve assimilation of elements to already established (and evolved) categories, then absolute reality, which pretends to be so different from objects of experience, cannot be so assimilated, and thus must remain uncognizable.[145] And second, since "Life in all its manifestations, inclusive of Intelligence in its highest forms, consists in the continuous adjustment of internal relations to external relations, the necessarily relative character of our knowledge becomes obvious."[146] In perception, whence stem immediate knowledge and (through racial experience) innate categories, we can only be aware of coexistences and sequences, the relations but not the things related. This constraint on the perceptual sources of knowledge also prevents us from knowing the nature of any reality beyond the human mind. Indeed, according to Spencer, we cannot even know the ultimate nature of the human mind itself, that is, its character as a *Ding an sich.*[147]

If this completely constituted Spencer's theory of the Unknowable, then one might snicker a bit with William James, who found the idea a metaphysical pretension. But Spencer moved beyond the conclusions of Hamilton and Mansel. First, he pointed out that our knowledge of extramental reality could not be merely negative. For "to say that we cannot know the Absolute, is, by implication, to affirm that there *is* an Absolute,"[148] a concession Kant himself made. The indefinite hypostatization of a reality beyond mind, Spencer took to be an incorrigible feature of thought—and thus it had the legitimizing sanction of the "universal postulate." Further, though we may only be aware in thought that perception *a* precedes perception *b,* and not know the realities behind *a* and *b,* we can know—as surely as we know anything—that something beyond the phenomena answers to this mental relationship.

145. Spencer, *First Principles,* p. 81.
146. Ibid., p. 85.
147. Spencer's doctrine of the "Unknowable" became progressively more knowledgeable over time. So that in a letter to Harald Hoeffding, the great historian of philosophy and psychology, Spencer explained that the concepts of "matter" and "mind" served as symbols "of the Unknowable Power of which they are both manifestations, and that the distinction between them is essentially this: that what we call our consciousness is a circumscribed portion, while that which we think of as unconscious or physical, is simply that which lies outside the circumscribed portion called consciousness" (Duncan, *Life and Letters of Herbert Spencer* 1:239). Spencer implied that the "Unknowable Power" that underlay mind and matter was one, not more. His own metaphysical stance thus became a double aspect identity theory, not unlike that of Wundt, Haeckel, James, and Baldwin.
148. Ibid., p. 88.

This last point can be made a bit clearer by reference to a chapter that Spencer added to the second edition of the *Psychology*. The chapter bears the title "Transfigured Realism,"[149] and presages in remarkable fashion the epistemological views of Wilfrid Sellars, the contemporary American neo-Kantian and, like Spencer, scientific realist.[150] Their respective modes of composition, though, undoubtedly differed: Spencer dictated his essay, as his amanuensis recalled, "in the intervals of a game at racquet in a court at Pentonville in the north of London."[151] It is nonetheless lucid; perhaps the rush of blood to the brain did nourish his thought, as Spencer believed. To illustrate the kind of knowledge we can have of reality, he offered the model of geometrical projection. A cube may be projected onto the surface of a cylinder from a variety of angles, producing a number of different representations on the surface of the cylinder. Since the projection might be highly distorted, if we perceived only the surface but did not know it was the surface of a cylinder, we would be unable to describe the object producing the figure. We would, however, know that the internal relations of the figure had their "counterpart relations" (as Sellars calls them) in the object and that the deformations were governed by certain, but unknown rules. The mind, Spencer contended, formed the surface of our intellectual cylinder, upon which reality impressed its relations of coexistence and sequence, but did not reveal the rules of projection.

Spencer's evolutionary Kantianism enforces epistemological modesty but certainly not skepticism in either scientific knowledge or morals. Though we may not know whether our ideas a and b resemble the realities x and y, we may be confident (in light of the universal postulate) that the sequence of our ideas answers to a counterpart sequence in reality. The evolution of scientific thought has, in Spencer's view, occurred precisely by replacing representations of the assumed intrinsic nature of things with representations of the order of their action.[152] And in the moral realm a comparable replacement is occurring. The evolution of society is, through the equilibration of adaptational forces, constantly shedding inadequate accommodations by individuals to changing social relations and replacing them with adaptations more finely

---

149. Herbert Spencer, *The Principles of Psychology*, 2d ed. (London: William and Norgate, 1872), 2:489–503 (chapter 19).

150. See in particular Wilfrid Sellars, "Being and Being Known," in his *Science, Perception and Reality* (London: Routledge and Kegan Paul, 1963), pp. 41–59.

151. The reminiscences of Spencer by James Collier, his personal secretary, are appended to Josiah Royce, *Herbert Spencer: an Estimate and Review* (New York: Fox, Duffield, 1904). The quotation is from p. 214.

152. Spencer, *First Principles*, p. 105.

reflective of the social state. The goal, as Spencer sighted it from the last pages of his *First Principles*, is final social balance: "Evolution can end only in the establishment of the greatest perfection and the most complete happiness."[153]

Mechanisms of Evolution

Spencer began working on his *First Principles* on 7 May 1860 in new lodgings taken in Bloomsbury. The publication of the final serial number took place in 1862, again at some cost to his health. The composition spanned the time during which Darwin's *Origin of Species* made its first dramatic impact, but there is no hint of this in *First Principles*. In the book Spencer never mentioned, for instance, natural selection as a possible mechanism of evolution; undoubtedly his disinclination for generous reading did not immediately allow him to perceive the distinctiveness of Darwin's doctrine of evolution. The adaptational device that Spencer relied on remained the principle of equilibration. Disturbing forces in the environment would cause an unbalancing of the rhythmic forces of life. Organisms in a changing environment would first reorganize their internal system of adjustments, and over many generations these would be gradually translated to offspring. Thus even after the publication of the *Origin*, Spencer relied exclusively on the inheritance of functionally acquired characters. He tinkered with this Lamarckian hypothesis, so that, in the network of his various theoretical considerations, it produced, when applied to human mental development, a final state of perfect equilibrium:

> the adaptation of man's nature to the conditions of his existence, cannot cease until the internal forces which we know as feelings are in equilibrium with the external forces they encounter. And the establishment of this equilibrium, is the arrival at a state of human nature and social organization, such that the individual has no desires but those which may be satisfied without exceeding his proper sphere of action, while society maintains no restraints but those which the individual voluntarily respects.[154]

When Spencer undertook the next project of his Synthetic Philosophy—the *Principles of Biology*, begun in autumn of 1862—he had, in light of the intellectual climate, to take heed of Darwin. He now distinguished two kinds of adaptational force: direct equilibration, the staple

153. Ibid., p. 486.
154. Ibid., pp. 470–71.

of his earlier theory, and indirect equilibration, which he defined as "the survival of the fittest," a term Darwin, at the behest of Wallace, adopted to characterize natural selection. Spencer admitted that indirect equilibration accounted for many instances of adaptation, particularly the features of plants and the simple traits of men and animals. But he observed that it could not explain the more complex systems of coadaptations found in higher organisms.[155] With the problem of coadaptations, Spencer lodged an objection in the throat of natural selection that at times during the latter part of the century threatened to choke it off as a major force in evolution.[156] He offered the Irish elk as a telling case. In order for its huge rack of antlers to have evolved, its skull must have been thickened, its neck muscles strengthened, its vascular network enlarged, and its nervous connections increased. It is highly improbable, however, that these coadaptations could have arisen spontaneously at the same time. Yet none of these traits would have had any selective value if they had appeared seriatim. Hence, according to Spencer's analysis—and that of many others during the century—such traits could have evolved only as a consequence of simultaneous functional adaptations that gradually altered heritable anatomical structures. Spencer pointed to other instances of characters that appeared to have little survival value (by common estimation); they too could only be explained by functional inheritance.[157]

In the second edition of the *Principles of Psychology*, Spencer appended to an example of a functionally altered instinct a note indicating that "Had Mr. Darwin's *Origin of Species* been published before I wrote this paragraph, I should, no doubt, have so qualified my words as to recognize 'selection,' natural or artificial, as a factor." He felt, however, no need for radical emendation of the text, since, "while holding survival

155. Herbert Spencer, *Principles of Biology* (New York: D. Appleton, [1866] 1884), 1:450–56.

156. Mark Ridley describes the late nineteenth-century debates over this objection in his well-wrought essay "Coadaptation and the Inadequacy of Natural Selection," *British Journal for the History of Science* 15 (1982): 45–68. Sewall Wright judged that the two most important objections to natural selection in the late nineteenth and early twentieth centuries were blending inheritance and complex coadaptations (cited by Ridley, p. 46). In the recent period the punctuationist school has resurrected the objection. See Stephen Jay Gould's "Introduction" to Richard Goldschmidt, *The Material Basis of Evolution* (New Haven: Yale University Press, 1981), p. xxxvi; see also Goldschmidt's elaboration of the objection in "Evolution, as Viewed by one Geneticist," *American Scientist* 40 (1952): 84–135 (in particular p. 93).

157. The objection to natural selection from coadaptation and evolution of nonadvantageous traits are discussed in Spencer's *Principles of Biology* 1:chap. 12, which is on "Indirect Equilibration."

of the fittest to be always a cooperating cause, I believe that in cases like these it is not the chief cause." [158]

Toward the end of the century, Spencer felt the increasing heat of arguments from the "ultra-Darwinians," as Romanes termed them, who would admit only natural selection as the mechanism of evolutionary change. In 1886, Spencer composed a long two-part article, "The Factors of Organic Evolution," in which he defended the role of functionally acquired modification in evolution, frequently citing Darwin's own employment of the device in the *Origin of Species* and *Descent of Man*. In the preface to the republication of the articles in book form the next year, he declared what was at stake in his defense of the mechanism of acquired modifications. It was the "indirect bearings upon Psychology, Ethics, and Sociology." The profound importance of these bearings on the social sciences, he confessed, was "originally a chief prompter to set forth the argument; and it now prompts me to re-issue it in permanent form.[159] Survival of the fittest was too crude a mechanism to yield up delicate mental structures, refined social adaptations, and a keen sense of justice—especially since these highly evolved traits had no survival value.

In 1890 Spencer first became aware of August Weismann's attacks on the inheritance of acquired characters. Feeling that the journal *Nature* had displayed some bias in favor of the ultra-Darwinians, Spencer launched a counterassault in the pages of the *Contemporary Review* against the presumption that natural selection alone could explain all the phenomena of evolution.[160] Attacking the views particularly of Weismann, he outlined a number of cases, highlighted by those of co-adaptation, that he believed demonstrated the inadequacy of natural selection for the task assigned it by the ultra-Darwinians. The controversy generated responses from Weismann and Romanes, and rebuttals from Spencer.[161] Some historians, knowing how the evolution story ends, have thought Spencer's arguments "fallacious" and have declared

158. Spencer, *Principles of Psychology*, 2d ed., 1:423.

159. Herbert Spencer, *The Factors of Organic Evolution* (London: Williams and Norgate, 1887), p. iii.

160. Herbert Spencer, "The Inadequacy of "Natural Selection," *Contemporary Review* 63 (1893): 153–66, 439–56.

161. August Weismann, "The All-Sufficiency of Natural Selection: a Reply to Herbert Spencer," *Contemporary Review* 64 (1893): 309–38, 596–610; George Romanes, "Mr. Herbert Spencer on 'Natural Selection,' " *Contemporary Review* 63 (1893): 497–516 "The Spencer-Weismann Controversy," *Contemporary Review* 64 (1893): 50–53; and "A Note on Panmixia," *Contemporary Review* 64 (1893): 611–12; and Herbert Spencer, "A Rejoinder to Professor Weismann," *Contemporary Review* 64 (1893): 893–912; and "Weismannism Once More," *Contemporary Review* 66 (1894): 592–608.

Weismann the winner.[162] A less hindsightful reading does not leave that impression. At one point Spencer forced Weismann to confess that "we must assume natural selection to be the principle of explanation . . . because all other apparent principles of explanation fail us, and it is inconceivable that there could be yet another capable of explaining the adaptations of organisms, without assuming the help of a principle of design."[163] In matters of logic, Spencer had few peers. He remonstrated: "The supposition that [a trait] has been produced by the inheritance of acquired characters is rejected because it presents insuperable difficulties. But the supposition that the [trait] has been produced by natural selection is accepted, though it presents insuperable difficulties. If this mode of reasoning is allowable, no fair comparison between diverse hypotheses can be made."[164] Weismann took it on faith that some explanation could be found for coadaptive variations, but Spencer was not of that religion (nor, for example was Ernst Haeckel, who complained to Spencer of Weismann's need "of somewhat mystical irrational dogmatic views").[165] Throughout the rest of his life, Spencer never quit the belief, as he summed it up in 1899, that "the inheritance of functionally-wrought modifications is the chief and almost exclusive factor in the genesis of all the more complex instincts and all the higher mental powers."[166] Willy-nilly, however, he made increasing use of Darwin's mechanism, especially in the further development of his moral theory; for it had special advantages in explaining the origins of altruism.

162. The quoted characterization is from James Kennedy, *Herbert Spencer* (Boston: Twayne, 1978), p. 85. See the like suggestion in James Moore, *The Post-Darwinian Controversies* (Cambridge: Cambridge University Press, 1979), pp. 172–73. Ridley gives a much fairer assessment of the controversy in "Coadaptation and the Inadequacy of Natural Selection," pp. 61–62.

163. Weismann, "The All-Sufficiency of Natural Selection," p. 328.

164. Spencer, "A Rejoinder to Professor Weismann," p. 894.

165. Ernst Haeckel to Herbert Spencer (10 November 1895), Athenaeum Collection of Spencer's Correspondence, MS. 791, no. 238, University of London Library.

166. Herbert Spencer, "The Filiation of Ideas," in Duncan, *Life and Letters of Herbert Spencer* 2:547.

7

Evolutionary Ethics:
Spencer and His Critics

After completing *Principles of Psychology* at his isolated French retreat, Spencer spent the next year and a half in slow recovery of his mental health. His *Autobiography* omits exact description of his symptoms, though indicates he succumbed to exhaustion and mental depression. The doctor he consulted advised him never again to live alone. "He thought, and no doubt rightly thought," Spencer recalled, "that my solitary days in lodgings had been largely instrumental in bringing on the physiological disaster which had already cost me so much of life and of work."[1] He took the advice, moving in with a family, whose two little girls were perfectly fitted, he said, "to serve as vicarious objects of the philoprogenitive instinct."[2] But this was only a palliative. While writing *First Principles,* he suffered a recurrent bout of emotional depression. His friend Huxley diagnosed the problem and cautiously prescribed the appropriate, but for Spencer, possibly toxic specific:

> I am grieved to hear such a poor account of your health. I believe you will have to come at last to the heroic remedy of matrimony, and if "gynopathy" were a mode of treatment that could be left off if it did not suit the constitution, I should decidedly recommend it. But it's worse than opium eating—once begin and you must go on, and so, though I ascribe my own good condition mainly to the care my wife takes of me, I dare not recommend it to you, lest perchance you should get hold of the wrong medicine.[3]

But no danger. By this time, in 1861, Spencer had given up the idea of marriage. In the early 1850s he had become, in his words, "very inti-

1. Herbert Spencer, *Autobiography* (New York: D. Appleton, 1904), 2:582.
2. Ibid., p. 583.
3. Thomas Huxley to Herbert Spencer (3 August 1861), in *Life and Letters of Thomas Huxley,* ed. Leonard Huxley (New York: Appleton, 1900), 1:249.

mate" with George Eliot.[4] Although she loved him and found it "impossible to contemplate life" without him,[5] he broke off. He judged her a fascinating intellect, but she evoked in him no sexual excitement. "The lack of physical attraction," he confided to a friend, "was fatal. Strongly as my judgment prompted, my instincts would not respond."[6] His friend George Lewes did, however, respond; and sometime after Spencer had introduced them, Lewes and Eliot lived openly together. Through the mid-1850s, he often took dinner with his two friends and immensely enjoyed their society. But during the intense concentration he bestowed on his major compositions, such as the *Psychology* and *First Principles,* he barely saw anyone. And despite the impression that we have of Spencer as an asocial, abstracted, and unemotional figure, he required companionship.

During his middle age, Spencer spent most every evening at his club, playing billiards and enjoying the company of fellow "X-Club" members Huxley, Tyndall, Hooker, and Lubbock.[7] In his later years he took special pleasure in the wit, conversation, and beauty of Beatrice Potter, whose parents he had known since the early 1840s. Beatrice, who later married Sidney Webb and became a leader of the Fabian Socialists, felt a deep intellectual and emotional attachment to her "old Philosopher." In her diary (for 4 January 1885), she related a story of one of their many trips together that reveals the character of their relationship:

> Royal Academy private view with Herbert Spencer. His criticisms on art dreary, all bound down by the "possible" if not the probable. That poor old man would miss me on the whole more than any other mortal. Has real anxiety for my welfare—physical and mental. Told him story of my stopping cart and horse in Hyde Park and policeman refusing to come off his beat to hold it. Want of public spirit in passers-by not stopping it before. "Yes, that is another instance of my first principle of government. Directly you get state intervention you cease to

4. The quotation is from a letter (3 February 1881) Spencer wrote to Edward Youmans, his American promoter. He described his relationship to Eliot to counter rumors that he never married her because he objected to matrimony. The letter and four of Eliot's letters to Spencer are transcribed in Richard Schonenwald, "G Eliot's 'Love Letters': Unpublished Letters from George Eliot to Herbert Spencer," *Bulletin of the New York Public Library* 79 (1976): 362–71.

5. Ibid., p. 370.

6. Ibid., p. 364.

7. John Fiske called this informal dinner society "the most powerful and influential scientific coterie in England," an estimate endorsed by Roy MacLeod in "The X-Club, A Social Network of Science in Late-Victorian England," *Notes and Records of the Royal Society of London* 24 (1970): 305–22.

Figure 7.1 Beatrice Potter Webb, 1858–1943, photograph from 1883.

have public spirit in individuals; that will be a constantly increasing tendency and the State, like the policeman, will be so
bound by red-tape rules that it will frequently leave undone the
simplest duties."[8]

On his deathbed Spencer called for this socialist flame, one of the dearest friends he had.[9] He could little tolerate books with which he dis-
agreed, but his strangled emotions often grasped at people with whom
he profoundly differed intellectually.

The Fortunes of Spencer's Evolutionary Theory

Huxley undoubtedly informed Spencer of the papers by Wallace and
Darwin read at the Linnean Society in October of 1858. Two months
later, Spencer sent Darwin a volume of his recently published *Essays,*
which included his evolutionary pieces. Darwin responded with considerable praise, which rings sincere. Darwin mentioned that he was "at
present preparing an Abstract of a larger work on the changes of species; but I treat the subject simply as a naturalist, and not from a general
point of view, otherwise, in my opinion, your argument could not have
been improved on, and might have been quoted by me with great
advantage."[10]

With the immediate success and notoriety achieved by the *Origin of
Species,* Spencer's own fame began its slow ascent. His *Psychology,* which
like Hume's *Treatise* fell dead-born from the presses, underwent a revaluation. Alexander Bain wrote him in 1863 to say that "I have gone
over your Psychology with great attention and I must add, with great
admiration, and nearly unqualified approval. You have certainly constituted yourself the philosopher of the doctrine of Development, notwithstanding that Darwin has supplied a most important link in the
chain."[11] John Stuart Mill also took up again the *Psychology,* but viewed
it this time in a different light, as he confessed to Spencer:

> There is much in it [the *Psychology*] that did not by any means
> strike me before as it does now: especially the parts which shew
> how large a portion of our mental operations consists in the

8. Beatrice Webb, *The Diary of Beatrice Webb: Volume one 1873–1892,* ed. Norman and
Jeanne Mackenzie (Cambridge: Harvard University Press, 1982), pp. 127–28.

9. David Duncan, *Life and Letters of Herbert Spencer* (New York: D. Appleton, 1909),
2:227.

10. Francis Darwin, ed., *Life and Letters of Charles Darwin* (New York: D. Appleton,
1891), 1:497.

11. Alexander Bain to Herbert Spencer (17 November 1863), Athenaeum Collection of
Spencer's Papers, no. 67, Senate House Library, University of London.

recognition of relations between relations. It is very satisfactory to see how you and Bain, each in his own way, have succeeded in affiliating the conscious operations of mind to the primary unconscious organic actions of the nerves, thus filling up the most serious lacuna and removing the chief difficulty in the association psychology.[12]

Spencer's publication of *First Principles,* beginning serially in 1860, captured the now quickened attention of the scientific public. The book set out the general laws of evolution, grounding them in a Kantian epistemology and a critical realism. With this volume he initiated his project for a thoroughgoing evolutionary science, his Synthetic Philosophy. The list of its first subscribers—Mill, Darwin, Huxley, Hooker, Lubbock, and the others whom I have already mentioned—indicate the growing respect for Spencer's evolutionary science.

Persistent Influence of Mill

While working on his *First Principles,* Spencer twice more received a stimulating jolt from his friendly antagonist John Stuart Mill. The consequences of the first of these escape precise measure, but the event itself must reorient our historical view of both Spencer and Mill.

During the late 1850s, Spencer had composed several essays on education, which he collected into a volume in 1860. He sent a complimentary copy to Mill, who undoubtedly felt a little odd in reading the last essay, particularly passages like the following:

It seems strange that there should be so little consciousness of the dangers of over-education during youth, when there is so general a consciousness of the dangers of over-education during childhood. Most parents are more or less aware of the evil consequences that follow infant precocity. In every society may be heard reprobation of those who too early stimulate the minds of their little ones. . . . [I]t is a familiar truth that a forced development of intelligence in childhood entails disastrous results—either physical feebleness, or ultimate stupidity, or early death.[13]

Spencer, of course, had no knowledge of Mill's early education, and more likely had in mind the beneficial effects of his own casual training.

12. John Stuart Mill to Herbert Spencer (3 April 1864), Athenaeum Collection of Spencer's Papers, no. 68, Senate House Library, University of London.
13. Herbert Spencer, *Education: Intellectual, Moral and Physical* (New York: D. Appleton, [1860] 1883), p. 267.

Mill responded to the gift in a letter that bespeaks a different side to his empiricism. The letter reads:

> I have been in the habit of attributing the diminished strength of constitution of the middle and higher classes (which I believe to be a *fact*) to a physiological cause not mentioned by you, being the same which explains the strong constitutions of many savage tribes. Formerly all the weakly children died, and the race was kept up solely by means of the vigorous specimens. Now, however, vaccination and improved bringing up of children by their very success keep alive to maturity and enable to become parents, a vast number of persons with naturally weak constitutions. This influence, diffused by intermarriage through the succeeding generations, must necessarily, unless countered by powerful causes of an opposite tendency, diminish the average vigor of constitution of the classes in which it occurs.[14]

It is difficult to judge the effect this letter had on Spencer. On the one hand, he had suggested much the same in his "Theory of Population" (in a passage already cited), though in a noninsistent way. Mill's proposal was redolent of natural selection, but as applied to man, something not yet undertaken by Darwin himself. The letter presaged a use of natural selection in Spencer's later works that won him the reputation of a "social Darwinist" (a term that would have inflicted a double-barbed injury on him). Likely the letter established a certain set of selective pressures that contributed to Spencer's gradual, but restricted, employment of a mechanism, natural selection, that was firmly associated with the name of his perceived rival.

Shortly after posting his 'natural selection letter,' Mill again intervened to foster the evolution of Spencer's ideas, this time in morals. The occasion was the publication of Mill's tract *Utilitarianism,* in which he threw Spencer in among the "anti-utilitarians."[15] Spencer wrote Mill to correct what he believed a misrepresentation of his moral philosophy. The letter, subsequently published in Bain's *Mental and Moral Science,* clearly sketched the basic features of his yet-developing ethical theory. Spencer first set Mill at ease: he did regard the greatest happiness as the moral end of man. He simply rejected felicity as the proximate end of man. Men do not, nor should they, tote up immediate pleasures and

14. John Stuart Mill to Herbert Spencer (30 July 1861), Athenaeum Collection of Spencer's Papers, no. 63, Senate House Library, University of London.

15. John Stuart Mill, "Utilitarianism," in his *Dissertations and Discussions,* vol. 3 (New York: Holt, 1882), p. 388–89. The essay originally appeared in *Fraser's Magazine* in 1861 and was issued as a volume in 1863.

pains before every moral judgment. As a matter of fact, they usually act from a nonrational, moral sense of right conduct. In a passage, which Darwin would later approvingly cite in the *Descent of Man,* Spencer expressed his evolutionary conception:

> there have been, and still are, developing in the race, certain fundamental moral intuitions; and that, though these moral intuitions are the results of accumulated experiences of utility, gradually organized and inherited, they have come to be quite independent of conscious experience. Just in the same way that I believe the intuition of space, possessed by any living individual, to have arisen from organized and consolidated experiences of all antecedent individuals who bequeathed to him their slowly developed nervous organizations—just as I believe that this intuition, requiring only to be made definite and complete by personal experiences, has practically become a form of thought, apparently quite independent of experience; so do I believe that the experiences of utility, organized and consolidated through all past generations of the human race, have been producing corresponding modifications, which, by continued transmission and accumulation, have become in us certain faculties of moral intuition—certain emotions responding to right and wrong conduct, which have no apparent basis in the individual experiences of utility.[16]

The position, then, is simply that we instinctively judge conduct under the unconscious guidance of the race's past experience, which necessarily has consisted of evolving patterns of altruistic social behavior. So the measures of utility are there, though mostly in our past; and because they are not now conscious calculations, they lack the selfish pleasure-seeking character that Bentham attributed to all human behavior. In short, the evolutionary perspective demonstrates that moral judgments are not vitiated by calculations of individual gain.

The second important clarification that Spencer provided was the distinction between an evolutionary psychology of moral sentiment and a moral science that draws from this evolutionary psychology but yet remains logically autonomous. He argued that moral science proper should be occupied with deriving from evolutionary science those laws and principles that as a matter of fact have produced human happiness and satisfaction. Having ascertained such laws and principles, moral science then postulates them as norms of human conduct. As he put it to Mill: "I conceive it to be the business of Moral Science to deduce,

16. Spencer to Mill (1862), quoted by Alexander Bain, in his *Mental and Moral Science* (London: Longmans, Green, 1868), p. 722.

from the laws of life and the conditions of existence, what kinds of action necessarily tend to produce happiness, and what kinds to produce unhappiness. Having done this, its deductions are to be recognized as laws of conduct; and are to be conformed to irrespective of a direct estimation of happiness or misery."[17] Spencer understood the difference between an evolutionary psychology of morals and moral science proper. The common charge that he committed the naturalistic fallacy thus appears weak, but, of course, appearances may be deceiving. In later sections of this chapter, I will explore more exactly the logic of Spencer's moral science and the views of those decrying it.

In autumn of 1862, Spencer commenced work on the second part of his Synthetic Philosophy, his *Principles of Biology,* the fascicles of which subscribers began receiving in 1863, the last installment arriving in 1867. The *Principles of Psychology,* logically the third part of his project, appeared in revised and expanded form in 1870 and 1872. In 1876 Spencer published the first of what would be an almost interminable series of his *Principles of Sociology.* But illness broke in upon him during winter of 1877–1878. He suspected he might not live to complete the Synthetic Philosophy, so he decided to turn immediately to what was to be the climax of the project, his *Ethics.* The reason for his change in plans affirms what has been the directive theme of these two chapters on Spencer. On 16 February 1878, he wrote a friend to explain his decision:

> I begin to feel that it is quite a possible thing that I may never get through both the other volumes of the *Principles of Sociology,* and that, if I go on writing them, and not doing anything towards the *Principles of Morality* till they are done, it may result in this last subject remaining untreated altogether; and since the whole system was at the outset, and has ever continued to be, a basis for a right rule of life, individual and social, it would be a great misfortune if this, which is the outcome of it all, should remain undone.[18]

Spencer finished the *Data of Ethics* the following year, in 1878. He lived another quarter of a century, time enough to complete his project and experience the rejection of his ethical theory by his best friend Huxley.

Spencer's Ethical System

The Spencer that evokes our moral indignation warned, in the *Study of Sociology* (1873), against legislation to aid the wretched, since such action would ignore fundamental biological law.

17. Ibid., p. 721.
18. Spencer, *Autobiography* 2 : 369–70.

> Besides an habitual neglect of the fact that the quality of a so-
> ciety is physically lowered by the artificial preservation of its
> feeblest members, there is an habitual neglect off the fact that
> the quality of a society is lowered morally and intellectually, by
> the artificial preservation of those who are least able to take care
> of themselves. . . . For if the unworthy are helped to increase,
> by shielding them from that mortality which their unworthi-
> ness would naturally entail, the effect is to produce, generation
> after generation, a greater unworthiness.[19]

In the face of such apparent callousness, it is easy to close the book
on Spencer's ethics. But we would then overlook the mitigating factors:
his quite general social principle that all action by the state, save to
protect individual freedom, could only violate natural laws of social evolution; and his balancing principle of personal responsibility, which
held that the "sympathetic care which the better take of the worst is
morally beneficial, and in a degree compensates by good in one direction for evil in another."[20] In judging the whole by its part and refusing
to take up the challenge of his moral system, we might easily conclude
that "Spencer's ethics were never philosophically reputable."[21] I believe,
on the contrary, that a fair estimate of his system shows it to be philosophically powerful and in its general structure perfectly sound, even
morally admirable.

Since the essential features of Spencer's ethical system fade into the
bulk of pages he devoted to its details, a brief sketch will, I hope, provide an orienting introduction. In his system, Spencer distinguished
among: (1) the goal of evolution, (2) the ultimate principle of ethical
evaluation, and (3) the moral sentiment guiding individual action. The
goal of evolution, toward which it ineluctably tends, is equal freedom—each individual within society exercising the full range and capacity of faculties without, however, infringing on the same freedom of
others. The ultimate criterion by which we judge actions to be morally
good or bad, according to Spencer, is the greatest happiness principle.
In separate arguments, he demonstrates (a) that happiness or pleasure
is the ultimate moral principle, and (b) that evolution of a society manifesting equal freedom will bring the greatest happiness. The principles
of social evolution thus become moral imperatives. The moral sentiments, which embody these imperatives, have, in Spencer's view,
evolved to lead individuals to promote equal freedom directly and
hence the greatest happiness indirectly. Now to the filling in.

19. Herbert Spencer, *Study of Sociology* (Ann Arbor: University of Michigan, [1873]
1961), p. 313.
20. Ibid.
21. James Kennedy, *Herbert Spencer* (Boston: Twayne, 1978), p. 69.

The End of Evolution and the End of Moral Behavior

One clear meaning for the concepts of good and bad is efficient or inefficient adjustment of means to ends. From the time he wrote *Social Statics,* in 1851, Spencer had defined good and evil in this way, as adaptation or nonadaptation to circumstances. In terms of his evolutionary conception, then, good and evil referred to "more evolved conduct" or "relatively less evolved conduct."[22] He measured adaptation against both the occurrent social environment and a more ideal environment, one toward which he believed progressive evolution was headed. Even though he took the ideal social environment as an absolute standard of adaptation or nonadaptation, he recognized that any assessments of behavioral fit with the ideal social environment remained empirical evaluations and did not immediately or necessarily translate into moral judgments. That would require an independent argument showing an identity between the ends of nature and moral ends. In the *Data of Ethics,* he constructed just this argument.

Spencer first discriminated three proximate evolutionary ends of individual human conduct, ends which furthered the ultimate goal of equal freedom.[23] The primary set of adjustments of behavior regarded one's own life, the goal being self-preservation. The second goal evolution set for conduct regarded offspring: behavior became biologically good or bad insofar as it advanced (or retarded) the health and safety of progeny. Finally, the welfare of society, of the community, formed an end of individual action. This last goal, in Spencer's judgment, derived its value from the function of society in preserving its units.[24] If it could be shown that these three ends of evolution were are also moral ends, then the evolutionary laws leading to these biological ends would become moral principles requiring our acquiescence.[25]

In the next stage of his argument, Spencer determined the ultimate moral goal and thus the criterion of ethical behavior. Despite Mill's suggestion to the contrary, Spencer provided an argument to show that the ultimate moral principle was the greatest happiness for the greatest number. But as a requisite for a scientific system of morals, he also attempted to demonstrate that happiness was the ultimate moral aim of behavior directed toward the individual, the family, and the society. His

22. Herbert Spencer, *Principles of Ethics* (Indianapolis: Liberty Classics, [1893] 1978), 1:58–61. For convenience, I will cite Spencer's *Data of Ethics* as it appears (unchanged) in the first part of the *Principles of Ethics.*

23. Spencer, *Principles of Ethics* 1:58–60.

24. Ibid., p. 167.

25. Ibid. 2:23.

strategy for establishing happiness as the ultimate moral criterion was twofold. First, he held that everyone in fact believed happiness or pleasure to be the ultimate moral goal of human action and that anyone who proposed another aim would simply be confusing means for ends. One needed only to ask: Would you feel obliged to take as the maxim of your behavior "Increase the misery in the world," or to toil for your self, your family, or your fellow man with the precise aim of leaving them no more happy than they happen to be? Such examination of conscience, Spencer was convinced, "quickly compels everyone to confess the true ultimate end."[26] His second tactic was to investigate other ethical systems to show that whether they invested the moral value in the agent, in the nature of his motives, in the quality of his deeds, or in the results—all such systems ultimately held pleasure to be the end governing moral behavior.[27]

While a critic might question the details of Spencer's demonstration, I do not think his general strategy for establishing an ultimate moral principle can be faulted. His argument was not a simple induction, since it not only surveyed individual judgments and other moral systems but also challenged them to provide another end of action. He attempted, in essence, to construct an abbreviated transcendental argument, which would demonstrate that moral life and any conceivable moral system could make sense only under the assumption of the greatest happiness principle. In light of his 'universal postulate,' he assumed that we had to accept the conclusion of such an argument as necessary.

An important corollary of the greatest happiness principle, for Spencer at any rate, was that one's pleasures could not be solitary. Genuine pleasure and happiness were found in altruistic behavior, behavior that sought the welfare of progeny and other members of society; hence the pure egoist who attended only to his own pleasure would enjoy less of it than the person who also acted to increase the happiness of others.[28] If the greatest happiness was the moral mandate, then altruism was a necessary consequence.

Spencer's moral system, as thus far described, appears to be a variety of hedonistic ethics; but it differs considerably from others of that sort, Benthamism, for instance. Unlike many other utilitarians (save perhaps Mill), Spencer denied that happiness could be distributed or even quantified. The only concrete meaning for the utilitarian injunction would be to distribute the means by which happiness could be pursued. And

26. Ibid. 1:66.
27. Ibid., pp. 66–80.
28. Ibid., pp. 241–43.

this had to be done equitably, that is, according to the deserts of each. Spencer perhaps failed to harden this last link, the joining of equity with the distribution of goods, against the weight of all objection. It is nevertheless clear, he believed that the proximate aim of moral conduct had to be justice; happiness flowed therefrom.[29] His essay on "Prison Ethics" (1860) illustrates this.

In that essay, Spencer directed his moral animus against the inhuman treatment given criminals held in various of the Empire's prisons. He described many cases in which the discipline was "carried to an extent of suffering, such as to render death desirable, and to induce many prisoners to seek it under its most appalling aspects."[30] His rebuke flowed from an outraged sense of justice, not, as he understood it, from mercy or empathy. He argued that the law of life, which required that natural connections not be interfered with, had its converse in the just measure of retribution. The morality of an evolved civilization demanded that "property stolen shall be restored, or an equivalent for it given" or that "any one injured by an assault shall have his surgeon's bill paid, compensation for lost time, and also for the suffering he has borne."[31] But, as he protested, "morality countenances no restraint beyond this—no gratuitous inflictions of pain, no revengeful penalties."[32] Justice, Spencer consistently held, formed the immediate criterion of our moral judgments. His moral system describes, therefore, the evolution not so much of greatest happiness as of greatest justice, whence comes happiness. The concept of justice had a cardinal role in Spencer's moral theory, for it united the ends of moral acts with the ends of evolution.

The next major step Spencer took in his argument was to show that the moral end of the greatest happiness (or greatest justice) constituted the end of evolution as well. The biological law of survival of the fittest implied, in Spencer's estimation, that "individuals of most worth, as measured by their fitness to the conditions of existence, shall have the greatest benefits, and that inferior individuals shall receive smaller benefits, or suffer greater evils, or both."[33] Those best adapted to their natural and social environments necessarily enjoyed their life more than those ill adapted. Moreover, organisms would generally be led to engage in life-sustaining acts through the allure of pleasure; those individuals who found that life-sustaining activity generally and consistently

29. Ibid., pp. 264–65.
30. Herbert Spencer, "Prison-Ethics" (1860), in his *Essays, Scientific, Political, and Speculative* (New York: D. Appleton, 1896), 3:49.
31. Ibid., p. 53.
32. Ibid., p. 54.
33. Ibid. 2:25.

produced misery would cease the struggle for existence.[34] On the other hand, activities that social life forced upon its members, to which they had to adapt, these, like most habitual acts, eventually would become a source of pleasure.[35] Evolution, therefore, led to ever greater adaptation and thus to ever greater happiness. The consummation of social evolution would be a state of equal freedom. But such a state would, by its nature, constitute a just social arrangement and furnish both necessary and sufficient conditions for the production of the greatest happiness. Spencer therefore concluded that moral conduct and evolution led to the same ends: a life fully adapted to the social state, a life that enjoyed the greatest happiness possible because it constituted the most just life possible.

It is worth lingering a moment on the role of justice in Spencer's ethics, since all major criticisms have ignored it. That great Apostle of Cambridge, G. E. Moore, who thought moral values as plain as the smirk on an undergraduate's face, overlooked the concept of justice in his dismissive analysis of Spencer's ethics. Moore focused one part of his scrutiny (I will discuss his remarks on the "naturalistic fallacy" below) on the relation Spencer asserted between highly evolved conduct and pleasure.[36] Moore urged two sorts of inconsistencies in Spencer's account that, he thought, pierced through the supposed relation between fit behavior and pleasure. First, it would seem quite possible that a smaller quantity of life (i.e., less evolved) might produce more intense pleasure than a greater quantity, hence the relation between highly evolved conduct and greater happiness lacked necessity. Second, if two states of life produced equal amounts of pleasure, but one was less evolved, surely Spencer would choose that which was more evolved as preferable: hence pleasure would not be the sole good, though Spencer so regarded it. Moore's analyses, while needling through some tender spots in Spencer's ethics, nonetheless ignored the role of justice. Moore presumed that Spencer believed happiness could be easily quantified; but as we have just seen, he did not. Moreover, since Spencer translated happiness into justice, or equal freedom, the relation between ethical and evolutionary goals held fast: both moral behavior and evolution aimed to produce as much freedom as circumstances would permit. Finally, if equal freedom served as the index of happiness, then Moore's proposed scenario of one individual being less evolved but more happy than another could never obtain. One might seriously object to Spen-

34. Ibid. 1 : 118.

35. Ibid., p. 216.

36. G. E. Moore, *Principia Ethica* (Cambridge: Cambridge University Press, [1903] 1929), pp. 51–52.

cer's evolutionary theory, as most now do, but the logical ties betweenit and his ethics remain secure.

If the ends of evolution conformed to the ends of moral conduct,then the principles of biological development could be reformulated asmoral imperatives; at least Spencer so constructed them. The primarylaw of evolution, survival of the fittest, thus became also the moral lawthat "each creature shall take the benefits and the evils of its own nature,be they those derived from ancestry or those due to self-producedmodifications." [37] Since this general law of evolution operated differ-ently in furthering subsidiary ends, those of progeny and of society,Spencer construed the moral law accordingly:

> During immaturity benefits received must be inversely propor-
> tionate to capacities possessed. Within the family group most
> must be given where least is deserved, if desert is measured by
> worth. Contrariwise, after maturity is reached benefit must
> vary directly as worth: worth being measured by fitness to the
> conditions of existence. The ill fitted must suffer the evils of
> unfitness, and the well fitted profit by their fitness. [38]

To these specifications of the moral law, he added a third: "If the con-stitution of the species and its conditions of existence are such that sac-rifices, partial or complete, of some of its individuals, so subserve thewelfare of the species that its numbers are better maintained than theywould otherwise be, then there results a justification for such sacri-fices." [39] I will consider this last derivation of a moral principle, the prin-ciple of altruism, from its biological counterpart in a moment.

Holding on to Spencer's dense and slippery argument is difficult,rather like trying to grab a fat eel. Let me therefore strip away the fleshto get to the logical skeleton. Spencer has argued, according to myreconstruction:

1. Progressive adaptation to social environment ≻— results in—→greatest freedom of individuals consistent with welfare of each≻—results in—→ greatest happiness for greatest number.
2. Moral end of conduct ≻— achieved through—→ greatest happinessfor greatest number ≻— achieved through—→ greatest freedom ofindividuals, that is justice for all.
3. But evolutionary end of greatest happiness = moral end of great-est happiness.
4. Therefore, evolutionary laws = moral principles.

37. Spencer, *Principles of Ethics* 1:218.
38. Ibid. 2:20.
39. Ibid., p. 23.

This spare representation might provoke objecting queries about the establishment of the first two premises, the interpretation of such concepts as freedom and justice, and the relation of evolutionary means to moral ends. That is why the tough meat of Spencer's conception must be restored to measure the beast.

In this chapter and the preceding one, I have argued that Spencer constructed his evolutionary theory to meet the demands of his moral theory, and not the reverse. His ethical model for the theory of evolution becomes quite evident in the *Principles of Ethics*. The moral ideal, as Spencer understood it, mandated behavior directed to the welfare, that is, the survival and happiness, of three classes of agents: the self, offspring, and others in society. But the process of evolution had, according to Spencerian biology, the same ends. Spencer thus proposed that the moral goals of happiness (with its instrument of justice) were achieved through evolutionary means. Darwin's ethical theory, by contrast, derived more directly from biological considerations. He maintained that the end of evolution was the health and vigor, though not necessarily the happiness, of the individual and community. Darwin constructed his idea of the moral end to conform with this biological end of health and vigor for individual and community. In reverse manner, Spencer, because of his antecedent ethical convictions, advanced the moral goals of happiness and justice (in the form of equal freedom) as the ultimate outcome of evolution. Spencer's procedure does not vitiate his moral theory, but it might suggest doubts about his biological theory. The weak joints in his moral theory reside not, I think, in the logical strategy of his argument but in the execution of the strategy, in the rather flexible connections (which G. E. Moore tried to slice through) holding together the ends of evolution and the ends of morality. For instance, one could imagine a highly evolved society of individuals, each of whom perfectly respected the freedom of others—yet who were not particularly happy, perhaps a society of Herbert Spencers. Spencer's synthetic vision, however, gains power as the moral end becomes justice rather than happiness—a redirection that actually catches the drift of much of his ethical thought.

Altruism

Another important test of Spencer's ethical system might focus on his conception of altruism, since therein the essence of morality is usually thought to reside. Spencer defined altruism[40] straightforwardly, as

---

40. Spencer acknowledged (*Autobiography* 1:517) borrowing the term "altruism" (as well as "sociology") from August Comte.

an action "unconscious or conscious," that "involves expenditure of individual life to the end of increasing life in other individuals."[41] Anonymous generosity serves up a clear example of altruistic behavior, the sort
that meets our moral intuitions as being prima facie good. In *Social
Statics*, Spencer had employed Adam Smith's theory of sympathy (a fundamentally selfish sentiment) to explain other-regarding behavior. But
in the second volume of his revised *Principles of Psychology*, completed
about a year after Darwin's *Descent of Man*, he enlarged his account. He
suggested that altruistic behavior would also originate in reciprocally
beneficial acts, but "thus commenced, and survival of the fittest tending
ever to maintain and increase it, it will be further strengthened by the
inherited effects of habit."[42] This explanation embodied elements of
Darwin's own; and no wonder, since by this time Spencer had made
natural selection part of his armamentarium, in the marshall uniform of
"survival of the fittest."

In the *Data of Ethics*, Spencer more carefully systematized his conception of altruistic behavior, giving it a three-layered explanation. The first
and most basic sort of altruism, to Spencer's mind, was family altruism,
which could be explained by natural selection of the family (or community selection as Darwin conceived it, or kin selection as we do).
Animals that displayed too much egoism would endanger the welfare
of their offspring and would thus occasion "disappearance from future
generations of the nature that is not altruistic enough."[43] For societal
altruism, Spencer had two accounts, group selection (in attenuated
form) and habitual accommodation to the requirements of social living.
When a society failed to produce sufficient numbers of altruists, it de-
cayed and thus inaugurated a "gradual decrease in the egoistic satisfactions of its members."[44] The idea seems to be that groups without
enough altruists would simply be selected against. Finally, the continual
accommodation to the social state through inherited habit would produce innate tendencies toward altruistic behavior.[45] Spencer thus believed that these processes of family and group selection and the adaptation to the social state would gradually yield highly refined altruistic
sentiments. What would characterize them as meeting the norm for
truly moral attitudes would be their nonselfish character: he conceived
evolved altruistic behavior as aiming directly at the welfare of another

41. Spencer, *Principles of Ethics* 1:231.
42. Herbert Spencer, *Principles of Psychology*, 2d ed. (London: William & Norgate,
1872), 2:561.
43. Spencer, *Principles of Ethics* 1:234.
44. Ibid., p. 235.
45. Ibid., p. 241.

Figure 7.2 Herbert Spencer, photograph from 1893.

and only incidentally at the happiness or benefit of the actor. For Spencer, as well as for Darwin, it could only be a gross sort of anthropomorphism to speak of selfish gemmules. But this consideration needs elaboration, since Spencer's ethics might appear to have been morally emasculated by its admittedly hedonic character.

The Reflective Science of Morals and the Innate Moral Sense

Spencer concluded that happiness or, more proximately, justice constituted the end of moral conduct. He did not thereby suppose that moral agents acted immediately from motives of happiness or even justice. In his letter to Mill, he distinguished between the inherited sentiment for right conduct and a reflective science based on recognition of such sentiments—the same distinction at which Darwin independently arrived. Spencer believed that evolutionary processes established, as we have just seen, sentiments of altruistic behavior and, due to the race's experience with pain produced by unfairness, a thirst after justice. Individuals acting under these motives respond instinctually: "the man who is moved by a moral feeling to help another in difficulty," Spencer observed, "does not picture to himself any reward here or hereafter; but pictures only the better condition he is trying to bring about."[46] In a cool hour of reflection, altruistic behavior would be judged as also meeting the moral criteria of justice and happiness. But reciprocally, reflection, he believed, could inform subsequent behavior, since "the relatively vague internal perceptions which men have of right human relations, are not to be accepted without deliberate comparisons, rigorous cross-examinations, and careful testings of all kinds."[47] Such testing, Spencer had held from the *Social Statics* on, should involve applying scientifically derived ethical principles (enlightened by evolutionary laws), or "absolute ethics", to particular situations, whence we might formulate maxims of practice, or "relative ethics." Spencer's complete ethical program, like Darwin's, united rational reflection with the wisdom of evolution.

One important aspect of Spencer's argument, which has already been touched upon, deserves final remark. This is his quasi-Kantian demonstration of moral values from evolutionary facts. In *Principles of Ethics,* Spencer considered what the ultimate authority might be for the moral principle of equal freedom.[48] Though all men must admit the principle as an intuitively valid moral axiom, we might still wish further justification. Spencer believed that he had demonstrated the essential soundness of his evolutionary tale—the story of the progressive development of human society toward the state of equal freedom. If we accept this empirical argument, then we have further to admit that the sentiment for equal freedom has been established as part of the race's innately determined repertoire of experience-forming principles. Evolution, Spencer supposed, has produced in human beings something like the Kantian categories. We would thus have as a necessary standard of moral judgment the principle of equal freedom. And since, according to the universal postulate, what we cannot even conceive as false we must accept as true, the principle stands justified. The moral principle of equal freedom (and the others Spencer discriminates) therefore has the sanction of the highest epistemological authority—the necessity of belief. In this way Spencer established moral values from evolutionary

46. Ibid., p. 154.
47. Ibid. 2:72.
48. Ibid., pp. 76–7.

facts, though in a fashion that escaped the notice of his critics. I will further explore the logic of this kind of argument in the second appendix.

Critics of Evolutionary Ethics

Huxley

A Break in Friendship

In their younger days, Spencer and Huxley wrangled constantly about evolutionary ideas. Even after Huxley became a convert to evolution, he continued to differ with Spencer on matters of substantial issue in biology. These differences, however, never sank below theory to damage the deep bonds of personal affection. On ethical questions, Huxley's rudimentary considerations (formulated in response to St. George Mivart's attack on Darwin's moral theory)[49] generally harmonized with Spencer's ideas in his letter to Mill. But Huxley's moral feeling reached depths Spencer seems not to have known. It was initially from a sense of morality rather than from higher theory that Huxley began to separate himself from Spencer. The first real strain in their relationship came with Huxley's essay "Administrative Nihilism" in 1871.[50] He defended state-sponsored education for the lower classes, opposing on the one hand the wealthy who complacently assumed that the poor could not (and should not) rise above their station, and on the other those laissez-faire thinkers who as a matter of principle denied that education was a proper function of the state. The objections of the first class he dismissed by recalling those noble lords who would have made good gámekeepers, "had they not been kept afloat by our social corks."[51] Against the second class, represented by Spencer, he prepared stronger powder. He lifted Spencer on his own metaphor. If the state were an organism, it must expire fed on laissez-faire. Suppose, he queried, that in the body physiological, "each muscle were to maintain that the nervous system had no right to interfere with its contraction, except to prevent it from hindering the contraction of another muscle," and individual cells protested that their unsanitary waste did not abridge the freedom and safety of neighbors? In such a liberal state, the biological organism would collapse, but just as surely would the political organism. If the end of government was the welfare of its citizens,

49. Thomas Huxley, "Mr. Darwin's Critics" (1871), in his *Collected Essays* (New York: D. Appleton, 1896–1902), 2:120–86. See the discussion in chapter 5.
50. Thomas Huxley, "Administrative Nihilism" (1871), *Collected Essays* 1:251–89.
51. Ibid., p. 255.

then restrictions on the freedom of some (e.g., through taxation) might be required for the benefit of all.[52]

Huxley returned to this theme in fall of 1887, but now motivated by more poignant circumstances. He began writing "The Struggle for Existence In Human Society"[53] shortly after his beloved youngest daughter, Marian, a woman of promising artistic ability, died after a lingering madness. The long ordeal gradually crushed the old man, driving him into fits of unexpressible sadness.[54] He felt nature's brutal indifference to human desire, and his article spoke of this. He wrote bitterly that few could take solace, in the face of human anguish, from "the reflection that the terrible struggle for existence tends to final good, and that the suffering of the ancestor is paid for by the increased perfection of the progeny."[55] Only the savage "fights out the struggle for existence to the bitter end, like any other animal."[56] Civilized man attempted to stay the struggle by morally responding to the needs of others in society. This effort to beat back nature required that industrial nations pay a living wage to workers, administer the benefits of sanitation to the poor, and provide the means for education to all.

52. Ibid., p 271.

53. Thomas Huxley, "The Struggle for Existence in Human Society" (1888), *Collected Essays* 9 : 195–236.

54. A few months before their daughter died, the Huxleys dined with Beatrice Potter and her father. In her *Diary* (Beatrice Webb, pp. 202–203), Beatrice described in touching, if a bit romantic, detail the decline of Huxley's emotional life:

"6 May [1887]

Huxley to dine. The old lion is broken down: he has only the remains of greatness. Has lost that delightful spring of mind and living energy of thought that charmed those who knew him. . . . Huxley, when not working, dreams strange things, carries on lengthy conversations between unknown persons living within his brain. There is a strain of madness in him; melancholy has haunted his whole life. 'I always knew that success was so much dust and ashes—I was never satisfied with achievement.' None of the enthusiasm for 'what is'—or the silent persistency in grasping truth; more the eager rush of the conquering mind, loving the fact of conquest more than the land conquered. And consequently achievement has fallen far short of capacity. Huxley is greater as a man than as a scientific thinker. The exact opposite might be said of Herbert Spencer. . . . [Huxley] suffers in his old age from the melancholy of true failure; from the wearing anxiety of unself-controlled children—his brilliant and gifted child has sunk into hysterical imbecility, another daughter, mad with restless vanity, has taken without a care to public singing. Two daughters have married mediocrities. One son is gaining a modest livelihood, has failed to distinguish himself from lack of ambition or purpose. Another son is worthy but dull, lives at home without understanding his father. And the little daughter gads about to balls, and flirts away with inferior young men. Ah! these great minds, seldom fit for everyday life."

55. Huxley "The Struggle for Existence in Human Society," p. 198.

56. Ibid., p. 203.

Huxley anticipated that his essay would put Spencer "in a white rage with me."[57] But his friend reacted unexpectedly. He wrote a long letter assenting to Huxley's basic premise.

> I have nothing to object and everything to agree to. In fact the leading propositions are propositions that I have myself enunciated either publicly or privately. It was but the day before leaving Bournemouth that I was shocking some members of the circle upstairs at Kildare by insisting on the non-moral character of Nature—immoral indeed I rather think I called it: pointing out that for 99 hundredths of the time life has existed on the earth (or one might say 999 thousandths) the success has been confined to those beings which, from a human point of view would be called criminal.[58]

Spencer, of course, dissented from the corollaries Huxley drew concerning state action. He rather maintained that social conditions had in fact steadily improved, and only our sharpened sensitivity to social evil suggested otherwise. His counsel remained the same: let evolution continue without bungling bureaucratic interference. But he did not wish to press the point, "being that my criticism might cause a coolness between us which I should greatly regret."[59]

Huxley misread the letter. His own weight of feeling pressed down on the first lines only, which he construed as a charge of plagiarism. He hastily replied: "I think that the first pages of your long letter are written under a considerable misapprehension as to the extent with which my views have been formed independently of anything you have written as being unacknowledged foolish [sic]."[60] Apparently detecting a patronizing attitude, Huxley also bristled at Spencer's refraining from further criticism. Surprised, Spencer answered the next day in a short letter saying that his friend had mistaken his meaning.[61] But too late. With this note, correspondence broke off between the two, except for a brief and nasty exchange, some nine months later, when they crossed letters in the *Times* on the question of land nationalization.[62] Spencer's

57. L. Huxley, *Life and Letters of Thomas H. Huxley* 2:199.

58. Herbert Spencer to Thomas Huxley (6 February 1888), Huxley Papers, VII, 209, Imperial College Library, London.

59. Ibid.

60. Thomas Huxley to Herbert Spencer (9 February 1888), Huxley Papers, VII, 211, Imperial College Library, London.

61. Herbert Spencer to Thomas Huxley (10 February 1888), Huxley Papers, VII, 212, Imperial College Library, London.

62. The controversy erupted when supporters of land nationalization sought the authority of Spencer's *Social Statics*. By this time, however, Spencer found reason to reject

agitation grew to such proportions that he resigned from the "X Club," the social pleasures of which he had once relished. The disaffection between Spencer and his old friend climaxed with Huxley's delivery of his celebrated Romanes lecture in 1893, "Evolution and Ethics."

The Romanes Lecture

Huxley's Romanes lecture, read to a jammed Sheldon Auditorium at Oxford, reverberated of earlier criticisms of evolutionary ethics as a normative discipline. He again asserted nature's indifference to human suffering: "grief and evil fall, like the rain, upon both the just and the unjust," he preached.[63] In the state of nature, the biologically fit survive, but not necessarily the morally fit. Civilized men must combat the savagery of the struggle for existence. "Social progress," he proclaimed, "means a checking of the cosmic process at every step and the substitution for it of another which may be called the ethical process; the end of which is not the survival of those who may happen to be the fittest, in respect of the whole of the conditions which obtain, but of those who are ethically the best."[64] Huxley willingly conceded that the moral sentiments, no less than the immoral sentiments, evolved as part of human character. But this knowledge provided no moral rules; it established no moral ends. "Cosmic evolution," he cautioned, "may teach us how the good and the evil tendencies of man may have come about; but, in itself, it is incompetent to furnish any better reason why what we call good is preferable to what we call evil than we had before."[65] With unmistakable reference to Spencer, he concluded that it was "from neglect of these plain considerations that the fanatical individualism of our time attempts to apply the analogy of cosmic nature to society."[66]

Huxley's critics, Spencer included, noted the apparent bifurcation he had introduced into nature, between natural processes and human activity, as if man could somehow lift himself out of nature. For a philosopher-physiologist who had earlier claimed that human consciousness flitted impotently above brain like a ghost in a machine,[67] the distinction seemed incoherent. Fifty years later, even his grandson in

---

the idea as incompatible with "relative ethics." See Duncan, *Life and Letters of Herbert Spencer* 2:26–9.

63. Thomas Huxley, "Evolution and Ethics" (1893), *Collected Essays* 9:60.
64. Ibid., p. 81.
65. Ibid., p. 80.
66. Ibid., p. 82.
67. Thomas Huxley, "On the Hypothesis that Animals are Automata, and Its History" (1874), *Collected Essays* 1:199–250.

Figure 7.3 Thomas Henry Huxley, portrait done in 1883.

his own Romanes Lecture found the distinction indefensible.[68] Huxley quickly became sensitive to the objection, especially when the Catholic philosopher and his one-time pupil St. George Mivart suggested that

68. See Julian Huxley, "Evolutionary Ethics" (1943), in Thomas Huxley and Julian Huxley, *Touchstone for Ethics, 1893–1943* (New York: Harper, 1947), pp. 115–16. John Greene probes the intellectual affinities between Julian Huxley and Spencer in his *Science, Ideology, and World View: Essays in the History of Evolutionary Ideas* (Berkeley: University of California Press, 1981), pp. 162–68.

the great Darwinist seemed to have himself undergone a mental evolution, from a position that regard man as a well developed orang to one
that now took him to be an authentic moral and spiritual being.[69] In
the printed version of his lecture, Huxley added a long "Prolegomena"
in which he quietly knitted together what he had dramatically rent before the assembly in Sheldon Auditorium. He now explained the ethical
process (as in fact both Darwin and Spencer had) as arising from group
selection: "I have termed this gradual strengthening of the social bond,
which, though it arrest the struggle for existence inside society, up to
a certain point improves the chances of society, as a corporate whole, in
the cosmic struggle—the ethical process."[70] Natural selection extin-
guishes the struggle *within* the group by selecting for altruistic sentiments and cooperative attitudes, though societies themselves continue
to do combat. For Huxley, this meant that within society "the kind of
evolution which is brought about in the state of nature cannot take
place."[71] He offered only the vaguest idea of the status and further development of ethical norms within a society.[72] His proposal for the gen-

69. St. George Mivart, "Evolution in Professor Huxley," *Nineteenth Century* 34 (1893):
198–211.

70. Thomas Huxley, "Evolution and Ethics: Prolegomena" (1894), *Collected Essays*
9:35. In the notes to his Romanes Lecture, Huxley had already pulled his punch. Note 20
(pp. 114–15) suggests the conciliation that he made more explicit in the "Prolegomena."
The note reads in part: "Of course strictly speaking, social life and the ethical process in
virtue of which it advances towards perfection, are part and parcel of the general process
of evolution, just as the gregarious habit of innumerable plants and animals, which has
been of immense advantage to them, is so. . . . Even in these rudimentary forms of society, love and fear come into play, and enforce a greater or less renunciation of self-will.
To this extent the general cosmic process begins to be checked by a rudimentary ethical
process, which is strictly speaking, part of the former, just as the "governor" in a steam-
engine is part of the mechanism of the engine." George Romanes, who saw an advanced
copy of the lecture, complained to Huxley just prior to delivery about the misleading
impression it would leave if the note were not included in the text. See George Romanes
to Thomas Huxley (27 April 1893), in Ethel Romanes, *Life and Letters of George John
Romanes,* 4th ed. (New York: Longmans, Green, 1897), pp. 325–26.

71. Ibid., p. 36.

72. Huxley continued to worry about the relation of ethical man in society to brute
nature existing outside of human intercourse. In the short interval between the publication of his Romanes Lecture in 1894 and his death in summer of 1895, he began a manuscript dealing with the relationship between "Natural History," which concerned "man
and the rest of the world in the state of Nature," and "Civil History," which concerned
"man in the state of art or civilization." The manuscript (Huxley Papers, XLV, ff 42–50,
Imperial College Library, London) is an incomplete first draft, but it indicates his struggle
to accord man a distinctive set of ethical norms while recognizing his naturally constituted
character. The following somewhat garbled excerpts will suggest the tenor of his
considerations.

"I have said that the Ethics of Evolution (I add 'so-called' because I do not admit the
[right?] & that which is so called is the title is applied Natural History because it supposes

eral course of ethical evolution simply did not substantially differ from Spencer's. Spencer, too, thought that group selection would establish the fundamental principles of altruistic action, while accommodation to social relationships (aided by inherited habit) would function as the main engine of progress within society. Moreover, Spencer did not simply accept without argument an identity between the ends of evolution and the ends of morality, which Huxley himself now came close to doing. The coda to the apparent about-face in Huxley's lecture was a reconciliation with Spencer, though surely more for reasons of the evolution of long friendship rather than because of agreement about evolutionary theory and social policy.

Other Critics of Evolutionary Ethics

Sidgwick

The Cambridge philosopher Henry Sidgwick became the most dogged and perceptive of the many critics of Spencer's ethical system. He initially took notice of the ethical ideas of *Social Statics* in the first edition (1874) of his magisterial *Methods of Ethics,* and he built up his criti-

---

that the struggle for existence on which progress in the 'Natural History' world depends, goes on & is the condition of progress in Civil History.

"The complete logical consequence of that doctrine is the ultra individualism of the philosophical anarchists—the half way to it is the [word illegible] of laissez faire philosophy—Practical results of it are seen in the ignoring of the value of the state; the denial of its authority & of the duties of the individual toward it; which occurs to me quite as mischievous as the antique error in the other direction—perhaps more so. I have desired to express my conviction that the fundamental assumption of this school are mistaken. True, that to begin with, social union gives an advantage "in the struggle for existence"—In respect of external relations the Natural History process and the dawning of the ethical impulse work in harmony—But from this truce that each individual becomes secured against robbery & murder by others;—from the truce that mutual help assumes the fulfilling of an obligation, the 'Natural history' struggle for existence is at an end. If social progress takes place it can owe nothing to the struggle for existence. And the whole foundation of modern individualism is cut away from under its feet.

"The first great step toward progress in the Civil history world is the establishment of security & life & property for all without reference to their adaptability or indeed to anything else than the [word illegible] that they are same human beings. Consequently when this stage has been reached—the struggle for existence of the natural history world is at an end. Moreover from this point ethical progress—the evolution of ethics—is antagonistic to any exercise of the machinery of Natural History Evolution. Unlimited self assertion is the foundation of the struggle for existence (& the cause of progress) in Natural History—Civil history beings with self renunciation—that is to say with the abstinence by the individual from executing some of his possibilities of action" (MS pp. 1–4).

cisms through subsequent editions.[73] His most sustained examination of Spencer came, however, in his lectures on ethics at Cambridge.[74]

Sidgwick followed Spencer through *Social Statics*, *The Data of Ethics*, and the completed system of *The Principles of Ethics*, exposing his misty formulations and congratulating him before the undergraduates on his several counsels, "judicious for the most part," but also "courageously commonplace."[75] Sidgwick's strongest objections focused on three areas.

The first important animadversion appeared in the original edition of the *Methods of Ethics* and occurred frequently, in various guises, in the lectures. Sidgwick observed that in *Social Statics* Spencer developed only the moral principles of the "straight man," the ideal principles of a perfect social organization. These principles, he complained, could have only an "indirect and uncertain . . . relation to the practical problems of actual life."[76] But Spencer had anticipated this criticism in his letter to Mill and explicitly answered Sidgwick in the *Data of Ethics* by emphasizing the distinction, already present in *Social Statics,* between "relative ethics" and "absolute ethics," the former consisting of pragmatic principles for realizing absolute ethical norms in a less-than-perfect society. Sidgwick remained of the opinion, though, that a perfect society was "so far distant that it cannot even afford us an ideal of any practical value"; he considered Spencer's maxims of relative ethics to be little better than moral bromides.[77] It would have been strange, however, if Spencer had not derived commonplaces of morality from his system: he intended, after all, to demonstrate that our intuitive moral feelings, often captured in the tritest of maxims, yet had validity, had moral force. Sidgwick sounded the usual protest made against the apparent remoteness of ethics done in the Kantian vein, which was also Spencer's. Spencer certainly did not intend his moral theory to cast up an indefinitely long list of injunctions for every contingent circumstance. He expected that the general principle of equal freedom would require practical intelligence for its application, much as Kant's categorical imperative relied on an individual's developing the appropriate maxims while wading through the pullulating affairs of life. Spencer believed that the discovery and promulgation of the absolute principles of the ideal society would more quickly bring men to their duty and hasten the attainment

73. Henry Sidgwick, *The Methods of Ethics,* 7th ed. (Indianapolis: Hackett, [1907] 1981).

74. Henry Sidgwick, *Lectures on the Ethics of T. H. Green, Mr. Herbert Spencer, and J. Martineau* (London: Macmillan, 1902).

75. Ibid., p. 310.

76. Sidgwick, *Methods of Ethics,* p. 18.

77. Sidgwick, *Lectures on Ethics,* pp. 165–67.

of the goal. Each man, he admonished, "may properly consider himself an agent through whom nature works."[78]

Sidgwick's second objection had more logical force. He wondered how Spencer could really justify the supposed necessary link between the goal of evolution and the goal of ethics: the coincidence of the full development of life and the greatest happiness. Further, whose life and whose happiness should prevail in a moral decision?[79] Spencer had attempted to work out the answers to these questions in absolute ethics, as we have seen above; and while I do not think he satisfactorily demonstrated the conjunction of the goal of evolution and the goal of ethics, the difficulty can be mitigated in the way I have already suggested. Yet one may still ask, as Sidgwick did, about the problem in relative ethics: "What is to be done if the two ends—Life and Happiness—do not coincide, in any particular case, here and now? and whose Life (or Happiness) is to be taken as standard by the individual, if circumstances arise in which a choice has to be made between action conducive to his own Life (or Happiness), and action conducive to the Life (or Happiness) of others?"[80] But again these are matters of practical intelligence; a priori resolutions simply cannot be offered.

Sidgwick probably expressed more inky irritation over Spencer than over any of the other moralists whom he considered. This seems strange, since his own ethical system undertook the same formal task as Spencer's: the reconciliation of moral sense intuitionism with utilitarianism. Sidgwick's theory discriminated several self-evident moral canons,[81] discovered them to require as the highest moral principle the

78. Herbert Spencer, *Social Statics* (London: Chapman, 1851), pp. 474–75: "[A man] must remember that whilst he is a child of the past, he is a parent of the future. The moral sentiment developed in him, was intended to be instrumental in producing further progress; and to gag it, or to conceal the thoughts it generates, is to balk creative design. He, like every other man, may properly consider himself as an agent through whom nature works; and when nature gives birth in him to a certain belief, she thereby authorizes him to profess and to act out that belief."

79. Sidgwick, *Lectures on Ethics,* p. 217.

80. Ibid.

81. Sidgwick variously formulated the moral axioms of his ethics (see his *Methods of Ethics,* chap. 13), but one set reads: (1) "whatever action any of us judges to be right for himself, he implicitly judges to be right for all similar persons in similar circumstances" (p. 379); (2) "a smaller present good is not to be preferred to a greater future good (allowing for difference of certainty)" (p. 381); (3) "each one is morally bound to regard the good of any other individual as much as his own" (p. 382); and (4) "as a rational being I am bound to aim at good generally,—so far as it is attainable by my efforts,—not merely at a particular part of it" (p. 382). See also J. B. Schneewind's extremely helpful book *Sidgwick's Ethics and Victorian Moral Philosophy* (Oxford: Clarendon Press, 1977), pp. 286–309.

maximizing of goodness,[82] and determined the highest good to be pleasure.[83] In skeleton it was rather like Spencer's system. From the perspective of an evolutionary historiography, however, Sidgwick's animus makes sense: for with conceptual systems, as well as with biological systems, the struggle will be keenest among those that are most similar. The difference that made the struggle meaningful lay below their morphological convergence.

Both Spencer and Sidgwick held that objective moral intuitions provided guidance for conduct and served as the raw principles which reason might refine into a coherent system. But each had a different source for the objectivity of these intuitions. For Sidgwick, the objectivity, truth, and certainty of moral intuitions did not derive from their being innate, nor from their origin in factual experience, but from their roots in the universal and necessary structure of rational action.[84] His methodological analysis established several conditions that were required to manifest the validity of such insights (e.g., clearness, care in ascertaining them, consistency among them, and agreement of experts),[85] though these conditions did not produce that validity. For Spencer, the instincts of sympathy for self, progeny, and society, the feeling for equality, and the rest—these also had conditions, but conditions producing validity: these moral instincts had their source in inherited mental structures and derived ultimately from the requirements of experience. This deep difference in their respective reconciliations between moral sense theory and utilitarianism led to Sidgwick's most important criticism.

Sidgwick asked of Spencer and his theory: "Why am I to seek the General Happiness?"[86] The "gap in ethical construction" that this question revealed was also exposed by Huxley, when he similarly queried why he should follow his moral rather than his immoral sentiments. For Sidgwick, the "oughtness" of moral principles depended upon no empirical conditions for its force, but only upon the general intention to maximize goodness, which intention such principles instantiated. For Spencer, their "oughtness" derived from the experience of ancestors who had adapted to the requirements of society and the consequent inherited mental structures that, as a matter of fact, impelled their descendants toward certain conduct. Both ethicians agreed that pleasure

82. Sidgwick, *Method of Ethics*, p. 105–15; and Schneewind, *Sidgwick's Ethics*, pp. 307–9.

83. Sidgwick, *Methods of Ethics*, pp. 391–407

84. Sidgwick, *Methods of Ethics*, p. 373–90; also Schneewind, *Sidgwick's Ethics*, chaps. 7 and 10.

85. Sidgwick, *Methods of Ethics*, pp. 338–42.

86. Sidgwick, *Lectures on Ethics*, p. 219.

constituted the highest good. But Sidgwick found that end to be revealed through insight into the *rational* requirements of action, while
Spencer concluded that pleasure was the coincident end reached by
means of the *empirical and evolutionary* requirements of action. The
structure of reason had independent and universal validity in Sidgwick's
founding epistemology; in Spencer's, it was determined by the common experience of evolving organisms. For Sidgwick, factual experience could not produce moral imperatives; for Spencer it did. And thus
in Sidgwick's estimation, Spencer's ethical construction left a "gap."
Though Sidgwick initiated this objection to Spencer's ethics, he unac-
countably did not press it home. His own student G. E. Moore, who
attended his lectures on Spencer, did; and it is to Moore that we owe
the term by which this objection usually travels. Spencer, Moore declaimed, had lapsed into fundamental moral error, he had committed
the "naturalistic fallacy."

G. E. Moore and the Naturalistic Fallacy

In the *Principia Ethica,* published in the year of Spencer's death, 1903,
G. E. Moore devoted considerable critical attention to Spencer's ethical
ideas, more than to any one else's, save those of his teacher and fellow
Apostle at Cambridge, Sidgwick. Moore is usually thought to have
demonstrated the fallacy of Spencer's position, and thus to have rendered any further efforts at constructing an evolutionary ethics sterile.
This is a mistaken evaluation, I believe. It supposes first that Spencer's
ethics (or any other) can be devastated by an objection that admits no
presuppositions and that the power of the attack is not a function of the
viability of its presuppositions. But Moore insinuated Bloomsburian
presuppositions, the universal validity of which one, with caution, may
doubt.

When first looking into *Principia Ethica,* Lytton Strachey announced
that "the Age of Reason has arrived." It was, however, a reason that
satisfied his own predilections. "It's madness of us to dream," he mused
as he further contemplated the book, "of making dowagers understand
that feelings are good, when we say in the same breath that the best
ones are sodomitical."[87] Moore's own tastes differed somewhat from
Strachey's. Moore conceived the highest ethical good, the "rational ultimate end of human action and the sole criterion of social progress,"[88]
to be not the health and vigor of the community, as Darwin did, nor

87. Strachey quoted by Leon Edel in *Bloomsbury: A House of Lions* (New York: Lippincott, 1979), p. 47.
88. Moore, *Principia Ethica,* p. 189.

the equal freedom and happiness of the community, as Spencer did, but rather "personal affection" and the "appreciation of what is beautiful in Art or Nature."[89] These, he felt, were self-evidently the ideal goods. In his own terms, he may not have committed the naturalistic fallacy. He certainly committed the Bloomsburian fallacy—the presumption that the comfortable intuitions of the Cambridge don fix truth and morality.

Moore condemned Spencer's moral system because it implied that the concept of "good" was definable, so that one might falsely believe it made sense to say "equal freedom constituted the moral good." This, according to Moore, was the naturalistic fallacy—"the fallacy which consists in identifying the simple notion which we mean by 'good' with some other notion."[90] Moore believed "good" was a simple, unanalyzable property attaching itself, more or less, to objects and actions. The perceptive Bloomsburian moralist could simply spy this property in things or behavior; Moore detected it in his Cambridge friendships, as did Strachey. Given such perceptions one might, however, well hesitate to endorse the moral theory that sanctioned them. Indeed several objections may be lodged against Moore's ethics. First, the belief that "good" or any other property is utterly simple provokes Parmenides's brief against Socrates in the *Parmenides:* an utterly simple property cannot even have meaningfully predicated of it that it is a property of this or that object. Second, individual judgments of the Bloomsburian sort have no guarantee of universal validity, no epistemological or psychological theory that would warrant their objectivity (that is, their public character)—we have no theory which would persuade us that all ethically acting individuals have the same intuitions as the Cambridge don. But every moral judgment requires the concomitant judgment (if only implicitly made) that anyone would evaluate a moral situation in a like manner. (Spencer's theory escapes both of these difficulties.) Moore's criticism of Spencer thus rested ill on weak presumption. Moreover, even in its hollow declamations, it only vaguely echoed the more logically telling, if nascent, criticisms of Huxley and Sidgwick. So, though Moore named the putative fallacy, his predecessors gave it force. Let me restate the straightforward claim underlying their objections.

Evolutionary ethics appears to identify what *is*—for example, a given trend of evolution—or what is *predicted to be*—for example, a final state of equal freedom—with what *ought to be*. But from factual descriptions one cannot, it is usually held, logically derive an imperative. While I will pursue this objection to evolutionary ethics—that is, to my historically descendent version—in the second appendix, let me here in sum-

89. Ibid., pp. 188–89.
90. Ibid., p. 58.

mary consider whether Spencer lapsed into this supposed error. I do not believe that he did.

Spencer did not simply identify the evolutionary process with the moral process. He rather attempted to construct three different arguments, which together led to the conclusion that certain trends in evolution (for instance, greater adaptation to the social state) were moral trends and thus "ought to be" promoted. The first argument, as we have seen, tries to demonstrate that the general end of evolution is complete adaptation to the social state, which creates equal freedom. Spencer added that such adaptation requires as means benefits bestowed on self, progeny, and community. The second and logically independent argument is an ethical one: it proposes that the general moral end of man is greatest happiness, but specifies it as a matter of distributive justice. Spencer attempted to establish this conclusion by the sound strategy of contending that all ethical systems really do have happiness as their ultimate moral value and that all men logically assert it, whether or not they verbally admit it. The mode of argument is not a mere induction; rather it derives its logical force from the challenge: produce any coherent ethical system, and I will show that its end is greatest happiness. In essence, Spencer had constructed an abbreviated transcendental argument, which proposed that moral life and any conceivable moral system could make sense only under the assumption of the greatest happiness principle. The third argument attempts to demonstrate that the end of the evolutionary process also establishes—as a matter of natural rather than logical connection—the end of morality. If that identity may be made, then one can regard the laws determining the evolutionary process in man (i.e., principles of adaptation to advancing social relations) also as moral principles (supposing that all principles intending a moral end are moral principles). In this sorites, a fresh set of moral concepts is introduced, so that the conclusion may quite legitimately contain moral terms. What is at stake in defending Spencer against the objections of Huxley, Sidgwick, and Moore is not that each of his individual arguments rings solid—they certainly do not—but rather that the strategy of deploying arguments in this fashion is sound, or at least does not commit the putative fallacy.

Conclusion

Spencer's work established an intellectual climate for several generations of theorists, who became imbued with his perception of the mental evolution of organisms and their social dependencies. In succeeding chapters, I will illustrate the specific legacies passed to Romanes, Morgan, James, and Baldwin. Let me now briefly offer an historical assessment of four areas of Spencer's thought and suggest their evolutionary trajectory beyond the nineteenth century.

Spencer's estimation of the role of government in human affairs certainly had an edge, which cuts out at least half of the truth. As a matter of course, committees of venal men, which is what governments frequently are, do a botched job. One's sympathy with Spencer's judgment also increases from historical awareness of the character of social institutions in early-nineteenth-century Britain. The Poor Laws of the early part of the century were, as Spencer claimed, often used to quarantine the French disease on the other side of the Channel. Upper-class landowners often manipulated the laws to keep wages of workers below subsistence—with the deficiency being made good by taxes on the local parish. We, however, recognize government as a necessary institution, so Spencer's hope that it might be restricted to protection and then completely eliminated in an evolutionary consummation—that is a hope we cannot rationally harbor. Indeed Spencer would have been warranted, in light of his own theory, to regard government as an intrinsic part of social development and thus bred into human nature.

Yet even Spencer's dyspeptic judgement about the liabilities of government lives on. In the mid-teens of this century, Truxtun Beale persuaded a number of eminent Americans to assess the contributions of Spencer to their own social philosophies. Senators Elihu Root, Henry Cabot Lodge, and William Howard Taft joined with university presidents Nicholas Murray Butler (Columbia) and Charles W. Eliot (Harvard), and other notables to write interpretative comments to the various essays of Spencer collected under the title *Man versus the State* (1884). Both the essays and the comments were largely occupied, as contributor David Jayne Hill observed, with "the extension of official control over the private life and activities of the individual, and the consequent subordination of personal liberty to the dictates of a ruling class acting in the name of the state."[91] The native Yankee soul resonated to Spencer's individualism, and continues to. American neoconservatives still find succor in Spencer. Liberty Fund has republished Spencer's *Principles of Ethics* and *Man versus the State* in high quality, but inexpensive, editions, certainly not simply to meet the needs of disinterested scholars.

Spencer's social philosophy, however, bore other progeny, rather different in appearance, but no less legitimate, I think. His conceptions

91. David Jayne Hill, "Introduction," *Man versus the State: a Collection of Essays by Herbert Spencer,* ed. Truxton Beale, with Critical Comments by William Howard Taft, Charles W. Eliot, Elihu Root, Henry Cabot Lodge, David Jayne Hill, Nicholas Murray Butler, E. H. Gary, Harlan F. Stone, and Augustus P. Gardner (New York: Kennerley, 1916), p. ix.

also promoted socialism. In his *Social Statics,* he envisioned as the terminus of evolution a socialistic utopia, a classless society in which government would have withered away, land would be held in common, and each citizen (including women and children) would enjoy the fullest freedom compatible with the freedom of all. This was a vision that initially appealed to the anarchistic socialist Prince Peter Kropotkin, who, though he dismissed Spencer's later exaggerated individualism, found in *Social Statics* the foundations for a socialist ethics.[92] The Italian socialist Enrico Ferri also discovered in Darwin and particularly in Spencer those scientific conceptions that led ineluctably to Marxism.[93] Ferri's *Socialism and Positive Science (Darwin-Spencer-Marx)* (1894) became the first volume that Ramsay MacDonald included in his series *The Socialist Library.* And Spencer's companion and dearest friend of later years, Beatrice Potter Webb, consumed his books, from *Social Statics* and *First Principles* to the volumes of psychology, biology, and sociology. While she broke in a direction that Spencer abhorred (and thus he requested she be his literary executor only in camera), she came to perceive under his tutelage that a scientific understanding of social evolution allowed us to take command of that evolution and direct it more proximately to moral ends.[94] Spencer's intellectual children, true to their paterfamilias, went in perversely different ways.

92. Prince Peter Kropotkin, *Ethics* (New York: Benjamin Blom, [1924] 1968), p. 290: "Both in his 'Social Statics' and 'The Principles of Ethics,' Spencer expounded the fundamental idea that Man, in common with the lower creatures, is capable of indefinite change by adaptation to conditions. . . . Gradually, under the influence of the external conditions of life and of the development of the internal, individual faculties, and with the increasing complexity of social life, mankind evolves more cultural forms of life and more peaceful habits and usages, which lead to a closer co-operation. The greatest factor in this progress Spencer saw in the feeling of *sympathy* (or *commiseration*)."

93. Enrico Ferri, *Socialism and Positive Science (Darwin-Spencer-Marx),* Socialist Library—1, ed. Ramsay MacDonald, M.P. (London: Independent Labour Party, [1894] 1905). When Spencer learned that the Italian socialist had made appeal to his ideas, he complained of the audacity. Ferri properly responded: "the personal opinion of H. Spencer is a different matter from the logical consequence of the scientific theories on universal evolution which he has developed farther and better than any other man, but of which he has not the official monopoly nor the power to prohibit their free expansion by the labour of other thinkers" (p. 153). Greta Jones discusses Ferri and other socialists who found comfort in evolutionary theory. See her *Social Darwinism and English Thought* (Atlanta Highlands, N. J.: Humanities Press, 1980), pp. 63–77.

94. In the first installment of Beatrice Potter Webb's autobiography, *My Apprenticeship* (Cambridge: Cambridge University Press, [1929] 1979), she made explicit both her intellectual and her personal debt to Spencer. In brief compass she sketched the way his ideas influenced her socialism:

"It was after Mother's death—in the first years of mental vigour—that I read the *First Principles* and followed his generalisations through Biology, Psychology and Sociology. This generalisation illuminated my mind; the importance of functional adaptation was,

   Two other central aspects of Spencer's thought, his evolutionary psychology and epistemology, and his evolutionary ethics, are indeed alive,
because they are essentially correct. These truths have survived the
struggle.
   Modern physiological and cognitive psychology is Kantian. Not in
the way Kant was a Kantian, but in the way Spencer was. The infant
comes into the world already outfitted with perceptual and cognitive
categories by which it organizes its experience. It does not encounter
Dingen an sich but objects that bear the marks of the races's evolutionary
history. Though we now, of course, reject the inheritance of acquired
characters, we still must agree with Spencer that human nature arises
out of experience—our own immediate experience, that which constitutes our individual history, and, most importantly, the adaptational
experiences of our ancestors. In subsequent chapters we will follow
some of the paths by which Spencer's evolutionary Kantianism worked
its way into twentieth-century cognitive theory.
   In ethics, it might be thought that Spencer, and anyone else who
attempted an evolutionary construction of moral behavior, was doomed
to commit the naturalistic fallacy. Huxley, in his famous Romanes lecture on "Evolution and Ethics," argued, in an only modestly veiled attack on his friend, that evolutionary development could not provide a
criterion for ethical judgment and behavior. We must rather, he urged,
struggle against the cosmic process. But Huxley misunderstood Spencer's evolutionary ethics. Spencer knew the distinction between a moral
psychology and a moral science. In moral psychology we attempt to
discover how men actually make moral decisions, and we might even
try, as Spencer did, to plot the evolutionary curve of human moral be-

---

for instance, at the basis of a good deal of the faith in collective regulation that I afterwards developed. Once engaged in the application of the scientific method to the facts of
social organisation, in my observations of East End life, of co-operation, of Factory Acts,
of Trade Unionism, I shook myself completely free from *laisser-faire* bias—in fact I suffered from a somewhat violent reaction from it. And in later years even the attitude
towards religion and towards supernaturalism which I had accepted from him as the last
word of enlightenment, have become replaced by another attitude—no less agnostic but
with an inclination to doubt materialism more than I doubt spiritualism—to listen for
voices in the great Unknown, to open my consciousness to the non-material world—to
prayer. If I had to live my life over again, according to my present attitude I should, I
think, remain a conforming member of the National Church. My case, I think, is typical
of the rise and fall of Herbert Spencer's influence over the men and women of my own
generation" [pp. 38–39].
   Webb records that in her visit with Spencer just before his death, he acknowledged
that his social theory and hers "'had the same ends . . . it is only in methods we have
differed'" (p. 37).

havior. But moral psychology, evolutionary or otherwise, cannot logically justify, according to Spencer, any particular criteria of judgment; this is the function of a moral science, whose logic is that of imperatives rather than factual propositions. Spencer believed that human happiness, the greatest happiness for the greatest number, was indeed a moral criterion; and its proximate imperative was to act so as to secure your own freedom and that of others in your society. When it was pointed out to him that his evolutionary vision of society as a kingdom of free individuals reconstituted precisely one of the formulations Kant gave the categorical imperative, he was gratified. The primary expression of Kant's imperative runs: act only on that maxim whose principle you can will to be a universal law of nature. Not only did Spencer will his moral imperatives to be universal laws of nature; he thought they actually were laws of nature, evolutionary laws. While today we cannot accept the letter of Spencer's evolutionary ethics, we can, as I hope to show in the second appendix, adopt its spirit.

Finally, we must consider the impact Spencer's conception of the evolution of social behavior had on subsequent theorists. The current discussions in biology concerning cooperative and altruistic behavior in organisms have many sources, though one strong line leads back to W. C. Allee, who at the University of Chicago studied the evolution of social behavior among lower organisms. In his *Social Life of Animals* (1938) and its revision, *Cooperation Among Animals* (1951), Allee traced his own interest in the evolution of sociality to his reading about the English moral-sense philosophers (particularly Shaftesbury) and his study of Alfred Espinas's *Des sociétés animales* (1878) and Kropotkin's *Mutual Aid* (1902).[95] He also imbibed Spencer's conception of the necessary balance between egoism and altruism in the evolution of society.[96] Cast into a neo-Darwinian framework, this conception of evolutionary balance dominated Allee's studies of animal evolution and his more speculative conclusions about human altruism. Moreover both Espinas and Kropotkin drew significantly on Spencer. Espinas, who had translated Spencer's *Principles of Psychology* into French, particularly liked the conception of sociality arising out of the general laws governing all evolutionary advance, correcting as it did Comte's assumption of unique social laws having no antecedents.[97] Kropotkin, also a translator of Spencer, found beneath the Englishman's sometimes exaggerated

95. W. C. Allee, *The Social Life of Animals* (New York: Norton, 1938), pp. 24–27; and *Cooperation Among Animals with Human Implications* (New York: Henry Schuman, 1951), pp. 8–12.

96. Allee, *Cooperation Among Animals,* p. 9.

97. Alfred Espinas, *Des Sociétés animales,* 2d ed. (Paris: Bailliere, 1878), pp. 83–155.

ideas about the struggle for existence a kindred, anarchical spirit. And Spencer also understood, in Kropotkin's view, the role that social instinct and societal cooperation played in evolutionary progress.[98] These are simply the other avenues by which Spencer's evolutionary thought has broken through to the latter part of the twentieth century.

Recent assessments of Spencer have usually done him grave injustice. Most historians of science believe that his evolutionary ideas died as he did, without issue. And they regard his ethics as a product of an unwarranted extension of the law of survival of the fittest into the moral realm. We will take a fuller measure of his impact on subsequent thought in succeeding chapters, though I believe we have seen enough to know that it was substantial. Concerning that further conclusion, my thesis through these two chapters has been that his evolutionary biology was formed to meet the demands of his ethics, which in its theoretical structure and ruling imperative must, I believe, be admired. It would be a foolish revisionist, however, who contended that recent evaluations have benightedly dismissed a thinker nonpareil. I believe that William James, in his obituary notice of Spencer, achieved the just balance:

> Rarely has Nature performed an odder or more Dickens-like feat than when she deliberately designed, or accidentally stumbled into, the personality of Herbert Spencer. Greatness and smallness surely never lived so closely in one skin together.[99]

98. In *Mutual Aid* (Boston: Extending Horizons, [1902] 1955), p. 65, Kropotkin recognized that Spencer's conception of functional adaptation to social conditions mitigated the need for "struggle for existence." See also the passage from Kropotkin in note 92.

99. William James, "Herbert Spencer's Autobiography" (1904), in *Memories and Studies* (New York: Greenwood Press, [1911] 1968), pp. 107–8.

8

Darwinism and the Demands of Metaphysics and Religion: Romanes, Mivart, and Morgan

Darwinian theory, as everyone knows, crushed nineteenth-century belief in a spiritually dominated universe and purged nature of intelligent design and moral purpose. "Natural-selection theory and physiological reductionism were explosive and powerful enough statements of a research program to occasion the replacement of one ideology—of God—by another: a mechanical, materialist science." So judge the scientists and cultural critics Richard Lewontin, Steven Rose, and Leon Kamin.[1] Gertrude Himmelfarb agrees. Darwinism produced its "traumatic effect" on humane sensibility by replacing "man by nature, moral man by amoral nature."[2] The historian of science Susan Cannon sustains this prevailing opinion: she is convinced that Darwin drained nature of moral significance, that he had shown nature to be "morally meaningless."[3]

These characterizations of Darwin's accomplishment control our perception of the late nineteenth and early twentieth centuries. But are they accurate? I do not think so. I believe they grievously distort historical reality. We have already seen that neither Darwin nor Spencer, certainly the iconic figures of evolutionary doctrine in the nineteenth century, rendered nature "morally meaningless." On the contrary, they scientifically reconstructed nature with a moral spine. And while Darwin's metaphysical views might have been materialistic—benignly and inconsequentially so, I believe—Spencer's were not. Though as he complained to a correspondent, he "had to rebut the charge of materialism

1. Richard Lewontin, Steven Rose, and Leon Kamin, *Not in Our Genes: Biology, Ideology, and Human Nature* (New York: Pantheon, 1984), p. 51

2. Gertrude Himmelfarb, *Marriage and Morals among the Victorians* (New York: Knopf, 1986), p. 79

3. Susan Cannon, *Science in Culture: the Early Victorian Period* (New York: Science History Publications, 1978), p. 276

times too numerous to remember"[4]—thus did the weight of words and hard logic fail to impress obdurate opponents.

Perhaps we ought to look rather to the successors of Darwin and Spencer in order to evaluate the impact of their ideas. These eponyms might be thought still too constrained by the theological conceptions of their youth to have worked out the full consequences even of their own ideas. After all, Darwin confessed that he cut natural selection against the mold of Paley's Creator, and Spencer's Unknowable seems to hover just behind the nonconformist's veil. That other founder of evolutionary theory, Alfred Wallace, might also be dismissed as too much the Victorian oddity, after his conversion to spiritualism, to serve as gauge for the philosophical and religious implications of evolutionary thought. We should, then, look to the next generation of Darwinian evolutionists. In subsequent chapters we will examine the views of the prominent American representatives of that generation, William James and James Mark Baldwin. Here, however, I would like to consider the ideas of George Romanes and Conwy Lloyd Morgan as a means of measuring up evolutionary thought in late nineteenth-century Britain.

Romanes was virtually anointed Darwin's successor by the old man himself; and he undertook the defense of Darwinism—and its further extensions into the evolution of mind and behavior—with a zeal that made Huxley look the model of Victorian reserve. Romanes did frequent battle with Mivart, a Catholic evolutionist and anti-Darwinian who sought to balance his religious orthodoxy with informed scientific judgment, though he succeeded only in destroying both. Mivart's story highlights the real difficulty, not to be minimized, of reconciling traditional faith with scientific rationality. For purposes of contrast, I will devote some time to his case. Morgan, the literary executor of Romanes, assumed leadership in English evolutionary biopsychology after the untimely death of his friend. He reformed evolutionary psychology's conceptual framework in terms of the new hereditary theory of Weismann. Through efforts of Romanes and Morgan the evolutionary biology of behavior emerged as a well-defined discipline at the turn of the century. Moreover both men attended to—even became obsessed in that peculiarly nineteenth-century way with—the implications of evolutionary theory for man's nature and aspirations. So they admirably serve to test the thesis that Darwinism established a materialistic, mechanistic, and amoral hegemony in late Victorian science.

In order to assess adequately this thesis, we must first get clear about

4. Herbert Spencer to Lewis Janes (4 May 1891), in the papers of Dr. Lewis G. Janes (private collection of Bradford Lyttle).

the meanings of "materialism" and "mechanism," and have ready to hand a steady idea of what a demoralized nature would be like. Mandelbaum, who has called the nineteenth century the "age of philosophical materialism," has constructed a definition of "materialism" with historical care.[5] I will follow his proposals, adapting them somewhat, and will understand by materialism a doctrine which holds: (1) that only material objects exist; (2) that God thus does not exist; and (3) that the human mind is a property of the material body. "Mechanism" would denote the strict application of this doctrine, maintaining that whatever the properties of mind or behavior, they are explicable by general laws governing all material manifestations. Mechanism so understood excludes any theories of emergentism.

A universe without moral character, one indifferent to the fall of sparrow or man, cannot provide any reasons to support ethical imperatives. I will take, then, as the nut of the moral-meaninglessness-of-nature proposition that nature offers no objective grounds for moral judgments. The two doctrines of mechanistic materialism and a morally meaningless nature lead logically and historically to the ethical theory of subjectivism, which claims that moral judgments are really only expressions of subjective preference. According to this view, an ethical imperative, such as "One ought not commit homicide, except to defend human life" must be translated, as A. J. Ayer and C. L. Stevenson have urged, as declarations of preference along with a recommendation—for example, "I don't like killing people who do no harm, and I hope you will feel the same."[6]

This chapter will have two purposes, first the portrayal of Romanes's and Morgan's theories of mind and behavior (with attention to those of Mivart as well), and second, an evaluation of their ideas in order to determine the extent to which these leading Darwinists might be classified mechanistic materialists, with all that seems to imply. The next two chapters will retain these historical interests, but situate them in rather different thematic contexts. The second appendix will take up the moral problem directly, turning from history (but not very far) to assay the logical implications of evolutionary theory for ethics, that is, to discuss the possibility of founding an objective ethics on evolutionary theory.

5. Maurice Mandelbaum, *History, Man, & Reason: A Study in Nineteenth-Century Thought* (Baltimore: Johns Hopkins Press, 1974), p. 22

6. The pertinent documents for a critical appraisal of ethical "emotivism" are conveniently gathered in Wilfrid Sellars and John Hospers, eds., *Readings in Ethical Theory* (New York: Appleton-Century-Crofts, 1952), pp. 391–440.

Prayer and the Imperatives of Scientific Reason

Romanes's Education and Early Career

George John Romanes was born under the sign of the pound sterling. On 2 May 1848, the day of Romanes's birth, his father inherited a fortune. This auspicious event permitted Dr. Romanes, then professor of Greek at Queen's University in Kingston, Ontario, to remove his family back to Britain. They settled in Regent's Park, London, where the family grew by two more sons and two daughters. Romanes's mother, Isabella Smith, was a Canadian Scot raised in the Presbyterian assembly, while his father held holy orders in the Anglican Church. A narrow sectarianism played no role in Romanes's upbringing; the family indifferently attended both rites. And when Romanes later expressed his intention of following his father into the clergy, his parents gave him no encouragement. The family's religious convictions appear to have been genteel and respectable, but hardly enthusiastic.[7]

Romanes, according to the report of his wife, had been regarded by his parents "as a shocking dunce."[8] Illness caused him to withdraw from preparatory school and to suffer an irregular education at home. Only after receiving some intense tutoring for over a year, was he ready to enter Gonville and Caius College, Cambridge, in October 1867. His original intention of taking holy orders seems to have faded as his interest in natural science grew stronger. His inadequate preparation, though, only carried him to a second class degree, a sore disappointment to his parents and an apparent confirmation of their early assessment of his intellectual abilities.

Romanes, however, was determined. He sought out Michael Foster and, under the young scientist's guidance, began a serious study of invertebrates in Foster's new Cambridge laboratory. In 1874 Romanes moved to London, where he continued his physiological studies with William Sharpey and John Burdon-Sanderson at University College. Together Foster, Sharpey, and Burdon-Sanderson brought British experimental physiology out of the dark shadow of its German counterpart. Romanes achieved distinction as an early contributor to the advancement of physiology in Britain.

While working in Foster's lab, Romanes began a study of the nervous system of Medusae. His precise anatomical descriptions and physiological experiments, conducted usually in his family's summer home at

7. For many of the details of Romanes's early years, I have relied on his wife's reverential biography. See Ethel Romanes, *Life and Letters of George John Romanes,* 4th ed. (London: Longmans, Green, [1896] 1897).

8. Ibid., p. 3.

Dunskaith, yielded numerous papers. The first, entitled "Preliminary Observations on the Locomotor System of Medusae," he communicated to the Royal Society in 1875. They selected it for the Croonian Lecture, an honor awarded the best biological paper each year. During the next two years he undertook to discover whether jellyfish had the rudiments of a nervous system, a question then undecided. On the basis of excision experiments, he concluded that jellyfish did have a primitive neural network, consisting of a grid of intercrossing "lines of nervous discharge." Over the next ten years several papers appeared, generally in the *Philosophical Transactions of the Royal Society* and in *Nature,* that continued this painstaking research into invertebrate physiology. Romanes regarded his work as further establishing an evolutionary link between primitive and higher organisms. Historically, however, it had a more important, if narrower, impact on theories of heart enervation, prompting Charles Sherrington to credit him with a direct influence on modern cardiology.[9] Romanes brought his research on Medusae to a wider audience in 1885, with the publication of *Jelly-Fish, Star-Fish, and Sea Urchins* in the International Scientific Series.[10] For his work on the physiology of Medusae Romanes was honored with membership in the Royal Society in 1879, at the very young age of thirty-one.

Darwin's Disciple

When Romanes first met Darwin at Downe in summer of 1874, the older man greeted him with outstretched hands, as Romanes liked to recall, and exclaimed "How glad I am that you are so young!"[11] Romanes had opened communication when he sent Darwin copies of some essays published in *Nature* earlier in the year.[12] In these pieces, he considered the causes Darwin had proposed for the atrophy of useless organs.[13] Darwin had postulated disuse as the principal reducing cause, with selection against deleterious organs, economy of growth (i.e., the disadvantage of nourishing a useless part), and free intercrossing pro-

9. See Gerald Geison's assessment of Romanes's contributions to nerve physiology in his *Michael Foster and the Cambridge School of Physiology* (Princeton: Princeton University Press, 1978), pp. 244–49.

10. George Romanes, *Jelly-Fish, Star-Fish, and Sea Urchins* (New York: D. Appleton, [1885] 1898).

11. E. Romanes, *Life and Letters of George John Romanes,* p. 14

12. George Romanes to Charles Darwin (10 July 1874), in the Darwin Papers, DAR 52 (series 4), Cambridge University Library.

13. George Romanes, "Natural Selection and Dysteleology," *Nature* 9 (1874): 361–62; "Rudimentary Organs," *Nature* 9 (1874): 440–41; "Disuse as a Reducing Cause in Species," *Nature* 10 (1874): 164.

viding additional sources of reduction. Romanes believed another factor was also at work: the cessation of selection. When selection, because of changed conditions, no longer operated to preserve and strengthen organs, then negative variations would accumulate, while economy of growth would continue to militate against any increases in such organs. Hence, according to Romanes, cessation of selection, along with economy of growth and disuse, would prove a powerful agent for removing no-longer-useful parts. Darwin generously responded to this as yet unknown Darwinian, and shortly thereafter invited him to visit Down House.[14] So began a brief, but psychologically intense relationship between Romanes and the man who would become his mentor, hero, paragon, and father substitute.

The relationship between Darwin and Romanes reached an intensity that seemed to have no rival. Their frequent meetings and correspondence bespoke the insinuating bonds of father and son. When Darwin died in 1882, Romanes grieved as he had previously done for no man. He wrote Francis Darwin:

> Even you, I do not think, can know all that this death means to me. I have long dreaded the time, and now that it has come it is worse than I could anticipate. Even the death of my own father—though I loved him deeply, and though it was more sudden—did not leave a desolation so terrible. Half the interest of my life seems to have gone when I cannot look forward any more to his dear voice of welcome, or to the letters that were my greatest happiness. For now there is no one to venerate, no one to work for, or to think about while working.[15]

A bit later Romanes expressed his feeling for Darwin in the way he was wont in times of high emotion; he composed a lyric in the grand Victorian style. It began:

> I loved him with a strength of love
> Which man to man can only bear
> When one in station far above
> The rest of men, yet deigns to share
> A friendship true with those far down
> The ranks . . . [16]

Romanes's personal relation with Darwin goes far to make intelligible both his preoccupation with those conceptually fragile hypotheses to

14. Charles Darwin to George Romanes (28 July 1974), in the Darwin Correspondence, no. 446, American Philosophical Society, Philadelphia.

15. E. Romanes, *Life and Letters of George John Romanes,* pp. 135–36.

16. Ibid., p. 138.

which the master had given special protection (e.g., pangenesis)[17] and his tenacious defense of "pure Darwinism" against the likes even of Wallace and Weismann.[18] His filial attitude toward Darwin also yields some explanation, I believe, for his most curious spiritual journey.

Agnostic Consequences of Evolutionary Theory

In 1873, Christ College set the topic for its annual essay contest as "Christian Prayer considered in relation to the Belief that the Almighty governs the World by General Laws." The subject for the College's Burney Prize appears to have been suggested by an article published the year before by Francis Galton. Galton examined, in his article "Statistical Inquiries Into the Efficacy of Prayer," what light science might shed

17. From the time he met Darwin in 1874, Romanes thought he could substantially aid the cause of evolutionary theory by providing experimental evidence for the hypothesis of pangenesis. Darwin, in his 1868 book *Variation of Animals and Plants under Domestication,* had postulated and elaborated a genetical theory to account for the hereditary transmission of characters, including those acquired during the lifetime of the organism. Darwin's cousin Francis Galton had attempted, during the months just prior to publication of the book, to demonstrate the validity of his cousin's theory by transfusing blood, presumably carrying the seeds of heredity—the gemmules—from one species of rabbit to another. Those rabbits that survived to give birth did not produce the desired half-breeds. (See Francis Galton, "Experiments in Pangenesis," *Proceedings of the Royal Society of London* 19 [1870–1871]: 393–410.) Romanes thought he could secure the needed evidence by doing hybridization and grafting experiments on both plants and animals. On and off from 1874 through the 1880s, he tried a large variety of different experiments, only some few, by his own estimation, showing any promise. He sent a notebook recording his proposed grafting experiments to Galton. They have the flavor of investigations carried out on the Island of Dr. Moreau. So for example:

"Graft a rat's tail on a male mouse. Do., do. on a female mouse, and see if, when grown up their progeny will have longer tails than normal mice. Graft old rat's tail on a young rat's body; and when that rat gets old, re-graft the tail on another rat's body; and so *ad infinitum,* in order to see how long the tail will live. Same exp., might be tried by grafting rat's tail on a cat's tail—or on any animal having a longer life than a rat."

The Notebook is held in the Galton Papers, Box 145, Manuscript Room of University College, University of London.

18. Throughout most of his career, Romanes remained persuaded of the inheritance of acquired characteristics, quite in opposition to Wallace and Weismann. He maintained to E. B. Poulton, in a letter of 11 November 1889 (E. Romanes, *Life and Letters of George John Romanes,* p. 229), that he had initially distrusted Lamarckian inheritance, but that Darwin persuaded him "that there was abundant evidence of Lamarck's principles apart from use and disuse of structures—e.g., instincts—and also on the ground of his theory of Pangenesis. Therefore I abandoned the matter, and still retain what may thus be now a prejudice against exactly the same line of thought as Darwin talked me out of in 1873" (misremembered, actually 1874). In his last years, Romanes came to regard Weismann's briefs against the inheritance of acquired characteristics as weighing heavily against the doctrine.

on the endemic belief that God suspended natural laws in order to grant petitions.[19] His study found that on average the clergy—certainly a prayerful group—did not live longer than physicians and lawyers; that ships bearing missionaries sank just as often as those carrying worldly goods; and that English society appeared to be blessed with leaders who were hardly more pious than the common run. Galton thought that if the devout enjoyed longer lives, surely the insurance offices would take note and adjust their rates accordingly. He concluded that the general laws of nature seemed to hold fast against the cries of the faithful.

One of those faithful undertook his own inquiry. While convalescing from typhoid fever in early 1873, Romanes decided to submit an essay for the Burney prize. To the astonishment of all, including himself, he won, beating out the odds-on favorite. The following year Romanes published the essay, along with additional material, under the title *Christian Prayer and General Laws.*[20] In his tract, he claimed that no logical or scientific barriers stood against the proposition that prayer had efficacy in a world governed by universal laws. His general defense of this thesis consisted in variously arguing that men were invincibly ignorant of the ultimate disposition of the universe, so that miracles—resulting either from the direct intervention of the Deity in answer to prayer or from God's prescient use of general laws to effect a petitioned outcome—might occur without possible human detection, including that of the statistician. Romanes drew powerful support for this argument from Spencer, whose theories of mind and the Unknowable brought the young metaphysician to his religiously orthodox conclusion. If human mind had evolved against the natural environment and was thus constructed for immediate and practical responses, it could hardly soar beyond nature to divine the ultimate arrangement of the universe and the Deity's movements therein. The scientific mind was likewise constrained:

19. Francis Galton, "Statistical Inquiries into the Efficacy of Prayer," *Fortnightly Review* 18 (1872): 125–133.

20. George Romanes, *Christian Prayer and General Laws, being the Burney Prize Essay for the Year 1873* (London: Macmillan, 1874). Since the rules of the essay contest restricted candidates to metaphysical considerations, Romanes could not address Galton's a posteriori arguments directly, even though he believed Galton's article "of greater argumentative worth than all the rest of the literature upon the same side put together" (p. ix). Romanes certainly thought his general argument about our ignorance would also rebut Galton, but he felt yet compelled to address specifically the statistical objections. In an addendum published with the Burney essay, he made the cogent, if not entirely persuasive, point that the answered prayers of the few truly pious might be masked by the averages for the clergy as a whole.

the only office of Science is the tracing back of phaenomena to the point at which they emerge from the ocean of the Unknowable, and the following of their course forward until they are again engulfed by its waters. And if such is the indisputable nature of that which underlies all science, it follows that even what we think we know we do not understand—that all our knowledge, absolutely considered, is merely another phase of our ignorance.[21]

Thus prayer could be efficacious, the laws of averages and nature notwithstanding.

Romanes candidly admitted that his conclusion was almost entirely negative. He had demonstrated that neither logic nor science nor evolved reason could show the Christian's *clamores ad Deum* to be swallowed by the wind. But he had no strong argument to prove that human cries did actually alter the course of the universe, only some hopeful surmises based on the presumption of God's goodness and moral purpose.[22] Nonetheless, the Burney committee selected Romanes's essay, undoubtedly for its acute and scientifically rigorous defense of orthodoxy.

Yet at the very moment of triumph, Romanes turned away. Shortly after the appearance of his book, he penned another essay contending that religious orthodoxy had no defense against reason and science. He preserved this piece unpublished for a few years. When *A Candid Examination of Theism* did come out in 1878, Romanes shielded his identity under the pseudonym "Physicus." Apparently in rendering his arguments for the Burney essay, he had felt the terrible weight of their further implication: if we had no knowledge of God, then how could we know he existed? In the *Candid Examination,* Romanes returned to the same arguments of the Burney essay, but now to admit that they actually gave no comfort to faith, but rather led to a bitter agnosticism. Though Romanes's arguments had for him the iron strength of cold reason, he mourned their conclusion:

> So far as I am individually concerned, the result of this analysis has been to show that, whether I regard the problem of Theism on the lower plane of strictly relative probability, or on the higher plane of purely formal considerations, it equally becomes my obvious duty to stifle all belief of the kind which I conceive to be the noblest, and to discipline my intellect with regard to this matter into an attitude of the purest skepticism. And forasmuch as I am far from being able to agree with those

21. Ibid., p. 38.
22. Ibid., p. 170–72.

who affirm that the twilight doctrine of the 'new faith' is a
desirable substitute for the waning splendour of 'the old,' I am
not ashamed to confess that with this virtual negation of God
the universe to me has lost its soul of loveliness.[23]

It would strain most theories of human psychology to conclude that
Romanes fell so rapidly from faith with a simple shove from rational
argument. As a student already schooled in the doctrines of Spencer,
Kant, and the opponents of religion (whom he vigorously attacked in
the Burney essay), Romanes was quite familiar with standard arguments about the abyss of ignorance that surrounded the human mind.
Some greater weight must have propelled him to disbelief. We must, of
course, take into account an awakened reason—the arguments for agnosticism are, after all, powerful—as well as the possibility that the
Burney essay itself might have quickly turned into an exercise in ingenuity rather than in rationally supported faith. Yet we must also recognize the crucial change in Romanes's personal and professional life that
occurred immediately prior to the composition of his second essay.

Though the *Candid Examination* was published only in 1878, evidence reveals that it was likely drafted sometime in 1875.[24] During the
summer of 1874, Romanes began his discipleship with Darwin. When
they met, the master had long since drifted into agnosticism and had
recently given a quite naturalistic explanation, in the *Descent of Man,* of
those essential traits of orthodox faith—religious feeling and moral
conviction. The very first argument for God's existence that Romanes

23. George Romanes [Physicus, pseud.], *A Candid Examination of Theism* (London:
Kegan Paul, Trench, Trübner, [1878] 1892), p. 114.

24. In the preface to *Candid Examination,* Romanes mentioned that "the following
essay was written several years ago; but I have hitherto refrained from publishing it, lest,
after having done so, I should find that more mature thought had modified the conclusions which the essay sets forth" (p. vii). He finally published it in 1878. But when did he
compose it originally? The clue to the date is given by his remark in the preface that Mill's
essay "Theism" had been published after the completion of his own essay and that references to Mill were subsequently inserted in the original text. Now Mill's essay first appeared in a posthumous collection in 1874. Hence, Romanes may have composed the
essay as early as winter of 1874–1875 or (granting him a little latitude in fixing a date) later
in 1875—thus not very long after seeing the Burney essay through the press. What appears
to be his first letter to Darwin is dated 10 July 1874; Darwin answered 16 July (*More Letters
of Charles Darwin,* ed. Francis Darwin [London: Murray, 1903], 2:352–54). The correspondence continued with Romanes writing again, apparently about Spencer. Darwin
answered on 28 July (Darwin Correspondence, no. 446, American Philosophical Society,
Philadelphia). Presumably Darwin invited Romanes to Downe shortly thereafter. In December, Romanes sent Darwin a copy of *Christian Prayer and General Laws.* On 16 December, Darwin sent an unenthusiastic acknowledgement (see note 28 below). Likely
Romanes adopted Darwin's own agnostic position shortly thereafter.

analyzed in the *Candid Examination*—that from the character of the human mind—he dissolved in the corrosive considerations of Darwinian evolution. That theory had provided a perfectly naturalistic account of human mind. And as for the moral sense, the pseudonymous Physicus averred: "Read in the light of evolution, Conscience, in its every detail, is deductively explained."[25] Romanes even anticipated the kind of self-justifying epistemological twist later executed by Freud. He contended that an opponent might be unable to appreciate the possibility that mind and the cosmos were simply the evolutionary result of matter and the persistence of force since a mind so evolved would not perfectly mirror external reality. It was because of the artifacts of mental evolution that we might "refuse to assent to the obvious deductions of our reason."[26] In a manuscript left unfinished at his death, Romanes expressly declared that the theory of evolution had caused him to abandon religion.[27]

But the agnostic implications of evolutionary theory were available to and even used by Romanes in his Burney essay. So Darwinian considerations alone seem insufficient to explain the fall from faith. The new factor that appeared in 1874 was the man himself. I believe it was Darwin's amiable good nature, parental solicitude, and scientific authority—along with intimations of professional support—that may well have tipped the unstable convictions of Romanes, so that a quick slide into disbelief resulted. When Darwin read parts of Romanes's Burney essay in December of 1874, he responded politely and unenthusiastically.[28] This contrasts with the delight later expressed by the older man when Romanes unmasked Physicus for him.[29] In 1876, Romanes inquired of Darwin what he thought of spiritualism, since the empirical evidence seemed so strong—and he himself had witnessed extraordinary occurrences (a hand appearing from nowhere, a head hovering over a table, and the like). Darwin poopooed the whole business, and almost immediately thereafter Romanes declared spirtualism to be grand fakery.[30] Darwin's personal influence on Romanes's scientific and

25. Romanes, *Candid Examination*, pp. 27–8.

26. Ibid., p. 87.

27. The manuscript was published posthumously. See George J. Romanes, *Thoughts on Religion*, ed. Charles Gore (Chicago: Open Court, 1895), p. 169.

28. In a letter dated 16 December 1874, Darwin thanked Romanes for the gift of the book on Christian prayer and indicated that he had read part of it. Uncharacteristically, however, he did not offer any kind words about its argument. The letter is held in the Darwin Correspondence, no. 455, the American Philosophical Society, Philadelphia.

29. See the exchange of letters between Darwin and Romanes over the book in late 1878, in E. Romanes, *Life and Letters of George John Romanes*, pp. 87–90.

30. The Romanes-Darwin interchange appears to be no longer extant. Wallace, who

philosphical opinions could hardly have been stronger. But whether or not Romanes's relationship with a revered father figure did contribute to his abandonment of faith, his subsequent immersion in an evolutionary study of mind and behavior surely held him fast in scientific skepticism and agnosticism—at least on the intellectual plane. Only after Darwin's death and when his own final illness began to take hold did he change his mind.

In his heart, however, Romanes never really left the fold. Despite his avowed agnosticism, he yet attended religious services with some regularity. He thoroughly enjoyed a good sermon and the aesthetic trappings of religious observance. He was ever sensitive to the moral encouragement that religion provided, even if he denied its theological rationale. He loved the Psalms, and attempted in his own verse to emulate their evocative qualities. He had among his friends, from early youth, churchmen of intellectual distinction, such as Francis Paget, bishop of Oxford; later he cultivated Dean Church, the Reverend Mr. Gore (who became bishop of Worcester and edited his posthumous writings), and the Reverend John Gulick (an American missionary in Japan and first-class naturalist). Romanes never cut the cord that joined his soul to the church. But throughout most of his career he warded off the intellectual advances of religion, always thrusting it back from the domain of science with the formidable instruments that Darwin had prepared. He publicly stepped out from the guise of Physicus to speak on his own, though, only in 1882, when he began a protracted debate on design in nature.

Natural Selection and Natural Theology

From the late 1830s through the publication of the *Origin* and the *Descent of Man,* Darwin reflected from time to time on the impact of his theory on traditional natural theology. He was hardly obsessed with the problem. But his disciple was. Guardedly at first, in the *Candid Examination,* and then more openly in essays of the 1880s and early 1890s, Romanes explored the implications of the scientific framework of evolution for theology. He took on all comers with the intellectual tenacity, if not the wit, of a Huxley. Like Huxley, he smashed hard against the conglomerate of science and religion defended by the avowed opponents and some of the friends of Darwinism; but in the end he

read the letters, recalled them to Romanes when he engaged the young scientist in a dispute about spiritualism. See Alfred Wallace, *My Life* (New York: Dodd, Mead, 1905), 2:333–40.

reached out for some of the larger chunks of religious belief and clung to them.

Romanes first openly animadverted on the traditional argument from design in a passing remark he made in a review of Wilhelm Roux's *Der Kampf der Theile im Organismus.*[31] He struck a glancing blow when he claimed that Darwin had broken up "the fountains of this great deep." The Duke of Argyle rose to defend the fountains by pointing out that the laws of evolution and their consequences supplied evidence of design no less compelling than the intelligent structure of the human hand.[32] Romanes responded by reiterating that Darwin's mechanism subverted all supposed cases of "special design," like the hand; and he consigned the larger question of whether a Mind stood behind the whole natural complex of laws to a plane beyond the stretch of natural science.[33]

Shortly after the interchange with Argyle, Romanes delivered to the Philosophical Institutions of Edinburgh and Birmingham a general lecture on evolution under the title "The Scientific Evidence of Organic Evolution." He prefaced the main part of his lecture, which recapitulated the argument of the *Origin of Species,* with a distinction he had made in his exchange with Argyle: "while Mr. Darwin's theory is thus in plain and direct contradiction to the theory of design, or system of teleology, as presented by the school of writers which I have named [i.e., Paley, Bell, and Chalmers], I hold that Mr. Darwin's theory has no point of logical contact with the theory of design in the larger sense, that behind all secondary causes of a physical kind there is a primary cause of a mental kind."[34] His remarks sparked a series of warm exchanges.

A doctor of divinity first objected. Mr. Eustace Conder brought two powerful sets of considerations to bear.[35] He first argued, playing upon Argyle's theme, that the lawlike structure of the universe, including the evolutionary process itself, bespoke design no less than instances of individual contrivance. Second, he insisted that natural selection alone was insufficient for its task; it needed intelligent assistance. Conder

31. George Romanes, "The Struggle of Parts in the Organism," *Nature* 24 (1881): 505–6.

32. Argyle, "The Struggle of Parts in the Organism," *Nature* 24 (1881): 581.

33. George Romanes, "The Struggle of Parts in the Organism," *Nature* 24 (1881): 604; and "The Struggle of Parts in the Organism," *Nature* 25 (1881): 29–30.

34. George Romanes, "The Scientific Evidence of Organic Evolution," in the Humboldt Library of Science, No. 40 (New York: Humboldt, 1882): 1–21.

35. Eustace Conder, "Natural Selection and Natural Theology: a Criticism," printed with Romanes, "Scientific Evidence of Organic Evolution," pp. 39–50.

maintained that natural selection only removed harmful or disruptive variations, but that it could not account for the positive and harmoniously coadaptive variations that gave rise to new species. The echo of this latter complaint reverberated with increasing strength as it rushed toward century's end.

In reply, Romanes concentrated his attack on the still-vivid assumption that individual adaptations provided evidence of the Creator.[36] He readily conceded that if a species were suddenly to appear we would be thrown back upon individual, intelligent creation as the only explanation. But we had strong evidence that species arose gradually, so natural selection could sufficiently account for their peculiar traits. The law of parsimony thus dictated that we should adopt a natural cause explanation for natural phenomena. Romanes then reaffirmed that Darwin's theory "has no point of logical contact" with the larger questions of design. Significantly, he did not deal forthrightly with Conder's objection that natural selection required harmonious coadapted variations but could not explain their origin. This last problem would remain a persistent irritant for Romanes and other Darwinians.[37]

A friend in common cause with the Darwinians, Asa Gray, the Harvard theologian and botanist, entered the controversy. He considered Romanes's reply to Conder to have slipped past the issues.[38] He returned to the question of design, pointing out that whether species appeared suddenly or gradually had no bearing on the question's resolution. Moreover, he thought Romanes had presumed nature writ large to be more than the sum of her parts, whereas the problem of the larger design in nature could only be raised in light of the design of the individual adaptations of which nature was composed. So the focus of inquiry had to be individual adaptations. Gray forcefully resurrected the original issue: Could natural selection fully account for individual contrivances? Gray thought not. For natural selection theory failed to explain the source of variations, those positive and harmonious traits that the mechanism of selection required in order to act.

Romanes retaliated. He admitted that if it could be shown that variations were directed, always occurring in anticipation of the vector of selection, then would design be deducible from the theory of natural

36. George Romanes, "A Reply," collected with Romanes, "Scientific Evidence of Organic Evolution," pp. 50–55.
37. For a discussion of the problem of coadaptation, see chapter 6 and Mark Ridely, "Coadaptation and the Inadequacy of Natural Selection," *British Journal for the History of Science* 15 (1982): 45–68.
38. Asa Gray, "Natural Selection and Natural Theology," *Nature* 27 (1883): 291–92.

selection. But no one had supplied patent evidence of this kind.[39] Gray responded with the tu quoque that natural selection theory assumed omnifarious variation, but that such variation was "no fact of observation."[40] Romanes countered with the demand for evidence that variation was directed.[41] And there the matter rested between the two friendly antagonists.

Romanes's consistent attitude in these debates over design in nature was that of the combative agnostic: no argument for a Creator could be mounted on the evidence of nature. Some eight years later, he again directly took up the question of design, but by that time his attitude had shifted. In a paper before the Aristotelian Society in 1891, he contended now against the agnostic, arguing that even if the universe were completely explicable in causal terms, this would still not settle the question of whether God, as *causa causarum*, used causation as an instrument. Romanes confessed that the spectacle of nature did not immediately yield persuasive evidence of design; but he maintained that if we first believed in God, then we might indeed perceive an intelligent pattern. In the Aristotelian lecture, he concluded by tracing the one route open to discover whether nature had a designing author: "it can only be determined in those mysterious depths of human personality, which lie beyond the reach of human investigation, but where it is certain that through processes as yet unknown to us, by causes—if they be causes—as yet unrevealed to us, there results for each individual mind either the presence or the absence of an indissoluble persuasion that 'God is.' "[42]

In this conclusion, Romanes let slip the veil that had shrouded an underlying motive for his earlier study of the evolutionary basis of human personality. He had begun his examination in the early 1880s rapt in a kind of militant rationalism, engendered, I believe, by his relation to Darwin, the man whose patient scientific investigations revealed to his young disciple a new earth and a new heaven. Romanes set out to explore the limits of human reason and to elaborate further a naturalistic account of morality. But this endeavor, in tandem with certain other

39. George Romanes, "Natural Selection and Natural Theology," *Nature* 27 (1883): 362–64.

40. Asa Gray, "Natural Selection and Natural Theology," *Nature* 27 (1883): 527–28; and "Natural Selection and Natural Theology," *Nature* 28 (1883): 78.

41. George Romanes, "Natural Selection and Natural Theology," *Nature* 27 (1883): 528–29; and "Natural Selection and Natural Theology," *Nature* 28 (1883): 100–101.

42. George Romanes, "Is There Evidence of Design in Nature?" *Proceedings of the Aristotelian Society* 1 (1891): 66–76; quotation from pp. 75–76.

considerations, brought him finally to recognize that an evolutionary causal analysis alone could not penetrate the hidden depths of mind. All along, this conclusion may well have been the desired end. Even during his Darwinian odyssey, Romanes felt the pull of religion and of a moral impulse past natural explanation. The evolutionary analysis of mind, which occupied him during the entire decade of the 1880s, seems to have revealed depths unfathomable, seems to have required a metaphysical hypothesis that it could not tame. The feelings that Romanes vented in poetry and in Sunday churchgoing appear to have directed him back to the convictions of his youth. So his plunge into evolutionary biopsychology might well have started out as the most important test area for his mentor's theories, but as Romanes's explorations progressed, and especially as the signs of his own mortality became more certain—the headaches, the intermittent eye problems, the failing memory—he seems to have sensed that the complete victory of an agnostic Darwinism would be Pyrrhic.

The Evolution of Mind

In "Scientific Evidence of Organic Evolution," his lecture of 1882, Romanes observed that a theory would receive strong confirmation if it could also explain classes of phenomena other than those for which it was first devised.[43] He mentioned instinct, reason, and moral sense as precisely those psychological traits which Darwin's theory was not originally designed to treat, but which the theory could explain if applied with a little imagination and enterprise. Romanes was not aware of the original context of his mentor's theorizing, which certainly included efforts to account for instinct, reason, and moral sense. But even if he were, he would not likely have been restrained by Darwin's pre-Malthusian jottings. For he believed he could reach beyond *Descent of Man,* where Darwin's earlier ideas had come to flower. Romanes felt assured that he could solidify evolutionary theory by giving a detailed and empirically grounded account of the psychological features of biological organisms.

Romanes conceived a series of volumes which would, as he expressed it in the first, trace "the principles which have been probably concerned in the genesis of Mind."[44] He completed three volumes of his projected series: *Animal Intelligence* (1881), *Mental Evolution in Animals* (1883),

---

43. Romanes, "Scientific Evidence of Organic Evolution," p. 17.
44. George Romanes, *Animal Intelligence,* 4th ed. (London: Kegan Paul, Trench, [1881] 1886), p. vi.

and *Mental Evolution in Man* (1888).[45] Though he intended to devote additional monographs to intellect, emotion, will, morals, and religion, he left only scattered essays touching on these more specialized topics.

The Empirical Evidence

In his published volumes, Romanes meant to examine more finely the question that had occupied Darwin in the *Descent of Man,* namely, whether the human mind could be explained as having evolved from animal mind. The question took on new urgency in the 1880s, since Wallace and Mivart had continued to mount argument and evidence to smother the question in doubt. Romanes, in response, collected numerous empirical reports, performed experiments, and undertook an analysis of various human and animal faculties to reinvigorate the thesis of evolutionary continuity, to show that "there is no difference *in kind* between the act of reason performed by the crab and any act of reason performed by a man."[46]

Romanes intended *Animal Intelligence* to serve as the repository for empirical evidence of the evolutionary bond between human and animal mind. In this volume he displayed widely accepted evidence about the memories, emotions, and general intellectual abilities of animals. Further, he surveyed large amounts of literature that dealt with the subject of animal mentality, beginning with accounts of various protozoa, and ascending the scale with coelenterates, echinoderms, ants, spiders, molluscs, fish, and reptiles, up through elephants, cats, dogs, and apes. He tried to avoid unsubstantiated and bizarre tales of the wonderful behavior of animals, of the kind that filled the pages of "anecdote-mongers."[47] His intellectual heir Conwy Lloyd Morgan would judge many of Romanes's own reports as only a cut above those of the "anecdote-mongers." Yet Romanes was sensitive to the liabilities of uncontrolled observation. In *Animal Intelligence,* he recounted at great length the more exacting experimental observations of specialists like John Lubbock, whose many studies of ants, bees, and wasps remain today examples of sophisticated analysis. He added to this large evidentiary store—20 percent of his five hundred pages were spent on Lubbock's careful inquiries about ants—his own experimental observations on coelenterates. He also devised a controlled situation in which the intelligence of a cebus monkey could be assessed.

45. George Romanes, *Mental Evolution in Animals* (London: Kegan Paul, Trench, 1883); and *Mental Evolution in Man* (London: Kegan Paul, Trench, 1888).
46. Romanes, *Mental Evolution in Animals,* p. 337.
47. Romanes, *Animal Intelligence,* p. vii.

Romanes borrowed a small cebus from the Zoological Society in December of 1880. At first he wanted to raise the young animal with his own newly born daughter, "but the proposal met with so much opposition that I had to give way."[48] Luckily his maiden sister was accommodating. She took in little "Sally" and roomed her with Romanes's invalid mother, for whom the monkey developed a strong attachment. Ethel Romanes kept a diary of the behavior of her charge from 18 December 1880 till 28 February 1881, when the animal was returned to the zoo.[49] Romanes printed his sister's diary in *Animal Intelligence,* offering it as a strong testimony of the humanlike rational abilities of a creature below man. Indeed, Romanes concluded that such abilities extended far below even monkeys. He discovered the roots of intelligent choice and conscious awareness in the social insects. These traits, he believed, grew in strength through the animal kingdom, so that dogs, monkeys, and apes could be shown to have as much reasoning ability as a year-old human child.

Romanes understood that the ascription of psychological faculties to an animal could only be made in light of well-defined standards. In *Animal Intelligence* he set down the two criteria he regarded as necessary to assign mind to lower creatures: consciousness and choice. He recognized that these were not easy traits to demonstrate unequivocally. Intentional choice could serve as evidence for consciousness, but we might be misled about an animal's intentions. Seemingly intelligent behavior might, after all, be the result of inherited reflex. Therefore the criterion for mind had to reduce to an organism's ability to learn—to make new adjustments of behavior on the basis of individual experience. Wherever, therefore, we find an animal altering its behavior through learning, "we have the same right to predicate mind as existing in such animal that we have to predicate it as existing in any human being other than ourselves."[50]

The Spencerian Psychological Framework

In his subsequent volumes, *Mental Evolution in Animals* and *Mental Evolution in Man,* Romanes refined his categories of mental activity in order to trace through more precisely the various stages of transformation joining animals to man. His discussion depended largely on Spencer's psycho-evolutionary theory. Like Spencer, Romanes distinguished an objective analysis of mind from a subjective postulation. For every

48. E Romanes, *Life and Letters of George John Romanes,* p. 110.
49. Romanes's sister and wife were both named "Ethel."
50. Romanes, *Animal Intelligence,* p. 7.

creature other than ourselves, we can be aware only of outward, objective manifestations, on the basis of which we postulate, by analogy to our own conscious experience, a subjective state. Mind or consciousness, on its subjective side, he thought reducible to sensation: "the elementary or undecomposable units of consciousness are what we call sensations."[51] Parallel, on the objective side, that which prompts us to ascribe sensibility to an organism is discrimination, choice.

Following Spencer, Romanes traced the objective manifestations of mind back to the most primitive forms of life, to plants and protozoa; indeed he conceived mind as an organic development out of the phenomenon of life. Since, however, plants and infusoria exercised only the rudiments of discrimination, we would stand on firmer ground, he conceded, if we postulated conscious feelings only of animals outfitted with sensory organs.[52] From the lowest forms of life through more advanced stages, organisms displayed increasing physiological and concomitant psychological complexity. So from merely sensory creatures, one could ascend the psychological scale to perceptive animals (i.e., those that interpreted sensations in terms of past experience), receptive animals (i.e., those whose perceptions coalesced according to primary laws of sensory association), and conceptive animals (i.e., those that actively reorganized abstract qualities into new combinations). All reasoning, properly so called, required the ability to form concepts, that is abstract, or general, ideas. Reasoning consisted, according to Romanes, in the inferring of unperceived qualities from perceived qualities, as when, for example, hearing a growl, one immediately formed the idea of a dog. In Romanes's estimation, all warm-blooded animals and some of the invertebrates had reached this level. The final evolutionary stage occurred when the process of reasoning itself could become an object of knowledge. This was the stage "at which it first becomes possible intentionally to abstract qualities or relations for the purpose of inference."[53] Here creatures began to employ symbols of ideas instead of the actual ideas themselves. Here human mind first appeared.

Romanes intended to sink a deeper and more elaborate argument for human descent than Darwin ever managed. His general strategy was directed to establishing the links of continuity between human mental faculties and their supposed roots in animal mind. In doing this, he drew from the resources of Spencer's psychology and generally from the sensationalist tradition, which encouraged him to slip easily from

51. Romanes, *Mental Evolution in Animals,* p. 72.
52. Ibid., pp. 78–79.
53. Ibid., p. 325.

sensation to ideas, from association of ideas to rational inference. Yet this kind of argument only showed, as he admitted, that an evolutionary transition was possible, that there was nothing in human mind that did not have rudimentary antecedents in animal mind. The strategy did not allow the demonstration of evolutionary transmission. This is a principal reason, I believe, why Romanes spent so much time on the question of instinctive behavior, devoting almost half the pages of *Mental Evolution in Animals* to tracing its origins and causes.

Instinct as the Critical Case

The instincts of animals seemed to provide instances of the heritability of mental faculties, the kind of hard evidence that would give the argument from continuity more bite. Romanes had additional reason, though, for dwelling on instinct. Darwin, just before he died, had given over to his protégé a large draft essay on animal instinct. This was to have been chapter 10 of his big book, *Natural Selection*. Romanes included as an appendix to *Mental Evolution in Animals* those sections of the draft that did not appear in the *Origin of Species*. Though he did this in homage to his recently deceased mentor, the published selections directed Romanes's own analysis in a more decidedly Spencerian way than even Darwin's intact draft might have; for Romanes omitted precisely those sections in his appendix that gave strongest support for the natural selection account of instinct, namely, those passages devoted to the wonderful instincts of neuter insects. The sections printed gave as much weight to inherited habit as to Darwin's chief instrument of transformation.[54]

Like Spencer, Romanes conceived instinct as a species of reflex that on its subjective side included consciousness.[55] In classifying instinct as a type of reflex, he furnished it with both a determinate structure and a concrete vehicle for its inheritance. He interpreted instinct, however,

54. Because of the significance given inherited habit by Romanes, the *Athenaeum*'s reviewer of *Mental Evolution in Animals* accused him of deriving his ideas, not from Darwin, but from Samuel Butler, who had claimed that inherited memory drove evolution. See the following exchanges by interested parties: George Romanes, "Mental Evolution," *Athenaeum* (January–June 1884), pp. 312–13; Samuel Butler, "Mental Evolution" (January–June 1884), p. 349; George Romanes, "Mental Evolution in Animals" (January–June 1884), pp. 411–12; and Herbert Spencer, "Mental Evolution in Animals (January–June 1884), p. 446.

55. Romanes, *Mental Evolution in Animals*, p. 159. Spencer had also described instinct as a type of "compound reflex action" and distinguished it from simple reflex action by reason of accompanying consciousness. See Herbert Spencer, *The Principles of Psychology*, 2d ed. (London: Williams & Norgate, 1872), 2: 432–34.

not as a simple reflex mechanically stimulated by blind sensation; rather he conceived of it as an action evoked by a mental perception. He thus construed instinctive behavior as a thoroughly cognitive act involving conscious interpretation of sensation in terms of inherited ancestral experiences and present aims.[56] Hence evidence for the inheritance of instinct would also constitute evidence for the inheritance of cognitive faculties —exactly the support he needed for his general argument from continuity.

As had Darwin, Romanes postulated two sources of instinct, one primary and the other secondary. The primary source was natural selection, concerning which he gave adequate account. The secondary source he thought consequent upon intelligence, as it were, going stale: "By the effects of habit in successive generations, actions which were originally intelligent became . . . stereotyped into permanent instincts."[57] His empirical proofs for the rise of instinct by way of lapsed intelligence amounted to assemblages of anecdotes of supposed instances of animals acquiring habits and transmitting them to their progeny—for example, a friend's cat had been taught to beg and had borne kittens which without prompting adopted the same habit.[58] As a concluding demonstration of the reality of hereditary transmission of instincts, whether primary or secondary, Romanes referred to experiments in the crossbreeding of distinct varieties in which the hybrids displayed a "blended psychology."[59]

In Romanes's conception, one also shared by Darwin, not only did some instincts evolve from originally intelligent habits, but all instincts, even those of primary origin, could be intelligently altered by the individual to meet the contingencies of existence.[60] Thus the instinct of the caterpillar to construct a web from which the chrysalis would hang was not unalterably fixed; if placed in a box covered with a muslin lid, the insect "perceives that his preparatory web is unnecessary, and therefore

---

56. Romanes, *Mental Evolution in Animals,* p. 125–31.

57. Romanes, *Mental Evolution in Animals,* pp. 177–78.

58. Ibid., pp. 194–97. In his posthumous second volume of *Darwin and After Darwin,* Romanes offered three kinds of evidence for a Lamarckian account of some classes of instinct: first, for very complex instincts, the mechanism of natural selection had to be aided by intelligent action; second, some instincts (e.g., howling of wolves at the moon) were completely useless and hence without selective value; finally, in our own species the religious and moral instincts were explicable by use inheritance, but hardly by natural selection. See George Romanes, *Post-Darwinian Questions, Heredity and Utility,* vol. 2 of *Darwin and After Darwin,* ed. C. Lloyd Morgan (Chicago: Open Court, 1895), pp. 87–90.

59. Romanes, *Mental Evolution in Animals,* p. 198.

60. Ibid., p. 203; and Romanes, *Darwin and After Darwin* 2:88–89.

attaches its chrysalis to the already woven surface supplied by the muslin."[61] Romanes believed such instances of intelligent modification of instinct to be normal occurrences, even among insects. His conviction was supported and prompted by the biological purpose he understood this flexibility to serve: "intelligent adjustment by going hand in hand with natural selection, must greatly assist the latter principle in the work of forming instincts, inasmuch as it supplies to natural selection variations which are not merely fortuitous, but from the first adaptive."[62]

Romanes's theory of instinct provided three supports for his general conception of the evolution of mind. First, the theory and its evidentiary base showed that behavior and associated psychological faculties could be transmitted by heredity—hence they were subject to the operations of natural selection. Second, the emphasis on the secondary mode of origin and the claim that intelligence could modify the transmissible character of all instincts made clear that cognitive faculties could also evolve. Finally, Romanes's treatment of instinct provided a scientific demand for recognition of a metaphysical reality, namely, that evolution had to be guided by something beyond blind mechanism.

In his posthumous *Darwin and After Darwin*, Romanes argued that in the case of complex instincts—such as the nine precisely located stings the Sphex wasp typically gave a caterpillar—it appeared "incredible that natural selection, unaided by originally intelligent action, could ever have developed such an instinct out of merely fortuitous variations."[63] While remaining the most orthodox of Darwinians, Romanes nonetheless detected an intelligent force beyond the veil of material reality, a force that science had to allow in its calculations. Romanes's fundamental metaphysical outlook thus distinguished him from his mentor. Darwin, of course, regarded use inheritance as a factor in evolution, but he did so principally because of his own intellectual inertia (acquired habit being his early evolutionary mechanism) and what he took as the requirements of empirical evidence. Romanes felt additionally constrained to admit the role of mind in evolution as the weight of his never-quite-abandoned religious convictions steadily pulled him back from Darwinian agnosticism. He began to give these convictions more rational scope in the later part of the 1880s, when he came gradually to pump life back into an interred natural theology. The argument that provided the intellectual afflatus was based on analogy: if the scientist had to appeal to finite intelligence in explaining particular cases

61. Romanes, *Mental Evolution in Animals*, p. 201.
62. Ibid., p. 219.
63. Romanes, *Darwin and After Darwin* 2:87.

of complex natural behavior, parity of reasoning would justify his recourse to infinite intelligence in an account of the transcendentally complex structure of the whole of nature. In metaphysical orientation Romanes was closer to evolutionists like Mivart than, say, to Huxley. Nonetheless his personal ties to Darwin made him a fierce opponent of expressly religious evolutionists, especially Mivart.

Controversy with Mivart

The Career of St. George Jackson Mivart

In biological evolution, the more similar the organisms, the more vigorously will they compete. We have seen analogous instances of this in the sphere of intellectual evolution, for example, Spencer's criticisms of Darwin and Huxley's of Spencer. St. George Jackson Mivart's harsh objections to Darwin's *Descent of Man* supply another example of this principle at work. Mivart was an evolutionist whose chief interest, like that of Romanes, lay in questions of mind; and also like Romanes, he was a scientist who sought to probe indelicately below the surface of scientifically describable phenomena. As he did he uncovered the God of a creed reviled in his native land. The Darwinians, like other English gentlemen, reacted with hostility to the ultramontanist cast of Mivart's theology, though no less to his particular style of argument. Mivart struck out against the Darwinians, treating them as his namesake had the fabled enemy. But in the end, the Darwinian dragon devoured this Catholic warrior and, adding to the tragicomedy of Mivart's life, his own coreligionists refused his remains sacred burial.

After a rather painful adolescent pilgrimage, Mivart joined the Roman Catholic Church in 1844 at age sixteen.[64] Being a member of the Roman communion barred him from university, so he chose a professional training that recognized no such disabilities: he became in 1846 a student at Lincoln's Inn. Mivart never practiced law, though his stewardship gave him that turn of argument identified by his later enemies as Jesuitical. His abiding interest had been zoology, and having some small means, he pursued a private course of study in that science under the tutelage of such men as George Waterhouse, the entomologist who aided Darwin, and Richard Owen, "the English Cuvier," as he was known.

Owen had a tremendous influence on the young Mivart. The great anatomist's leading ideas in morphology—that the organism displayed

64. The details of Mivart's conversion and his long odyssey within the church are provided in Jacob Gruber, *A Conscience in Conflict: the Life of St. George Jackson Mivart* (New York: Columbia University Press, 1960).

serial and lateral homologies (e.g., vertebrae and limbs respectively) and that homologies permitted systematizing the animal kingdom into types—these conceptions bespoke a design in nature that could be explained only by reference to a supermundane intelligence. Owen's transcendental anatomy became the framework for Mivart's later evolutionary theory. Mivart would argue that a purposeful force directed the development of species within the confines of certain very general types, which themselves had independent and transnatural origins. Natural selection had only the mundane task of sweeping out unfit variations. Though Mivart's mature science came to rest at some distance from Darwin's, the young scholar rose to the evolutionary perspective under the wing of Darwin's stalwart, Huxley.

Mivart attended Huxley's lecture series on the principles of biology at the Royal Institution in early 1858. Then, encouraged by some casual conversation, he enrolled as Huxley's student in a course of lectures at the School of Mines in autumn of 1861. Mivart became a demonstrator as well as auditor for Huxley's subsequent lecture series, and profited greatly thereby: he recalled in his memoir of Huxley that "I learnt more from him in two years than I had acquired in any previous decade of biological study."[65] During the early 1860s, Mivart formed an intimate acquaintance with his teacher, occasionally strolling with him, discussing the questions of the day, and dining with the family. He thus found it difficult to balance his friendship with Huxley against his considerable admiration for Owen, especially during the great debate between the two anatomists on the origin of the vertebrate skull.[66] Mivart apparently achieved the judicial measure, since both Huxley and Owen helped launch him professionally, supporting his successful candidacy as lecturer of comparative anatomy at the medical school of St. Mary's Hospital in London. Mivart remained at St. Mary's until his retirement as professor in 1884.

Mivart's dissent from full-fledged Darwinism ostensibly derived from

65. St. George Mivart, "Some Reminiscences of Thomas Henry Huxley," *Nineteenth Century* 42 (1897): 985–98; quotation from p. 991.

66. In pursuit of substantiation for his doctrine of serial homologies, Owen had argued that the skull was merely a modification of terminal vertebrae in backboned animals. Huxley, always suspicious of high philosophy imported into biology, made a careful study of the stages of osteological development in the embryo. His Croonian Lecture of 1858 and his book *Man's Place in Nature* laid that little piece of transcendental anatomy to rest. For a brief history of the debate, see Thomas Huxley, *Evidence as to Man's Place in Nature* (London: Williams & Norgate, 1863), pp. 133–38. Adrian Desmond follows out the complex personal and professional relationship between Huxley and Owen in his *Archetypes and Ancestors: Palaeontology in Victorian London, 1850–1875* (Chicago: University of Chicago Press, 1984).

his classificational studies. The primate order, which he investigated intensely from the 1860s through the early 1870s, appeared to him a group sharply separated from other mammalian types. Moreover, structurally similar classes and orders seemed to have arisen independently of one another. From such evidence he concluded: "the notion that 'similarity of structure' necessarily implies 'genetic affinity' can no longer be ranked as a biological axiom."[67] But this Owenite conclusion, declared in 1873, was really more an axiom determining acceptable evidence. By this time Mivart had rejected Darwinism for other reasons, specifically moral and religious. In his memoir of Huxley, he mentioned the crucial considerations that brought him to oppose the Darwinian hypothesis:

> It was in 1868 that difficulties as to the theory of Natural Selection began to take shape in my mind, and they were strongly reinforced by the arguments of one who became . . . a highly valued friend, whose acquaintance I made at Professor Huxley's lectures, at which love of science had also made him a regular attendant. This was the Rev. W. W. Roberts. . . . The arguments he again and again urged upon me were the difficulties, or rather the impossibilities, on the Darwinian system, of accounting for the origin of the human intellect, and above all for its moral intuitions—not its moral *sentiments*, but its ethical *judgments*.[68]

In 1869 Mivart undertook a discreet examination of Darwin's principle. He published anonymously a set of three essays, "Difficulties of the Theory of Natural Selection," in the Catholic periodical *The Month*.[69] And in this same year he confessed his scruples to Huxley.

> After many painful days and much meditation and discussion my mind was made up, and I felt it my duty first of all to go straight to Professor Huxley and tell him all my thoughts, feelings, and intentions in the matter without the slightest reserve, including what it seemed to me I must do as regarded the theological aspect of the question. Never before or since have I had a more painful experience than fell to my lot in his room at the School of Mines on that 15th of June, 1869. As soon as I had made my meaning clear, his countenance became transformed as I had never seen it. Yet he looked more sad and surprised

67. St. George Mivart, "On Lepilemur and Cheirogaleus and on the Zoological Rank of the Lemuroidea," *Proceedings of the Zoological Society of London* (1873), p. 506; quoted in Gruber, *A Conscience in Conflict*, p. 33.

68. Mivart, "Some Reminiscences of Thomas Henry Huxley," p. 994.

69. [St. George Mivart], "Difficulties of the Theory of Natural Selection," *Month* 11 (1869): 35–53; 134–53; 274–89.

than anything else. He was kind and gentle as he said regret-
fully, but most firmly, that nothing so united or severed men as
questions such as those I had spoken of.[70]

Mivart's rejection of Darwinian theory had much the same emotional
character as his rejection of the Church of England; moreover, he embraced his new scientific faith with the zeal of a convert. As before, he
estranged himself from a loved and admired father, and then immediately set out to take his stand openly against a hostile world, now with
a publication whose title left little doubt as to the spiritual battle he
would wage.

In his *On the Genesis of Species,*[71] which appeared in January 1871, Mivart directed a sustained attack on the theory of natural selection. He
argued his brief by attempting to show the mechanism insufficient to
produce a variety of zoological phenomena. Darwin's principle could
not explain the incipient stages of subsequently useful adaptations; it
could not account for independently derived homologous structures
within different animal groups, or serial and lateral homologies in the
same species; and it supposed minute gradual evolution over long periods of time, whereas the fossil evidence spoke of long periods of stasis
followed by abrupt transitions occurring in a time constrained by the
physicists' calculations.

Mivart, in this his best known book, piled the negative evidence high
and in so doing attempted to secure space for the operations of another
principle to explain what natural selection could not. He conceived this
principle as an "internal law or 'substantial form,' moulding each organic being, and directing its development," a principle which would
account "at the same time for specific divergence as well as for specific
identity."[72] Though Mivart could only vaguely specify this law (or, perhaps, laws) as operating in the conceptual space vacated by natural selection, it had the positive virtue of pointing beyond to its supernatural
author, as he suggested in the final paragraph of his book:

> The aim has been to support the doctrine that these species
> have been evolved by ordinary *natural laws* (for the most part
> unknown) controlled by the *subordinate* action of "Natural Se-
> lection," and at the same time to remind some that there is and
> can be absolutely nothing in physical science which forbids
> them to regard those natural laws as acting with the Divine

70. Mivart, "Reminiscences of Huxley," p. 995.
71. St. George Mivart, *On the Genesis of Species* (New York: D. Appleton, 1871).
72. Ibid., p. 201.

concurrence and in obedience to a creative fiat originally imposed on the primeval Cosmos, "in the beginning," by its Creator, its Upholder, and its Lord.[73]

Mivart devoted the longest chapter of his book to the Darwinian theory of ethics. His task was difficult, since up to this time Darwin had not published anything of substance on the subject. Nonetheless, Mivart—by lumping the views of J. S. Mill and Spencer together with a passing remark of Darwin's in *Variation of Animals and Plants*—thought he could divine what the Darwinian position must be. It would consist of a theory that confused the useful with the ethically right and traced the moral faculty to animal antecedents. But a theory having these features simply could not sustain the necessary and obvious distinction that had to be made in ethics between the 'materially' moral and the 'formally' moral. Some acts that produced useful consequences might, perchance, be materially good, but they would deserve moral approbation only if executed with the formal intention to effect that good. Darwinian theory, which had to ignore this distinction, could only be, for those "who accept the belief in God, the soul and moral responsibility, . . . utterly unendurable."[74]

When the *Descent of Man* came out, within a few days of *Genesis of Species,* Mivart found his extrapolation confirmed. Darwin's explicit treatment of morality had all those failings that condemned it in scholastic eyes. Mivart avenged true morality in a review of the *Descent* (discussed in chapter 5) that added invective and sarcasm to some modestly telling objections. Though Mivart shielded his attack in anonymity, Darwin thought he recognized the Jesuitical style, and Huxley unleashed a mordantly sardonic rejoinder that signaled total war between the two opposing evolutionary camps.[75] It was in the light of these exchanges that the new recruit to the Darwinian side, George Romanes, suspected that it was "probable that Mivart and I shall have a magazine battle some day on Mental Evolution."[76]

The Rights of Reason

Romanes drew first blood in what became a protracted engagement. The wound was inflicted in a review of a new book by Mivart, *Nature*

73. Ibid., pp. 306–307.
74. Ibid., p. 277.
75. See chapter 5.
76. George Romanes to Charles Darwin (5 November 1880), in E. Romanes, *Life and Letters of George Romanes,* p. 102.

and Thought: An Introduction to a Natural Philosophy (1882).[77] Mivart
had composed the book as a dialogue between two friends, Maxwell
(the shadow of Mivart himself) and Frankland. Maxwell, through
shrewd but commonsensical argument, convinces Frankland that the
intellect and senses are generally reliable and that skepticism, whether
of a phenomenalistic, Kantian, or evolutionary variety, is unreasonable.
He likewise demonstrates to his friend that the Darwinian analysis of
morality fails to make the necessary distinction between material good
and formal good. Romanes, in an extended review, forcefully countered
the naive realism underpinning the dialogue with the simple argument
that the world could be known to us only in idea and that we were thus
never in a position to compare our thoughts with things in themselves.[78] All claims to knowledge about the natural world must therefore
be corrigible. Romanes also dissented, as one might expect, from Mivart's representation of the man-animal relationship. But he conceded
the importance of the ethical distinction that his opponent constantly
urged. Mivart, though, could draw no comfort from this, since Romanes believed the distinction could easily be accommodated in an evolutionary ethics. And then, in what seemed almost a passing observation
on the last chapter of Mivart's book, which took up the question of
theism, Romanes thrust home: "In this chapter the most novel feature
which we observe is that of systematic plagiarism."[79] Mivart had borrowed, without benefit of quotation marks or attribution, several lines
from an obscure little book published several years before under the
title *A Candid Examination of Theism,* by Physicus. Rather unhappily
for Mivart, his reviewer knew the book intimately.

Certainly the charge of plagiarism damaged Mivart's position more
than any deficiencies in argument. But it is difficult now to judge
whether the charge should be sustained. Mivart indeed lifted some lines
from *Candid Examination,* though not as many as Romanes suggested.
Moreover he put the skeptical lines in the mouth of Frankland, so that
Maxwell could demolish their argument. Mivart nonetheless felt the
blow. He immediately wrote the editor of the *Contemporary Review* to
explain that

> I happen to be personally acquainted with Physicus, who,
> when he confided to me the secret of his authorship, earnestly
> requested me to be most careful in no way to betray that se-

77. St. George Mivart, *Nature and Thought: An Introduction to a Natural Philosophy*
(London: Kegan Paul, Trench, 1882).
78. George Romanes, "Nature and Thought," *Contemporary Review* 43 (1883): 831–41.
79. Ibid., p. 840.

cret. . . . As it happens, part of my last chapter was expressly written for the sake of Physicus himself, and it was my very desire to represent his old arguments with perfect accuracy, which made me employ his own *ipsissima verba* as the expression of certain views opposed to my own, and which I deemed unreasonable and foolish.[80]

Since Romanes did not reply to this letter, perhaps Mivart really did know the identity of Physicus; but such knowledge hardly explains why Mivart or his mouthpiece did not simply attribute the words to Physicus. And the lame excuse that a desire for "perfect accuracy" required verbatim transcription must have squeezed a chuckle from every schoolmaster who read the letter. This harsh personal exchange between Romanes and Mivart set the tone for the several encounters to follow.

The two antagonists next met over the question of instinct. In 1885 Mivart published a long two-part article, "Organic Nature's Riddle."[81] In this piece for the *Fortnightly,* he undertook a study of instinct, the phenomena of which flashed out intelligent design he thought and so provided the surest evidence against the gospel of irrationality preached by some evolutionists. He identified Haeckel as the archsinner in maintaining the doctrine of the ultimate purposelessness of life; but in that insidious way so often inflaming the Darwinists, he aimed his haranguing objections squarely against their recently deceased leader.

Certain instincts, in Mivart's view, could not have arisen through lapsed intelligence or chance variation, since they displayed intelligent foresight far beyond an animal's capacity and a complexity that defied mere lucky occurrence. This "innate mysterious rationality" must have a source beyond the animal itself and thus gives testimony to "the existence of a constant, pervading, sustaining, directing, and all-controlling but unfathomable Intelligence which is not the intelligence of irrational creatures themselves."[82]

Romanes again took Mivart on, responding within the month to the *Fortnightly* article. He wished, significantly, not to try the question of whether the ultimate order of nature were due to mind, but to indict Mivart's particular effort at solving the riddle of instinct. In Romanes's estimation, Mivart misemployed his scientific knowledge when he argued that our inability to explain certain phenomena required immediate recourse to the "agency of final causes." The very logic of this procedure was "essentially unscientific." Moreover the several cases of

80. St. George Mivart, "Letter to the Editor," *Contemporary Review* 44 (1883): 156.
81. St. George Mivart, "Organic Nature's Riddle," *Fortnightly Review* 43 (1885): 323–37; 519–31.
82. Ibid., p. 531.

instinct Mivart examined either had been or could be explained by Darwin and his followers.[83]

Mivart fired back, but attacked the man more than the argument.[84] Referring to Romanes's recent article, "Mind and Monism," he claimed that his adversary also considered the universe intellectually ordered; at least that appeared to be the clear implication of an obscure metaphysics. Mivart scrutinized Romanes's monistic account of the mind-brain problem, according to which mind and brain were expressions of an underlying substance that, though we had no proper conception of it, yet had to be considered mental in character. Mivart complained that this substance, since it was not sensible, could not be known or characterized. Romanes himself thus "solved" a scientific problem through the dark agency of the unknowable, so that his last position was no different than the theist's first. Mivart concluded his attack with a burst of sweetly smelling invective. He excused himself for not mentioning any of Darwin's supposed explanations of instinct, because in respect of the memory of a great scientist and friend, he could never recall to criticism statements "unlikely . . . to do him honour."[85]

Romanes surely choked on these cloying last remarks, but with icy restraint observed that "Mr. Darwin lived to see the 'criticisms' of his 'friend and opponent' become matters of merely historical interest."[86] Romanes also pointed out that Mivart had typically misquoted him, so that he appeared to know more about the unknown than he claimed. His article "Mind and Monism" had merely exposed the deficiencies of the rival doctrines of spiritualism and materialism and concluded that the only live possibility left was monism. While admitting monism as a reasonable metaphysical hypothesis, Romanes reiterated that no logically defensible conclusions could be drawn about the existence of God.

The crashing denouement to the controversy between Mivart and Romanes came in 1889 with the publication of Mivart's *Origin of Human Reason,* a large volume written specifically in response to Romanes's *Mental Evolution in Man.*[87] Mivart subjected Romanes's views to a minute inspection and an Aristotelian criticism. From Mivart's stand-

83. George Romanes, "Professor Mivart on Instinct," *Fortnightly Review* 44 (1885): 90–101.

84. St. George Mivart, "The Rights of Reason," *Fortnightly Review* 45 (1886): 61–68.

85. Ibid., p. 68.

86. George Romanes, "Mr. Mivart on the Rights of Reason," *Fortnightly Review* 45 (1886): 329–38; quotation from p. 337. See also Mivart's response "An Explanation," *Fortnightly Review* 46 (1886): 525–27.

87. St. George Mivart, *The Origin of Human Reason* (London: Kegan Paul, Trench, 1889).

point, animals gave no outward sign of having rational intelligence, though they could manipulate sensory images for practical advantage. And if the evidence indicated a gulf in kind between man and animal, then no gradual progressive development could bridge it. Mivart, as was his style, did not failed to color his philosophic-scientific conclusions with remarks about his opponent's frequent flights into "bathos" and stumbles "over the edge of an abyss of absurdity."[88]

Romanes did not confront Mivart's arguments directly. He received them in the middle of his three-year lecture series on Darwinism, which he delivered at the Royal Institution (1888–1890), and at the outbreak of those boding symptoms that signaled the onset of his terminal illness. During the last years of his life, however, Romanes took up the several metaphysical and religious issues which the two antagonists contested. The results were posthumously published as *Mind and Motion and Monism* and *Thoughts on Religion*. They represented not a volta faccia, as suggested by his wife,[89] but a careful drawing out of a metaphysics upon which Romanes had based his evolutionary science and a growing personal need for the consolations of religion. The problem of the bearing of biological science on religion likewise continued to occupy Mivart, but with opposite consequence.

Happiness in Hell

In 1864, Pius IX issued the encyclical "Quanta cura," to which was appended an attack on modernism, the "Syllabus errorum." This reactionary afterthought asserted the Church's teaching authority even in scientific matters. The doctrine was reinforced a few years later when the Vatican Council convened to declare, in 1870, the Pope infallible when teaching on faith or morals. These pronouncements aroused great suspicion and hostility in Protestant England and America, provoking John William Draper, a chemist and physiologist at the City University of New York, to identify Rome as the archvillain in *The History of the Conflict between Religion and Science,* as the title of his famous polemical tract of 1874 had it.[90] These events also kept Mivart on the defensive, not only with his scientific associates, but with his coreligionists.[91] In 1885 an Irish ecclesiastic challenged him to disavow evolution, since it

88. Ibid., p. 25.

89. E. Romanes, *Life and Letters of George John Romanes*, p. 372.

90. See John William Draper, *History of the Conflict between Religion and Science,* 4th ed., International Scientific Series (New York: D. Appleton, [1874] 1875).

91. For an account of Mivart's slow departure from the Catholic Church, see Gruber, *Conscience in Conflict,* pp. 141–213.

flew in the face of the Church's common and recently reasserted authoritative interpretation of scripture. Mivart responded, in his essay "Modern Catholics and Scientific Freedom," that the Church was simply mistaken if she condemned evolutionary theory, just as she had been mistaken in condemning Copernican theory.[92] History made it plain, he argued, that in the seventeenth century the Church had "founded its erroneous decree affecting physical science, which was *not* its province, upon an erroneous judgment about the meaning of Scripture, which was universally supposed *to be* its own province."[93] He concluded that in the Galileo case, "God has thus taught us that it is not to ecclesiastical congregations but to men of science that He has committed the elucidation of scientific questions."[94] Though Mivart's "reconciliation" of science and religion brought howls from the Catholic press, especially from the Jesuits, Rome and the English hierarchy did not raise cudgels against this still-faithful son of the Church. Galileo's ghost cautioned forbearance. But a few years later, as he approached his own end, Mivart took an unorthodox stand on eschatology, a province into which even his scientific predecessor Galileo had not dared to trespass.

Mivart examined the Catholic Church's doctrine that souls without baptism or those baptized but dying in mortal sin would suffer eternal punishment in hell. He did not believe that the usual image of the damned undergoing profound mental agony and extreme physical torture could be squared with "right reason, the highest morality and the greatest benevolence."[95] Dante's inferno surely could not be the abode of innocent children or primitive adults who died without baptism, or even of those whose social or intellectual circumstances would mitigate responsibility for sin. Mivart reviewed statements of the early Fathers, medieval theologians, and contemporary writers; and he weighed what reason could demand of a good and just God. He concluded that although souls in hell would be deprived of the blissful vision of God, yet most would enjoy a measure of natural happiness, depending on the degree of their culpability. He even proposed that punishment for less-heinous sins would be finite, so that a duly punished soul might thereafter be released to a pleasant existence.[96] For many of the damned, there might be happiness in hell.

92. St. George Mivart, "Modern Catholics and Scientific Freedom," *Nineteenth Century* 18 (1885): 30–47.

93. Ibid., p. 39.

94. Ibid., p. 41.

95. St. George Mivart, "Happiness in Hell," *Nineteenth Century* 32 (1892): 899–919; quotation from p. 899.

96. See also Mivart's subsequent articles, "The Happiness in Hell: a Rejoinder," and "Last Words on the Happiness in Hell," *Nineteenth Century* 33 (1893): 320–638, 635–51.

Mivart had entered his last decade during a time of retrenchment in the English Catholic Church. John Henry Cardinal Newman, who had fought against the doctrine of papal infallibility, died in 1890; Cardinal Manning two years later. The new prelate of Westminister, Cardinal Vaughan, was a man of no intellectual distinction. With his approval, Mivart's articles on hell were placed on the *Index librorum prohibitorum* by the Holy Office in 1893. That same year, Pope Leo XIII issued an encyclical that appeared to require a literal interpretation of scripture. Those priests who had harbored evolutionary ideas similar to Mivart's were called to Rome, where they had to recant. And finally, at century's close came the Dreyfus affair. Neither the French clergy nor the Roman Pope did anything to curb the anti-Semitic attacks during the trials of Dreyfus in 1899. For Mivart this represented the clearest example of the dogmatic blindness and complete fallibility of the Bishop of Rome. In a last effort, a few months before his death, he wrote to the *Times* to state his repudiation of the attitudes of the clergy. He also composed three articles, two for the *Nineteenth Century* and one for the *Fortnightly,* reaffirming his heterodoxy.[97] Vaughan responded by demanding Mivart sign a profession of the Catholic Faith that emphasized precisely those doctrines the scientist held in doubt. Mivart, ill and dying, refused to sign. Vaughan, who would die shortly himself, took the ultimate step. He prohibited his priests from administering extreme unction to the heretic. Mivart died on 1 April 1900 without the consolation of those rituals he had once so cherished.

Mivart's religious convictions bent his science and rational judgment along lines that diverged from other evolutionists; finally, however, there could be no more give. His intellectual odyssey beached him outside the institutional Church. Romanes navigated in just the opposite direction. As a young man his scientific views caused him to abandon formal religion. But just as surely as Mivart, his deeply felt religious desire guided his metaphysics and science to a harbor of rational belief in God. He died being received back into the Church.

Evolution, Metaphysics, and the Return to Religion

Natural Selection and Monistic Metaphysics

In his book *Animal Intelligence,* Romanes traced the phylogeny of mind from its first glimmer in protozoa, through reflex creatures, to

97. See St. George Mivart, "The Continuity of Catholicism" and "Scripture and Roman Catholicism," *Nineteenth Century* 47 (1900): 51–72, 425–42; and "Some Recent Catholic Apologists," *Fortnightly Review* 67 (1900): 24–44. These articles and Mivart's final break with the Catholic Church are discussed by Gruber in *Conscience in Conflict,* pp. 188–213.

animals exhibiting instinct and intelligence. He used complexity of nervous development as one index of the growth of mind; grades of learned behavior, expressive of the development of consciousness, provided another. These measures, he held, were in fact used when we predicated mind and conscious states of other human beings. If such objective standards justified our attribution of mind to our fellows, then parity of consideration ought to allow us to use them in making similar ascriptions of nonhuman animals. Romanes thus dismissed Descartes's theory of animal automatism. But he did not consequently endorse Cartesian interactionism in the case of animals, or of man; for, as he believed, the doctrine of conservation of energy militated against the supposition that mental force might propel the physiological machinery.[98] With this perplexity of the mind-body relation introduced, Romanes reserved further discussion of the problem for another occasion. His mission in *Animal Intelligence* was only to set out the evidence of a hierarchy of mental development in the animal kingdom.

Romanes did not wait long to tackle the problem of mind in nature. In 1882, the year after *Animal Intelligence* was published, he authored an article that began to lay the ground for a systematic resolution. In "The Fallacy of Materialism," he examined several hypotheses concerning the mind-body relationship; rejected completely the leading contenders—materialism and spiritualism; lingered over monism; but concluded in agnosticism. Over the next few years, however, Romanes's doubts began to dissolve; he recorded his new conviction in two more essays: "Mind and Motion" (1885) and "The World as an Eject" (1886).[99] At decade's end, he had started on a book that would thoroughly examine the metaphysical relation between mind and body, and that would set out the place of mind in the universe; but illness and a protracted controversy with Wallace[100] prevented his manuscript from appearing. By July of 1893, Romanes knew his projected book would not leave his own hand. He entrusted Lloyd Morgan, his literary executor, to publish his earlier metaphysical essays, along with his unfinished manuscript. These appeared posthumously in 1895 under the title *Mind and Motion and Monism.*[101]

I have earlier argued that Romanes abandoned his religious beliefs

---

98. Romanes, *Animal Intelligence*, p. 7.

99. Both of these essays were reprinted in Romanes's posthumous *Mind and Motion and Monism* (London: Longmans, Green, 1895). My citations will be from these reprints.

100. See the appendix at the end of this chapter for a description of the Romanes-Wallace debate over physiological selection.

101. See note 99.

not entirely for the intellectual and scientific reasons he gave, but also because of the impact of the personality of Charles Darwin. Darwin exuded quiet wisdom, scientific eminence, and paternal solicitude for his young disciple. Romanes, I think, responded to Darwin's intellectual authority and personal concern by zealously advancing evolutionary theory and its apparent materialistic implications. Yet the secret strings of his heart remained tied to the Church. And shortly after Darwin died in 1882, Romanes began, I believe, a slow retreat back along the path to faith. He took the first hesitating steps in his essay the "Fallacy of Materialism," published in December of 1882, a few months after death severed the strong personal bond between master and disciple.

In his essay, Romanes investigated six possible hypotheses about the relationship between mind and body, with the aim of eliminating all but one.[102] He first dismissed the view that mind could operate directly on body, since that would inject new energy into a physical system and thus violate the conservation laws. The supposition that a divine planner preestablished a harmony between the two realms of mind and body remained a logical possibility, but a hypothesis perfectly incapable of demonstration. He rejected radical idealism (i.e., a denial of the existence of matter) with a wave, since it was "illusive of argument."[103] This left materialism (i.e., the idea that brain utterly determined mind), monism (i.e., the idea that brain and mind expressed two aspects of one underlying substance), and agnosticism (i.e., the admission that no answer was possible).

Romanes interpreted as the most cogent form of materialism the doctrine advanced by Huxley under the name "conscious automatism": the theory that mental events were completely determined by cerebral events, but that mental events themselves merely dangled—that they were causally inefficacious.[104] Romanes opened a wedge against this position with the instruments of epistemological idealism. He argued that our idea of causal force in the external world derived from our experience of it in the internal world, when we caused an alteration in our own thoughts; he further maintained that what we called physical causation could only be a sequence of mental modifications—a Humean train of ideas. Therefore we not only had no evidence but "we can have no evidence of causation as proceeding from object to subject."[105]

102. George Romanes, "The Fallacy of Materialism," *Nineteenth Century* 12 (1882): 871–888.

103. Ibid., p. 884.

104. Thomas Huxley, "On the Hypothesis that Animals Are Automata and Its History," *Fortnightly Review* 22 (1874): 555–89.

105. Romanes, "Fallacy of Materialism," p. 873.

Romanes undoubtedly realized that epistemological objections of this kind could not hold against the strong current of scientific empiricism and mechanism. But he spiked another argument that he thought could sustain the weight of scientific reason, since it was drawn from science itself, from that theory Darwinians like Huxley could not abandon, namely natural selection theory. Romanes argued that "on the principles of evolution, which materialists at least cannot afford to disregard, it would be a wholly anomalous fact that so wide and general a class of phenomena as those of mind should have become developed in constantly ascending degrees throughout the animal kingdom, if they are entirely without use to the animals."[106] Darwinian theory required complex and general traits exhibited by animals, such traits as constituted conscious mind, to have evolved under the aegis of natural selection. But for them to have been selected, they must have been useful in their environments. Hence mind could not be merely the inert product of brain. Precisely on those principles Huxley held most dear, mind had to be causally efficacious.

This transposition of natural selection theory into a new key by an *echt* Darwinist demonstrates rather conclusively, I believe, the wild inaccuracy of the presumption that Darwinism led inevitably to materialism and agnosticism. Romanes understood the power of this new argument from natural selection, and repeated it in his Rede Lecture, "Mind and Motion," given in 1885 at Cambridge, as well as in his unfinished manuscript on metaphysics. This use of the Darwinian principle, however, was not original with Romanes, and he may have borrowed it. William James advanced virtually the same argument—and also brought it to bear against Huxley's theory—in his essay "Are We Automata?" which he published in *Mind* in 1879,[107] a few years before Romanes's article appeared. The link between James's argument and Romanes's might have been forged by Darwin himself, since the older man, in 1878, had recommended "strongly" to Romanes that he read an important article by James ("Brute and Human Intellect").[108] So both the title of James's essay and the author's name would have caught Romanes's attention as he perused the pages of *Mind*. There is no clear

---

106. Ibid. p. 880.

107. William James, "Are We Automata?" *Mind* 4 (1879): 1–22. See chapter 9 for a discussion of James's discovery and deployment of the natural selection argument for mind.

108. In a note to Romanes on 27 December 1878, Darwin "strongly" recommended that the young naturalist read James's "Brute and Human Intellect" in the *Journal of Speculative Philosophy,* 1878. The postcard from Darwin to Romanes is held in the Darwin Correspondence, no. 556, American Philosophical Society, Philadelphia.

evidence, however, indicating that Romanes did in fact get the argument directly from James. We may have here another case of convergent conceptual evolution. Certainly the intellectual milieu of both Romanes and James would have fostered such convergence: each was a committed Darwinist; each expressed deep religious feelings, while retaining a theologically skeptical attitude; each was aware of the epistemological constraints the theory of evolution placed on the human knower; though each sensed that man was more than a passive mechanism; and finally, each focused on the Huxleyan version of materialism.

While the argument from natural selection had a surprisingly negative impact on the doctrine of materialism, Romanes thought that it also suggested an alternative resolution of the mind-body problem—monism. The conclusion that mind had evolved to fit an animal into its environment prompted two directive questions, for which monism seemed the only satisfactory answer. First, why should there be consciousness at all? And, second, why should a particular neural sequence that made coherent physical sense be always accompanied by a particular mental sequence that made coherent logical sense? The doctrine of monism held that 'mind' and 'matter' only expressed the phenomenal aspects of a single, underlying substance. So if brains evolved, then minds must have as well, since they both represented two aspect of the same thing. Moreover any coherent neural sequence ought simply be the obverse of a coherent logical sequence. The advantage of the monistic doctrine, however, did not reside merely in answering these two questions. More importantly, monism explained how natural selection could operate on intelligence: "intelligence being, not a result of matter in motion, but itself matter in motion, natural selection working upon the movements (functions) of organs, may thereby at the same time be working upon intelligence."[109]

Though monism greatly attracted Romanes, it yet had two consequences about which he remained wary. First, it seemed to lead to William Clifford's thesis that each atom of matter had its concomitant atom of mind.[110] But the doctrine did not explain the brute fact of such concomitance. Just why did reality and its elements have this dual character? In posing such a question, Romanes got caught up in the logical conundrum of explanation: all explanation requires an unexplained explainer. Science and most philosophy can, when pressed to the limits, offer only certain unexplained facts as explanations for other facts. Ro-

109. Romanes, "Fallacy of Materialism," p. 886.
110. William Clifford, "Body and Mind," in *Lectures and Essays,* ed. L. Stephen and F. Pollock (London: Macmillan, [1879] 1901), 2:39.

manes was perfectly aware of this logical slide from fact to more remote fact. Yet he moved easily down the slope, apparently hoping to fall into the hands of the ancient and self-justifying faith. But since, in 1882, he could not bring himself to appeal to God as the final disposer of natural facts, he stumbled back on the last possible solution to the mind-body problem, which was to admit no solution: "The Fallacy of Materialism" concluded that "the association between mind and matter is one which is beyond the reach of human faculties to explain."[111]

Romanes, however, soon began to probe beyond the apparent barrier of ignorance. As his own empirical and theoretical studies of animal mind continued—with *Mental Evolution in Animals* in 1883—he became more convinced of the Jamesian argument that natural selection meant mind functioned in nature. In his Rede Lecture in 1885, he elaborated this conviction that "not without a reason does mind exist in the frame of things."[112] He also returned to the monistic hypothesis. While admitting that empirical evidence could not demonstrate monism, he yet argued that reason urged us to accept it as the only logical possibility. And whereas before he had shied from monism because it seemed to lead to panpsychism and even pantheism, he now welcomed universal mind as an acceptable consequence. Romanes concluded his lecture expressing an attitude which undoubtedly provided the initial direction for his argument: that "if a little knowledge of physiology and a little knowledge of psychology dispose men to atheism, a deeper knowledge of both, and still more, a deeper thought upon their relations to one another, will lead men back to some form of religion, which, if it be more vague, may also be more worthy than that of earlier days."[113]

Romanes continued to ponder the metaphysics of the mind-body problem through the latter part of the 1880s. In 1886 he composed an essay entitled "The World as Eject," in which he explicitly extended the monistic view to the whole of nature. He proposed that the ever so-complex physical universe be regarded as an outward appearance having a subjective side, a suprapersonal consciousness of which the thoughts of individuals might be understood as the many elements. This would mean that rigidly necessary causal law would be identical to rational volition, both in the universe at large and in our own personal world of apparently determined behavior.[114] In a few years, then, Romanes trav-

111. Romanes, "Fallacy of Materialism," p. 887.
112. Romanes, "Mind and Motion," *Mind and Motion and Monism,* p. 25.
113. Ibid., p. 38.
114. Romanes's hypothesis that causal laws might be the objective side of supreme rational will was hardly original with him. In the closing passages of *Contributions to the Theory of Natural Selection,* first published in 1870, Alfred Wallace developed a similar

eled from regarding monism as interesting but without rationale, to being persuaded logically of its merits, finally to extending it as a reasonable hypothesis suggesting a mindful universe.

At the end of the decade, in 1889 or 1890, Romanes began a work that would knit up the various strands of his loose metaphysics. The unfinished manuscript appeared in 1895, under the editorship of Lloyd Morgan, as *Mind and Motion and Monism*. In the manuscript, Romanes expanded the conclusions of his earlier essays. He argued that monism had reason on its side. It explained both the existence and the function of mind in nature (the Jamesian argument powerfully put). But especially it made intelligible man's moral faculty. Materialism denied human freedom and morality. But according to Romanes, so did spiritualism, since spiritualism had to recognize the constraints of physical necessity on human behavior. Monism, by contrast, alchemized physical force, making it the same as psychic force. According to this doctrine, then, "the human mind is itself a causal agent, having the same kind of priority within the microcosm as the World-eject has in the macrocosm."[115] Monism thus situated the human organism in a familiar world, one which did not sterilize human values or mechanically foreclose our deepest aspirations: "the moral sense no longer appears as a gigantic illusion: conscience is justified at the bar of reason."[116] And finally, monism restored theism. For in Romanes's estimation, if human mind were due to anything other than something like itself, the necessary proportion between cause and effect would be destroyed. Monism thus supported the doctrine that the universe stood as the embodiment of an ultimate personality.[117]

Through the 1880s, Romanes worked out a series of arguments, largely dependent upon Darwinian theory, whose consequence was to reestablish an order that Darwin seemed to have overturned: a universe in which human freedom could be exercised, moral choices could be made, and divine purpose could govern. But these intellectual conclusions, as most critical readers have already discerned, were not bound

hypothesis, perhaps even suggesting it to his friend: "If, therefore, we have traced one force, however minute, to an origin in our own will, while we have no knowledge of any other primary cause of force, it does not seem an improbable conclusion that all force may be will-force; and thus, that the whole universe is not merely dependent on, but actually *is*, the will of higher intelligences or of one Supreme Intelligence." See Alfred Wallace, *Contributions to the Theory of Natural Selection* (London: Macmillan, 1870), p. 368.

115. Romanes, "Monism," *Mind and Motion and Monism*, p. 140.
116. Ibid., p. 149.
117. Ibid., p. 149–70.

to their premises with the logical steel of anything like a Euclidean demonstration. The required connections in Romanes's argument crept up from his emotional life. As a young researcher overpowered by the force of Darwin's personality, he had stumbled into the stark world of the new science. When Darwin died in 1882, his supportive hand could no longer guide his protégé along the unmarked path of an emotionally spare agnosticism. The reality of Darwin's personality, however, did not pass away. That spiritual presence seemed, however, to have a force different from that of its living incarnation. Romanes became ever more entranced by the wonder of human personality itself, especially its power to transform and mold cold rational considerations. Paradoxically the lingering reality of Darwin's personality worked to overcome Romanes's original—and conventional—acceptance of the apparent implications of evolutionary theory for moral judgment and religious belief. The justification of morality and religion could not be found in cool empirical reasoning of the kind that his friend so well exemplified, but only, as he declared in his Aristotelian Society Lecture of 1889, in "those mysterious depths of human personality, which lie beyond the reach of human investigation."[118] In the realm of personality he thus began to uncover a less dreadful road, one that brought him back, not exactly to the religious world of his natural father, but to one enough like it to assuage the mounting anxiety produced by his final illness.

Return to Religion

In the early 1890s, Romanes's symptoms became more ominous—worsening headaches, increasingly blurred vision, more fearful lapses of memory, all apparently due to a growing brain tumor. Yet he continued with anxious determination to work on a variety of scientific problems, recapitulated in the three volumes of *Darwin and After Darwin*, the last two of which required the finishing touches of Lloyd Morgan. During this time of ever-more-difficult and hesitating scientific effort, his thoughts moved frequently and finally incessantly back to religion. His correspondence with the Reverend John Gulick, an American missionary to Japan and a field naturalist, indicates the eruptive force of his growing concerns.

The two correspondents shared many scientific interests, especially in overcoming Wallace's objections to "physiological selection," a mechanism Romanes had proposed as an aid to natural selection.[119] But on

118. Romanes, "Is There Evidence of Design in Nature?" p. 75.
119. See the discussion in the appendix to this chapter. Gulick's contribution to the

Figure 8.1 George John Romanes, 1848–1894,
photograph from ca. 1890.

Christmas day of 1890, Romanes changed abruptly the course of their discussions. He wished to ask of his friend, whom he thought scientifically accomplished and calmly rational, "How is it that you have retained your Christian belief?" He confessed that "years ago my own belief was shattered—and all the worth of life destroyed—by what has ever since appeared to me overpowering assaults from the side of rationality."[120] Gulick responded from his missionary post in Osaka, Japan,

---

controversy is described by John Lesch, in "The Role of Isolation in Evolution: George J. Romanes and John T. Gulick," *Isis* 66 (1975): 483–503.

120. George Romanes to John Gulick (25 December 1890), in Correspondence of John Gulick, no. 78, Academy of Natural Sciences of Philadelphia (film 839 in American Philosophical Society, Philadelphia).

the following March. He sent Romanes a small, specially composed tract, entitled "Christianity and the Evolution of Rational Life."[121] He argued that Christianity gave direction to the altruistic instincts that biological evolution had bestowed on the human race. Science, he maintained, displayed the means for achieving our ends, but only Christian wisdom and love declared what ends we ought to pursue. Gulick then urged a consideration that would ultimately tell on Romanes:

> But is it wise, is it rational, to act on these assumptions [i.e, the doctrines of Christianity] before we prove that they are in accordance with fact? I believe it is. This is, it seems to me, just what rational man has always done in some degree and I believe he will always have to unless he abandons rational life. . . . They are necessary to the continuance of rational life. They give vigor, enthusiasm and joy to life and they bring all parts of our knowledge into a harmonious whole.[122]

This suggestion—that we need exercise a will to believe in order to make real the possibility of rational action and, ultimately, to discover the truth of religion—immediately evoked from Romanes only a list of the conundrums that seemed to embarrass Christian doctrine in the light of advanced science.[123] He objected that "the entire structure of Pauline theology has had its formulation undermined by Darwinian science: the 'first-man' having been politely removed, there is no longer any logical justification (according to this theology) for the 'second man.'" It was precisely the doctrine of Christ's divinity—the pouring of God into that "second man"—that simply could not be squared with Darwinian theory, German higher criticism, or careful common sense. After this response, the two friends returned to less emotionally searing matters.

In fall of 1893, Romanes suffered a paralytic stroke, but no apparent diminishment of the hard focus of his intelligence. Though he kept up his scientific correspondence during the following months, his reading turned to religious books, especially the *Pensées* of Blaise Pascal, which he kept by his bedside.[124] The *Pensées* appeared to justify Christian assent for reasons other than the scientific, for reasons hidden in the heart

---

121. John Gulick to George Romanes (7 March 1891), in Correspondence of John Gulick, nos. 106–11, Academy of Natural Sciences of Philadelphia (film 839 in American Philosophical Society, Philadelphia).

122. Ibid.

123. George Romanes to John Gulick (19 May 1891), Correspondence of George Romanes, American Philosophical Society, Philadelphia.

124. E. Romanes, *Life and Letters of George John Romanes,* p. 371.

of human nature. Recognition that deeper sources in personality needed to have their say supplied Romanes with a "new and short way with the Agnostics," as he expressed it in a letter to his dear friend Dean Francis Paget of Christ Church, Oxford.[125] He explored this new way through a reexamination of the positions he had taken in his earlier *A Candid Examination of Theism.* He called his new manuscript *A Candid Examination of Religion* and intended to publish it under the name "Metaphysicus."[126]

Romanes admitted that his new way derived not so much from any "purely logical processes of the intellect," as from "the subconscious (and therefore more or less unanalyzable) influences due to the ripening experience of life."[127] Nonetheless, he set out to offer reasons for harkening to these profounder resources of a mature personality. His justifying argument bears striking resemblance, again, to one of William James, though without, I think, likelihood of borrowing.[128] The argument recognizes that in any fundamental demonstration employing first principles, whether in science, philosophy, or religion, the premises of the demonstration cannot themselves be justified within the same sphere of discourse, since there can be no recourse beyond *first* principles. Hence our only way of justifying larger frameworks of thought is by appeal to argument and evidence outside the area of demonstration. This sort of consideration led Romanes to affirm a purer agnosticism than before in the realm of reason. Reason alone thus cannot give us God, but neither can it give us the world of natural law and physically determinable consequences. Both science and religion, according to Romanes, must seek ultimate justification in a kind of trust, in an intuition enriched by experience, in short, in faith. The belief in universal causality, after all, is a belief in things not seen.[129] His effort, in fine, was to begin with a purified agnosticism, with the hope of ending in justified religious conviction.

Romanes thought religion, like science, was also capable of a kind of pragmatic justification. Were religion true, particularly a humanized Christianity (shorn of the troubling doctrine of the Incarnation), it would give meaning to life and significance to suffering.[130] Moreover the existence of a divine volitional side to causal regularity would allow

125. Ibid., p. 375.

126. Romanes's manuscript of *A Candid Examination of Religion* was posthumously published in *Thoughts on Religion.*

127. Ibid., p. 100.

128. See chapter 9.

129. Romanes, *Thoughts on Religion,* p. 146.

130. Ibid., p. 152.

reconciliation between natural law and moral purpose[131]—something he had earlier argued in advancing his doctrine of monism.

The only real justification of religion, however, would be through intuition and faith. Not an intuition or faith arrived at through reason, whether pragmatic or theoretical, for religious conviction was more than intellectual prudence; nor an intuition or faith passively received, since Romanes had little patience with Calvinism. Only through an intuition struggled for and a faith exercised. But even in the winter of his illness, Romanes could not yet will himself to believe:

> For assuredly the strongest desire of my nature is to find that that nature is not deceived in its highest aspirations. Yet I cannot bring myself so much as to make a venture in the direction of faith. . . . Even the simplest act of will in regard to religion—that of prayer—has not been performed by me for at least a quarter of a century, simply because it has seemed so impossible to pray, as it were, hypothetically, that much as I have always desired to be able to pray, I cannot will the attempt.[132]

In time, Romanes did take Pascal's wager. During spring, 1894, he regularly attended church services, and on Easter Monday he received Holy Communion. His strength ebbed, but returned sufficiently so that he could attend the third in the series of annual lectures on science that he had funded. On May 3, August Weismann gave the Romanes Lecture at Oxford, and dined with Romanes and his wife afterward. On Thursday during Pentecost he again took Communion. His wife related that after services he told her, "I have now come to see that faith is intellectually justifiable." Somewhat later he added, "It is Christianity or nothing."[133] On May 23, George Romanes fell into a coma, and he died five days later at age forty-six.

With the publication of Romanes's *Thoughts on Religion,* many Christian writers rejoiced at the return of a sheep that had been lost to Darwinism, while some Darwinists blanched at Romanes's effort "to slaughter his reason on the alter of faith," as Paul Carus, editor of the *Monist,* put it with exiguous sympathy.[134] But Romanes neither abandoned evolutionary theory for religious orthodoxy, nor sacrificed his

131. Ibid., p. 121.

132. Ibid., pp. 132–33.

133. E. Romanes, *Life and Letters of George John Romanes,* p. 379.

134. Paul Carus, "The Late Professor Romanes's Thoughts on Religion," *Monist* 5 (1894–1895): 385–400; quotation from p. 398. See also Frank Turner's prudent evaluation of Romanes's religious trajectory and what his friends made of it, in Turner's *Between Science and Religion* (New Haven: Yale University Press, 1974), pp. 134–63.

reason. The faith in which he died had few of the doctrinal filigrees of traditional Christianity; his was precisely the kind of spare faith that could be harmonized with the reign of natural law and evolutionary processes. Moreover, his faith demanded not a murder of reason, but a lively and dexterous use of it: Romanes's thesis, and William James's as well, that fundamental argument could only be set in the concrete of intuition and experience—this thesis could not be more reasonable, as I hope to show in the second appendix.

As Darwin conferred a legacy on his heir apparent, so Romanes bequeathed an intellectual inheritance to his. The more superficial aspect of the heritage required of Lloyd Morgan that he see through the press Romanes's final two volumes in the *Darwin and After Darwin* series and the unfinished collection of metaphysical essays, *Mind and Motion and Monism*. But this professional debt was incurred because of the peculiar intellectual attachments binding the two evolutionary psychologists together.

Morgan versus Romanes on the Status of Comparative Psychology

Morgan's Career

The paper hardly stirred the multitudes, but George Romanes paid attention. The 28 September 1882 number of *Nature* printed a lecture on "Animal Intelligence," originally delivered in Cape Town, South Africa.[135] The lecturer challenged Romanes's presumption that a dog might have "abstract ideas." He argued that while an animal could form general ideas by focusing on one aspect of an object or by letting the differences among similar objects fade, it could isolate such qualities as "goodness" or "whiteness" only if it had the use of words; for only language could bind up the qualities of objects in a stable symbol. Animals, being bereft of language, could not therefore have any such abstract ideas. About the time Romanes received this criticism, he also received a book manuscript from the lecturer, seeking aid in finding a publisher.[136]

Perhaps his former teacher Thomas Huxley assured Conwy Lloyd Morgan sufficiently of Romanes's professionalism and kindness so that the young colonial felt little reluctance to criticize gently the man he importuned. Romanes, in any case, agreed to help. He also conceded

---

135. C. Lloyd Morgan, "Animal Intelligence," *Nature* 26 (1882): 523–24.
136. George Romanes to C. Lloyd Morgan (11 March 1883), in the Papers of C. Lloyd Morgan, 128/13, University College Library, Bristol.

that Morgan was right, that certainly no animal could perform higher-level, language-dependent abstractions.[137] By his generosity, Romanes secured a friend who shared his interests in biopsychology, a future executor, and an incisive and persistent critic.

Morgan's background and concerns guaranteed that Romanes would patiently listen to his objections. Born four years after Romanes, in 1852, Morgan, like his new friend, came from a professional, middle-class family. His father, a solicitor with business interests in mining, advised his second son about the financial advantages of becoming an engineer. Balking at the alternative—an office job—Morgan entered the Royal School of Mines in London, where Huxley reigned as professor of natural history. While pursuing a degree in metallurgy and mining, Morgan cultivated a growing interest in a philosophy of a distinctly unearthly sort. The rector of his Weybridge parish introduced him to Berkeley's *Principles* and *Dialogues,* which left a transforming impression on the young scholar—enough so that when he came to write a biographical essay fifty years later, he recorded his rector's tutelage as the initiating episode of his recollections.[138] The cleric led him from Berkeley through the philosophic greats, thus joining Morgan's professional training in science with a deepening interest in the nature of mind.

Morgan mentioned his avocational study, as well as his lamentable ignorance of biology, to a chance dining companion one evening at a dinner for graduates. He sat next to Professor Huxley. Huxley's reputation among the students at the time could not have been more lustrous: a few years before, he had conquered Richard Owen with his superior knowledge of the vertebrate skeleton and then had upended the guardian of orthodoxy, Bishop Samuel Wilberforce, by dexterous intelligence and wit. More recently St. George Mivart had been crushed by the weight of his scholastic learning. Bloodthirsty undergraduates took delight in this champion who would smite the Amalekites whenever they dared rise up against Darwinian theory. Morgan remembered that in his dinner conversation, Huxley "gave me of his riches without emphasizing my poverty."[139] The great man kindly suggested that the young engineer might well spend another year at university under his instruction. Morgan pondered the opportunity while traveling through the rough bywaters of the Americas—acting as tutor to the scion of a

137. George Romanes to C. Lloyd Morgan (21 July 1883), in the Papers of C. Lloyd Morgan, DM 612, University College Library, Bristol.
138. C. Lloyd Morgan, "Autobiography," in *A History of Psychology in Autobiography,* ed. Carl Murchison (Worcester, Mass.: Clark University Press, 1932), pp. 237–64.
139. Ibid., p. 241.

wealthy Chicago family. On his excursion he tackled the *Origin of Species* and the *Descent of Man,* which churned up a resolve to ask his father to support him a year longer so that he might accept Huxley's invitation. Huxley's lectures on evolutionary theory, as well as the stimulation of essays on science by William Clifford and the continued fascination with Berkeley's problem of the place of mind in nature—these all conspired to alter the path of Morgan's career. After his tenure with Huxley, he obtained a position for which only a madly dedicated teacher could aspire, as lecturer in physical science, English literature, and constitutional history (all three!) in the Diocesan College at Rondebosch, near Cape Town.

Despite his considerable instructional duties, Morgan nonetheless managed during his five years there, from 1878 to 1883, to complete three book manuscripts—only the beginning symptoms of what he called his "cacoethes scribendi." The publications *Water and Its Teachings* (1882) and *Facts Around Us* (1884) undoubtedly recommended his appointment as professor of geology and zoology at the small University College at Bristol. His book *Springs of Conduct: An Essay in Evolution* waited until Romanes interceded with his own publisher; the book appeared in 1885. Morgan moved rapidly to the office of Principal of University College in 1887, and in 1909, when it received its university charter, he agreed to assume temporarily the role of vice-chancellor. He happily relinquished administrative duties a few months later, but remained as a teacher, he said, "till I was placed on the shelf of superannuation as Emeritus Professor of Psychology (1920)."[140] During the time he was transforming University College into a first-class institution, he became the chief British spokesman for evolutionary biopsychology, being the first fellow elected to the Royal Society for work in psychology (1899). He initially directed his efforts in the science to turn it away from Romanes's casual projections of large mental capacities into the small minds of animals.

The Inference to Lower Minds

Around the time of his return to England, Morgan drafted an essay, never published, that detailed his opposition to Romanes's procedures.[141] In his attack, he made sport of the comparative psychologist's excesses—for instance, Romanes's assigning mental images of "home"

140. Ibid., p. 245.
141. C. Lloyd Morgan, "Mental Evolution in Animals," in C. Lloyd Morgan Papers, DM 612, University College Library, Bristol.

to the molluscan limpet that clung to its rock.[142] But the hardened engine from which he launched his objections—and which drove most of his subsequent efforts in biology of behavior—was an epistemological theory he constructed from resources drawn from Berkeley and especially William Clifford. The specific argument he exploded under Romanes's comparative science was one used generally by idealist philosophers to lift most of their opponents: namely, that "all my knowledge of consciousness in others, is a knowledge of my consciousness, or is built up out of that knowledge." Hence when I would attribute a conscious state to other human beings—or to animals—"it is an eject, an image of my own consciousness which I throw out from my self."[143] But what was the constitution of the self thus ejected? Plucking off a metaphor from Clifford—one that William James would soon expand into an influential psychological theory—Morgan likened the conscious self to a stream,[144] with deep channels of social, personal, and physical "subconsciousness."[145] So we perceived ourselves, indirectly, through the eyes of our fellows, through the weight of our past experiences, and even through our bodily feelings—that weariness in the loins and dull ache behind the eyes also constituted the ego. Morgan claimed that upon the waters of self-consciousness floated those bits of psychological states that we flung onto other human beings, though with caution, and onto animals, only in careless moments of scientific abandon. Morgan, of course, recognized that lower creatures did have nervous systems bearing similarity to ours and that they sometimes acted in ways at least reminiscent of the human tribe. So he quickly admitted we might justly grant animals subjective states, though only of the most general and unspecified sort. We simply had no way of verifying specific attributions and had every reason to be cautious in assigning even the most generic

142. Romanes, *Mental Evolution in Animals,* p. 153; cited by Morgan in "Mental Evolution in Animals," MS p. 10.

143. Ibid., MS p. 2. Morgan studied the collection of Clifford's essays published in 1879, shortly after the Cambridge mathematician's death. In "Body and Mind" (originally published in 1874), Clifford maintained that one's awareness of another person's subjective states was really a consciousness of one's own feelings 'ejected' into the other person. See William Clifford, "Body and Mind," *Lectures and Essays* 2:54.

144. Clifford described the mind as "a stream of feelings which runs parallel to, and simultaneous with, a certain part of the action of the body" ("Body and Mind," *Lectures and Essays* 2:34). William James began developing his theory of the "stream of thought" in "On Some Omissions of Introspective Psychology" (*Mind* 9 [1884]: 1–26), where he spoke variously of "the stream of our feeling," "the subjective stream," "thought's stream," and the like. James had already read and taken exception to what he took to be the Huxleyan cast of Clifford's "Body and Mind." The full expression of the Jamesian theory is in chapter 9 of James's *The Principles of Psychology* (New York: Henry Holt, 1890).

145. Morgan, "Mental Evolution in Animals," MS p. 2.

psychological traits. Morgan thus concluded that uncertain projections onto the primitive minds of animals provided no solid base upon which to conduct a science.[146] No comparative psychology, such as Romanes attempted to institute, had any warrant. At best a careful scientist might develop a comparative physiology of nervous systems or a comparative study of the adjustive behaviors of animals. These latter alone, Morgan allowed, might constitute "objective psychology."[147]

Morgan's unpublished essay served as the source for a more particularized attack on Romanes. In his article "Instinct," Morgan sketched the argument just described—concluding that comparative psychology was impossible—and then turned to Romanes's conception of instinct.[148] Romanes defined instinct as "reflex action into which there is imported the element of consciousness."[149] But without any inkling of the nature of animal consciousness, so Morgan argued, the definition had no clear meaning. Moreover, Romanes held (as did Morgan at the time) that instincts might evolve not only through natural selection but also through intelligent acts that had become automatic and unconscious—the Spencerian idea that instincts arose from "lapsed consciousness." But if that were so, then such instincts obviously need have nothing of consciousness appended to them.

Romanes protested that he had already cautioned in *Mental Evolution in Animals* that a reliable ejective predication would depend on a close analogy between a human being and another organism, so that indeed it "ceases to be trustworthy in the ratio in which the analogy fails."[150] He further reminded Morgan that such objection to comparative psychology would also tell against the possibility of human psychology. The skeptical critic, he thought, must retire in the face of the actual accomplishments of scientists in these areas. Romanes concluded that it was not therefore hopeless to ascribe subjective states as concomitants to instinctive acts, since some difference was needed to distinguish instincts from simple reflexes.

In 1885 Morgan's *Springs of Conduct* was published.[151] Romanes gave it a strange and rather patronizing review in *Nature*.[152] While observing that "there is not much in it that is strikingly original," he nonetheless

146. Ibid., MS pp. 6–7.
147. Ibid., MS p. 15.
148. C. Lloyd Morgan, "Instinct," *Nature* 29 (1884): 370–74.
149. Romanes, *Mental Evolution in Animals*, p. 159.
150. George Romanes, "Mr. Lloyd Morgan on Instinct," *Nature* 29 (1884): 379–81.
151. C. Lloyd Morgan, *The Springs of Conduct; an Essay in Evolution* (London: Kegan Paul, Trench, 1885).
152. George Romanes, "The Springs of Conduct," *Nature* 33 (1886): 436–37.

thought its style and compendious organization rendered it "a most interesting epitome of modern thought upon the subjects of which it treats." He further remarked that Morgan had moderated his views about the possibility of comparative psychology, probably, he added, because of their previous interchange. But Romanes must have read a different book (or remembered only vaguely the manuscript he had earlier received from Morgan). Actually, Morgan explicitly held his ground: "I have elsewhere stated my opinion that no science of comparative psychology from the ejective standpoint is possible. And I see no cause to change that opinion."[153] What Romanes may have been reacting to, though, was the fact that in 1885, about the time *Springs of Conduct* appeared in the stores, Morgan did relent. In a lecture given at the Bristol Naturalists Society in October, which Romanes may have attended or at least heard report of, Morgan considered again the problems of ejective psychology, and began shifting his position.[154] "I must here add," he concluded for his audience,

> that I am a believer not only in the parallelism but in the identity of neuroses & psychoses [i.e., states of nerves and mind]. Hence I believe it possible that, in the far distant future, we may attain to
> 1. A sufficiently exact knowledge of neuroses and psychoses in the human subject to enable us to infer the one from the other.
> 2. A sufficiently exact knowledge of neuroses in animals to enable us to correlate them with human neuroses.
> and 3. By the combination of 1 and 2, an indirect method by which to infer the psychoses of animals from the nature of their neuroses.[155]

While the philosophical problem of other minds initially led Morgan to pronounce comparative psychology stillborn, the metaphysics of monism—"the identity of neuroses & psychoses"—brought him to detect the possibility of revival. As he pursued monistic philosophy over the next decade, he became ever more sanguine about comparative psychology, not only declaring it alive and well, but vigorously exercising it in a series of books and articles during his most creative period. To insure the health of the science, he prescribed a prophylactic for its conduct, one that preserved the spirit of his earlier strictures. His famous canon, formulated in 1892, proclaimed: "That in no case is an

153. Morgan, *Springs of Conduct,* p. 164.

154. C. Lloyd Morgan, "On the Study of Animal Intelligence" (Bristol Naturalists Society, 1 October 1885), in the papers of C. Lloyd Morgan, DM 612, University College Library, Bristol.

155. Ibid., MS p. 11.

animal activity to be interpreted as the outcome of the exercise of a higher psychical faculty, if it can be fairly interpreted as the outcome of the exercise of one which stands lower in the psychological scale."[156] To understand the transformation in Morgan's views about comparative psychology, as well as the force binding the empirical results of that science to its conceptual framework, we need to consider the character and development of his metaphysical theory of monism.

The Monistic Framework of Morgan's Science

Throughout his career Morgan maintained, as he argued during his Lowell Lectures in Boston in 1904, that "a complete and satisfactory interpretation of nature is, so far as it is attainable by man, partly scientific and partly metaphysical."[157] The metaphysical part provided a framework for the more particular scientific interpretations: it determined the permissible areas of investigation, justified the methods, and led science to speak to issues of larger concern. From the mid-1880s through his last works on emergent evolutionism, Morgan continued to develop his brand of metaphysics, that of monism.

Monistic doctrine formed the heart of Morgan's *Springs of Conduct,* though it beat with the pulse of his youthful engagement with Berkeleyan idealism. In this early work, he augmented his Huxleyan vocabulary of "neuroses" (referring to activity of nerves) and "psychoses" (referring to accompanying mental activity) with the term "hypopsychoses," by which he designated mental states below the threshold of consciousness. He proposed that in both the individual and the race, parallel evolutionary processes led from elemental neural states to more complex cerebral states, on the one hand, and from elemental hypopsychotic states to complex conscious states, on the other. The monistic psychologist would regard these processes as simply two expressions of the same evolutionary development. There would then be no question of an obscure communication between two separate entities of mind and brain or of a mysterious harmony synchronizing them. The eco-

156. Morgan apparently first announced his canon in a paper, "Limits of Animal Intelligence," read at a session of the International Congress of Experimental Psychology, 2 August 1892. This version is recorded in the proceedings of the meeting. See *International Congress of Experimental Psychology,* second session (London: Williams & Norgate, 1892), p. 44. Morgan promulgated his canon in virtually this same form in his *An Introduction to Comparative Psychology* (London: Walter Scott, 1895), p. 53.

157. Morgan gave the Lowell Lectures in Boston in 1904 and published them, with additions, as *Interpretation of Nature* (New York: Putnam's Sons, 1906). Quotation is from p. 105.

nomics of explanation recommended monism as a resolution of the chief problem of evolutionary biopsychology.[158]

But in the enthusiasm of his first major work in evolutionary science, another consideration led Morgan to peer below the surface of his pragmatic justification of monism. In an admittedly pale light, he speculated in the fashion of Berkeley and Clifford: "Thought is the one absolute reality that we know. The elements out of which thought is built up we may call mind-stuff. And it is conceivable that just as the mind is the true reality which underlies that phenomenal mass of matter we call the human organism, so too is mind-stuff the true reality which underlies all phenomenal masses of matter."[159] Morgan granted that this was nothing but idealism, but thought that it at least remained faithful to experience.

This idealist brand of monism lost its savor for the mature scientist; only in old age did this youthful variety of metaphysics again become attractive. In his address to the Bristol Society of Natural History, given about the time *Springs of Conduct* appeared in 1885, he had already suggested a version of monism that escaped idealism. He developed this suggestion a year later in his paper "On the Study of Animal Intelligence,"[160] and set it out in some detail in his most innovative and important book, *Animal Life and Intelligence*, which appeared in 1890.[161]

Alfred Wallace judged *Animal Life and Intelligence* "worthy of all praise."[162] Romanes lauded its biological sections, marvelled at its psychology (especially in that he thought it mirrored his own), and commended the author as a "gifted philosopher."[163] I will discuss in a moment Morgan's treatment of animal psychology, particularly instinct, and his ingenious ideas about mental evolution. But I would like first to consider the fuller development of his monistic theory as he elaborated in *Animal Life* and in subsequent works.

Notwithstanding his generous estimate of Morgan's book, Wallace took issue with its monistic thesis that matter and mind expressed the different traits of an underlying neutral stuff. He likely objected because that thesis had been designed by Morgan to cut through a dilemma the older evolutionist used to comfort his own spiritualistic assumptions. Wallace had insisted that human consciousness could not be the mere product of brain, otherwise we would be compelled to acknowledge

158. Morgan, *Springs of Conduct*, pp. 188–94.
159. Ibid., p. 209.
160. C. Lloyd Morgan, "On the Study of Animal Intelligence," *Mind* 11 (1886): 174–85.
161. C. Lloyd Morgan, *Animal Life and Intelligence* (Boston: Ginn, 1890–1891).
162. Alfred Wallace, "Modern Biology and Psychology," *Nature* 43 (1891): 337–41.
163. George Romanes, "Animal Life and Intelligence," *Mind* 16 (1891): 262–67.

that the constituent molecules of the nervous system were themselves conscious. But rejecting the absurd idea that atoms could think meant we had to admit, or so Wallace believed, that consciousness was added to the material organization of the brain. This admission, he pointed out, recognized the possibility that higher spiritual powers might exist independently of material nature.[164]

Morgan accepted the basic proposition of Wallace's dilemma; he simply did not balk at the idea that even the smallest speck of matter might have a mental side, as long as we did not suppose that such an atomic mind exhibited consciousness. As in *Springs of Conduct*, he argued, with yet more Greekized jargon, that "parallel to the evolution of organic and neural kinesis [i.e., material phenomena] there has been an evolution of metakinetic [i.e., mental] manifestations culminating in conscious thought."[165] He again justified the monistic assumption from economic principles. Monism allowed us to ignore Wallace's plea for higher powers, as well as Romanes's projection of a supraconsciousness: human mind did not need to look to higher powers for justification, only to the humbler neutral stuff which mental traits expressed; nor did the universe of material bodies require the supposition of a supraintelligence, only the recognition of a constellation of lesser mental lights. And finally, of great economic advantage, the metaphysics of monism sanctioned what was already increasingly fruitful work in comparative psychology, while avoiding the excesses of some researchers (like Romanes).[166]

During 1893 and 1894, Morgan distilled the scholarship of *Animal Life and Intelligence* into the more modest *Introduction to Comparative Psychology*. The book may have surprised the naturalist who casually browsed in its pages when it appeared in 1895. For it began with a formidable discussion of monism—a measure of the importance Morgan placed on a philosophic framework for science, as well as of the caliber of serious reader he expected. In the "Prolegomena" to his book and in the article from which it derived,[167] Morgan further elaborated his doctrine. He distinguished its three mutually supportive features: a monistic epistemology, a monistic interpretation of nature, and an analytic monism. The monistic theory of knowledge held that the necessary starting point of science and philosophy was common experience. Beginning with an undivided flow of experience, we separated out the objective and the subjective sides. According to Morgan, our abstract-

164. Alfred Wallace, *Contributions to the Theory of Natural Selection*, p. 365.
165. Morgan, *Animal Life and Intelligence*, p. 467.
166. Ibid., p. 476.
167. C. Lloyd Morgan, "Three Aspects of Monism," *Monist* 4 (1894): 321–32.

ing and generalizing thought constructed out of experience both a cosmos of objects and a self that observed them.[168] The monistic interpretation of nature kept this theory of knowledge from sliding into idealism. Morgan protected against the enthusiasm of his youth by grounding his monism on a pragmatic but critical realism, of the kind he found emerging from William James's *Principles of Psychology*. The ultimate justification, then, of a particular metaphysical interpretation of nature would be its utility for and harmony with experience. He thus felt warranted in endorsing our ineluctable natural attitude that "the world which forms the objective aspect of knowledge continues somehow to exist quite independently of its being sensed or perceived."[169] Finally, the analytic side of monism recognized that the psychologist confronted an independent human (or animal) organism, which for special purposes could be analyzed into physical and mental components. Neither component had priority; neither determined the existence of the other. Rather each expressed a feature of the one natural organism. And since biological science had as its subject the whole organism, we were completely justified in applying evolutionary theory not only to the physical nature of the organism but to its mental nature as well.

Morgan shed his early Berkeleyan idealism for a naturalistic monism.[170] I believe there were three converging considerations that encouraged this change. First, he took to heart Romanes's initial criticism that severe idealistic strictures must undermine not only comparative psychology but all psychological science. Second, he found the economic principles that allowed him to dismiss ontological excess, particularly in Romanes and Wallace, might also be applied to his own idealism: it was simpler to suppose that alterations in sensation had a source in an independent, external nature. These two considerations,

168. Morgan's version of radical empiricism may owe something to the similar conceptions of William James and Wilhelm Wundt, both of whom Morgan read. See James, *Principles of Psychology* 1 : 284–86; and Wilhelm Wundt, *Vorlesungen über die Menschen- und Thierseele*, 2d ed. (Leipzig: Voss, 1892), pp. 491–95.

169. Morgan, *Introduction to Comparative Psychology*, p. 4.

170. During the 1890s, Morgan refined his theory of monism and spread its message into many areas of biology and psychology. Most of his important papers on the subject were published in the *Monist* while that journal was under the editorship of Paul Carus. In addition to "Three Aspects of Monism," see C. Lloyd Morgan, "Mental Evolution," *Monist* 2 (1891–1892): 161–77; "The Doctrine of Auta," *Monist* 3 (1892–1893): 161–75; "Naturalism," *Monist* 6 (1895–1896): 76–90; "Animal Automatism and Consciousness," *Monist* 7 (1896–1897): 1–18; "Causation, Physical and Metaphysical," *Monist* 8 (1897–1898): 230–49; "The Philosophy of Evolution," *Monist* 8 (1897–1898): 481–501; and "Biology and Metaphysics," *Monist* 9 (1898–1899): 538–62.

though, did not cause him to abandon all restraint in comparative psychology. He remained convinced that the investigator's individual consciousness could not immediately penetrate to external objects and other minds. This conviction along with other concerns led him, in his *Introduction to Comparative Psychology,* to promulgate his famous cautionary canon that "in no case may we interpret an action as the outcome of the exercise of a higher psychical faculty, if it can be interpreted as the outcome of the exercise of one which stands lower in the psychological scale."[171] Morgan's desire to redress the grievances caused by Romanes's looser approach brought him to concentrate on that aspect of animal behavior that had an objective character and could be studied systematically in evolutionary terms: animal instinct. So finally any residual tincture of idealism faded in his active research into the behavior of chickens and other perversely independent creatures.

Morgan's research on animal instinct keeps his name still alive, at least in the introductory chapters of textbooks in ethology and biopsychology. His studies became models for subsequent investigators in England and America (such as E. L. Thorndike) and his methods were employed on the continent (by the likes of Konrad Lorenz). Several of his theories in the area of mental evolution at the time caused large ripples, but they were quickly swamped in the new wave of behaviorism during the 1920s and 1930s. This is unfortunate, since they displayed considerable creativity and have, I believe, continuing significance. But let us first consider the several features of his theory of instinct.

Morgan's Theory of Instinct

Empirical Research and Philosophic Conviction

In his earliest discussions of instinct, Morgan sought to replace Romanes's psychological definition, which referred to animal consciousness, with an objective definition, which referred to specific behaviors and neurological states. He first suggested that we distinguish reflexive, instinctive, and intelligent acts from one another by their respective

171. Morgan, *Introduction to Comparative Psychology,* p. 53. Morgan's canon is, of course, but the psychologist's application of Occam's razor. Even Romanes had observed this economical principle when, in a lecture in 1882 ("Scientific Evidence of Organic Evolution," p. 3), he cited Sir William Hamilton's "law of parsimony—or the law which forbids us to assume the operation of higher causes when lower ones are found sufficient to explain the observed effects." Wilhelm Wundt, in the same year as Morgan first urged his canon upon psychologists, wrote of the "lex parsimoniae, which allows recourse to more developed principles of explanation only when the simpler ones have proved insufficient." See Wundt, *Vorlesungen über die Menschen- und Thierseele,* p. 380.

Figure 8.2   Conwy Lloyd Morgan, 1852–1936, photograph from ca. 1900.

places in the nervous organization of the genus, the species, and the individual.[172] Thus behavior common to a genus would be regarded as reflexive; that typical of a species would be taken as instinctive; and acts peculiar to the individual would be classified as intelligent. Romanes quickly deflated this effort by pointing to the patellar reflex in human beings, which on Morgan's terms had to be classified as instinctive.[173] Morgan returned to the problem of definition in 1888, now specifying the three classes of behavior causally: reflexes responded to definite stimuli and were confined to particular organs in a group of animals; instincts were inherited habits (stemming from natural selection or

172. Morgan, "Instinct," p. 373.
173. Romanes, "Mr. Morgan on Instinct," p. 380.

lapsed intelligence) that uniformly characterized a class of animals; and intelligent acts arose in an individual's adaptations to special circumstances.[174]

Morgan sought objective criteria for instinct in order to remove animal psychology from the hands of dilettante naturalists and establish it scientifically. But he seems to have had another, deeper motive as well, originating, I think, from his philosophic training. Morgan, while an evolutionary naturalist with impeccable credentials, yet always insisted on the difference in kind between animal and human consciousness: our species enjoyed bright reason, perceived abstract relationships, and pursued moral and aesthetic ideals; animals lived on a darkling plane of sensory associations, chased after fleshy objects, and had no knowledge of the goals of their instinctive behavior. Morgan, directed by these professional and philosophical motives and armed with the epistemological weapons of ejective monism, attacked Romanes's comparative studies, which supposed particular conscious states, much like our own, to be a part of animal instincts. But this was not simply an ideological battle; Morgan was a sensitive scientist and empirical evidence counted, in some measure at least. As the result of his own experiments on instincts, he was forced to consider with greater refinement the role of consciousness in animal instinctive behavior. Yet, though the evidence spoke, he was still able to interpret it so as to preserve his deeper conviction that animal mind and human mind fundamentally differed. His canon for conducting comparative psychology became both the instrument to preserve this philosophic faith and an expression of it. Morgan's empirical research in the early 1890s followed the lead provided by a remarkable and tragic predecessor, Douglas Spalding.

The Influence of Spalding

Spalding, a poor barrister who had studied with Alexander Bain at Aberdeen, took a position as tutor in the household of John Russell, Lord Amberley. The domestic assembly of mother and children (with the youngest, Bertrand Russell, looking on) aided the young naturalist with his investigations of animal instincts. The family often helped him conduct observations on broods of chicks and young ducklings in the dining room, greatly upsetting visiting relatives and acquaintances.[175] Lady Amberley, with her husband's permission, reciprocated lessons in

174. Morgan, "On the Study of Animal Intelligence," p. 184.
175. Bertrand Russell recounts in his family history the delight his mother and other siblings took in Spalding. See Bertrand Russell and Patricia Russell, *The Amberley Papers* (London: Allen & Unwin, 1937), 2:533–67.

natural history with private instruction in human reproductive biology.[176] But in those experiments for which he is remembered, Spalding performed deprivation studies on young animals. In one line of investigation, he opened chicken eggs just before they hatched and hooded the chicks to prevent them from learning correct responses. When he removed the hoods one to three days later, after they were able to walk around, the chicks immediately "pecked at some speck or insect, showing not merely an instinctive perception of distance, but an original ability to judge, to measure distance, with something like infallible accuracy."[177] Isolation experiments like this (now a standard technique of ethologists) showed that instinctive behavior resulted from inherited nervous organization, which Spalding, following Spencer, believed to have arisen principally from the acquired habits of progenitors. Spalding's small fame rested on such original experiments demonstrating the evolved behaviors of animals.[178] When his consumption grew worse, the Amberleys released the young naturalist, with pension, to southern France, where he died in 1877 at about age thirty-seven.[179]

In *Animal Life and Intelligence,* Morgan cited Spalding's investigations as evidence that animals came into the world prepared to engage in complex activity, with consciousness playing little or no role.[180] But a few years later, in 1892 and 1893, when he performed experiments similar to Spalding's, the situation began to appear a bit different.[181] Morgan hatched chicks in a controlled environment, so that he could assess what was innate and what learned in the young animals' attempts to grab and swallow grain pellets, insects, and worms. He found that though chicks would peck at small objects soon after birth, they were clumsy and completely promiscuous: the instinct to pick up and consume objects of a certain size came built in, but the animals had to

176. Bertrand Russell, *The Autobiography of Bertrand Russell,* vol. 1 (Boston: Little, Brown, 1967), p. 10.

177. Douglas Spalding, "Instinct, with Original Observations on Young Animals," *Macmillan's Magazine* 27 (1873): 282–93; quotation from pp. 283–84.

178. See also Douglas Spalding, "On Instinct," *Nature* 6 (1872): 485–86; "Instinct and Acquisition," *Nature* 12 (1875): 507–508.

179. "Douglas A. Spalding," *Nature* 17 (1877): 35–36.

180. Morgan, *Animal Life and Intelligence,* pp. 423–25.

181. C. Lloyd Morgan, "The Limits of Animal Intelligence," *Fortnightly Review* 60 (1893): 223–39; "The Scope of Psycho-Physiology," *Nature* 49 (1894): 504–5. These experiments are discussed at some length in his *Introduction to Comparative Psychology.* Morgan continued experimental work through the 1890s. He corresponded with other protoethologists on questions of inherited behavior and even supplied C. O. Whitman pigeons for the latter's work on instinct. See C. O. Whitman to C. Lloyd Morgan (January 1897), in the Papers of C. Lloyd Morgan, 128/88, University College Library, Bristol.

improve their aim with practice and learn what objects were edible. Morgan's chicks, for instance, stuffed themselves with numerous strips of worsted wool before learning that worms were more satisfying. He perceived that "the role of consciousness on the matter of pecking is to select the adequate responses and to steady the muscular mechanism to its work."[182] Most instincts were exhibited in varying contexts, to which an animal had after their first occurrence intelligently to accommodate itself.

When naturalists such as Romanes observed animals acting instinctively, they saw an already integrated skein of behaviors, one in which consciousness played a biologically vital role. In his isolation experiments, Morgan could tease out the respective contributions of the unconscious automatism of instinct and the subsequent adjustive response of conscious intelligence. In later work, influenced especially by the ablation studies of Charles Sherrington,[183] Morgan postulated that beneath the psychologically distinct faculties of instinct and intelligence existed physiologically separate substructures. The intelligent behavior of an animal, including the human animal, was "the function of the cerebral cortex with its distinguishing property of consciousness," while "the coordination involved in instinctive behavior, and in the distribution of physiological impulses to the viscera and vascular systems, is the primary function of the lower brain centers."[184]

Origins of Instinct

In his early studies, Morgan disputed Romanes's definition of instinct as conscious reflex action, but he embraced his colleague's faith in the two sources of instinctive behavior: natural selection and "the inheritance of habitual activities intelligently acquired."[185] Orthodox Darwinism sanctioned a limited role for the inheritance of acquired characters, and Morgan initially defended that view with rather traditional arguments, for instance, by appeal to the evidence of vestigial organs and coordinated variations.[186] But he was still very tentative about the La-

182. Morgan, "Limits of Animal Intelligence," p. 227.

183. Sherrington's experiments demonstrated that in decerebrate and spinal animals complex reflex and instinctive responses could be elicited. The frog whose spinal cord was severed above and below the brachial region still emitted the sexual clasp when the skin of the sternal area was stimulated. Even children born without cerebrum and midbrain displayed normal instinctive reactions to stimuli during early infancy. See Charles Sherrington, *Integrative Action of the Nervous System* (New Haven: Yale University Press, 1906).

184. C. Lloyd Morgan, *Instinct and Experience* (London: Methuen, 1912), p. 7.

185. Morgan, *Animal Life and Intelligence*, p. 434.

186. Ibid., pp. 197–213.

marckian thesis, basically, I believe, for two reasons. The first concerned an emendation he had made to the principle of natural selection. His refinement insinuated a role for mind even in the operation of Darwin's basic mechanism of evolution. In 1888 Morgan read a paper before the Bristol Naturalists' Society in which he distinguished natural selection proper from natural elimination.[187] In the latter process, unfit variations were eliminated as the result of struggle and competition. In the former, intelligence, appetency, and individual choice functioned: these faculties led an animal to select and shape adaptations—when, for instance, an insect selected the brightest flower to pollinate, or when a hen chose the gaudiest or most tuneful mate. In Morgan's view, natural elimination only dispatched relatively unfit traits, but did not touch neutral traits; while selection proper worked against both harmful and neutral variations. If natural elimination were the most prevalent force in evolution, then a principal defense of Lamarckian inheritance—that only it could explain the existence of neutral traits—crumbled: for natural elimination left neutral traits intact.

The second reason for Morgan's wariness of the Lamarckian hypothesis was that there seemed to be no convincing mechanism to explain the inheritance of acquired characters. In *Animal Life and Intelligence,* he examined Darwin's theory of pangenesis and found it wanting. Haeckel's theory of perigenesis and Spencer's theory of physiological units—each holding that external bodily acquisitions inscribed their effects on the molecular structure of heritable elements—and Nägeli's speculation that germplasm and body plasm were convertible, all seemed unsupported by the evidence and intolerably vague on cardinal points.[188] Paradoxically, Morgan thought the strongest theory to be Weismann's. Weismann proposed that the developing embryo consisted of two distinct sets of cells: somatic cells that gave rise to muscle, nerve, bone, and so forth, and germinal cells that issued both germ cells and body cells of the next generation. In this model of heredity, no communication from body cells to hereditary substance would be possible; only chance variation of the germ could provide the raw material of evolution. In his evaluation, Morgan balanced the persuasiveness of the Weismannian theory against the apparent observational evidence for Lamarckian inheritance, and inclined toward the empirical evidence.

By the time *Habit and Instinct* appeared in 1896, however, the weight of Morgan's considerations shifted against acceptance of the Lamarckian hypotheses. He endorsed Weismann's basic thesis that the germ and

187. C. Lloyd Morgan, "Natural Selection and Elimination," *Nature* 38 (1888): 370.
188. Morgan, *Animal Life and Intelligence,* pp. 197, 212, 213, 447.

Figure 8.3 August Weismann, 1834–1914, photograph from 1882.

somatic lines were separate. One might suppose that this introduction of an ultra-Darwinism into biopsychology, a bold move that won the admiration of Weismann himself,[189] resulted from the accumulation of empirical evidence on the other side. Weismann had, after all, experimentally demonstrated that snipping the tails of several generations of mice never produced progeny with shorter tails.[190] But what seems really to have convinced Morgan was his own formulation of a powerful theory, compatible with ultra-Darwinism, that described a process which simulated the Lamarckian. Evidence that bespoke the inheritance

189. August Weismann to C. Lloyd Morgan (26 November 1896), in the C. Lloyd Morgan papers, 128/82, University College Library, Bristol.

190. August Weismann, "The Supposed Transmission of Mutilations" (1888), in *Essays on Heredity,* vol. 1, trans and ed. E. Poulton, et al. (Oxford: Clarendon Press, [1889] 1891), pp. 444–45.

of acquired characters could thus be neutralized as an objection to the Weismannian theory of hereditary variations. Morgan's discovery came first to be known as "organic selection," and then later the "Baldwin effect." I will discuss it in a moment.

Opposition to Morgan's Mechanistic Conception of Instinct

Morgan strove to analyze instinctive behavior into its components and to provide objective, physiologically assignable causes to account for it. In the early part of this century, his effort was perceived as representing the further encroachment of Darwinistic mechanism into psychology. In 1910 a joint meeting of the Aristotelian Society, the British Psychological Society, and the *Mind* Association sponsored a symposium on instinct and intelligence. The participants turned quickly to Morgan's theory of instinct. G. F. Stout sounded a criticism that echoed, with some variations, through the papers of Charles Myers and William McDougall.[191] Stout delineated the consensus that instinctive activity, even on its first occurrence, was not merely automatic but involved intelligence, else the animal would not know how to modify it the second time around.[192] Morgan considered this objection in his book *Instinct and Experience,* published in 1912. Earlier, in *Animal Behavior* (1900), he had defended the view that natural selection would tend to favor those animals which derived emotional satisfaction from performing species-preserving acts.[193] If such were the case, then it would be reasonable, he thought, to postulate that cortical centers of emotion might be hereditarily linked and jointly activated with subcortical motor centers of instinct. But he insisted that any conscious conation accompanying instinctive behavior, even on first occurrence, could only be a vague emotional aura rather than true cognition—unless we were prepared to grant innate ideas to animals.[194] Morgan regarded his proposal as consistent with the supposition that animals might be subsequently aware of their behavioral patterns and attendant feelings, and that this experience would suffice for the intelligent modification of their imperfect instincts thereafter.

The differences between Morgan and his critics might seem slight—whether intelligent awareness initially accompanied instinct or

191. G. F. Stout, "Instinct and Intelligence," *British Journal of Psychology* 3 (1909–1910): 237–49; Charles Myers, "Instinct and Intelligence," *British Journal of Psychology* 3 (1909–1910): 209–18; William McDougall, "Instinct and Intelligence," *British Journal of Psychology* 3 (1909–1910): 250–66.

192. Stout, "Instinct and Intelligence," p 238.

193. C. Lloyd Morgan, *Animal Behaviour* (London: Arnold, 1900), pp. 293–94.

194. Morgan, *Instinct and Experience,* pp. 46–48, 104–14.

only subsequently recalled it. Beneath the dispute, however, lay a much larger theoretical and methodological gulf, which Stout's criticism exemplifies. G. F. Stout—fellow of St. John's, Cambridge (1884), later Wilde Reader in Mental Philosophy at Oxford (1898), and finally professor of metaphysics at St. Andrews (1903)—was probably the most prominent British systematic psychologist of the period. But his mental science owed much more to the traditional, philosophical psychology of Mill, Bain, and Ward than to the newer approaches of the evolutionists and experimentalists.[195] His objection was long-standing; he had developed the basis for it in a series of letters and manuscripts communicated to Morgan in the late 1890s. Stout argued that all perceptual activity, including the chick's first spying of a worm, had to be regarded as meaningful, in the sense that it involved elements that were synthesized into a whole, such that one element would reproduce the others.[196] But in Morgan's view, this kind of philosophical focus on the supposed contents of consciousness and the postulation of their synthetic unity inhibited an experimental analysis of behavior and an evolutionary explanation of its components. Stout's approach to mind simply ignored the exigencies of the new scientific methods for studying psychological subjects: not inspection of the private mental world of organisms, but dissection of their public physiology and behavior. Morgan did not lack sympathy for the philosophic concerns of Stout; he simply thought them uncongenial to scientific work.

In respect to McDougall, the eminent Oxford (and later Harvard) social psychologist, Morgan's attitude was almost the reverse. McDougall had a keen interest in experimental work—he held a degree in medicine and had studied sensory psychology at Göttingen—but his science paraded to different pipes: the vitalism of Driesch and the hereditary theory of the neo-Lamarckians.[197] Morgan attacked these ideas as retrogressive.[198] The doctrine of monism provided a more economical conception than vitalism of the forces shaping development. And while Morgan, like most orthodox Darwinians, had initially accepted the inheritance of acquired characters as empirically justified, several discoveries made about mental evolution in the 1890s had already led

195. Stout's systematic treatises on psychology had an enduring impact on British psychology. See G. F. Stout, *Analytic Psychology*, 4th ed. (London: Allen and Unwin, 1914); *A Manual of Psychology*, 5th ed. (London: University Tutorial Press, [1899] 1938). The last volume was reprinted through 1949.

196. G. F. Stout, "A Genetic Scheme" (MS sent to Morgan 4 October 1897), in the papers of C. Lloyd Morgan, DM 612, University College Library, Bristol.

197. See especially William McDougall, *Body and Mind* (New York: Macmillan, 1911).

198. Morgan, *Instinct and Experience*, pp. 241–92.

him to reinterpret the evidence according to the texts of of the ultra-Darwinian Weismann.

Mental Evolution and the Theory of Organic Selection

Animal Intelligence and Human Reason

In his early skirmishes with Romanes, Morgan granted animals intelligence but refused them reason. Intelligence, he supposed, acted at the level of concrete sensory images, which could be chained together through associative bonds. Learning from trial and error as well as through imitation of specific behavior forged these links. Human reason, on the other hand, could focus on abstract relationships to draw rational inferences. Morgan maintained that he did not have to postulate of animals an awareness of the means-ends relationship in order to explain, for example, how his favorite chick, "Blackie," escaped from its pen. One could account for such apparently rational acts by showing that through trial and mistake the animal would finally hit on the solution; one could then reasonably suppose that the consequent pleasure associated with success would encourage an animal to repeat the puzzle-solving behavior in the appropriate circumstances.[199]

Not willing to rest with casual observation, Morgan attempted to test experimentally his thesis about the limits of animal intelligence. During a period of some days, he had his pet terrier, Tony, try to fetch a stick through the gaps between slats of a fence. Only after many trials did the animal gradually learn to turn its head sideways when carrying the stick back through the narrow opening. Whenever the conditions of the test were modified, the animal failed to recognize the similarity of relations and had to relearn the task again through trial and error. So, for instance, Morgan changed the situation slightly by using a crooked stick, which would usually get caught on the bottom of the fence. After repeated efforts to get through, the dog happened to break the crook and then easily made it. A man who had chanced to wander by at that moment exclaimed: "Clever dog that, sir; he knows where the hitch do lie." This episode summed up for Morgan the whole literature on the marvelous reasoning ability of animals.[200]

A good scientist will often choose a set of experiments to confirm a conclusion that theory demands. Edward Thorndike, while preparing his doctoral dissertation with Cattell at Columbia, built puzzle boxes to further demonstrate Morgan's thesis that animals learned only through

199. Morgan, "Limits of Intelligence," pp. 234–35.
200. Morgan, *Introduction to Comparative Psychology*, pp. 255–58.

trial and error. But other psychologists, notably the Gestaltists, just as prudently constructed experimental situations to show that animals could insightfully comprehend relationships[201] Even in the same experimental conditions, scientists of discordant theoretical persuasions will often, though not inevitably, interpret results differently. The great Cambridge idealist F. H. Bradley suggested that Morgan's lack of intimate acquaintance with the mind of Canidae led to misinterpretation. He wrote to his friend:

> When the dog hears the words "a cat in the garden" he, as you remark, probably understands only "cat." But I would submit to you that he probably doesn't understand even "cat" in abstraction from his own relation to it. "Me chasing a cat," "me being beaten about by a cat," is how I should interpret his idea. . . . I never could see any difference at bottom between my dogs & me, though some of our ways were certainly a little different.[202]

No dog, though, ever did metaphysics like Bradley's. Bertrand Russell, always perceptive on such issues, understood the delicate balance between theoretical assumption and empirical observation: he remarked that "animals studied by Americans rush about frantically with an incredible display of hustle and pep, and at last achieve the desired result by chance," while those "observed by Germans sit still and think, and at last evolve the solution out of their inner consciousness."[203]

Morgan was certainly prepared, at the very least, to have his long-held reservations about animal reason confirmed empirically. And his canon—the admonition not to ascribe to animal mind more than was required to explain the behavior exhibited—served to foster just the right conclusions. For he devised his principle as a methodological restraint on experimental observation and interpretation. The canon did not arise out of experience, but scientific experience arose out of it. Morgan formally justified it by appeal not to observation but to evolutionary theory. He argued that behavior exhibited within an animal's typical environment had to be the gauge for estimating its various mental powers: for a greater faculty than the behavior and environment required could not be explained by natural selection.[204]

201. Robert Boakes offers a thorough account of Thorndike's debt to Morgan and of the views of the Gestaltists in his *From Darwin to Behaviourism* (Cambridge: Cambridge University Press, 1985), pp. 68–78, 184–96.

202. F. H. Bradley to C. Lloyd Morgan (16 February 1895), in the C. Lloyd Morgan Papers, DM 612, the University Library, Bristol.

203. Bertrand Russell, quoted by Boakes in *From Darwin to Behaviourism*, p. 202.

204. Morgan, *Introduction to Comparative Psychology*, pp. 55–59.

Mental Evolution, the Analogue of Biological Evolution

Morgan was fully persuaded that natural selection operated on animal mind to raise it through successive levels of consciousness and intelligence; natural selection also produced the discontinuous phase shift that transformed our animal ancestors into rational beings. Other evolutionists of his time—notably Herbert Spencer, Ernst Haeckel, Samuel Alexander, and Benjamine Kidd—held that not only did selective forces (as well as inherited habit) raise animals into men, those agents continued to improve the human mind, transforming benighted savages into cultured Englishmen and Prussians. Morgan, while certainly recognizing differential intellectual abilities, denied the continued operation of natural selection on the mind of men.[205]

Much like Wallace, he believed that modern man differed little in mental capacity from even remote ancestors. He proposed that once human brains achieved the plasticity of rational response, they could adjust to their environment without the necessity of those adaptations being carved into the nervous system. Rationality relieved the selective pressures on human beings to alter their physiology to accommodate new situations.

Morgan certainly did not deny that civilized man stood above his savage cousins. The Victorian gentleman was the product of considerable mental evolution. The deposit of these changes, however, lay not in a more puffed-up brain, but in a more enriched social environment. Morgan argued, beginning in *Animal Life and Intelligence,* that the social environment retained and served as the transforming base for the products of art, great scientific ideas, inventions, and the ideals that served to advance the mental lives of successive generations. The social environment continued to evolve, while the human frame remained constant.[206]

Morgan's monism, which precluded any direct interaction between mind and brain, furnished the foundation for his theory of conceptual evolution. "The environment of an idea," he claimed, "is the system of ideas among which it is introduced."[207] Human thought could exist and become modified only through the agency of other thought, not, he insisted, through any imposition from physical objects. When I perceive an object that calls to mind a previous experience, both perception and memory reside entirely within my consciousness. The nervous system, to be sure, might be altered by impingements from the natural

205. Morgan, *Animal Life and Intelligence,* p. 488.
206. Ibid., pp. 480–503.
207. Ibid., p. 485.

environment, but the metakinetic sphere of consciousness could not be causally influenced by the kinetic sphere of physical occurrences.

In *Animal Life and Intelligence* and in a subsequent article, "The Law of Psychogenesis" (1892),[208] Morgan worked out the principles of the evolution of ideas within the mental-social realm. He proposed that learning by experience was the analogue of natural selection in the biological world. Natural selection operated to eliminate traits that did not fit into their particular surroundings; just so, ideas that were incongruous with the host of other ideas, feelings, beliefs, and attitudes forming the mental environment of an individual consciousness would be likewise rejected. In this theory, true ideas—true for the individual—would be those that survived the selective process. Similarly, true moral ideas would be those that fitted into the niche of beliefs that constituted the internal moral environment. "This is accepted as right, that is rejected as wrong," Morgan contended, "according as each is congruous or incongruous to our moral nature. The sense of congruity or incongruity is what we term the voice of conscience."[209]

Morgan's Darwinism of mind would appear to endorse an unseemly epistemological relativism. But he had three anchors for mental systems that restrained the evolution of chimeras. First, any system of ideas that utterly failed to represent in some way the natural environment would be eliminated by having its biological carriers dispatched.[210] Those protohumans that conceived of the sabertooth as an edible cabbage would not have had opportunity to leave an intellectual deposit for the next generation. But second, a belief system would have to harmonize with intransigent sensory perceptions, the mechanisms of which would been honed through generations of biological evolution.[211] Finally, the traditions of social evolution would provide common patterns of ideas against which individuals, during the processes of enculturation, might accommodate their particular conceptual systems. This deposit, in its own reflexive fashion, would rely for its stability on previous stages in social evolution and on the first two anchors mentioned.

Morgan's doctrines of monism and conceptual evolution obviously did not obscure all the shadows that still hung over the mind-body problem. For example, he really failed to expose the lines of communication that presumably radiated from external nature, through the physical nervous system, dipping down to the underlying neutral substance, and terminating in the mode of mind. Romanes's vexing ques-

208. C. Lloyd Morgan, "The Law of Psychogenesis," *Mind* n.s. 1 (1892): 72–93.
209. Ibid., pp. 86–87.
210. Ibid, p. 81.
211. Ibid., p. 88–91.

tion about the harmony between mental and physical traits of the one underlying entity remained unanswered. But Morgan sensed, as did Romanes himself, that monism offered the last best hope for understanding the biopsychology of organisms. And they were hardly alone. At century's end, a host of other scientists came to adopt some form of monism—Spencer, Haeckel, Mach, Wundt, Bergson, James, and Baldwin, just to mention those with deep interests in evolution. It was an arresting doctrine, which even captured the period's leading scientific philosopher, Bertrand Russell, though only after James charmed him into it.[212] The Darwinian biopsychologists also found congenial the idea that mental evolution operated according to the same principles as organic evolution. That happy conception has recurred in our own time, though not, I hope, as comedy. The first appendix to this volume will examine some of these recent developments. I will also detail there how a Darwinian approach might be taken to the evolution of scientific ideas—since my historiographic practice undoubtedly requires defensive argument. Morgan himself made some suggestive remarks in this direction. But he really designed his theory for a somewhat different use. He showed how the processes of mental evolution might simulate the inheritance of acquired characteristics, and thereby eliminate any need for Lamarckism. The theory also led him to formulate a powerful biological principle, one that involved him in a politically tangled priority dispute.

Morgan's Discovery of the Baldwin Effect

In the winter of 1895–1896, Morgan toured the United States, lecturing in several cities. In Boston he stayed for a week in the house of William James while he gave the Lowell Lectures.[213] George T. Ladd invited him to Yale for a talk and dinner.[214] He then made his way into the heartland, where he had traveled as a young tutor and companion to an offspring of Chicago wealth. George Peckham brought him to the University of Illinois in Urbana.[215] And he spent some time with a coexperimenter in instinct theory, Charles Otis Whitman, who was a professor at the University of Chicago and directed the University's marine

212. Bertrand Russell, *A History of Western Philosophy* (New York: Simon and Schuster, 1945), p. 812.

213. William James to C. Lloyd Morgan (21 January 1896), in the C. Lloyd Morgan Papers, 128/66, the University Library, Bristol.

214. George T. Ladd to C. Lloyd Morgan (25 January 1896), in the C. Lloyd Morgan Papers, 128/67, the University Library, Bristol.

215. George Peckham to C. Lloyd Morgan (28 January 1896), in the C. Lloyd Morgan Papers, 128/68, the University Library, Bristol.

biological station at Woods Hole, Massachusetts.[216] In New York, where his lecture tour began, he delivered a paper before the Academy of Science, which met just after Christmas. He shared the platform with another eminent evolutionary psychologist, James Mark Baldwin. Morgan described in his lecture an interesting new principle that he thought would remove support for the doctrine of the inheritance of acquired characteristics. Baldwin, who spoke next, quite surprising to all, delivered a lecture proposing just this same principle, which he had apparently derived independently. A few months later Henry F. Osborn announced again virtually the same principle—a thrice discovered idea.[217]

By dint of shrewd political maneuvering, the principle at first came to be known by the title that Baldwin chose, "Organic Selection"; but later it traveled simply under the name "The Baldwin Effect." The idea was spare and elegant. If animals entered a new environment—or their old environment rapidly changed—those that could flexibly respond by learning new behaviors or by ontogenetically adapting would be naturally preserved. This saved remnant would, over several generations, have the opportunity to exhibit spontaneously congenital variations similar to their acquired traits and have these variations naturally selected. It would look as though the acquired traits had sunk into the hereditary substance in a Lamarckian fashion, but the process would really be neo-Darwinian. For instance, some ground-feeding birds that happened to enter a new swampy terrain might learn in each generation to wade out onto a pond to feed off the bottom. Those that were flexible enough to acquire such responses would survive. In time, congenital variations might begin slowly to replace acquired traits, with natural selection molding them into instincts for wading and pecking at the right-sized objects. So what began as learned behavior and acquired modification might in time become innately determined and part of the hereditary legacy of the species. Organic selection thus imitated Lamarckian inheritance but remained strictly neo-Darwinian.[218]

In chapter 10, I will discuss how Baldwin arrived at this principle and sketch his political strategy for claiming priority in its discovery. I would like now to concentrate on how Morgan came to formulate it and suggest why he lost the priority dispute with Baldwin.

Some six years before hitting on the principle, Morgan indicated, in

216 C. O. Whitman to C. Lloyd Morgan (January 1897), in the C. Lloyd Morgan Papers, 128/88, the University Library, Bristol.

217. See chapter 10 for the details of Baldwin's and Osborn's versions.

218. Morgan first described the principle of organic selection in *Habit and Instinct,* pp. 307–22; this part of the book was preprinted as an article, C. Lloyd Morgan, "Of Modification and Variation," *Science* 4 (1896): 733–40.

Animal Life and Intelligence, that his theory of mental evolution, although it was not a biological theory, nevertheless had implications for organic processes. He claimed that conceptual evolution would have the effect of segregating the population of a society into different breeding classes and consequently directing the flow of hereditary traits. He thought that individuals who had intellectually evolved along certain lines would more likely choose marriage partners that met special ideals and exhibited complementary cultural adaptations. So, for instance, men and women of refined intellect would more probably select one another for marriage and thus be able to pass on undiluted the organic cerebral structures that allowed their mental refinements.[219] Another consequence of the theory of mental evolution for biology was that acquired traits would give the appearance of being inherited. This would occur because the progeny of such selective unions as Morgan described would come endowed with the intellectual ability to quickly adopt the mental culture of their parents.[220] Now both of these features of his earlier theory were essential elements in his later conception of what became known as organic selection: both theories proposed that acquired mental traits might have a real biological effect, and both attempted to remove support for Lamarckian heredity. Only a slight shift was required to transform the earlier theory of conceptual evolution into the theory of organic selection. Two alterations in Morgan's own mental environment seem to have provided just the selective pressure needed to produce the new discovery.

In May 1894, August Weismann came to Oxford to deliver the Romanes Lecture. Likely Morgan was in the audience, since he had become Romanes's literary executor, and the event would have drawn all the leading evolutionary biologist in the vicinity. In any case, Morgan read the lecture in its published form a few months later.[221] In his talk, Weismann described a process that he called "intra-selection." This process seemed to void the Spencerian objection that coadaptive parts required functional inheritance for their mutual evolution. Weismann suggested that environmental influences might cause a struggle of anatomical parts for nourishment, and so under stress the parts would develop into a harmoniously functioning network. To illustrate, he offered Spencer's favorite example of the deer's development of antlers. Should a somewhat larger rack of antlers spontaneously appear as the result of a congenital variation in the germ, the other anatomical parts—neck

219. Morgan, *Animal Life and Intelligence,* p. 498 "The Law of Psychogenesis," p. 84.
220. Ibid., p. 92.
221. August Weismann, *The Effect of External Influences upon Development* (Oxford: Clarendon Press, 1894).

muscles, vertebrae, and so forth—would adjust by consuming more nutriments, though without any adaptive transmission of greater size to the next generation. But the animal and its coadaptations would be preserved, so that later spontaneous variations of the germ plasm might gradually be selected to further perfect the system. As Weismann put it in a paragraph from which Morgan would later quote:

> as the primary variations in the phyletic metamorphosis oc-curred little by little, the secondary adaptations would prob-ably as a rule be able to keep pace with them. Time would thus be gained till, in the course of generations, by constant selec-tion of those germs the primary constituents of which are best suited to one another, the greatest possible degree of harmony may be reached, and consequently a definitive metamorphosis of the species involving all the parts of the individual may occur.[222]

When Morgan first published his version of the principle of organic selection—in *Habit and Instinct* and in a prepublication excerpt from the book in the journal *Science*—he cited Weismann as having first really established the principle.[223] Yet to our eye some ambiguity clouds the first sentence of the Weismann quotation, where it appears that ac-quired variations would follow congenital variations instead of leading the way for them. Baldwin seized on this vagueness to deny that Weis-mann really had the idea of organic selection.[224]

Though the Romanes Lecture seems to have played an important role in Morgan's formulation of the principle, he appears not to have cited Weismann in his first announcement of it in New York; for Baldwin, who was on the same platform, only learned of the Weismann connec-tion after reading a revision of Morgan's lecture that appeared later in *Science*.[225] Morgan's discussion of the theory in the printed version sug-gests that, though Weismann's proposal was undoubtedly part of the shaping intellectual environment for his own formulation, he high-lighted it more in retrospect. Another event likely led him to bring

222. Ibid., p. 19.

223. Morgan, *Habit and Instinct,* pp. 312–15;

224. See chapter 10.

225. See note 218. Baldwin wrote Morgan in November 1896, after the excerpt from *Habit and Instinct* describing the principle appeared in *Science*. He suggested that they mutually accord one another credit for having independently discovered the principle of organic selection, and mentioned that "your quotations from Weismann I find very inter-esting & I shall look up his content at once." James Mark Baldwin to C. Lloyd Morgan (20 November 1896), in the C. Lloyd Morgan Papers, 128/81, the University Library, Bristol.

together his theory of mental evolution with the relevant passages from Weismann. This was his job as editor of Romanes's second volume of *Darwin and After Darwin*.

Morgan undertook the task of editing Romanes's volume in 1895. While most of the manuscript was complete in typed format, Morgan actually had to compose chapters 5 and 6 from Romanes's notes (at his friend's request). In chapter 1, Romanes sketched a theory of conceptual evolution much like Morgan's (with a footnote acknowledging the similarity). In his version, Romanes compared the "intellectual trans-mission of acquired experience" with its Lamarckian counterpart, sug-gesting that while the latter might not be real, the former assuredly was. The whole effort of the volume, however, was to show that Lamarckism could not yet be rejected. This perhaps served as a goad, as Romanes's work often did, for Morgan to come to a contrary conclusion. But even more of a stimulus may have come in chapter 6, which Morgan carefully redacted. One passage, which refers to the Lamarckian principle, needs only minor adjustment to read as a description of organic selection:

> if functionally produced changes, and changes produced in adaptive response to the environment, are ever transmitted in a cumulative manner, a time must sooner or later arrive when they will reach a selective value in the struggle for exis-tence—when, of course, they will be rapidly augmented by natural selection. . . . Thus, if in any degree operative at all, the great function of these principles must be that of supply to natural selection those incipient stages of adaptive modifica-tions in all cases where, but for their agency, there would have been nothing of the kind to select.[226]

One creative reading of this passage might run in the following way. In a particular environment, individually acquired adaptations would "reach a selective value in the struggle for existence." These adaptations would not be biologically transmitted (according to ultra-Darwinism), but they would preserve the organism. Organisms so selected would subsequently have the individually acquired adaptations replaced by spontaneous variations. Those traits, then, would be "augmented by natural selection." Morgan, in the context of his earlier theory of mental evolution, just might have read Romanes in this way. It is of some cor-roborating interest that in Baldwin's New York lecture, where he too announced the principle, he had commented on Romanes's volume.

The principle of organic selection was perceived by the scientific community as an extremely important addition to evolutionary theory. Alfred Wallace judged that in light of the principle, "all the theoretical

226. Romanes, *Darwin and After Darwin* 2: 153.

objections to the 'adequacy of natural selection' have been theoretically answered."[227]. His approval made it a significant item in the science of the time. Osborn, even though a Lamarckian, heralded the principle because it united the two great tradition in evolutionary biology.[228] And Baldwin claimed the principle as his own, for which reason Osborn asked Morgan to print an advance excerpt from *Habit and Instinct,* where the principle was described. Osborn somewhat maliciously suggested the exerpt be printed in *Science,* the journal edited by Baldwin's antagonist James McKeen Cattell. Baldwin responded to the November *Science* article by immediately writing Morgan. He asked his colleague to mention his independent discovery of the principle, since he had "taken pains to refer to you [Morgan] as having reached similar views." Moreover, as he politely pointed out, his own version in the previous March issue of *Science* constituted "the first full statement [of the principle] of mine in print."[229] Morgan accommodated this request by inserting a footnote in the page proofs of *Habit and Instinct* to give Baldwin—and Osborn—proper due.

In the next few years, a struggle for intellectual hegemony ensued—to whom belonged the real credit for discovering the principle? That we now refer to it as the "Baldwin Effect" answers the question. There are, I believe, some sociologically important reasons for Baldwin's success, which suggest lessons to be drawn from the history of science. First was the simple question of controlling the name for the principle. Baldwin had immediately suggested it be called "Organic Selection," a title he had already bestowed on his theory of nonbiological conceptual evolution (a theory like Morgan's and Romanes's). This tied the principle to Baldwin's earlier ideas. Morgan, though he usually refrained from using Baldwin's term, yet never christened the principle with any convenient name. This sapped a claim not only to parentage but even to having recognized something real. Without signature, it could exist in his writings only as a set of shifting possibilities, not as an insistent creature. A name would have given it flesh, made it solid. Second, Baldwin never ceased to lobby, implicitly and explicitly, for rights to the principle. Morgan, by contrast, maintained that it could really be found in Weismann—so by his own admission, he never really owned it. Finally, Baldwin wrote several articles devoted exclusively to the principle, and even brought them altogether in a book. Morgan always buried his discussion of the principle in the middle of some gen-

227. Alfred Wallace, "The Problem of Instinct," a review of Morgan's *Habit and Instinct* reprinted in Wallace's *Studies, Scientific & Social* (London: Macmillan, 1900), 1:508.
228. H. F. Osborn, "Organic Selection," *Science* 4 (1897): 583–87.
229. James Mark Baldwin to C. Lloyd Morgan (20 November 1896), in the C. Lloyd Morgan Papers, 128/81, the University Library, Bristol.

eral treatise or made it only part of a broader lecture. In Morgan's hands, the principle never stood out as a well-defined entity. I will detail in chapter 10 other twists in Baldwin's campaign to land the principle as his own. It suffices here to summarize Baldwin's triumph by observing that success in science requires both intelligence and craft.

Conclusion: Science, Metaphysics, and Religion

Morgan retired in 1920. In his active period at Bristol, he helped transform the institution into a major university. He also presided over the tenuous establishment of experimental psychology in England, serving as the first president of the psychological section of the British Association the year after his retirement. The election to president indicates both his stature and, because of his age, the middling prospects for the future of experimental and biological psychology in England. Only after the Second World War did British psychologists really begin to think of their science in terms other than those fixed by the philosophers of mind.[230]

In the decade after his retirement, Morgan kept scratching the writer's itch. His books *Emergent Evolution* (1923), *Life, Mind, and Spirit* (1926), *Mind at the Crossways* (1929), and *The Emergence of Novelty* (1933) exposed and further developed the metaphysics that framed his biological psychology. In these books, he proposed a monistic philosophy that in its maturity hardly differed from Romanes's. He argued that reality was essentially one, but that it had diverse modes. Body and mind expressed corresponding features of the one reality, best described as spirit, that evolved through higher emergent stages both in its physiological and mental modes. The whole natural world, constituted by the diverse modes of spirit, gradually revealed, in Morgan's judgment, a progressive, rational, divine plan. "The world-plan," he lectured to his St. Andrews audience in 1923, "through and through, from its lowest to its highest expression, is manifestation of God; in you and me—in each of us severally—God as Spirit is partially revealed."[231]

George Romanes's evolutionary considerations absorbed the high emotion aroused by his religious and personal needs. Morgan, more philosophically detached, coolly extended the reach of evolution from the instincts of animals to the achievements of advanced civilization.

230. Edwin Boring describes the Edwardian insouciance that characterized the formation of British experimental psychology in his *A History of Experimental Psychology*, 2d ed. (New York: Appleton-Century-Crofts, 1950), pp. 488–95.

231. Morgan delivered the Gifford Lectures at St. Andrews in 1922–1923. They were published as *Emergent Evolution* (London: Williams & Norgate, 1923) and *Life, Mind, and Spirit* (London: Williams & Norgate, 1926). The quotation is from *Life, Mind, Spirit*, p. 32.

Despite their very different personal attitudes, virtually the same metaphysics underlay their evolutionary ideas. And at the end of their respective careers, they both found monism to support a conception of the Divine. They conceived evolutionary progress to mirror darkly God's purpose.

What Are the Metaphysical and Religious Implications of Darwinian Science?

Mandelbaum has isolated three defining proposals of nineteenth-century materialism: that only material objects exist; that God therefore does not exist; and that mind is merely a property of body. Mechanistic materialism would further insist that the activities of mind are fully explicable through the laws governing all matter. By these criteria neither Romanes's nor Morgan's science could be classified as materialistic. The philosophy that gave foundation and direction to their theories was pitted against materialism, and eventually emerged as a monism of spirit, of personality; it was a metaphysics that led to theism, not perhaps to religious orthodoxy, but certainly to a conception of God at work in the universe. In one sense, though, the definition of mechanism does describe their science: they both believed that the properties of mind could be explained by the same principles as those governing organic evolution. But divorced from materialism, this sort of mechanism, even if it could rightly be called that, was a completely different beast.

After the publication of such recent historical studies on the relation of science to religion as Frank Turner's *Between Science and Religion* and James Moore's *Post Darwinian Controversies,* it might be thought anomalous that any scholars would still cultivate the vintage belief that Darwinism and religious conviction were fundamentally opposed.[232] But that belief can be easily sustained if one is persuaded that Darwinism essentially incorporated a metaphysics of mechanistic materialism, which subsequent religious thinkers either ignored or distorted. Lewontin, Rose, and Kamin, as well as Himmelfarb and Cannon, seem to be of this opinion. John Greene perceives the Darwinian world as inclining toward mechanism, agnosticism, and positivism.[233] Garland Allen finds Darwinism pervaded by materialism in its several varieties

232. Frank Turner, *Between Religion and Science* (New Haven: Yale University Press, 1974); James Moore, *The Post-Darwinian Controversies: a Study of the Protestant Struggle to Come to Terms with Darwin in Great Britain and America, 1870–1900* (Cambridge: Cambridge University Press, 1979).

233. John Greene, "Darwinism as World View," in *Science, Ideology, and World View* (Berkeley: University of California Press, 1981).

(e.g., mechanistic, holistic, dialectical).[234] And Ernst Mayr suggests, with unintended irony, that Darwinism essentially opposes essentialism, as well as any element of divinely imposed teleology.[235] So the theologians whom Moore calls "Christian Darwinists" can be admitted, as long as we recognize that they have sugarcoated the unpalatable hard kernel of evolutionary theory.

The effort to distill the essential nature of the Darwinian revolution could be sanctioned by the historiographic theory of Imre Lakatos.[236] Lakatos maintains that research programs—such as the Darwinian—have a hard core of principles and concepts that remain immutable through the life of the program and that this hard core has definite logical consequences. It can occur, and Lakatos provides examples, that even the principal architect of a research program will not perceive all of its implications. Hence in the case of Darwinism, it could well be that some religiously minded thinkers simply failed to understand the message of evolutionary theory—that God died in 1859. The historians and cultural critics employing this Lakatosian model, however, will be sure to write the obituary.

The history examined in this and other chapters should, I hope, undermine both the historiographic assumption of essentialism and the particular belief that Darwinism implies materialism and atheism. Logically the Lakatosian presumption should collapse after any passing acquaintance with the history of science. After all, what thinkers shall we include as establishing the Darwinian research program? Greene selects Spencer, Darwin, Huxley, and Wallace—the earlier Wallace, of course. But why not Haeckel, Romanes, Morgan, James, Baldwin, or a host of others who certainly identified themselves as Darwinians, as did their contemporaries? Perhaps it is just Darwin who defines Darwinism. But the early Darwin, the middle Darwin, or the later Darwin? Darwin's thought about central matters also evolved through the different periods of his career. But even if we choose the later, agnostic Darwin as defining the program, that Darwin certainly did not eviscerate the universe of moral purpose—or so has been the argument in previous chapters.

Lakatos admonishes the historian of science to include in a description of a research program all that is really contained therein. This

234. Garland Allen, "The Several Faces of Darwin: Materialism in Nineteenth and Twentieth Century Evolutionary Theory," in *Evolution from Molecules to Men,* ed. D. Bendall (Cambridge: Cambridge University Press, 1983).

235. Ernst Mayr, "The Nature of the Darwinian Revolution," in *Evolution and the Diversity of Life* (Cambridge: Harvard University Press, 1976).

236. I compare Lakatos's model of scientific research programs with the evolutionary model that I advance in the first appendix to this volume.

would be a formidable task. The historian must discern the "real," that is, for Lakatos, the logical implications of a research program, despite the historical noise in the system. But the job is even more difficult. Since any set of premises—not to mention a large, inchoate network of concepts that constitutes most scientific programs—since any set of premises logically has an infinite number of conclusions, great forests of deductive tree systems would have to sprout from these roots. Lakatosian histories would give new meaning to the *longue durée*.

If, however, we adopt another historiographic model, the one I have been employing, then deriving the "real" implications of nineteenth-century evolutionary theory becomes less formidable. The model that I prefer, a natural selection model, suggests first that the thought of a scientist can indeed be captured over a span of time, though there is no guarantee that even its central elements will remain unchanged. We discover its implications not by hiring a Laplacean demon to do a logical calculation but by tracing out its descending branches, the ones that actually live in the historical record. Along one of these branches the historian can find, to be sure, Huxley's materialistic epiphenomenalism, but along an even more sizeable limb, the monism of Romanes, Morgan, James, and Baldwin (with intertwining branches holding the similar metaphysics of Haeckel and Spencer).

The biopsychologists Spencer, Romanes, Morgan, James, and Baldwin all sought to understand the evolution of mind and behavior. Monism became a necessary condition for their explanatory efforts. Since they focused precisely on those evolutionary problems whose solution required the right metaphysical turn, who is to say that their choice does not more adequately represent the implications of Darwinism?

Appendix: The Romanes-Wallace Debate on Physiological Selection

On 6 May 1886, Romanes read a long paper before the Linnean Society entitled "Physiological Selection: An Additional Suggestion on the Origin of Species." He introduced his hypothesis of physiological selection to account for three difficulties he perceived in Darwin's original theory: first, that the crosses of very similar species were inexplicably infertile, while those of very different domestic breeds of the same species were not; second, that swamping out of incipient differences ought to prevent divergence of species; and finally, that species displayed many apparently useless traits, which therefore could not be explained by natural selection. Romanes proposed that these difficulties could be removed were we to recognize that a kind of physiological selection preceded divergence of species. The principle of physiological selection, which he urged as a necessary auxiliary to natural selection, supposed that if alterations in the reproductive system of varieties would occur for whatever reason, such modifications would isolate the varietal from the parent type, with the consequence that infertility would initiate speciation, that other incipient differences would be preserved, and that accidentally useless traits would be perpetuated by heredity.

Alfred Wallace rose to challenge what he took as an attack on Darwinism by a previously faithful disciple. He defended the sufficiency of natural selection to account for evolution. He contended that most supposedly useless traits in fact had uses; bizarre coloration in birds, for instance, would be seen, on more careful consideration than Romanes offered, to be extremely useful as camouflage in natural environments. Romanes's presumption that incipient traits should be swamped out failed to recognize the empirical fact that traits within a local variety often varied simultaneously in the same direction; moreover natural selection would immediately preserve individuals having the same sort of useful traits, so that similar organisms would have greater opportunity of meeting and mating. Further, according to Wallace, numerous closely allied species showed themselves to be mutually fertile, so that speciation might frequently occur without sterility. In Wallace's view, Romanes had constructed his theory on a series of faulty assumptions; but even if they had been sound, the proposed principle, Wallace argued, could not work. For were the reproductive organs of some animals of a species to vary, rendering them infertile with the parent species but mutually fruitful, the odds against them meeting and mating would be so large as to constitute an impossibility.[237]

237. The dispute between Romanes and Wallace—and then their respective supporters—devoured many a pulp-producing tree during the later part of the 1880s. The controversy may be followed through these essays: George Romanes, "Physiological Selection: an Additional Suggestion on the Origin of Species," *Journal of the Linnean Society: Zoology* 19 (1886): 337–411; "Physiological Selection," *Nature* 34 (1886): 314–16, 336–40, 362–65; Alfred Wallace, "Romanes versus Darwin," *Fortnightly Review* 46 (1886): 300–316; Francis Darwin, "Physiological Selection and the Origin of Species," *Nature* 34 (1886): 407; George Romanes, "Physiological Selection and the Origin of Species," *Nature* (1886): 407–8, 439; "Mr. Wallace on Physiological Selection," *Nature* 35 (1887): 247–48, 390–91; "Physiological Selection," *Nineteenth Century* 21 (1887): 59–80; "Definition of the Theory of Natural Selection," *Nature* 38 (1888): 616–18; Alfred Wallace, "Dr. Romanes on Physiological Selection," *Nature* 43 (1890): 79, 150; George Romanes, "Mr. Wallace on Physiological Selection," *Nature* 43 (1890): 127–28, 197–98. Romanes had, as it were, the last word in his posthumous *Post-Darwinian Questions, Isolation and Physiological Selection*, vol. 3 of *Darwin and After Darwin*, ed. C. Lloyd Morgan (Chicago: Open Court, 1897).

9

The Personal Equation in Science: William James's Psychological and Moral Uses of Darwinian Theory

To be an intellectual in the mid-nineteenth century required that one suffer a severe spiritual crisis or mental breakdown. At least the lives of the more famous thinkers of the period suggest this. John Stuart Mill felt profound emotional emptiness when he realized that the measure of his happiness would not be increased even if his Benthamite reformist desires were satisfied. As he confided in his *Autobiography,* he "seemed to have nothing left to live for." [1] Charles Darwin's immobilizing digestive and cardiac problems began when he started work on his theory; and five years after the publication of the *Origin of Species,* his health and spirits reached their nadir. [2] Herbert Spencer, in the great effort to finish his *Principles of Psychology,* said his "nervous system finally gave way"—he languished for eighteen months. [3] Francis Galton, who at Cambridge failed to meet his father's expectations, complained of obsessive ideas, along with "intermittent pulse and a variety of brain symptoms of an alarming kind." [4] A student of Wilhelm Wundt described his teacher's tenure as Helmholtz's assistant as "seventeen years of depression." [5] And while a medical student traveling in Germany—and plagued by professional doubts, metaphysical insecurities, and women—William James fell into a depressive abyss, trailing thoughts of suicide after him.

The historian of nineteenth-century thought usually notes these in-

---

1. John Stuart Mill, *Autobiography of John Stuart Mill* (New York: New American Library, [1873] 1964), p. 107.

2. See chapter 5.

3. Herbert Spencer, *An Autobiography* (New York: D. Appleton, 1904), 1:543. See chapter 7.

4. Francis Galton, *Memories of My Life* (London: Metheun, 1909), pp. 79–79.

5. Edward Titchener, "Wilhelm Wundt" (1921), in *Wundt Studies,* ed. W. Bringmann and R. Tweney (Toronto: Hogrefe, 1980), p. 324.

stances of mental collapse as decorative episodes in the biographies of men whose outward pursuits paled in contrast to the romantic exploits of many of their contemporaries. Little effort has gone into assessing the impact of private crises on the philosophical and especially the scientific ideas of these men and others considered in this volume.[6] The presumption seems to be that generally little return would be paid to the historian who tried. After all, these philosophers and scientists did not produce great imaginative works fired in passion. Gerard Manley Hopkins's profound acedia and Franz Kafka's lingering illness and traumatic relations with his father rightly concern the literary historian, since these conditions help explain aspects of their poetry and prose. But the thinkers discussed in this volume were empiricist philosophers and tough-minded scientists; their emotional lives seem not to have altered the shape of their theories. The occasional attempts to explain, for instance, even Darwin's hesitation in publishing the *Origin of Species*—in the opinion of some psychoanalysts the book symbolized the killing of the old Adam and Darwin's real father[7]—have not been embraced by historians of science. But the case is manifestly different for William James. To understand his psychological science, his epistemological, metaphysical, and moral ideas, James's emotional life must be considered. So if the previous chapters have failed to demonstrate the importance of the psychology of scientists for assessing the evolution of their theories, I bring the case of William James.

This chapter, accordingly, will constitute a sustained argument against the proposition of one historian of James's thought, namely that "to provide a proper perspective for the study of James . . . attention must be diverted from his life, however interesting, to his published philosophy."[8] James himself, in his early article "Quelques considérations sur la méthode subjective," made subjective preference, even in the face of contrary objective evidence, reason to accept or reject a scientific hypothesis.[9] We are authorized, then, to recover James's subjective state in an effort to explain his adoption and use of certain scientific

6. A notable exception is Bruce Mazlish's *James and John Stuart Mill* (New York: Basic Books, 1975). This psychohistory convincingly shows ties between J. S. Mill's crisis and his conception of utilitarianism, though a fair amount of psychoanalytic speculation must be passed through to appreciate the connections.

7. See, for instance, Rankin Good, "Life of the Shawl," *The Lancet* (9 January 1954), pp. 106–107. See also Ralph Colp, Jr., *To Be an Invalid* (Chicago: University of Chicago Press, 1977), pp. 122–26.

8. William Earle, "William James," in *The Encyclopedia of Philosophy,* ed. Paul Edwards (New York: Macmillan, 1967), 4:241.

9. William James, "Quelques considérations sur la méthode subjective," *Critique philosophique* 2 (1878): 407–13.

Figure 9.1 William James, 1842–1910,
photograph from ca. 1873.

ideas. Indeed, his suicidal despair and the metaphysical remedy he chose
to stanch it help explain why he found Darwin's theory of evolution so
attractive.

Most James scholars acknowledge a connection between his pragmatism—also perhaps his psychological functionalism—and evolutionary ideas.[10] Some critics, though, regard the influence of evolutionary

10. See, for instance, John Wild, *The Radical Empiricism of William James* (New York:
Doubleday Anchor, 1970), pp. 16–17; Earle, "William James," pp. 241–49; and Andrew
Reck, "The Philosophical Psychology of William James," *Southern Journal of Philosophy* 9
(1971): 293–312. Philip Wiener, Bruce Kuklick, Marcus Ford, and Daniel Bjork perceive
but do not measure nearly the full force of evolutionary ideas on James's psychology. See
Philip Wiener, *Evolution and the Founders of Pragmatism* (Cambridge: Harvard University
Press, 1949), chap. 5; Bruce Kuklick, *The Rise of American Philosophy* (New Haven: Yale
University Press, 1977), pp. 51–52, 160–61, 170–71; Marcus Ford, *William James's Philoso-
phy* (Amherst: University of Massachusetts Press, 1982), pp. 26–29; and Daniel Bjork,
The Compromised Scientist (New York: Columbia University Press, 1983), pp. 7–9. Don
Browning, in his insightful *Pluralism and Personality: William James and Some Contempo-
rary Cultures of Psychology* (Lewisburg, Pa.: Bucknell University Press, 1980), pp. 52–58,

theory on James's science as negligible.[11] Both of these attitudes inhibit attempts at a deeper understanding of James's intellectual development. It was Darwin's theory that provided the essential structure and objective justification for James's scientific and philosophical conceptions about the nature of mind, the acquisition of knowledge, and the possibility of moral action. To comprehend fully James's achievement, then, demands that we follow the careful construction of his psychology against the framework of Darwinian theory. But we must view his use of that theory principally in light of his spiritual crisis.

James's Depressive Period, 1865–1878

James's educational and early professional pursuits were a continued *cursus interruptus*. Against the wishes of his father Henry James, Sr., he studied painting till he finally admitted that his talent was insufficient. He then entered the Lawrence Scientific School at Harvard University to take up chemistry, an endeavor the elder James hoped might enable the son to defend rationally and empirically the father's Swedenborgian religious beliefs. But William was only a fair chemist. He quickly perceived that a lack of both desire and mathematics recommended a switch to comparative anatomy, a choice congenial to the artist's interests, but also one with a more practical consequence—it opened the way to medical school. James enrolled in the Harvard Medical School in 1864, though not from any secure vision of himself in clinical practice. The following spring he left off studies to sail up the Amazon with a contingent of students led by Professor Louis Agassiz, the American Cuvier and fierce opponent of Darwinism. While in South America James contracted a mild form of smallpox and a more serious case of depression. In very low spirits, he wrote his father that the zoologist's collecting and classifying "work was not in my path" and that his excur-

---

insists on the importance of Darwinian theory in the development of James's thought, as does William Woodward, in his "Introduction to William James's Essays in Psychology," in *The Works of William James: Essays in Psychology,* ed. Frederick Burkhardt (Cambridge: Harvard University Press, 1983), pp. xx–xxiv.

11. Bruce Wilshire, in *William James and Phenomenology: A Study of "The Principles of Psychology"* (Bloomington: University of Indiana Press, 1968), pp. 50–51, even supposes that James rejected the "darwinizing" of human mind. Other scholars, by omission, suggest that Darwinian theory had no strong influence on James's thought. So, while Ralph Barton Perry does occasionally refer to James's Darwinian interests in his masterful *The Thought and Character of William James* (Boston: Little, Brown, 1935), he fails to mention evolution or Darwin in his distillation of James's views, *In the Spirit of William James* (New Haven: Yale University Press, 1938).

sion "was so much a waste of life."[12] The depression passed as his physical health returned. Back in Boston in March 1866, he again resumed the medical curriculum; but in spring of the next year, he interrupted his studies and set out for Germany.

James believed the travel to Germany necessary to preserve his health. It would also give him an opportunity to work up his German and perhaps to study some physiology. During the summer, while he was in Dresden and Bohemia, his health deteriorated; he suffered insomnia, digestive difficulties, headaches from reading, and serious back problems. Accompanying these physical distresses, a great pitch colored his moods. In November he began attending physiology lectures at Berlin, with the intention of continuing the following summer with "Helm holtz and a man named Wundt at Heidelberg."[13] But that winter his ill health and lack of scientific preparation produced a deep frustration, which he poured out to his friend Tom Ward: "my habits of mind have been so bad that I feel as if the greater part of the last ten years had been worse than wasted, and now have so little surplus of physical vigor as to shrink from trying to retrieve them. Too late! Too late! If I had been *drilled* further in mathematics, physics, chemistry, logic, and the history of metaphysics."[14] Two months later, in January 1868, James attempted to bolster Ward's slipping spirits with the confession of his own suicidal despair: "I fancy you have always given me credit for less sympathy with you and understanding of your feelings than I really have had. All last winter, for instance, when I was on the continual verge of suicide, it used to amuse me to hear you chaff my animal contentment."[15]

During his eighteen months in Germany, James's emotions ebbed and flowed. One episode of melancholy gave off mists of a faint hypothesis that eventually would become a more firm biopsychological theory. This occurred while he was listening to the piano playing of Katherine Havens, an American toward whom he felt a strong attraction. In his diary for 22 May 1868, he wrote:

> Tonight while listening to Miss H's magic playing & the Dr. and the Italian Lady sing my feelings came to a sort of crisis. The intuition of something here in a measure absolute gave me such an unspeakable disgust for the dead drifting of my own life for some time past. I can revive the feeling perhaps hereaf-

12. William James, *The Letters of William James,* 2nd ed., ed. Henry James (Boston: Little, Brown, Brown, 1926), 1:63.
13. Ibid., p. 119.
14. Ibid.
15. Ibid., p. 129.

ter by thinking of men of genius. It ought to have a practical effect on my own will—a horror of wasted life since life can be *such*—and Oh god! an end to the idle, idiotic sinking into *Vorstellung* disproportionate to the object. Every good experience ought to be interpreted in practice. Perhaps actually we cannot always trace the effect, but we won't lose if we try to drop all in wh. this is not possible. Keep [one word illegible] all the while—and work at present with a mystical belief in the reality interpreted somehow of humanity.[16]

Five days later, on 27 May, James again brooded on unrealized representations and unrequited love: "About 'Vorstellungen disproportionate to the object' or in other words ideas disproportionate to any practical application—such for instance are emotions of a loving kind indulged in where one cannot expect to gain exclusive possession."[17]

James's experience with Miss Havens was repeated with many women whom he knew before finally marrying Alice Gibbens in 1878. In his letters, he would sing the delights of a pretty woman, but he could not bring himself to reduce his *Vorstellungen* to action.[18] In the long passage just quoted from his diary, James resolved not to dwell on conceptions that he could not act upon. Later he made this resolve a biopsychological principle. He came to argue that the function of cognition—its evolutionary purpose—was to produce action, to allow the will to be effective in the world. "Cognition, in short, is," as he explained in a later article, "incomplete until discharged in act."[19]

James's Mental Collapse

James returned from Germany in November 1868 and took up residence again in his father's house. In June of the next year he received his medical degree, but he harbored no intention of practicing medicine. He continued to be depressed. In December 1869, James wrote to Henry Bowditch, a friend from medical school, to excuse a lapse in correspondence: "I have been prey to such disgust for life during the last 3 months as to make letter writing almost an impossibility."[20] Shortly thereafter, on 1 February 1870, he seems to have completely

16. William James, *Diary* MS p. 55, James Papers, Houghton Library, Harvard University.

17. Ibid., MS pp. 56–57.

18. See, for instance, *Letters of William James* 1: 93–94, 113, 116; and Perry, *Thought and Character of William James* 1: 240.

19. William James, "Rationality, Activity and Faith," *Princeton Review* 2 (1882): 66.

20. William James to Henry Bowditch (29 December 1869), James Papers, Houghton Library, Harvard University.

broken down and for a time underwent treatment in the McLean Asylum near Boston.[21] He suffered the breakdown, as was characteristic of the James family, with pen in hand.

> Feb. 1. A great dorsal collapse about the 10th or 12 of last month has lasted with slight interruption until now, carrying with it a moral one. Today, I about touched bottom, and perceived plainly that I must face the choice with open eyes: shall I frankly throw the moral business overboard, as one unsuited to my innate aptitude, or shall I follow it, and it alone, making everything else merely stuff for it?—I will give the latter alternative a fair trial. Who knows but the moral interest may become developed. Hitherto I have given it no real trial, and have deceived myself about my relation to it, using it in reality only to patch out the gaps which fate left in my other kinds of activity, and confusing everything together.[22]

By moral interest James appears to have meant exercising the will in pursuit of definite goals. In his diary, he associated it with "attaining certain difficult but salutary habits."[23]

James's spiritual crisis had three major components—professional, interpersonal, and psycho-metaphysical.[24] Its gradual remission during

21. James Anderson followed up certain rumors about James's commitment to the McLean Asylum to discover that the hospital refused to confirm or deny his stay there. See James Anderson, "William James's Depressive Period (1867–1872) and the Origins of his Creativity: A Psychobiographical Study" (Ph.D. diss., University of Chicago, 1979). I also contacted McLean and met with a similar response. Afterward, however, I spoke with someone who had worked in the hospital archives in an official capacity, and she confirmed James's stay as a patient at McLean.

22. James, *Diary,* entry for 1 February 1870.

23. Ibid.

24. Cushing Strout argues that James's professional insecurities betrayed a more fundamental crisis of identity, which led to his deep despondency in these early years. See Cushing Strout, "William James and the Twice-Born Sick Soul," *Daedalus* 97 (1968): 1062–82. Gay Allen, in his biography of James, suggests that James's emotional distress resulted primarily from taking too seriously, as the Victorians were wont, the question of free will. See Gay Allen, *William James* (New York: Viking Compass, 1969), pp. 164–70. James Anderson's thesis, in his "William James's Depressive Period," is that any extrinsic cause of James's illness can only be understood as fragmenting an already fragile self-system. Howard Feinstein and Daniel Bjork both maintain that the root of James's trouble lay in his father's demand that he give up painting for science. James's emotional problems thus stemmed, they suppose, from an unresolved Oedipus complex. See Howard Feinstein's sensitive and highly insightful biography of James's early years, *Becoming William James* (Ithaca, N.Y.: Cornell University Press, 1984), pp. 117–45; and Bjork, *The Compromised Scientist,* pp. 15–36. My own interpretation fishes up a somewhat different tangle of causes.

Figure 9.2   William James, self-portrait done ca. 1873.

the decade of the 1870s required specific but related therapies. James had despaired over his professional prospects, feeling that his education had left him unprepared for serious scientific work and that he had wasted his years in desultory study. In 1872, however, President Eliot of Harvard inquired of his neighbor whether he would be interested in filling a vacancy in the physiology department. Thus began James's teaching career at Harvard, a career that carried him from an appointment in physiology to one in psychology and finally to a professorship in philosophy, from which he retired in 1907.

During his early adulthood, James appears not to have been able to bring himself to translate his intentions concerning women into action. His diary portrays the morbid condition to which this brought him. But in 1876 he was introduced to Alice Howe Gibbens, who pulled him back from his frustrations and bachelorhood. A day after the encounter, he wrote to his brother Wilkie that he had met "the future Mrs. W. J."[25] James believed that Alice had resurrected his interred soul. He wrote to her in June of 1877: "Last fall and last winter what pangs of joy it sometimes gave me to let you go! to feel that acquiescing in your unstained, unharnessed freedom I was also asserting my deepest self, and cooperating with the whole generous life of things.!"[26] Alice supplied James that emotional elixir which sparked his sense of self-possibility and infused him with a zest for life. Indeed, Schwehn has forcefully argued that Alice redeemed James, for in her he found a living embodiment of that religion which he had earlier rejected when his father preached it in dreamy Swedenborgian periods.[27]

The third dimension of James's spiritual crisis was psycho-metaphysical. During the period of the 1860s, he enthusiastically embraced Herbert Spencer's scientific philosophy, which proclaimed an eternally predestined evolution of matter and mind out of primal stuff.[28] And even as Spencerian doctrine began to evaporate under James's scrutiny, its accompanying determinism yet crusted over his convictions. The attitude that modern science revealed an inexorability that even mind could not escape was also urged on James by his study of German physiology, especially that of Du Bois-Reymond, whose lectures he attended in Berlin. The skeptical positivism of his friends Chauncey Wright and Oliver Wendell Holmes undoubtedly also exerted a strong force. The precise origins of James's belief in determinism may be a bit uncertain, but his own persuasion was not, as he recounted to Ward in 1869: "I'm swamped in an empirical philosophy. I feel that we are nature through and through, that we are wholly conditioned, that not a wiggle of our will happens save as the result of physical laws; and yet, notwithstanding, we are *en rapport* with reason.—How to conceive it? Who knows?"[29] James's own lack of purpose, his inability to reduce *Vorstellungen* into action, his "palsied" will (as he described in it his diary)[30] could all be understood and even justified if mind were a puppet to nature's laws. Then "the task," as he concluded in late 1869, would be

25. *Letters of William James* 1: 192.
26. William James to Alice Gibbens (June 1877), James Papers, Houghton Library, Harvard University.
27. Mark Schwehn, "Making the World: William James and the Life of the Mind," *Harvard Library Bulletin* 30 (1982): 426–54.
28. William James, "Herbert Spencer," *Atlantic Monthly* 94 (1904): 99–108.
29. *Letters of William James* 1: 152–53.
30. James, *Diary*, entry prior to 21 December 1869.

"to act without hope."[31] But the task was beyond him, and in February of the following year he "about touched bottom."[32]

Renouvier and the Subjective Method

A few months after sinking into the depths, James chanced to read a book by the French Kantian Charles Renouvier, his *Traité de psychologie rationnelle*.[33] It gave James his first great lift out of despair. On 30 April 1870, he recorded the decisive experience in his diary:

> I think that yesterday was a crisis in my life. I finished the first part of Renouvier's 2nd Essay, and saw no reason why his defi-nition of free will—the sustaining of a thought *because I choose to* when I might have other thoughts—need be the definition of an illusion. At any rate I will assume for the present—until next year—that it is no illusion. My first act of free will shall be to believe in free will. For the remainder of the year, I will abstain from the mere speculation & contemplative *Grübelei* in which my nature takes most delight, and voluntarily cultivate the feeling of moral freedom, by reading books favorable to it, as well as by acting . . . . Hitherto, when I have felt like taking a free initiative, like daring to act originally, without carefully waiting for contemplation of the external world to determine all for me, suicide seemed the most manly form to put my dar-ing into; Now, I will go a step further with my will, not only act with it, but believe as well; believe in my individual reality and creative power. My belief to be sure can't be optimis-tic—but I will posit life (the real, the good) in the self govern-ing *resistance* of the ego to the world.[34]

In the chapters of the *Traité* that captured James's attention (chapters 13 and 14), Renouvier analyzed two opposing doctrines of will, that of determinism and that of the liberty of indifference. He found both un-acceptable. Determinism implied that authentic moral behavior, which assumed the agent could have done otherwise, was a delusion. Deter-minism thus undermined our primitive experience that men did make

31. Ibid.
32. Ibid., entry for 1 February 1870.
33. Charles Renouvier, *Essais de critique générale, Deuxième essai: Traité de psychologie rationnelle,* 2d ed. (Paris: Librarie Armand Colin, [1875] 1912). James read the original edition of 1864 during his emotional crisis. In his later essay "Bain and Renouvier" (*Na-tion* 22 [1876]: 367–69), he used the second edition of the *Psychologie rationnelle.*
34. James, *Diary,* entry for 30 April 1870.

valid moral judgments. Although Renouvier recognized that the determinist would regard this as a weak objection, one based on an illusion, he thought that the determinist could not so easily dismiss two further consequences of his doctrine. First, if all men were determined, then so would be their philosophical assertions: each of their decisions, including the acceptance of determinism, had to result from coercive causal processes. In practice, then, truth and falsity would have to merge in a system that permitted no judgments freely executed for good reasons. Moreover the determinist had to face the antinomy of his position, that an actually infinite series of causes existed, a series requiring a beginning but having none. According to Renouvier, those advocating the liberty of indifference stood no more securely. They endowed man with a pure will, indifferent to and uninfluenced by motives, intellectual convictions, or passions. Their theory of freedom, however, would actually deny that men—those bundles of hopes, fears, and fluctuating beliefs—could be assigned responsibility for their acts. Freedom in this sense became identical with chance.

Renouvier's own theory accorded man a will enmeshed in the thicket of judgment and motives, a will that did not simply react to pressing needs, passions, and desires, but one that antecedently reflected on plans leading to alternative motives for behavior. Will actively selected interests as well as responding to them.[35] In this conception, liberty would be "that character of human acts, reflective and voluntary, in which consciousness joins in close union the motive and the drive identified with it, and affirms that other acts different from the first are possible at the same moment."[36] From this footing, James launched his own theory of human liberty. He would come to define the free agent as one who chose what interest to pursue. Much in the world potentially beckons the individual, but he must decide which interest to cultivate, which to reject. He actively examines plans of action in order to evoke new motives within himself. And when he deliberately acts, he does so, as James expressed it in his diary, with a belief in freedom, with the conviction that he could have done otherwise.[37] In this respect, James came to regard mind neither as a mere passive recorder of events nor as a simple calculator of efficient means. Mind was preeminently a "fighter for ends."[38]

35. Renouvier, *Traité de Psychologie rationnelle* 1: 318–19.
36. Ibid., p. 317.
37. James, *Diary,* entry for 30 April 1870.
38. William James, *The Principles of Psychology* (New York: Henry Holt, 1890), 1: 141.

But some might still insist on the doctrine of causal necessity, claiming with John Stuart Mill that induction proved it. To them Renouvier responded that the inductive method actually assumed causal regularity in order to demonstrate it: induction would fail were nature not constant and our perceptions not stable and reliable.[39] But if an ultimate postulation of a first principle must occur without demonstration—as it must to avoid circular reasoning—then why not the principle that forms an inevitable part of our conception of human acts, one that makes sense of the epistemology of truth and error and that makes moral behavior meaningful—the principle of liberty? This principle, in Renouvier's Kantian consideration, had the unyielding support of our practical nature: men instinctively predicated freedom of their own acts and those of others. Renouvier believed this thoroughly human attitude brought a moral certitude about freedom. And moral certitude, he argued, was the only kind available to man.[40] For anyone who pretended his certitude was logical, based on rational principles, could always be asked whether he was certain of his certitude. Since there was no fixed point (*aliquid inconcussum*) that certified itself logically, the only recourse would be to some more ultimate guarantee of certitude. But in that direction lay the devouring chasm of infinite regress. We must begin with moral certitude, which for Renouvier was a certitude about human freedom.[41]

James adopted the subjective method of Renouvier in his first major essays, published in the late 1870s. He insisted that we were right to confirm a theory which met with our natural preferences, because all ultimate foundations for philosophical or scientific theories rested, not on ineluctable reason—since first principles could not be demonstrated—but on belief and conviction. If our taste ran to determinism, we had to recognize that such a choice nullified significant aspects of apparently valid experience. But if we decided for freedom, then we at least insured that moral action would not wither, but could be vigorously tested. In a review of Bain and Renouvier for the *Nation* in 1876, James wrote: "If this be a moral world, there are cases in which any indecision about its being so must be death to the soul. Now, if our choice is predetermined, there is an end of the matter; whether predetermined to the truth of fatality or the delusion of liberty, is all one for us. But if our choice is truly free, then the only possible way of getting at that truth is by the exercise of the freedom which it implies."[42]

39. Renouvier, *Traité de Psychologie rationnelle* 1 : 321–22.
40. Ibid., p. 328.
41. Ibid. 2 : 97–98.
42. James, "Bain and Renouvier," p. 369.

The philosophical cure Renouvier offered seemed to have had a strong and cumulative effect on James's emotional life. His spirits waxed after the spring of 1870, though they again slid during the next year. But in late 1872 he seems to have dosed himself with the right strength of Renouvier. He wrote on 2 November to introduce himself to Renouvier and express his gratitude to his healer: "Thanks to you I possess for the first time an intelligible and reasonable conception of freedom. I accept it almost entirely. On other points of your philosophy, I still have doubts, but I can say that through that philosophy I am beginning to experience a rebirth in the moral life; and I assure you, Monsieur, that it is no small thing."[43] Gradually, then, James's spirits lifted. His father, noticing the difference in him, asked about it, and then relayed the news to his other son, Henry, in March 1873:

> He came in here the other afternoon when I was sitting alone, and after walking the floor in an animated way for a moment, exclaimed "Dear me! What a difference there is between me now and me last spring this time: then so hypochondriacal" (he used that word, though perhaps in substantive form) "and now feeling my mind so cleared up and restored to sanity. It is the difference between life and death." He had a great effusion. I was afraid of interfering with it, or possibly checking it, but I ventured to ask what specially in his opinion had promoted the change. He said several things: the reading of Renouvier (specially his vindication of the freedom of the will) and Wordsworth, whom he has been feeding upon now for a good while; but especially his having given up the notion that all mental disorder required to have a physical basis. This had become perfectly untrue to him. He saw that the mind did act irrespectively of material coercion, and could be dealt with therefore at first-hand, and this was health to his bones.[44]

43. The letter to Renouvier can be found in Perry, *Thought and Character of William James* 1:661–62. Perry was the first to argue that reading Renouvier's philosophy resuscitated James's emotional life. While Renouvier cannot be credited as James's lone savior, the letter to the Frenchman, as well as the next letter quoted in the text (from Henry James, Sr., to Henry, Jr.), testifies to the powerful effect Renouvier had. This evidence can be contrasted with the conclusion that Feinstein, his biographer and a psychoanalyst, has reached. Feinstein asserts that reading Renouvier neither dramatically changed James's philosophical convictions, nor had any real impact on his health. But a psychoanalyst by professional disposition must assume that a patient's intellectualization of his problems disguises more profound Oedipal difficulties—an assumption the historian of science need not make. See Feinstein, *Becoming William James*, pp. 311–12.

44. Henry James, Sr., to Henry James, Jr. (March 1873), quoted in Perry, *Thought and Character of William James* 1:339–40.

The elder James's letter suggests another important aspect of the cure Renouvier wrought, though not, I think, without the help of Darwin. The doctrine of freedom meant that mind was not identical with brain, nor its slave. Hence, any mental or emotional disturbances, any signs of insanity, need not be attributed to an incurable organic disorder. Direct spiritual therapy could be effective. And James administered this to himself with strong doses of Renouvier, plus, it would seem, an anodyne that J. S. Mill also used, the poetry of Wordsworth.

James's protracted spiritual crisis climaxed in the early 1870s, and probably really subsided only after his marriage. It was virtually on his honeymoon that he produced his first important scientific and philosophic papers. The emotional consolation of a wife and the security of a teaching position helped considerably. But the remedy of Renouvier, taken alone, was not potent enough for a lasting cure. The French Kantian demonstrated that the determinist position was not more logically persuasive than the libertarian; yet he failed to counter the full strength of Victorian science, which seemed to support determinism. James required objective evidence to compound with his subjective preference for freedom. This he found, oddly enough, in the ideas of one usually credited with introducing a pervasive mechanism in biology—Charles Darwin. But to understand exactly what Darwin offered and how James adapted it, we must first consider his intellectual relation to that other great nineteenth-century evolutionist, Herbert Spencer.

The Psychological and Moral Uses of Darwinism

Spencerian Evolutionism

Spencer, like Darwin, believed that the various extant species had descended from simpler, more primitive forms over long periods of time. Also like Darwin, he initially formulated his theory with ideas drawn from Lamarck, especially the French zoologist's notion that acquired habits could produce heritable adaptations in animals and men. Darwin, however, after he had worked out the principle of natural selection in summer and early fall of 1838, gradually reduced the role of Lamarckian mechanisms, though without denying their existence (see chapters 2 and 3). Spencer, on the other hand, continued to recognize an essentially Lamarckian device as the chief engine for species alteration: habits acquired in response to environmentally produced needs, he believed, would eventually transform simpler organisms into more complex ones by molding their structures against external environmental relations (see chapter 6). Spencer offered the simple organic adaptation of the cuttlefish's sucker as illustrative:

The established relation between the tactile and muscular changes in the sucker and its ganglion is parallel to the uniform relation between resistance and extension in its environment—the inner cohesion of psychical states is as persistent as is the outer relation between attributes. And if we remember that in the actions of the cuttle-fish this inner relation is perpetually being repeated in response to the outer relation, we see how the organization of its species answers to the infinitude of experiences received by the species.[45]

According to Spencer's theory, when an organism's equilibration is upset by a change in the environment, the creature naturally seeks to reestablish balance by altering its behavior. Such alterations themselves disrupt anatomical relations, which in turn move toward a new equilibration at a higher level of complexity. These individually acquired adaptations, Spencer argued, would be inherited by subsequent generations of a species.[46]

In *Principles of Psychology,* Spencer proposed that the "law of growth of intelligence," a special formulation of the principle of adaptive equilibration, directed the evolution of consciousness in the animal kingdom.[47] The law operated initially to establish reflex connections between perceptions and adaptive responses in lower organisms. The first glimmerings of consciousness, in Spencer's view, mirrored these primitive nervous linkages. More complex neural organization carried in tandem correspondingly higher states of consciousness—those associated with instinct, memory, and reason. On this interpretation, conscious reasoning became merely the feeling of deliberation over different courses of action; the neural realities behind the sentiment of reason were competing nascent motor reactions to a complex environment. Thus he explained the apperance of rationality: "As the groups of antagonistic tendencies aroused will scarcely ever be exactly balanced, the strongest group will at length pass into action; and as this sequence will usually be the one that has recurred oftenest in experience, the action will, on the average of cases, be the one best adapted to the circumstances. But an action thus produced is nothing else than a rational action."[48]

Spencer's theory of evolution declared that the chief mechanism

45. Herbert Spencer, *The Principles of Psychology,* 2d ed. (London: Williams & Norgate, 1872), 1:428.
46. Ibid., pp. 407–26.
47. Ibid., p. 455.
48. Ibid., p. 462.

of evolution was the internalizing of external relations and that this mechanism progressively drove anatomical forms and conjoint mental structures from more generalized adaptive states to more definite correspondences with the environment, from simpler, more homogeneous patterns to more complex and heterogeneous configurations. Or, as James liked to rephrase Spencer's theory for his students: "Evolution is a change from a no-howish untalkaboutable all-alikeness to a somehowish and in general talkaboutable not-all-alikeness by continuous stick-togetherations and somethingelseifications."[49]

Darwin Pitted against Spencer

James unfailingly yielded to the allure of Spencer's philosophical science, though the temper of his considerations changed dramatically from his adolescence to early manhood. He recalled reading Spencer's *First Principles* "as a youth when it was still appearing in numbers," which would have been when he was between eighteen and twenty years old. At that time, as he later confessed, he was "carried away with enthusiasm by the intellectual perspectives which it seemed to open."[50] His ardor for Spencer's evolutionism was, however, dashed in the cooler reflections of his friend Charles Sanders Peirce, who surgically exposed what he considered to be Spencer's vagueness, vacuity, and pretension. Thereafter, James never lost a fascination for—almost sadistic pleasure in—dismembering Spencer's speculations at every opportunity.

James's formal introduction to Darwinian theory probably came in the comparative anatomy and physiology courses of Professor Jeffries Wyman, a defender of evolution against Agassiz's Cuvierian criticisms. For two academic years (1863–1865), James studied with Wyman, whom he came to regard as a paragon of "quiet wisdom."[51] His mentor's dispassionate discussions, though, must have seemed thin beer in comparison to the wranglings of his friends Chauncey Wright and Charles Sanders Peirce over the philosophical implications of Darwinism. For Wright, the *Origin of Species* testified to the scientific power of British

---

49. Quoted in Perry, *Thought and Character of William James* 1:482. James Moore has identified this very Jamesian parody as actually borrowed from Thomas Kirkmann's *Philosophy without Assumptions* (London: Longmans, Green, 1876), p. 292. See James Moore, *The Post-Darwinian Controversies* (Cambridge: Cambridge University Press, 1979), p. 375 n. 39.

50. James, "Herbert Spencer," p. 104.

51. *Letters of William James* 1:48.

empiricism; for the Kantian Peirce, it provoked a search after the logical flaws in the theory.[52] James reflected the attitudes of both friends in some of his earliest pieces, two reviews, written in 1868 while in Germany, of Darwin's *Variation of Animals and Plants under Domestication*.[53] James commended Darwin's "painstaking and conscientious industry in the accumulation of fact," but recognized that the British naturalist's interpretation "has just so much of the hypothetical element in it, in all the cases, that a sceptic who should refuse to accept it would have no trouble in presenting a legal and logical justification for his conduct." Yet James thought the value of Darwin's hypotheses could not really be settled by logic, but by "the learned tact of experts, which alone is able to weigh delicate facts against each other, and to decide how many possibilities make a probability, and how many small probabilities make an almost certainty."[54]

James's still fluid ideas about evolutionary theory seeped into the courses he offered as a lecturer in the anatomy department at Harvard. In January 1873, his professional career was inaugurated with "Natural History 3: Comparative Anatomy and Physiology," Wyman's old course and one James taught until he went over to the philosophy department in 1880. A student who took the class remembered that James, unable to keep to the more pedestrian aspects of his subject, "launched out, on almost any occasion, into a lecture which took shape gradually in a course on evolution."[55] His enthusiasm seems to have breached standards of scientific restraint (at least at Harvard), since another student complained that in the course "Darwinism is to be treated metaphysically, that is to say . . . precisely as Darwin and his followers say it should not be treated."[56] In 1876, James legitimated his concern with the deeper aspects of evolutionary theory, particularly its implications for mind and behavior, by introducing a new course into the anatomy department. This was "Natural History 2: Physiological Psychology," for which Spencer's *Principles of Psychology* served as the textbook. James's surviving lecture notes and the marginal annotations in his copy

52. Wiener discusses Wright's and Peirce's views of Darwinian theory in *Evolution and the Founders of Pragmatism,* pp. 31–96.

53. James's reviews of Darwin's *Variation of Animals and Plants Under Domestication* appeared in *The North American Review* 107 (1868): 362–68; and in *Atlantic Monthly* 22 (1868): 122–24. Each review emphasized different aspects of Darwin's work.

54. Ibid.

55. "A member of the Class of 1878," *Harvard Graduates' Magazine* 39 (1920): 324; quoted in Perry, *Thought and Character of William James* 1:469.

56. Ibid., p. 476.

of Spencer indicate the scope of his objections to the evolutionary psychologist and his growing reliance on Darwin. Most of his criticisms of
Spencer focused on the philosopher's view of the mind as passive and
fixed by natural forces. It was this sort of conception that had plagued
James during his spiritual crisis and that now could be scientifically
assuaged only by a very different sort of evolutionary hypothesis, the
Darwinian.

In a class lecture entitled "Spencer's Law of Intelligence," James set
out to refashion Spencer's idea that mind was passively molded against
the external environment:[57] "There might be in the mind," he cautioned, "Principles quite as natural as those of the outer world which
nevertheless alter the shape taken by the outer facts in thought."[58] One
need not, therefore, be forced to choose between Spencer and the cate-
chism. There was another way, a decidedly scientific way, of construing
the relationship between mind and the environment—the Darwinian way.

James contended that Spencer "repeats the defects of Darwin's predecessors in biology."[59] That is, the pre-Darwinians supposed anatomical adaptations to be direct responses to environmental relations,
whereas Darwin showed them to have two different sources: spontaneous variations, which did not mirror their causes; and a selection by
external circumstances of fit variations, which if retained would indicate
a kind of correspondence with the environment. The main point of a
Darwinian analysis, James insisted, was that "the variation or inner relation does not 'correspond' with its *cause* but with some environing
relation entirely removed from its cause. This outward relation has a
perfectly definite function: to take the variation once made and preserve
or destroy it."[60]

In applying the Darwinian perspective to the mental realm, James did
not deny that immediate experience often shaped ideas. He simply
could not swallow Spencer whole, as he explained to his friend Charles
Eliot Norton at this time: "My quarrel with Spencer is not that he
makes much of the environment, but that he makes *nothing* of the glar-
ing and patent fact of subjective interests which cooperate with the environment in molding intelligence. These interests form a true spontaneity and justify the refusal of a priori schools to admit that mind was

57. William James, "Spencer's Law of Intelligence," James Papers, 4493, Houghton
Library, Harvard University.
58. Ibid., MS pp. 2–3.
59. Ibid., MS p. 5.
60. Ibid., MS p. 8.

pure, passive receptivity."[61] James argued that categories of thought acquired over our long evolutionary history and the novel ideas that are produced by men of genius—and ourselves on occasion—were not due to direct adaptations, to immediate environmental coercion. He proposed, instead, that new modes of thought and conceptual innovations sprang up in the mind as spontaneous mental variations, and that we would come to accept them as representations of the environment only if they continued to meet the test of survival. James of course recognized that natural selection theory was usually interpreted as deterministic. Darwin himself had tried to specify the causes of "spontaneous" variations in the *Variation of Animals and Plants,* though, as James suggested in his reviews, not without appeal to unconfirmed hypotheses. In the lecture to his class, James intimated that our inability to give adequate scientific account of these causes might possibly be a result of mental variations erupting freely. At least the strength of natural selection theory did not depend on specifying the causes of variation.

Darwin's conception of the principles of evolutionary change thus appeared to James not only theoretically powerful but scientifically cautious as well; for Darwin segregated the mechanism of selection from the causes of variation and discussed these latter in the appropriate language of hypothesis. By contrast, Spencer characteristically overreached himself. In *Principles of Psychology,* he presumptuously declared psychological science had demonstrated that, through inherited habit and acquired associations, the motives of individuals were fully determined. In the margins of his copy of Spencer's text, James disciplined the errant psychologist:

> Nonsense yourself! Psychology don't pretend to be a quantitative science. Free will is solely a question of <the> *quantity* <of> in motives. *The* motives are as to their possible *kinds* always determined. Out of several possible liberty chooses one. But psychology is the science of the possibles, of the *classes* of representations. To claim more wd be to make it knowledge not of general laws but of all the particular details of all future history.[62]

61. William James to Charles Eliot Norton (22 November 1878), James Papers, Houghton Library, Harvard University. I am indebted to Deborah Coon for calling this letter to my attention.

62. The annotation is on p. 503 of Spencer's *Principles of Psychology,* vol. 1. The words in wedge quotes were crossed out. James probably penned this in preparing for his "Spencer elective," Natural History 2: Physiological Psychology. He began teaching the course in the academic year 1876–1877.

Spencerean psychology erred not in supposing that experiences fixed the range of possible motives operating on the will but in presuming that introspection could discriminate the fine weightings which tipped the balance in favor of one sort of motive over another. The Darwinian approach allowed James to elaborate a psychology that, while recognizing the role of ancestral experience (conveyed by natural selection) in establishing categories of motivational response in the individual, also permitted mental variations spontaneously to invest interest in one motive out of the several entertained. Natural selection, for James, thus worked on two levels: the phylogenetic, in fitting the species out with modes of response to different environments; and the ontogenetic, in spontaneously electing one motive over the others available. This two-level analysis of willful behavior is only adumbrated in James's class notes, but becomes more distinct in his later writings. At this point (in the late 1870s), it is clear, however, that his psychology sanctioned general laws but not a playground for a Laplacean demon of will.[63]

In the class lecture from which I have quoted, James brought his Darwinian scheme to bear on the historical propositions of Spencer's disciple Grant Allen. Allen had undertaken the Laplacean task. In a series of articles in 1878, he attempted to formulate a deductive history of nations.[64] He argued that "every national character must necessarily be due to the special physical characteristics of the country in which it is developed."[65] The external environment must alter the hereditary traits of people and, during a long evolutionary incubation, mold the distinctive personalities that national groups exhibit. Allen's historical determinism meant that, as he expressly avowed, "there is no caprice, no spontaneous impulse in human endeavors. Even taste and inclinations *must* themselves be the result of surrounding causes."[66] This was a hard saying for James, who with virtually every stroke of his pen proclaimed and displayed the spontaneous, the unexpected varieties of active thought.

Against this abstract and for James morally distasteful evolutionary tale, he urged what he believed, in an Emersonian vein, to be the evidence of history: that social evolution was due to the work of great

63. Darwin himself, of course, would not have approved use of his theory to support the idea of free will. He was fully persuaded that human mental behavior was completely determined. See chapter 2.

64. See Grant Allen, "Hellas and Civilization," *Popular Science Monthly—Supplement* 13–20 (1878): 398–406; and "Nation-Making," *Popular Science Monthly—Supplement* 13–20 (1878): 121–27.

65. Ibid., p. 121.

66. Ibid., p. 126.

thinkers and magnetic leaders in a society. "We can note with our eyes," he lectured to his class, "the way in which the great man works and to abandon this solid foundation for the emptiness of an unknown ultimate cause is in the highest degree unscientific."[67] Here again the Darwinian idea served. The great man could be understood as a spontaneous variation in the social organism. Unknown physiological causes conspire to fashion his brain; the dice roll and up comes a Napoleon, a Goethe, or a Bismarck. If the social and physical environments are receptive, if they select and preserve the great man and his ideas, then society will become adapted to a new mode of existence. If the genius and his schemes are out of time, if the environment proves hostile, well, undoubtedly there have passed many a rejected and inglorious Newton.

When in his lecture and the article based on it ("Great Men, Great Thoughts, and the Environment"),[68] James pitted his version of Darwinian social evolutionism against Spencerian historical determinism, he filled a need to reconcile two ideals of human development that warred in his breast. The one stemmed from his father, the other from an intimate of the James dining table, Ralph Waldo Emerson. Through the mists of the elder James's several religious writings—W. D. Howells said of *The Secret of Swedenborg* that Henry James kept it—divine creation was depicted as continuous and progressive. First God becomes other in the guise of the natural man, whose self-consciousness atheistically rejects the idea that it has been created; and then He returns to Himself by overcoming individual selfishness in the redemptive unfolding of an altruistic society. Henry James's spiritual evolutionism bore resemblance to the deterministic evolutionism of the Spencerians; as his son remarked, "according to both doctrines, man's mortality and religion, his consciousness of self and his moral conscience, are natural products like everything else we see."[69] Moreover both views submerged individual self in an inevitable process. William strenuously objected to the absolutism and determinism of these conceptions. He seems rather to have thought Emerson touched the right cord in praising "the sovereignty of the living individual."[70] Emerson and James (whose article "Great Men" resonates with the Concord sage's essay "Uses of Great Men") both perceived men of genius as shaping their

67. James, "Spencer's Law of Intelligence," MS p. 12.
68. William James, "Great Men, Great Thoughts, and the Environment," *Atlantic Monthly* 46 (1880): 441–59.
69. William James, Introduction to *The Literary Remains of the Late Henry James* (Boston: Osgood, 1885), p. 20.
70. William James, "Address at the Emerson Centenary in Concord (1903)," in F. O. Mathiessen, *The James Family: A Group Biography* (New York: Vintage, 1980), p. 457.

own destinies and redirecting the course of their societies. James's Darwinian scheme thus allowed him to preserve his father's idea of the spiritual development of society, but to regard this evolution as the chancy affair of morally independent selves.

The Darwinian Argument for the Independence of Mind

For James, 1878 was an annus mirabilis. In June he signed a contract with Henry Holt to write a textbook in psychology—though he delivered the manuscript of his two-volume *Principles of Psychology* in 1890, ten years over schedule. In July, after a frustrating year of courtship, with tears, departures, and nobility in suffocating abundance, he married Alice Gibbens. In autumn he began the second year of the "Spencer elective," his course in physiological psychology. During the year he saw published his first three major articles, and the while worked at high speed on several others. In February he received an invitation from D. C. Gilman, president of the recently founded Johns Hopkins University, to give a series of lectures. Gilman, who was very impressed with the performance, hoped to lure James into the psychology department at Hopkins; the invitation succeeded in getting James an advance in rank to assistant professor at Harvard. In his ten lectures, which the financial worries of marriage caused him to repeat at Boston's Lowell Institute that autumn, he explored the relations between mind and brain. It was in these lectures that he elaborated an extremely powerful evolutionary argument, one which would objectively and firmly ground his subjective desire to postulate an active and independent mind.

In the first five of his Hopkins lectures, James sought to demonstrate his mastery of brain physiology.[71] He recounted the latest experiments from Germany and wove them into a coherent pattern. In the sixth lecture, he finally confronted the question of the relationship between mind and brain, taking as his point of departure Huxley's essay "On the Hypothesis that Animals are Automata."[72] Huxley advanced an extreme form of the passive view of mind—epiphenomenalism. The brain, he claimed, received stimulation from the environment and issued motor acts as a result; the engine of the central nervous system simply transformed one kind of energy into another, without consciousness playing any mediating role at all. Rather, conscious mind hovered over brain activity like mists of steam coughed up from the dynamo actually doing

71. William James, Ten Lectures, untitled, James Papers, 4397 and 4469, Houghton Library, Harvard University. Hereafter referred to as "Hopkins Lectures."

72. Thomas Huxley, "On the Hypothesis that Animals Are Automata and Its History," *Fortnightly Review* 22 (1874): 555–89.

the work. James, in his subsequent Hopkins lectures, indicted this materialistic theory with an argument as powerful as it was elegant.

James sketched the argument in his lectures,[73] but rendered it more explicitly in the article "Are We Automata?"[74] Its full and compelling form is in chapter 5 of his *Principles of Psychology*.[75] The argument has an a priori part and an a posteriori part. The former can be reconstructed by the following sorites: Consciousness is a manifest trait of higher organisms, most perspicuously of man; like all such traits it must have evolved; yet it could have evolved only if it were naturally selected; but if naturally selected, it must have a use; and if it has a use, then it cannot be causally inert. Mind therefore must be more than an excretion of brain; it must be (at least in some respects) an independently effective process that is able to control some central nervous system activity. Here then, as James put it in his lecture, was "objective evidence" for our "aesthetic demands."

In the remaining lectures, and more fully in his article and in *Principles of Psychology*, James laid out the second part of his argument, the empirical evidence for the actual effectiveness of consciousness in the natural economy. The first sort of evidence came from his analysis of cerebral physiology. Animals higher in the evolutionary scale have hemispheres adapted for response to minute features of their complex environments. The delicately balanced cortex of these animals has, in James's terms, a "hair-trigger"; the slightest jar or accident could set it firing erratically. Its organization makes it "happy-go-lucky," "hit-or-miss." "Caprice is its law," he claimed. Yet, "if consciousness can load the dice, can exert a constant pressure in the right direction, can feel what nerve processes are leading to the goal, can reinforce & strengthen these & at the same time inhibit those which threaten to lead astray, why, consciousness will be of invaluable service."[76] Consciousness, in short, could serve to stabilize the machinery of the brain.

The most important function of consciousness, though, was that it established goals and selected interests. Man and higher organisms, in James's judgment, clearly revealed purpose in their behavior; they became fascinated by certain interests—from seeking food to seeking beauty—to the exclusion of others. This could not result from a passive accommodation to the occurrent environment, since goals and ideals were precisely those things beckoning from the future, and interests often transcended the commonplace and the present time. In James's

73. James, "Hopkins Lectures," MS pp. 56–57.
74. William James, "Are We Automata?" *Mind* 4 (1879): 3–4.
75. James, *Principles of Psychology* 1:138.
76. James, "Hopkins Lectures," MS pp. 60–61.

view, goals, ideals, and interests could be understood only as spontaneous mental variations that, in the life of higher creatures, had been
selected to steer them through their natural and social terrain. "Consciousness," in James's pugnacious metaphor, "is a fighter for ends."[77]

In his lectures James itemized other instances of the potential effectiveness of consciousness, all ultimately based on its presumed capacity
for selecting interests and focusing attention. So, out of the "swarming
continuum, devoid of distinction or emphasis" that nature presented to
experience, consciousness might, by attending to this motion and ignoring that, carve out a coherent world of related objects. Exercising
this same sort of selective ability, consciousness might also facilitate
some nerve processes and inhibit others, thus allowing a person to act
freely and with moral responsibility.

James also mentioned three other kinds of evidence for the utility of
consciousness that have an appeal independent of his particular psychological assumptions. First, if animals consciously intended to preserve
themselves and used an evolved reasoning power to do so, then the time
required for evolution might be shortened. This possibility, James calculated, would mitigate the potent objection that geologic time was
insufficient for evolution to have occurred.[78] Second, the empirical connection between subjective feelings of pain and objective injury on the
one hand, and between feelings of pleasure and life-enhancing activities
on the other, could only be explained if evolution had rendered subjective states effective in adapting animals to their environments. Finally,
dramatic instances of brain-damaged people slowly recovering intellectual functioning seemed to bespeak an autonomous agent regaining
control over its instrument—for machines cannot repair themselves.[79]

James thus buttressed his a priori argument with weighty empirical
evidence and reached the unavoidable conclusion that mind must exist,
in some of its activities at least, as an effective reality independent of
brain. The Darwinian analysis vindicates the strongest urgings of our
inmost selves—and James's preferred moral conception of the universe.[80]

77. James, *Principles of Psychology* 1 : 141.

78. In fact, there is good evidence for the more rapid speciation of organisms with
higher nervous centers. See, for instance, Leigh Van Valen, "Two Modes of Evolution,"
Nature 257 (1974): 298–300; and A. Wilson, G. Bush, S. Case, and M. King, "Social
Structuring of Mammalian Populations and Rate of Chromosomal Evolution," *Proceedings of the National Academy of Science* 62 (1975): 5061–65.

79. The great neurophysiologist Wilder Penfield made basically this same argument in
his *Mystery of the Mind* (Princeton: Princeton University Press, 1974), pp. 67–72.

80. In their otherwise insightful article, "William James and Gordon Allport: Parallels

James's Discovery of the Darwinian Argument for
Independent Mind

I believe that James hit upon this evolutionary argument sometime in late 1872 or early 1873, and that it helped heal his emotional sickness. James's spirits had noticeably picked up by March of 1873, when his father wrote Henry, Jr., about William's improvement. In the letter, quoted above, the elder James claimed that William's mood was elevated not only through reading Renouvier (and Wordsworth), but also because he had become convinced that "mind acted irrespectively of material coercion." The Darwinian argument was James's most powerful demonstration of that autonomy.

Other evidence also suggests that the discovery was made in the early 1870s. In his review of Wundt's *Grundzüge der physiologischen Psychologie*, published in the July 1875 issue of the *North American Review*, James, objecting to the epiphenomenalism of Shadworth Hodgson and William Clifford, succinctly countered:

> Taking a purely naturalistic view of the matter, it seems reasonable to suppose that, unless consciousness served some useful purpose, it would not have been superadded to life. Assuming hypothetically that this is so, there results an important problem for psycho-physicists to find out, namely, how consciousness helps an animal, how much complication of machinery may be saved in the nervous centres, for instance, if consciousness accompany their action. . . . In a word, is consciousness an economical *substitute* for mechanism?[81]

The evolutionary argument for the independence of mind is clearly sketched here; but James probably had it even earlier, in 1874, when he penned in his copy of Wundt's *Grundzüge*, next to a section on the origin of self-consciousness, the following remark: "This 'Bewusstsein' seems then with him to mean the element of spontaneity as distinguished from receptivity."[82] James's contrast of spontaneity with receptivity smacks of his evolutionary hypothesis.

---

in their Maturing Conceptions of Self and Personality," in R. Rieber and K. Salzinger, eds., *Psychology: Theoretical-Historical Perspectives* (New York: Academic Press, 1980), pp. 57–70, Richard High and William Woodward propose that James, "educated as a scientist in an era shaken by Darwin's evolutionary theory, feared the moral implications of an entirely mechanistic universe" (p. 60). It is, of course, my argument that Darwin's evolutionary theory freed James from such fears.

81. William James, "Review of *Grundzüge der physiologischen Psychologie* by Wilhelm Wundt," *North American Review* 121 (1875): 201.

82. William James's annotation is on p. 463 of vol. 2 of Wilhelm Wundt's *Grundzüge der physiologischen Psychologie* (Leipzig: Engelmann, 1874).

In his review of Wundt, James referred to an earlier article published in the *North American Review* of April 1873, his friend Chauncey Wright's "Evolution of Self-Consciousness." In this article Wright argued that Darwinian evolution was opportunistic. He showed how already established structures in animals, mental as well as anatomical, might be put to new and unexpected uses. Evolution therefore need not produce a continuous gradation of functions. "The truth is, on the contrary," Wright explained, "that new uses of old powers arise discontinuously both in the bodily and mental natures of the animal, and in its individual developments, as well as in the development of its race, according to the theory of evolution, although, at their rise, these uses are small and of the smallest importance to life."[83] Wright was a thoroughgoing positivist and scientific reductionist. In his article, he contended that human self-consciousness was not a supernatural endowment but had evolved quite naturally and accidentally out of ordinary animal awareness. Yet the passage just quoted could be read somewhat differently, as indicating that specifically mental activities were accidentally introduced into their environments, but that they might have had adaptive uses there, and that because of their causal efficacy, they might have continued to evolve. This, of course, is the nut of James's argument. Since Wright and James were close friends at this time, they may have discussed the article before it was published. If so, this would perhaps place the time of James's discovery of the argument prior to his father's letter, that is, sometime between late 1872 and early 1873.

A more direct source for James's argument, however, might have been Darwin's *Descent of Man*. In April 1871, soon after the book's appearance, James wrote from Cambridge to Bowditch, then in Germany, about the stir it was causing: "Darwin's new book appears to make a good deal of noise in the papers. I suppose it makes hardly less in Germany notwithstanding the war. I have not yet seen it. I had the pleasure of hearing Agassiz call it 'rubbish' the other day."[84] Probably, James read the *Descent* a bit after this, about the time his spirits revived; his first printed reference to the work, though, came in 1878.[85] In the book, Darwin clearly argued for the effectiveness of "the intellectual and moral faculties of man." "These faculties," Darwin wrote, "are variable; and we have every reason to believe that the variations tend to be inher-

83. Chauncey Wright, "Evolution of Self-Consciousness," *North American Review* 116 (1873): 246.
84. William James to Henry Bowditch (8 April 1871), James Papers, Houghton Library, Harvard University.
85. William James, "Brute and Human Intellect," *Journal of Speculative Philosophy* 12 (1878): 260

ited. Therefore, if they were formerly of high importance to primeval man and to his ape-like progenitors, they would have been perfected or advanced through natural selection."[86] Darwin, rather like Huxley, considered mental faculties to be completely determined by brain patterns. But James could not have known this, since Darwin expressed this interpretation only in his private notebooks.[87] What tells against James's forthright derivation of his argument for the causal independence of mind from a passage such as this is simply his failure to identify the source, and James usually gave credit where he thought it was due. His reading of the *Descent,* of course, along with conversations with Wright, might have caused the argument to bob to the surface of his thoughts, without his being reflectively aware of its origins

The best source for recovery of James's ideas during late 1872 and early 1873 would be his diary. Unfortunately, entries between April 1870 and February 1873 are missing. The evidence, then, that James formulated his argument before 1874, when he was preparing his review of Wundt's *Grundzüge,* must remain circumstantial.

My reason for trying to fix an early date for James's discovery of the Darwinian argument for mental autonomy is simply to sink a strong anchor for the thesis of this chapter: that Darwinian evolutionary theory played a fundamental role in James's mental development, emotional as well as intellectual, and thus that the historian cannot ignore the possible play of personality factors in the development of science. I think the several parts of this thesis are sufficiently sustained even without fixing an early date for James's discovery, but I would feel better about it if James had indeed seized upon Darwinism immediately as an instrument of his own emotional therapy.

The Reach of James's Darwinian Psychology

The argument of James's Hopkins lectures and the ensuing article formed a central nerve radiating out into the many other papers he composed during 1878 and the years following. Finally, after some ten years, he began systematically to incorporate most of these essays into the chapters and sections of his *Principles of Psychology,* which Holt with great relief published in 1890. The article "Are We Automata?" became chapter 5 of the book, while the Darwinian analyses of the selective capacities of consciousness and its spontaneously generated interests provided the framework for those several other chapters and related

86. Charles Darwin, *The Descent of Man and Selection in Relation to Sex* (London: Murray, 1871), 1 : 159.

87. See chapter 2.

essays that developed James's characteristic theses in epistemology, metaphysics, religion, psychology of self, and ethics. Let me indicate the ways in which James employed Darwinian ideas for solving central problems in these areas. I will first dwell a bit on his evolutionary epistemology, since that functions as the principal control over conceptions in the other areas; then I will more briefly follow the ramifications of his Darwinian considerations through the body of his biopsychology.

Evolutionary Epistemology and Its Foundations

In Jamesian epistemology, the evolutionary perspective explains how fundamentally new ideas might be introduced both to an individual consciousness, which preserves those that accord with its already-established interests, and to a larger society, which will select or reject them. Basic conceptions in science, for instance, are not, as Spencer thought, forced on the thinker by stern nature. The booming buzzing confusion of unorganized experience could hardly reveal the aesthetically structured world of, say, Newtonian mechanics. Rather, spontaneous conceptions put the hodgepodge of experience in order, emphasizing some aspects and ignoring others. The final test of such ideas is their ability in the long run to survive the rigors of experience.[88]

This summary description of James's evolutionary epistemology leaves obscure his understanding of the relation between inherited and individually acquired patterns of thought. This is one of the hazier regions of his Darwinian psychology. Nonetheless a path can be made through the several considerations that bear on the question. The idea that mental structures evolved in the animal kingdom was completely bound up in James's original view with Spencer's notion that fixed mental habits, which were molded directly on the physical environment, could be transmitted to progeny. James initially wedged this idea in as a distinction between animal and human mind. In animals, he thought, fixed habit was the rule: "The brain grows to the exact modes in which it has been exercised, and the inheritance of these modes—then called instincts—would have in it nothing surprising." But with human beings, the case was different: "in man the negation of all fixed modes is the essential characteristic.[89] James urged, in his 1878 essay "Brute and

---

88. James, *Principles of Psychology* 2 : 633–40.

89. James, "Brute and Human Intellect," p. 275. James left this passage unchanged when he incorporated "Brute and Human Intellect" into chapter 22 of *Principles of Psychology* (2 : 368). James's belief in the extreme flexibility of the human mind dimmed as he worked through his emerging theories of emotion, will, and instinct—or so I argue below.

Human Intellect," that men, unlike animals, might easily crack patterns of thought and rearrange the pieces anew. The human mind, in short, was chancy and spontaneous, the animal mind fixed and predetermined.

During the 1880s, however, James began gradually to uncover more native veins in the human mind. These harder streaks appeared as he worked through the theories of will, emotion, and instinct that would give ballast to the loftier speculations of *Principles of Psychology*. In his essay on psychology of will, "The Feeling of Effort" (1880), which became the basis of chapter 26 of *Principles of Psychology*, he advanced his "ideo-motor" hypothesis, the pieces of which he pulled from Lotze, Carpenter, Renouvier, Darwin, and his own early experiences in Germany.[90] The hypothesis proposed that "every representation of a motion awakens the actual motion which is its object, unless inhibited by some antagonistic representation simultaneously present to the mind."[91] Simply concentrating on a plan of action releases the action itself. The human mind hesitates, James supposed, only when several courses of action vie for attention.

This conception of willful behavior supported James's new theory of emotion, announced in 1884 in his article "What is an Emotion?"[92] Here he portrayed emotions as feelings of bodily response. We perceive certain objects or events (e.g., the crazed killer stalking toward us); our body reflexively and instinctively reacts (e.g., knees buckle, face grows pale, hands shake); and we feel this physical reaction (e.g., we are afraid). James's theory, independently arrived at by the Danish physiologist Carl Lange, asserts, then, that cognition does not cause emotion, but rather that emotion is a direct response to the instinctive wisdom of the body. Emotions are a stage in the release of innate instinctive reactions.[93] The only difference between the two is that true

90. William James, "The Feeling of Effort," *Anniversary Memoirs of the Boston Society of Natural History* (Boston: Boston Society of Natural History, 1880).

91. Ibid., p. 17. For a lucid discussion of the origins of the "ideo-motor" hypothesis in W. B. Carpenter's physiological theory, see Kurt Danziger, "Mid-Nineteenth Century British Psycho-Physiology," in *The Problematic Science: Psychology in Nineteenth-Century Thought,* ed. William Woodward and Mitchell Ash (New York: Praeger, 1982), pp. 119–46.

92. William James, "What is an Emotion?" *Mind* 9 (1884): 188–205. This article became the basis for chapter 25 of the *Principles*. Gerald Myers skillfully discusses James's theory in "William James' Theory of Emotion," *Transactions of the Charles S. Peirce Society* 2 (1969): 67–89.

93. James, *Principles of Psychology* 2:442. Browning, in his *Pluralism and Personality,* p. 163, argues that James disconnected emotions from meaning by interpreting them as direct responses to the body. But Browning's objection supposes that bodily reactions themselves are meaningless, which James's evolutionary theory of instinct denies.

instincts not only evoke emotions but also spill over into overt and more elaborate behavior. Though experience subsequently alters innate patterns, instincts and concomitant emotions nonetheless form the original base for all human behavior. Indeed, man's emotional and instinctive endowment functions as the motor for willed behavior: on any occasion incipient instincts, colored by emotion, clamor for attention; and that which consciousness exclusively selects becomes automatically executed. Thus James's conception of will and emotion led him to find in human mind a native legacy. The character of that deposit was spelled out in his theory of instinct.

The third step in James's advance toward a more nativistic conception of human mind was reached in a series of papers he published on instinct in 1887.[94] He derived his ideas about instinct largely from sources that harmonized with his theories of will and emotion. The *Thierische Wille* (1880) and *Menschliche Wille* (1882) of Wundt's student G. H. Schneider, *Mental Evolution in Animals* (1883) by Darwin's own disciple George Romanes, and the articles of Spencer's advocate Douglas Spalding, as well as the work of their mentors, provided James the conceptual foundations and many examples for his own theory of instinct.[95] He defined instinct, in accord with this literature, as "the faculty of acting in such a way as to produce certain ends, without foresight of the ends, and without previous education in the performance."[96] Essentially, for James and for those whom he read, instincts were complexes of reflex actions released by appropriate environmental stimuli. Since they were initially executed without directive experience, instinctive behaviors could only be fixed innately in the organism. The above-mentioned writers, particularly Schneider, also helped him to see that man did not differ from brutes by reason of fewer instincts. On the contrary, man, according to James, "is more richly endowed in this respect than any other mammal."[97] James counted over thirty classes of human instinct, a list large enough to choke the behaviorist John Watson.[98] Only

94. William James, "What Is an Instinct?" *Scribner's Magazine* 1 (1887): 355–65; and "Some Human Instincts," *Popular Science Monthly* 21 (1887): 160–70, 666–81.

95. See Georg Schneider, *Der thierische Wille* (Leipzig: Abel, 1880) and *Der menschliche Wille vom Standpunkte der neueren Entwicklungstheorien* (Berlin: Dummlers, 1882); George Romanes, *Mental Evolution in Animals* (London: Kegan Paul, Trench, 1883); and Douglas Spalding, "Instinct and Acquisition," *Nature* 12 (1873): 507–508; "Instinct, with Original Observations on Young Animals," *Macmillan's Magazine* 27 (1873): 282–93; and "On Instinct," *Nature* 6 (1872): 485–86.

96. James, "What Is an Instinct?" p. 355.

97. James, "Some Human Instincts," p. 666.

98. In his *Behaviorism*, rev. ed. (Chicago: University of Chicago Press, 1930), John Watson listed the many instincts James regarded as the endowment of man. He then

the jumble of different instincts (e.g., anger, pugnacity, sympathy, and love, when a child has dropped his jelly sandwich on your open volume of Kant's *First Critique*) shrouds the inbred grains of human decision. In such situations, the strongest instinct will usually out. But, in James's view, reason has its role: it can neutralize a particular instinct by mobilizing another (so the hand ready to thrash is stayed by one's attending to the angelic smile of an offspring). And "thus, though the animal richest in reason might be also the animal richest in instinctive impulses too, he would never seem the fatal automaton which a *merely* instinctive animal would be."[99]

Guided by Schneider, Romanes, Spalding, and Darwin, James reluctantly granted that some instincts probably resulted from inherited habits. He thought, though, that in man natural selection produced complex instincts and the higher mental faculties. A short time before the publication of his *Principles of Psychology*, however, he read of August Weismann's experiments, which demonstrated that succeeding generations of mice with clipped tails failed to bear progeny with any shortening of their hind members.[100] James gratefully concluded, in the last few pages of his book, that Darwinian chance variation and selection could be the only agents of evolutionary change in human and animal behavior.

James's elaboration of theories of will, emotion, and instinct thus altered his perspective on the inheritance of mental categories in man. In the last chapter of *Principles of Psychology*, he summarized his conclusions and carefully specified those mental traits he believed were products of evolution. Certainly sensations of color, taste, sound, pleasure, and pain were evolved mental responses. Representations of space and time relations were also heritable; he agreed with Spencer (before reading Weismann) that they were forged through habitual exposure to real space and time connections in nature. But the two fundamental abilities that made rational thought possible—the discernment of differences and the ability to hold in consciousness a series of objects—these could not be the result of impressed experience, as a Spencerian approach would re-

---

responded: "The behaviorist finds himself wholly unable to agree with James and the other psychologists who claim that man has unlearned activities of these complicated kinds. We have all been brought up on James or possibly even on a worse diet, and it is hard to run counter to him" (p. 110).

99. James, "What Is an Instinct?" p. 360.

100. See August Weismann, "On Heredity" (1883) and "The Supposed Transmission of Mutilations" (1888), in his *Essays upon Heredity,* 2d ed., ed. and trans. E. Poulton et al. (Oxford: Clarendon Press, [1889] 1891). James cites this translation in *Principles of Psychology* 2:686.

quire. Rather they were necessary for coherent experience in the first place. These faculties must have spontaneously flashed in our ancestors, permitting them to survive and prosper. With such preformed equipment, then, along with evolved emotional capacities and instinctive responses, the child could initially organize experience and render it intelligible. Higher mental categories—causality, logical principles, necessary truths of mathematics, and ideal relations of aesthetics and morality—derived naturally from empirically acquired concepts that had been structured through our inherited cognitive framework.

James's epistemology thus rested on Darwinian principles. The mind comes already outfitted with fixed sensory and emotional responses, instinctive reactions, and basic rational abilities; these constitute our evolutionary legacy. But the acquisition of new ideas is also Darwinian; spontaneous hypotheses, guesses, and notions erupt in our pedestrian and scientific encounters with the world; those that survive the pitiless force of reality live for another day. But this Darwinian epistemology, in both its phylogenetic and ontogenetic phases, implied a peculiar sort of metaphysics. Indeed, James's epistemology led to a notion of reality as provisional and in the making—a conception that congenially accorded with his ultimate religious and moral purposes.

Metaphysics and Religion

The spontaneous and selective aspects of consciousness explain our postulation of a world of natural objects in the first place. What we decide is real, James maintained, will depend on the nature of the intellectual environment. Our interests, beliefs, and conceptual framework select mental variations. The objective worlds of common sense and of science ultimately have a subjective foundation: our ego, our self selects what will be real for us. *"The fons et origo of all reality, whether from the absolute or the practical point of view, is,"* James proclaimed, *"thus subjective."* "Reality, starting from our Ego," he continued, "thus sheds itself from point to point—first, upon all objects which have an immediate sting of interest for our Ego in them, and next, upon objects most continuously related with these. It only fails when the connecting thread is lost."[101] The objects we regard as most real are those that touch our senses.[102] Our ancestors who chanced to test their conceptions against sensations were able to avoid the saber-toothed tiger lurk-

---

101. Ibid., pp. 296–97.
102. Ibid., p. 306, 311–12. Richard High considers the character and sources of James's perceptual realism in "Shadworth Hodgson and William James's Formulation of Space Perception," *Journal of History of the Behavioral Sciences* 17 (1981): 466–85.

ing in their paths; and we have inherited their mental penchant. As for those protomen who fancied other criteria for real objects, their lines have not prospered.

In chapter 21 of *Principles of Psychology,* James described the various orders of reality that men were apt to postulate: the manifestly natural world primarily, but also the world of scientific objects (of atoms, molecules, and the rest), of abstract truths, of madness, and of prejudice and superstition. We rank these orders from most real to least real by their relation to our sensory, emotional, and active lives. "In this sense," according to James, "what ever excites and stimulates our interest is real." [103] But if this is the scale for measuring reality, then for James the world of religion was decidedly real.

During his young manhood, James felt the sting of religious demands from several quarters. His father unleashed upon him baroque missiles of Swedenborgianism, from which he defended himself with logical objection and retaliatory skepticism.[104] Yet the father's feeling for religion, his profound conviction that an ideal reality hovered just above the horizons of human life, that there would be a "final evolution *of human nature itself* into permanent harmony with God's spiritual perfection" [105]—these sentiments penetrated the son's defenses. James became a champion himself not of orthodox theology nor even of heterodox faith but of the warm and energizing feeling of divinity. He marshalled this hopeful energy against the scientific materialism of his friends Wright and Holmes. In his wife he found an ally. Alice enshrined for James the sacred presence that his father declaimed, and he constantly turned to her to stoke his own religious fervor. As he got older, James developed a growing sense of a divinity beyond theological formulation.[106] It was this feeling he sought to vindicate empirically in his Gifford Lectures, *The Varieties of Religious Experience* (delivered in 1901 and 1902). Before his Edinburgh audience he stacked the hard data of religion at work in the lives of exceptional men against the evidence upon which the sciences rested. He found religion to have as much

103. James, *Principles of Psychology* 2:295.

104. See the exchange of letters between father and son on the topic of the elder James's religious beliefs in Perry, *Thought and Character of William James* 2:705–16.

105. Henry James, "Spiritual Creation," in *The Literary Remains of the late Henry James,* p. 457.

106. James had a minimalist definition of religion. Religion meant essentially the belief that "the so-called order of nature, which constitutes this world's experience, is only one portion of the total universe, and that there stretches beyond this visible world an unseen world of which we now know nothing positive, but in its relation to which the true significance of our present mundane life consists." See William James, "Is Life Worth Living?" in his *Will to Believe* (New York: Dover, [1897] 1956), p. 51.

empirical grounding as science. Religious belief makes a difference in the lives of men, it alters their attitudes, changes their behavior, and thus does work in the world. The unseen region toward which religious sentiment arches has a claim on reality. James preached to his Scots congregation:

> Yet the unseen region in question is not merely ideal, for it produces effects in this world. When we commune with it, work is actually done upon our finite personality, for we are turned into new men, and consequences in the way of conduct follow in the natural world upon our regenerative change. But that which produces effects within another reality must be termed a reality itself, so I feel as if we had no philosophic excuse for calling the unseen or mystical world unreal.[107]

In his earlier essays and in *Principles of Psychology,* James transfused the blood that gave life to his Gifford Lectures. The richest part flowed from Darwin.

In the essays and *Principles of Psychology,* James utilized Darwinian theory much as Romanes and Morgan had, to provide scientific support for religion. He did so principally in three ways.[108] First, he proposed that religious faith could be regarded as a spontaneously generated set of beliefs not yet rejected by the winnowing hand of reality. "We know so little about the ultimate nature of things, or of ourselves," he warned, "that it would be sheer folly dogmatically to say that an ideal rational order may not be real. The only objective criterion of reality is coerciveness, in the long run, over thought."[109] Religious belief might yet be confirmed, as James liked to put it, *ambulando,* in the great by and by. Second, the reality that religion postulated might require our continued acquiescence in order to evolve. Divinity itself might well be, as his father held, an organic reality fed on our constant belief in it.[110] Lastly, though James regarded the spiritual world as something ultimately verifiable only when we passed beyond, he harbored no doubt about the effective reality of religious belief itself. Religion as purely a belief system gave natural advantage to those who possessed it. He thus cautioned that a nation which succeeded in suffocating the religious senti-

107. William James *Varieties of Religious Experience,* ed. Frederick Burkhardt et al. (Cambridge: Harvard University Press, [1902] 1985), p. 406.

108. In his essay "William James and the Culture of Inquiry," *Michigan Quarterly Review* 20 (1981): 264–83, David Hollinger describes the problematic character and the implications of the concept of science in James's later thought.

109. William James, "Remarks on Spencer's Definition of Mind as Correspondence," *The Journal of Speculative Philosophy* 12 (1878): 17.

110. James, "Is Life Worth Living?" p. 61.

Figure 9.3   William James and Josiah Royce, photograph from 1903.

ments of its citizens, replacing venerable beliefs, for instance, with the sterile doctrines of empirical science, "that nation, that race, will just as surely go to ruin, and fall prey to their more richly constituted neighbors, as the beasts of the field, as a whole, have fallen prey to man."[111]

By the time of the Gifford Lectures, James had moved from regarding the unseen world as temporarily preserved from natural elimination to believing it had been positively selected by natural experience. Through the aid of Darwinian theory, then, James met his filial obligation to find scientific evidence for his father's deeply felt conviction that "religion is real."[112]

111. William James, "Reflex Action and Theism" (1881), in *The Will to Believe*, p. 132.
112. On 9 January 1883, after learning of his father's death, James wrote his brother

Psychology of the Self

James's evolutionary approach also sanctioned his psychology of the self. In the famous chapter of *Principles of Psychology* "Consciousness of Self," a chapter ripe with descriptions that have appealed to many phenomenologists,[113] James distinguished the several selves with which we become identified on different occasions: the material self, our clothes and possessions, but particularly our body; the social self, our image reflected in the attitudes of others; the spiritual self, those inmost memories, feelings, and beliefs we associate with our deepest being; and the transcendental ego, a self usually found only in the souls of philosophers. In James's view, the foundation for our recognizing all of these selves as us is ultimately an emotion, a feeling of familiarity, a self-love. We appropriate our body—rather than that of the person next to us—and say "Here, this is me," because we are comfortable with it; it elicits our love and protection. Moreover, we love our selves in our friends, because we like what we see of us in them. Finally, we love our spiritual dispositions, our perishable powers, our passions and hates, our willingnesses and sensibilities. These three selves, according to James, "must be the supremely interesting objects of each human mind."[114]

Henry from England of his intention to give expression to their father's conviction: "I must now make amends for my rather hard non-receptivity of his doctrines as he urged them so absolutely during his life, by trying to get a little more public justice done them now. As life closes, all a man has done seems like one cry or sentence. Father's cry was the single one that religion is real. The thing is so to 'voice' it that others shall hear,—no easy task, but a worthy one, which in some shape I shall attempt" (Perry, *Thought and Character of William James* 2:165).

113. The interest that John Wild and Bruce Wilshire (see notes 10 and 11) have in James rests precisely in those aspects of his thought most closely resembling the conceptions of Husserl and other phenomenologists. See also the remarks of Browning in *Pluralism and Personality*, pp. 64–86, and Herbert Spiegelberg in his *The Phenomenological Movement*, 2d ed. (The Hague: Nijhoff, 1965), 1:111–17.

114. James, *Principles of Psychology* 1:323. James's depiction of the several selves with which we identify owed a good deal to the social metaphysics of his brother Henry. One might compare the psychologist's discussion of the material and social selves with the novelist's, as in this passage from *The Portrait of a Lady* (New York: New American Library, [1881] 1963), p. 186: "we see that every human being has his shell, and that you must take the shell into account. By the shell I mean the whole envelope of circumstances. There is no such thing as an isolated man or woman; we are each of us made up of a cluster of appurtenances. What do you call one's self? Where does it begin? Where does it end? It overflows into everything that belongs to us—and then it flows back again. I know that a large part of myself is in the dresses I choose to wear. I have a great respect for *things!* One's self—for other people—is one's expression of one's self; and one's house,

With this last assertion James moved from phenomenological description to scientific explanation. Every human mind must resonate to these interests, because they have been naturally selected.

> All minds must have come, by the way of the survival of the fittest, if by no directer path, to take an intense interest in the bodies to which they are yoked. . . . And similarly with the images of their person in the minds of others. I should not be extant now had I not become sensitive to looks of approval or disapproval on the faces among which my life is cast. . . . Were my mental life dependent exclusively on some other person's welfare, . . . then natural selection would unquestionably have brought it about that I should be as sensitive to the social vicissitudes of that other person as I now am to my own. . . . My spiritual powers, again, must interest me more than those of other people, and for the same reason. I should not be here at all unless I had cultivated them and kept them from decay. And the same law which made me once care for them makes me care for them still.[115]

James regarded these basic interests—in the material, social, and spiritual selves—as the inherited products of a long evolutionary history. These fundamental interests were instinctive; they were given a variety of expression in the multitude of instincts that human beings inherited, and which James so lovingly and (to those not recognizing the structure of his evolutionary psychology) indulgently recounted in the long chapter on instinct in *Principles of Psychology*.

The Moral Will

In several early essays and in various chapters of *Principles of Psychology*, especially the chapter "Will," James poured out the scientific foundations for moral choice.[116] His construction was in two parts: first, an account of the evolutionary source of moral interests, and second, an explanation of the way the will operated so as to make free choice possible. In tackling the first part, James's strategy was, of course, to argue that moral interests, like other ideal standards, were spontaneous mental

---

one's clothes, the books one reads, the company one keeps—these things are all expressive."

115. James, *Principles of Psychology* 1:324.

116. For an excellent study of theories of will in late nineteenth-century psychology, see Lorraine Daston, "The Theory of Will *versus* the Science of Mind," in *The Problematic Science: Psychology in Nineteenth-Century Thought*, pp. 88–115.

variations that had been selected. Like Darwin, he conceived these interests, in his early essays and lectures, as instincts having survival value, and thus selectively perpetuated in intelligent organisms.[117] But also like Darwin, he encountered the basic objection that altruistic behavior, promoted by moral inclination, often appeared not to benefit the individual exercising it. James's solution to this conundrum, sketched in his essay "Spencer's Definition of Mind" (1878), seems to have been formulated with only vague understanding of Darwin's discussion of altruism in *Descent of Man*.[118] Early tribal communities, James argued, had an interest in promoting the hero, the martyr, the gallant warrior, since, as he macabrely put it, "it is death to you, but fun for us."[119] That is, individual altruistic action would advance the welfare of the whole community, which then would become the selecting force for those organisms manifesting such behavior. Unlike Darwin, who proposed that altruistic behavior resulted from the selection of the whole tribe in competition with other tribes not having altruistic individuals, James conceived such behavior as a consequence of selfish individuals being selectively eliminated by their own communities. This explanation, of course, presupposes an already existing community of moral individuals. James offered no account of the evolutionary origins of their initial altruistic behavior. Nor, I think, could he have. He really did not fully understand something that Darwin saw straight through to the bottom, that to explain altruistic behavior the unit of selection cannot be the individual, since moral acts usually offer him no advantage. The unit of selection must be the whole tribe or community. James did appreciate, however, that the altruistic instinct in man would be delicately balanced against the instinct for self-preservation.

The second part of James's analysis, his account of the operation of the will, required acceptance of his particular evolutionary construction

117. In his early essays, notably in "Remarks on Spencer's Definition of Mind as Correspondence," pp. 8–9 and 14–17, James suggested (rather obliquely) that moral principles or attitudes might be directly inherited. This suggestion, however, did run counter to his presumption of the great flexibility of human mind. These antitheses were synthesized in the last chapter of the *Principles of Psychology*, where he argued that moral ideals arose from acquired concepts which had been sifted through an inherited mental framework. See the discussion of evolutionary epistemology, above.

118. See Darwin, *Descent of Man* 1:161–67; and see chapter 5.

119. James, *Principles of Psychology* 1:325: "If the zoological and evolutionary point of view is the true one, there is no reason why any object whatever *might* not arouse passion and interest as primitively and instinctively as any other . . . . I might conceivably be as much fascinated, and as primitively so, by the care of my neighbor's body as by the care of my own. The only check to such exuberant altruistic interests is natural selection, which would weed out such as were very harmful to the individual or to his tribe."

of ideas. Thought, James supposed, originally evolved to facilitate action. "It is far too little recognized," he observe in an early essay, "how entirely the intellect is built up of practical interests. The theory of Evolution is beginning to do very good service by its reduction of all mentality to the type of reflex action. . . . Cognition, in short, is incomplete until discharged in act."[120] This hypothesis of the motor function of ideas, which I have already mentioned above, served to make free choice a scientific possibility for James. In deciding what to do in a situation, the mind, he believed, becomes the playground for competing plans of action. When one idea finally dominates our attention to the exclusion of rivals, action follows automatically. In a moral decision, selfish, pleasure-preserving proposals vie with altruistic intentions. A free act, in James's estimation, consists in the mind selectively attending to one idea over others, becoming interested in it, while letting the others fade. Such attention, the active entertaining of the moral idea, for instance, would put a thumb on the scale. With the moral intention the most weighty, moral action would follow as a matter of course. Natural selection within the social environment of the primitive tribe may have outfitted modern man with moral ideas, but his behavior merits moral approbation only if he pursues those ideas freely.

James admitted that psychological introspection could never really decide whether the interest invested in an idea was merely a function of the idea's own attractive force or a mental variation spontaneously bestowed on it.[121] Empirical investigation and sound theory—that is, Jamesian evolutionary psychology—proved to him that determinism, at least, was not a mandate of modern science. But what science alone could not demonstrate, the subjective method of Renouvier allowed him to postulate: as a first principle we could choose freedom.[122]

Conclusion

The subjective method of Renouvier served to boost James's declining spirits during his mental crisis. Yet it did not really lighten the weight of modern science, which seemed to press toward determinism as a necessary conclusion. James needed an objective counterbalance to give the subjective method real force. He found this in Darwinian evo-

120. James, "Rationality, Activity and Faith," pp. 65–66.
121. James, *Principles of Psychology* 2 : 569–74.
122. On 28 September 1882, James wrote Renouvier: "I believe more and more that free will, if accepted at all, must be accepted as a postulate in justification of our moral judgment that certain things already done might have been better done. This implies that something different was possible in their place" (James Papers, Houghton Library, Harvard University).

lutionary theory, which in its application to cognition and behavior seemed to require a mind independent of the machinery of brain and one thus capable of free choice. James's Darwinian approach extended far into the workings of his scientific psychology, as we have seen, and supplied the unity of conception and power of explanation that made it a significant influence on the course of early modern psychology in America.

Good history should have those qualities, I believe, that Horace attributed to good poetry. It should be *dulcis et utilis*. Even a moderately faithful representation of James's intellectual development could not be less than sweetly fascinating, since an extraordinary mind, a lively wit, and a deeply emotional personality dance on virtually every page of his essays and books. The history of James's development is also usefully instructive. He formulated four different arguments that for me and many others are still intellectually coercive, as the reader of this history will have gathered.

The first is the elegantly simple argument for the mind being at least partly independent of brain machinery: if conscious mind is an evolved trait, which it certainly seems to be, then it could have been naturally selected only if it added some utility to the material of the brain, that is, to that organ as it is usually described in physiology textbooks. Karl Popper, for one, has employed this argument, though apparently without recognizing its originator.[123]

The second contribution is equally persuasive. At least, I am persuaded, enough so that it has ensouled the historiographic model used in the composition of this volume. The contribution is James's proposal

123. See, for instance, Karl Popper, "Natural Selection and the Emergence of Mind," *Dialectica* 32 (1978): 339–55. For a dissenting opinion about the cogency of the Jamesian argument for the independence of mind, see William Bechtel and Robert Richardson, "Consciousness and Complexity: Evolutionary Perspectives on the Mind-Body Problem," a paper presented at the Eastern Division Meeting of the American Philosophical Association, December 1979. The theory of mind-body relation adopted most frequently by tough-minded philosophers of science is identity theory. (For the classic statement, see J. J. C. Smart, "Sensations and Brain Processes," in *The Philosophy of Mind*, ed. V. Chappell [Englewood Cliffs, N.J.: Prentice-Hall, 1962], pp. 160–72.) This theory holds that certain conscious processes are identical with specific brain processes, that introspection picks out aspects of the same process that neurophysiology also (but more reliably) describes. But this theory is open to the objection that if conscious mind has evolved, as it must if carried by an evolved brain, then it must be useful, and therefore . . . well, the rest follows the Jamesian path. Nor is it plausible that the consciousness of higher organisms should be dismissed as a pleiotropy—that is, a side effect of a gene selected for other reasons. Explaining traits as pleiotropic really makes sense only if they are single gene effects and thus simple traits. It is hardly likely that the complex mind of higher organisms is a one gene effect. Only a Leibnizian geneticist believing in soul monads might propose that.

that ideas be regarded as comparable to chance variations. Their truth value—in a correspondence sense—becomes then a function of their survival value in the various intellectual environments into which they are plunged. This epistemological hypothesis is grounded in a compelling argument adumbrated by James: if novel ideas are not innate, and are not simply logically induced from observation, then only a kind of blind or unjustified variation could first introduce them; and they will be retained only if they are adapted to the intellectual problem conditions to which they are applied. There are now several epistemologists for whom this evolutionary theory of knowledge has struck home.[124]

Third is James's approach to moral judgment, particularly his recognition of the instinctual base for other-regarding virtues. He thought that man, as an evolved animal, could not shed his biology upon being civilized. Rather, those instincts that constitute the possibility of society, such as parental affection and altruism, he took to be evolved traits, selected for over our long evolutionary history. James did not blush to admit, nor I believe should we, that our moral character is as much a hereditary product as our ability to use language and think great thoughts—a proposition that informs the discussion in the second appendix of this history.

Finally, there is the subjective method itself. Here again, the argument is disarmingly simple. No system of thought can demonstrate its own first principles, something Aristotle long ago recognized. Foundational principles must result from custom, deep preference, or personality—all ultimately shaped by our ontogenetic and phylogenetic evolutionary history. This conclusion has for me two important lessons, one philosophical, the other historiographical. First, if the postulation of freedom and a spiritual world would enrich life and competitively exclude no more important considerations, then why should we hesitate to endorse their reality? Second, the historian of science must attempt to reconstitute the actual intellectual environment that spawned scientific ideas, and not rest satisfied with some rationally sanitized reconstruction of that environment.[125] Those advocating what is usually

124. See, for instance, Donald Campbell, "Blind Variation and Selective Retention in Creative Thought and in Other Knowledge Processes," *Psychological Review* 67 (1960): 380–400; and "Evolutionary Epistemology," in *The Philosophy of Karl Popper,* ed. P. Schilpp (LaSalle, Ill.: Open Court, 1974), pp. 413–63; David Hull, "Altruism in Science: A Sociobiological Model of Co-operative Behavior among Scientists," *Animal Behavior* 26 (1978): 685–97; and "Central Subjects and Historical Narratives," *History and Theory* 14 (1975): 253–74; Robert Richards, "The Natural Selection Model of Conceptual Evolution," *Philosophy of Science* 44 (1977): 494–501; Karl Popper, *Objective Knowledge: An Evolutionary Approach* (Oxford: Oxford University Press, 1972).

125. Such as offered by Imre Lakatos, in "Falsification and the Methodology of Scien-

called the "strong program" in the sociology of science have perceived this.[126] And while historians who seek to portray the selecting environment cannot be as antilogistical as are some enthusiasts for the strong program, it is clear they can no longer pursue the kind of history of science in which ideas inexorably unfold, bound together only by internal logical chains. This is neo-Spencerianism. Historians, rather, must give due weight to social influences and, as James's own history powerfully shows, to the psychology of individual personality.

---

tific Research Programs," in *The Methodology of Scientific Research Programmes: Philosophical Papers of Imre Lakatos,* vol. 1 (Cambridge: Cambridge University Press, 1978).

126. For example, David Bloor, *Knowledge and Social Imagery* (London: Routledge and Kegan Paul, 1976); and Barry Barnes, *Interests and the Growth of Knowledge* (London: Routledge and Kegan Paul, 1977).

10

James Mark Baldwin: Evolutionary Biopsychology and the Politics of Scientific Ideas

A system of scientific ideas is comparable to a biological species. Such ideas, as realized in the work of a given scientist, vary, compete with rivals, become adapted to interlocking intellectual, cultural, and social environments, and as a result slowly evolve over time. This conception has only recently emerged as a historiographic model,[1] but its founding epistemological assumptions were formulated in the late nineteenth century. William James, it will be recalled, argued that the creative thinker spontaneously generated ideas which, if robust enough, would be naturally selected within their various environments. During the decade of the 1890s, this Darwinian approach to ideas seemed to be in the air, or rather the intellectual environment fostered a convergent evolution of scientific and philosophic thought. At the beginning of the decade, Conwy Lloyd Morgan advanced a Darwinian epistemology, as did the psychiatrist Theodor Ziehen, who taught at Jena, and the Berlin sociologist Georg Simmel.[2] In the 1890s and through the next two decades, James Mark Baldwin refined this theory of ideas both empirically and conceptually, and wove it into the fabric of his developmental and social psychology. Baldwin's work, however, must capture the interest of historians and philosophers of science for two additional reasons.

First, Baldwin's conceptions gave dramatic expression to an aspect of evolutionary theory which had always been a part of its development and which grew in intensity toward the end of the nineteenth century: the view that biological evolution operated not as a blind mechanical process but as one governed by mind. Baldwin pursued this notion in formulating a principle that met the requirements of neo-Darwinian

1. See the first appendix.
2. See chapter 8 for a discussion of Lloyd Morgan's evolutionary epistemology; reference to Ziehen is given below. See Georg Simmel, "Über eine Beziehung der Selectionslehre zur Erkenntnistheorie," *Archiv für systematische Philosophie* 1 (1895): 34–45.

assumptions and indeed appeared initially to rival the importance of Darwin's own principle of natural selection. Baldwin called it "organic selection."

The second compelling feature of the history of Baldwin's thought, at least for those already inclined toward an evolutionary historiographic model, might be termed the politics of origin and extinction. When an evolutionary systematist arranges the lineages of animal groups, he does so in view of their supposed descent. Though similarity of species is an important organizing criterion, what really matters is the causal relationship of descent, as David Hull never tires of reiterating.[3] In arranging his groups into taxonomic categories and tracing their connections, however, the systematist on occasion engages in the politics of science. After all, if he discovers a new species, his own name gets attached, and he thereby achieves that second-best kind of immortality. So too with the evolution of ideas. Descent counts, and politics often influence it.[4] This is a round-about way of indicating one of the reasons why today biologists refer to organic selection as the "Baldwin effect," despite its simultaneous discovery by Baldwin, Lloyd Morgan, and H. F. Osborn. Politics also play a role in biological extinction. Some animal species do not die off; they are erased by rival systematists. Many of Baldwin's important theoretical concepts, such as "social heredity," vanished from psychological science partly because he suffered the collapse of a reputation and foreign exile, due in no small measure to the politics of science.

Training in the Old Psychology for the New

The Civil War and James Mark Baldwin arrived in Columbia, South Carolina, during the same year, 1861. He was the third of five children born to Cyrus Hull Baldwin and Lydia Ford Baldwin. Because of his father's Unionist sympathies the family was spared loss of property when Sherman's armies burned their way through the Carolinas. After the destruction in 1865 of Columbia (save some houses of sympathizers, including the Baldwins'), the family moved north to Norfolk, Virginia. Upon returning to Columbia after the war, Cyrus Baldwin became mayor by military appointment. His son judged that the father's politi-

---

3. See, for example, David Hull, "The Naked Meme," in *Learning, Development, and Culture,* ed. H. C. Plotkin (New York: John Wiley, 1982), pp. 273–327.

4. By the term "politics" I mean that set of activities with which most academics are familiar—generally artful activities designed to secure power and prestige within a community, activities that usually involve manipulation, rhetoric, and the suggestion of reason, rather than reason itself.

cal moderation and personal integrity, united with his business interests in the South, insured exemption from the persecution to which many northerners were then exposed in South Carolina.[5]

James Mark was raised a Presbyterian of the strict observance. "I became a convert in my extreme youth," he recalled in his autobiography, "and united with the church pure in heart and devoted in life."[6] The faith of his father, though strict and based on a Biblical morality, was nevertheless sweetened by a "liberal Congregationalist" attitude. Baldwin described his family as a happy one in which "the austerities of Calvinistic faith [were] tempered by a genuine piety and a large charity." He remembered his early religious feelings as "normal, sincere, and happy."[7]

Baldwin received his primary education in the private schools of Columbia. After he graduated, his father installed him as a clerk in a dry goods store for two years, an experience in commerce he later valued. In 1878, at age eighteen, he entered the Salem Collegiate Institute in New Jersey. In this two-year preparatory school, Baldwin fell under the guidance of its genial founder, Colonel Harlan P. Davidson, a southern gentleman he held in great affection. At the school, Baldwin's religious faith quickened, and in 1881, with the intention of becoming a minister, he enrolled as a sophomore at the College of New Jersey, as Princeton was then called.[8]

At college Baldwin followed the "academic" course, based on classical studies, rather than the "scientific." The dominating personality in the college was its president, James McCosh, a Scots realist philosopher who taught many of the courses Baldwin took and who eventually directed his Ph.D. dissertation. Baldwin recalled McCosh as a formidable presence, who "had a very humane and withal Presbyterian conception of his relation to the undergraduates. If taken ill the student was in daily fear that 'Jimmy' would come and pray with him."[9]

McCosh's thought evolved like an archaeopteryx. Its heavy reptilian skeleton of Scots metaphysical and epistemological realism prevented graceful flight into the newer areas of psychology and biology. But at least it did get off the ground. McCosh studied Darwin and Spencer and integrated their views into his theistic cosmology. He dismissed the fear that evolutionary ideas led to irreligion, maintaining rather that

5. James Mark Baldwin, *Between Two Wars, Being Memories, Opinions and Letters Received* (Boston: Stratford, 1926), 1:4–8.
6. Ibid., p. 10.
7. Ibid., pp. 10–12.
8. Ibid., pp. 13–18.
9. Ibid., p. 21.

454   JAMES MARK BALDWIN

"development by causation is the plan by which God carries on his works." He granted that natural selection did operate to prune species, and he endorsed Spencer's theory of progress, holding that the "tendency of animal life is generally upward, from all fours to the upright position, from which men can look up to heaven." [10] McCosh, however, rejected the mechanistic interpretation often given evolution: he insisted that internal, vital forces having a divine origin were the true causes of development. These superadded powers directed the evolutionary process toward ever-greater perfection in nature and explained what the Darwinian mechanical causes could not: specifically, life itself, sensation, instinct, intelligence, and morality. [11] Among contemporary theorists, McCosh sided with the neo-Lamarckians, such as E. D. Cope, who found a place for design and creative mind in evolution. [12]

In his senior year (1883–1884), Baldwin wrote a bachelor's thesis that undoubtedly gave root to what would become an abiding interest. It was entitled "The Nature and Extent of *a priori* Principles, with Special Criticism of the Evolutionary Theory of Conscience." [13] The title of this (apparently no longer extant) work suggests that he took potshot at a favorite target of McCosh, namely Herbert Spencer's evolutionary theory of mind. Whatever the intellectual worth of that senior essay, it had the immediate value of winning Baldwin a traveling fellowship—the Green Fellowship in Mental Science—for a year's study in Germany.

10. James McCosh, "Development: What It Can Do and What It Cannot Do" (1883), reprinted in his *Realistic Philosophy* (New York: Scribner's, 1887), 1:157 and 162.

11. James McCosh, *The Religious Aspects of Evolution*, 2d ed. (New York: Scribner's Sons, 1890), pp. 47–57. James Moore would make McCosh out as a Darwinian "save for his views on human descent." Moore's conclusion appears more a reflection of his revisionist thesis that among Christian theologians, "Darwin's theory of evolution by natural selection could be accepted in substance only by those whose theology was distinctly orthodox." See James Moore, *The Post-Darwinian Controversies* (Cambridge: Cambridge University Press, 1979), pp. ix and 248–49.

12. McCosh ("Development: What it Can Do and What it Cannot Do," p. 190) illustrated the compatibility of his views with current science by drawing this quotation from Cope: "In living things the powers display design, having direct reference to consciousness, to the satisfaction of pleasure and the avoidance of pains. Mind also controls structure: the evolution of mind has a corresponding effect on organism, a view which is confirmed by palaeontology. The mind producing struggles of animals has led to machines for grinding, cutting, seizing, digging; for running, swimming, and flying." For a discussion of the development of McCosh's views about evolution, see J. David Hoeveler, Jr., *James McCosh and the Scottish Intellectual Tradition* (Princeton: Princeton University Press, 1981), pp. 180–211.

13. Karin Wetmore discovered the title of Baldwin's senior essay, but in spite of extensive search in various archives, not the manuscript itself. I am grateful to her for calling the title to my attention.

Baldwin's year in Germany held certain portents of his future. In the summer before his serious study began, he toured the countryside, and quickly became romantically involved with a young German girl. Her father misread both Baldwin's intentions and financial status, and the green American had to retreat hastily to the safety of Wundt's laboratory in Leipzig. Baldwin stayed only one term in the lab but nonetheless became "an enthusiast for the new psychology, and took back . . . the full outfit of ideas—Fechner's and Weber's laws, the technique of reaction-time experiments, theories of mind and body, and cognate points of view as propounded by Lotze, Fechner, and Wundt." [14] From Leipzig he moved to Berlin and enrolled in Friedrich Paulsen's seminar, where he was introduced to the system of the "God-drunk" philosopher Spinoza. Baldwin found Spinozistic metaphysics congenial to his incipient desire to reconcile the possibility of scientific knowledge with apparently obdurate nature. When Baldwin returned to Princeton he intended to do his dissertation on Spinoza. McCosh, however, caught wind of his sympathetic reaction to the Jewish lens grinder. "No," he remonstrated, "Spinoza will not do, you must refute materialism." [15] Later, when buffered against his former teacher, Baldwin heeded this command with residual undergraduate perversity. He wrote an article attempting to show that correctly interpreted, the Spinozistic system yielded an absolute realism and a direct epistemological intuitionism—a precursor of Scots philosophy no less. [16]

Baldwin spent the academic years 1885 to 1887 in the Princeton Theological Seminary and as an undergraduate instructor of elementary French and German in the college. His language classes, he remembered, included two French boys whose snickers made him poignantly aware of his linguistic deficiencies. To further his grasp of French, and not incidentally to feed his growing interest in the new psychology, he translated for publication Théodule Ribot's *Psychologie allemande contemporaine.* The translation appeared in 1886 with a preface by McCosh. [17] The book and a satisfactory refutation of materialism brought Baldwin a call to his first teaching position.

In the summer of 1887, Baldwin set out for Lake Forest University, a small Presbyterian school that was principally concerned with preparing missionaries for China. He took with him a new wife, the daughter of

14. Baldwin, *Between Two Wars* 1:32.

15. Ibid., p. 20.

16. James Mark Baldwin, "The Idealism of Spinoza," *Presbyterian Review* 10 (1889): 65–76.

17. Théodule Ribot, *German Psychology of To-day,* trans. James Mark Baldwin, with an introduction by James McCosh (New York: Scribner's, 1886).

the Princeton Seminary president, and a resolve to make philosophy and psychology his vocation. The first order of business, then, was to get out of that small Illinois college on the western shore of Lake Michigan. Baldwin accomplished this by exercising a talent he must have acquired during his short stay with Wundt—the ability to write quickly and in quantity. His several essays and *Handbook of Psychology*[18] brought him the professional advancement he desired.

The Foundations of Baldwin's Psychological Science

Spiritualistic Metaphysics

Baldwin's philosophy and psychology evolved together, though with the philosophic structure often obscured (at least in the middle years) by psychological preoccupations. His early work at Lake Forest, however, reveals the spiritualistic metaphysics that gave backbone to his science. He nourished his nascent philosophic convictions on the spiritualism and idealism brewing in the latter part of the nineteenth century as well as on more traditional fare, the monism of Spinoza.

The "new spiritualists" whom Baldwin mentioned in his review, "Contemporary Philosophy in France" (1887), included such thinkers as Victor Cousin, Paul Janet, Charles Renouvier, Théodule Ribot, and Jean Charcot.[19] Their theories were variously competitive, but they agreed in a pursuit also undertaken by Baldwin (and his teacher McCosh)—the reconciling of philosophy with science. The new spiritualism (unlike the older theological version) required that philosophy temper its claims in light of empirical science. Baldwin, however, insisted on a reciprocal relationship. Philosophy had the office of "rationalizing science."[20] His own efforts at such rationalizing uncovered for him a congruent principle structuring both the new spiritualism and the new psychology—the principle that only nature *as known* could be an object of science. Though a directive idea of apparent innocence, it nevertheless had Spinozistic and Kantian depths, for in Baldwin's interpretation it meant that "nature is intelligent and that the laws of thought are the laws of things."[21] This conclusion would later yield the belief that life

18. James Mark Baldwin, *Handbook of Psychology: Senses and Intellect* (New York: Holt, 1889).

19. James Mark Baldwin, "Contemporary Philosophy in France," *New Princeton Review* 3 (1887): 137–44.

20. James Mark Baldwin, "The Postulates of Physiological Psychology," *Presbyterian Review* 8 (1887): 427–40; reprinted in James Mark Baldwin, *Fragments in Philosophy and Science* (New York: Scribner's Sons, 1902). The quotation is from the latter, p. 140.

21. Ibid., pp. 140–41.

and mind were coextensive, so that the principles of mental growth would find their functional equivalents in those of organic evolution.[22]

The particular metaphysical postulate that underlay Baldwin's epistemology of science and later his evolutionary biopsychology had its roots in Spinoza but its ramifications in French spiritualism and Wundtian idealism. This was the double aspect identity postulate, which held that mind and body displayed different attributes of a common substrate. From this postulate, Baldwin, in an early essay, developed two corollaries: "First, complexity of organism is the reflection and not the cause of complexity of thought, the opposite of the position of materialistic evolution. Wundt has actually drawn this inference. And, second, since thought must logically precede its realization, the inner must precede the outer aspect, and our monism is, after all, a monism of mind."[23]

Baldwin believed a monism of mind to be a happier metaphysical postulate upon which to ground a psychological science than, say, the dualism of James.[24] James's evolutionary psychology seemed to suggest that mind could spontaneously vary independently of any causal relation with brain. The Spinozistic perspective, by contrast, required a harmony of mental and physical laws, so that, as Baldwin argued in his *Handbook of Psychology:*

> The ultimate laws of psychology must, therefore, find their completion in the psychophysical connection, since a complete explanation of a phenomenon must include its cause and essential conditions. . . . With any other supposition, we destroy the unity of mind, since, with the lower operations governed by laws of mind and body in their relation, and the higher by laws of mind without relation to body, how could the two systems of laws be held in harmony?[25]

In his work of the next decade, Baldwin would transform this metaphysical directive into a natural law. It became the principle of *organic selection,* which supposed that an organism's mental and physical aspects together formed the unit upon which natural selection operated: "In

22. See Baldwin's assertion of the principle of coextension in his *Mental Development in the Child and the Race* (New York: Macmillan, 1895), p. 213.

23. James Mark Baldwin, "Recent Discussion in Materialism," 1 (1890): 357–72; reprinted in *Fragments in Philosophy and Science.* Quotation is from the latter, p. 53.

24. The dualist stance of James's *Principles of Psychology* later collapsed into the radical empiricism and spiritualism foretold in Baldwin's review of the *Principles.* See his "Review of James's *Principles of Psychology,*" *Educational Review* 1 (1891): 357–71; reprinted in *Fragments in Philosophy and Science,* pp. 371–89.

25. Baldwin, *Handbook of Psychology: Senses and Intellect,* pp. 32–3.

other words, *it has been the psychophysical, not the physical alone, nor the mental alone, which has been the unit of selection in the main trend of evolution.*[26]

Since in the preface to the *Handbook* Baldwin expressed gratitude to McCosh for his training and to Wundt and Elie Rabier, a French spiritualistic philosopher, for their intellectual contributions, one might be led with Wozniak, one of the subtler historians of Baldwin's thought, to regard his psychology as "an integration of Wundt with McCosh."[27] That impression is encouraged by the topics of the last chapter of the *Handbook,* in which Baldwin sketched the several kinds of intuition operative in rational mind. McCosh, after all, regarded intuition as the fundamental source of our knowledge about reality; so the presumption is easily formed that Baldwin's thought came half out of McCosh's brain. But Baldwin was not so docile a student as that. His metaphysical monism of mind would have offended Scots common sense, which could certainly intuit the clear difference between matter and spirit. Moreover intuition as Baldwin described it reconstructed the external world using subjective categories, while for McCosh intuition opened a direct highland vista onto the real world.[28] Baldwin owed McCosh a debt as an esteemed teacher who had whetted his metaphysical appetite but who could not satisfy it. Only the more critical provisions drawn from Spinoza and French neo-Kantianism would sustain him.

Toronto, 1889–1893

Due to the initial success of his *Handbook of Psychology* and the timely death of the professor of metaphysics and logic at the University of

26. James Mark Baldwin, *Development and Evolution* (New York: Macmillan, 1902), p. 26.

27. Robert Wozniak, "Metaphysics and Science, Reason and Reality: the Intellectual Origins of Genetic Epistemology," in *The Cognitive-Developmental Psychology of James Mark Baldwin,* ed. John Broughton and John Freeman-Moir (New York: Ablex, 1982), p. 26. Though Wozniak exaggerates the influence of McCosh on Baldwin's early work, he yet recognizes that the stronger currents of Baldwin's thought ran in channels trenched by Spinoza and the French neo-Kantians.

28. Baldwin's discussions in the *Handbook* have an indefinite and elusive quality that marked all of his later writing. The critical difference between his theory of intuition and McCosh's, though, becomes apparent in his account of the unity we attribute to an object of sense intuition: "Its unity is first an ideal unity, through which the unity of the external thing of perception is interpreted and reconstructed: but this ideal unity is in so far concealed in the potential unfolding of the process of perception, that it seems to arise consciously by abstraction from the unity of the thing" (Baldwin, *Handbook of Psychology: Senses and Intellect,* p. 320).

Toronto, Baldwin had opportunity to leave his outpost north of Chicago. He was the leading candidate to fill the position at Toronto. But the president of the university, Sir Daniel Wilson, wished to keep the faculty in the English tradition and favored the student of the previous incumbent. Considerable debate erupted over the chair, which the minister of education finally becalmed, in Solomonic fashion, by creating two chairs; Baldwin received one and the Canadian professor James Hume the other.[29]

Shortly after Baldwin arrived at Toronto in 1889, funds were secured for him to create a laboratory for experimental psychology, "the first ever opened in the British Empire," he boasted in his autobiography.[30] Among the important studies conducted there, and in the nursery of his own home, were those on handedness, which showed that the establishment of speech centers on one side of the brain determined that the dominant hand would be on the contralateral side.[31] Baldwin also completed, in 1891, the second volume of his *Handbook*, subtitled *Feeling and Will*[32]

Baldwin's two volumes were initially conceived as a continuous theoretical discussion, more or less following the pattern of Bain's similarly titled *The Senses and Intellect* (1855) and *The Emotions and the Will* (1859). (McCosh also adopted this traditional division of the psychological faculties in his *The Cognitive Powers* [1886] and *The Motive Powers* [1887].) While volume 2 of the *Handbook* offered some brief report of experimental work conducted at Toronto, its agenda were nonetheless set, as they had been in the previous volume, by the work of Wundt, Bain, Spencer, James, Mill, and Lotze, though now also by the ideas of Schneider, Romanes, and Darwin. Yet certain theoretical differences emerged in the second volume. Baldwin signaled this change of direction in the preface by observing that "the phenomena of emotional and volitional life have not been worked over for purposes of philosophical system, as intellectual phenomena have." For that reason he thought "the psychologist has in this field greater freedom of treatment and a larger scientific opportunity."[33] In the second volume of the *Handbook*, Baldwin seized the opportunity to make good on the promise of his

29. Baldwin, *Between Two Wars* 1:42.

30. Ibid., pp. 43–4.

31. Baldwin provided a brief report of these studies in "Origin of Right or Left Handedness," *Science* 16 (1890): 247–48, and a much fuller discussion in his *Mental Development in the Child and the Race*.

32. James Mark Baldwin, *Handbook of Psychology: Feeling and Will* (New York: Holt, 1891).

33. Ibid., p. iii.

identity postulate, that is, he began to treat mind as fully integrated into the psychophysical system.

He did this under the aegis of the law of mental dynamogenesis, his version of the ideomotor hypothesis (also advanced by Carpenter, Spencer, and James). The law simply asserted that "every state of consciousness tends to realize itself in an appropriate muscular movement."[34] For Baldwin this implied that all states of consciousness involved feelings of muscular movement; all consciousness essentially became motor consciousness. The meaning, therefore, of a particular pattern of sensations, or even the conclusions reached in rational judgment, could be interpreted in motor terms. This construction of consciousness echoed James's proposal that thought originally evolved to produce action in the natural world.

Genetic Psychology and the Theory of Imitation

Sources of Baldwin's Genetic Psychology

In 1895 Baldwin published *Mental Development in the Child and the Race,* a book that laid the foundations for his genetic psychology and set the program that would occupy him for the next thirty years. His methods in genetic psychology stood in contrast to those he employed in the two volumes of the *Handbook.* In these latter, conscious reflection on conscious processes in the mature mind constituted the principal technique. Even at the beginning of his systematic theorizing, though, Baldwin acknowledged the danger of this method, especially when it was used to formulate the first principles of mental activity. "The very existence of 'first principles,'" he cautioned in the opening pages of the *Handbook,* "the determination of the barest woof and warp of thought itself, is a matter of origins, as the evolutionists claim, and the problem should be approached, as well from the side of infant and comparative psychology, as from the side of the observation of developed reason."[35] Wundt had offered a similar admonition over two decades earlier in his *Vorlesungen über der Menschen- und Thierseele,* and had acted upon it.[36] Baldwin, in the *Handbook,* did not. What then induced him, in the early 1890s, to change course and undertake a study of the ontogenetic and phylogenetic evolution of the human mind?

Baldwin's new approach was perhaps stimulated by a publication of

34. Ibid., p. 281.
35. Baldwin, *Handbook of Psychology: Senses and Intellect,* p. 14.
36. See Robert Richards, "Wundt's Early Theories of Unconscious Inference and Cognitive Evolution in their relation to Darwinian Biopsychology," in *Wundt Studies,* ed. Wolfgang Bringmann and Ryan Tweney (Toronto: Hogrefe, 1980), pp. 42–70.

Figure 10.1   James Mark Baldwin, 1861–1934,
photograph from ca. 1895.

Wilhelm Preyer, a colleague of Ernst Haeckel at Jena and like Haeckel an enthusiastic Darwinian. Preyer's *Die Seele des Kindes* (1882) was a psychogenetic study of his own infant son's first three years of life. It probably first came to Baldwin's attention in the English translation (1888), which contained an introduction by G. Stanley Hall,[37] one of the most influential of the Americans importing the new psychology. Preyer conceived his study as demonstrating a natural, but not reductively mechanical, origin for human mind[38]—a scientific perspective

37. Wilhelm Preyer, *The Mind of the Child*, trans. H. Brown, with introduction by G. Stanley Hall (New York: D. Appleton, 1888–1889).

38. The idealist strains of Preyer's evolutionary psychology are discussed by Siegfried Jaeger in "Origins of Child Psychology: William Preyer," in *The Problematic Science: Psy-*

Baldwin must have appreciated. Yet mere availability of a theoretical resource cannot really explain the adoption of a particular line of research. After all, the period flowered with exciting scientific possibilities. An evolutionary historiographer must, like his counterpart in biological ecology, discover what specific pressures and what combinations of forces directed the evolution of scientific thought along certain paths. Baldwin certainly read Preyer at the right time. But the changes in his personal environment that more proximately pushed him to adopt the methods of genetic analysis were, I believe, the births of his two children and his own experience in rearing them.

Baldwin's real laboratory at Toronto was his home nursery. He doted on his two daughters, Helen, born in 1889, and Elizabeth, in 1891. Uncharacteristically for a man of his station, Baldwin devoted considerable time to helping his wife and their Canadian nurse in the feeding, changing, and general care of the children. A university usually pays its professors in coin that banks do not recognize; but a flexible schedule and freedom to pursue interests, both of which allowed Baldwin to spend so much time with his children, are its most generous compensation. He seems to have begun systematic study of Helen in order to test certain hypotheses about the origin of handedness. These hypotheses had been suggested by the discussions of his colleague (and president of the university) Daniel Wilson, in the latter's monograph *The Right Hand: Left-handedness*. Baldwin tested hand preference in his infant daughter from her fourth to tenth months under a variety of circumstances. In her ninth month he also began a study of some hypotheses about color perception initially formulated by Preyer; and he undertook a variety of experiments concerning the development of other specific behaviors in his children. During the years 1890 to 1893, Baldwin published some of these studies in *Science,*[39] but he provided a considerably more complete account of his experimental work in *Mental Development in the Child and the Race,* which appeared in 1895.

The advantage Baldwin claimed for the study of his children (aside from the obvious pleasure he took in it) was that their consciousness lacked the entangling obscurity of an adult's; it was innocent, moreover,

chology in Nineteenth-Century Thought, ed. William Woodward and Mitchell Ash (New York: Praeger, 1982), pp. 300–321.

39. James Mark Baldwin, "Origin of Right or Left Handedness," *Science* 16 (1890): 247–48; "Right-handedness and Effort," *Science* 16 (1890): 302–3; "Infant Psychology," *Science* 16 (1890): 351–53; "Suggestion in Infancy," *Science* 17 (1891): 113–17; "Infants' Movements," *Science* 19 (1892): 15–16; "Origin of Volition in Childhood," *Science* 20 (1892): 286–87; "A New Method of Child Study," *Science* 21 (1893): 213–14; "Distance and Color Perception by Infants," *Science* 21 (1893): 231–32.

of the theoretically contaminating influence of self-reflection. Any idea, because of its dynamogenic force, tended to realize itself immediately in the child's action.[40] Through careful observation and test, the experimenter could therefore chart the gradual appearance of the different mental powers and determine their necessary sequences. Baldwin did not ignore, however, the liabilities of using children as subjects of genetic analysis. Since their minds were not blank slates, but from the beginning deeply grooved by specific and individually varying heredity, he knew that it would often be difficult to distinguish the acquired from the native, the peculiarities of one child from the traits possessed by all children. He also recognized the dangers of generalizing from just a few cases. Yet he understood that certain hypotheses prevalent in psychology textbooks could be falsified by observation of only one subject.[41] And the testing of his fellow scientists, along with his children, was a major aim of the experiments reported in his *Science* papers.

Suggestion as a Developmental Principle

The theoretically most important of Baldwin's *Science* papers dealt with the dynamogenic property of consciousness. In his article "Suggestion in Infancy" (1891), he urged that the phenomenon of hypnotic suggestion demonstrated that ideas functioned as a fundamental kind of motor stimulus. It was this phenomenon, he said, that prompted the theory of suggestion which he used in discriminating stages in the developing behavior patterns of his daughter Helen.[42] But for suggestion to serve as an all-purpose explanatory principle, Baldwin had to define it very sparely: "from the side of consciousness," he stipulated, "suggestion in general is the tendency of a sensory or ideal state to be followed by a motor state."[43] This minimal definition, however, sacrificed the usual notion that a suggestive sensory or ideal state furnished the behavioral plan for the motor response. The definition really only amounted to a reiteration of the law of dynamogenesis.

Baldwin identified three basic stages of suggestion in the development of the child's mind. First was "physiological suggestion." This earliest achievement of mental growth consisted in particular sensory

40. Baldwin, *Mental Development in the Child and the Race*, p. 5.

41. Ibid., pp. 10–11.

42. Baldwin, "Suggestion in Infancy," p. 113. Baldwin thought the theoretical implications of hypnotic suggestion so important that in 1892 he travelled to France, first to Paris to visit Charcot and then to Nancy to see Bernheim; he wished to study the marvelous phenomenon firsthand. See James Mark Baldwin, "With Bernheim at Nancy," *Nation* 55 (1892): 101–103.

43. Baldwin, "Suggestion in Infancy," p. 113.

states (e.g., perception of bed, nanny singing, low lights) becoming integrated with motor reflexes or habits (e.g., child falling asleep), such that subsequent occurrences of those particular sensations would tend to reproduce the conjoined behavior. Baldwin referred to the second stage of mental growth as that of "sensori-motor suggestion." It differed from the previous stage in that the child now attained the use of consciousness, whereas before it merely reacted "unconsciously or subconsciously by means of an extra-organic stimulus."[44] The final step in early infant development, "ideo-motor suggestion," had two sublevels, the deliberative and the imitative. "Deliberative suggestion" arose in Helen at the end of her first year, when conflicting suggestions would vie with each other until the strongest prevailed. The final level of development Baldwin discriminated was that of "imitative suggestion," of which there were two kinds: simple and persistent. In simple imitation, some sensory suggestion stimulated behavior, which the child would repeat without being really able to make it conform to the model—for example, continually parroting a phrase overheard without correcting mispronounciations. Persistent imitation marked the transition, in Baldwin's view, from suggestion to will, from the reactive to the voluntary consciousness. Here the child attempted to improve its imitative behavior, bringing it closer to the model. Baldwin defined this stage as "the tendency of a sensory process to maintain itself by such an adaptation of its re-actions that they become in turn new stimulations." The essential feature of both simple and persistent imitation, which Baldwin would soon come to regard as central to all levels of mental development, was that imitative behavior tended to reproduce its initiating stimulus and thereby itself again.[45] (Jean Piaget's theory of mental development in the child resonates of Baldwin's earlier conceptions, and for good reason, as Jacques Vonèche relates.)[46]

44. This characterization is taken from Baldwin's *Handbook of Psychology: Feeling and Will*, p. 297, where he offered an analysis similar to that of his *Science* paper.

45. Baldwin, "Suggestion in Infancy," p. 117. One should compare Baldwin's stage analysis with that offered by Preyer, in *The Mind of the Child* 1, chaps. 8–9. Preyer recognized four early stages of development: the impulsive, characterized by random motions due to internal stimuli; the reflexive, in which external causes stimulate fixed actions; the instinctive, consisting of complex actions having a determinate aim; and the imitative, wherein will is first clearly evinced. Preyer conceived these stages as largely overlapping, but with succeeding ones occurring generally a bit later in the maturation process.

46. Jacques Vonèche interviewed Piaget, just before the great psychologist's death, about Baldwin. Piaget mentioned that he had carefully read and annotated the French translations of *Mental Development in the Child and the Race, Social and Ethical Interpretations,* and the first volume of *Genetic Logic.* Piaget's mentor, Pierre Janet, was a close

From Suggestion to Imitation: the Reconstruction of an Ideational Lineage

In the early years of the 1890s, Baldwin's stature in the scientific community assumed larger proportions. He had completed a two-volume *Handbook of Psychology,* highly enough regarded that, like James, he was encouraged to undertake a one-volume abridgment (*Elements of Psychology,* 1893) for use in the colleges. He held a prestigious university position at Toronto, and his experimental work there bore fruit. In 1893 he was called back to his alma mater, Princeton, to occupy the new Stuart chair in experimental psychology. His book *Development in the Child and the Race* (1895), prepared in his seminar of the previous year, marked a dramatic change in the direction of American psychology, heading it decisively toward evolutionary biopsychology. It was followed two years later by the second volume in his series on development, *Social and Ethical Interpretations,* the first book to describe itself (as part of the subtitle) "A Study in Social Psychology." Baldwin solidified these intellectual accomplishments in an institutional way by starting a new journal in partnership with James McKeen Cattell, the editor of *Science.* The *Psychological Review,* still the most important journal in the field, came to birth in 1894. At the end of the decade, Oxford University recognized Baldwin's eminence by awarding him one of the first of two new honorary doctor of science degrees. The scientific community of the 1890s perceived within its midst a powerful and original intellect, and Baldwin did not wish cloud that view.

In order to ensure his originality, Baldwin reconstructed the historical lineage of an important concept in his emerging evolutionary psychology. I do not mean to suggest there was any deceit on Baldwin's part, that is, beyond the frequent self-deceit of men in his position. Most creative intellects incorporate the ideas of others and turn them into the flesh of their own theories; it is just that in science, and particularly in science, the social pressures to make original contributions can seduce a thinker to redress a natural lineage to the advantage of his own reputation. He may have good reasons for taking a new position or advancing an apparently novel idea, but an equally strong need to proclaim, explicitly or implicitly, that his intellectual child is a virgin birth. Baldwin, I believe, expressed this need in a relatively minor way when he adopted 'imitation' in place of 'suggestion' as the cardinal prin-

---

friend of Baldwin and, according to Piaget, cited Baldwin constantly in his courses. See Jacques Vonèche, "Reflections on Baldwin," *The Cognitive Developmental Psychology of James Mark Baldwin,* pp. 80–86.

ciple of his genetic psychology; but he set the pattern for a significantly more artful reconstruction a short time later.

A reviewer of *Social and Ethical Interpretations,* a book which extended the analysis of imitation worked out in *Mental Development in the Child and the Race* to social institutions, suggested that Baldwin owed an unacknowledged debt to Gabriel Tarde's *Les lois de l'imitation* (1890). Baldwin defended his independence in the preface to the second edition (1899) of *Interpretations* by pointing out that the theory of imitation had formed the conceptual foundation for his 1895 book *Mental Development,* but that the manuscript for this book was "finished before my attention was called to M. Tarde's *Lois de l'imitation,* and the allusions to him were then made in it as it went to print."[47] Baldwin included as part of his defense a small paragraph of a letter from Tarde in which the Frenchman granted that *Mental Development* had come to rest at that point from which his own work had commenced. But Tarde could not have reached the second part of the book, wherein Baldwin had already begun to use the principle of imitation for analyzing social institutions. Perhaps, though, he simply stifled discontent, since Baldwin held translation rights to *Les lois de l'imitation* and *Les lois sociales.*[48]

A Tarde supporter, Gustavo Tosti—the Italian consul general of New York and frequent contributor to Baldwin's *Psychological Review*—did not remain so reticent. He had detected the similarity of Tarde's ideas and those found in the latter part of *Mental Development.* In a general review of Tarde's theories—which did not appear in Baldwin's journal—Tosti had politely suggested that the French sociologist had made a real discovery about the influence of imitation in social life, but that Baldwin in *Mental Development* merely "generalizes and completes Tarde's statements."[49] Tosti became considerably less polite after the appearance of Baldwin's defense in the second edition of *Interpretations.* On the occasion of another priority dispute—this time between Baldwin and the American sociologist Franklin Giddings—Tosti wrote a letter of indictment to *Science.* Cattell, who had begun to nurse a growing antipathy for Baldwin, gladly printed it. Tosti called Baldwin's *Interpretations* only "a transcription of Tarde in another key," asserting

47. James Mark Baldwin, *Social and Ethical Interpretations in Mental Development, A Study in Social Psychology* (New York: Macmillan, 1897), p. xiii.

48. Ian Lubeck, "Histoire de psychologies sociales perdues: le cas de Gabriel Tarde," *Revue française de sociologie* 22 (1981): 382.

49. Gustavo Tosti, "The Sociological Theories of Gabriel Tarde," *Political Science Quarterly* 12 (1897): 507. I am grateful to Ian Lubeck for calling this and other articles by Tosti to my attention. See his "Histoire de psychologies sociales perdues" for a discussion of further details of the relationship between Baldwin and Tarde.

that aside from work on imitation in individual psychology, "Professor Baldwin has never brought to light any fact in the line of social evolution that had not been previously intimated or actually mentioned by Tarde."[50] In private correspondence with Tarde, Tosti went beyond insinuation. He threatened that if Baldwin dared to respond to his *Science* letter, "I am going to reply by dotting the 'i's, that is, I will indicate those passages of the 'Laws' and 'Social Logic' that Baldwin has simply translated in bad English in his chief work 'Ethical Interpretaions.'"[51] The community of psychologists had begun to grow suspicious and irritated by Baldwin's constant and agressive claims to scientific originality.

And, it seems, with some justice Baldwin had actually discussed Tarde's *Lois de l'imitation* in his essay "Imitation: the Natural History of Consciousness," published in *Mind* in January 1894 and thus finished at least a year before his "attention was called to M. Tarde's *Lois de l'imitation.*" In the essay Baldwin affirmed that "the theory [of imitation] so far advanced, with extreme brevity, is in accord with that first announced (obscurely I think) by Tarde."[52] Moreover the first several chapters of Tarde's book were published as articles beginning in 1882 in *Revue philosophique,* a journal Baldwin kept up with; indeed Baldwin also cited two of these articles in his *Mind* paper. To understand the nature of his likely debt to Tarde—and the validity of his own claims to originality—a consideration of Baldwin's concept of imitation is required.

For the purpose of adapting the concept of imitation as a fundamental principle of ontogenetic (and phylogenetic) development, Baldwin had to pare its usual meaning, just as he had previously done with the concept of suggestion. In *Mental Development,* he proposed that "an imitative reaction is one which tends normally to maintain or repeat its own stimulating process."[53] Under Baldwin's definition, the infant reaching for or holding on to a cookie is imitating, since it attempts to maintain a stimulus (perception of the cookie) that reinforces the original behavior (of reaching and holding). This definition of imitation had previously served for "imitative suggestion," but now Baldwin intended the conception to take over the role of suggestion in his stage analysis of the child's mental growth. The stages of mental development to be

50. Gustavo Tosti, "Baldwin's Social and Ethical Interpretations," *Science* 15 (1902): 552.

51. Gustavo Tosti to Garbriel Tarde (2 May 1902), quoted in Lubeck, "Histoire de psychologies sociales perdues," p. 382.

52. James Mark Baldwin, "Imitation: A Chapter in the Natural History of Consciousness," *Mind* 19 (1894): 30.

53. Baldwin, *Mental Development in the Child and the Race,* p. 350.

explained by imitation were precisely the same as the previously discriminated stages of suggestion. In *Mental Development,* Baldwin called them "biological imitation" (recall his "physiological suggestion"), "cortical imitation" (recall "sensori-motor suggestion"), and, finally, "simple and persistent imitation" (carrying the same labels as before).

In the preface to *Mental Development,* Baldwin claimed the concept of imitation had an eruptive significance for his burgeoning genetic psychology. He said the importance of imitation struck him in 1892, while he was working on his paper "The Origin of Volition in Childhood."[54] He continued in the preface to his book: "The further study of this subject brought what was to me such a revelation of the genetic function of imitation that I then determined—under the inspiration, also, of the small group of writers lately treating the subject [including Tarde?]—to work out a theory of mental development in the child incorporating this new insight."[55]

Both in "Origin of Volition" and in *Mental Development,* Baldwin recounted, with great delight, several instances of Helen persistently attempting to imitate skillful behaviors she had observed, as well as the "mother-child" charades of his two daughters. In these persistent imitative performances, he found "the child's first exhibition of will."[56] Such experiences, reinforced by his earlier theory of suggestion stages (of which the final was that of imitative suggestion), undoubtedly constrained the evolution of Baldwin's ideas about the role of imitation in development. But so did, I think, the views of Tarde. For in his *Mind* paper of 1894 (where he admitted cultivating Tarde's insights) and in his 1895 book, Baldwin used the concept of imitation to argue, just as Tarde previously had, that the habits of the individual, his accommodations to novel situations, and the ideas he acquired—all resulted from social models. Moreover Baldwin generously expanded his theory of individual mental growth through imitation to account for the growth of cultural traditions within a society—the very phenomenon that Tarde was most concerned to explain through his own "Laws of Imitation." Tarde apparently moved Baldwin to perceive the contribution that the social environment made to individual mental development, and he undoubtedly provided the right hints on how to extend the theory of development to the mental growth of societies. The desire for originality seems, however, to have subtly eased Baldwin away from acknowledging his debt to the French sociologist. Baldwin's genius was

54. See note 36.
55. Baldwin, *Mental Development in the Child and the Race,* p. vii.
56. Baldwin, "Origin of Volition in Childhood," p. 286.

nevertheless secure enough, as demonstrated by his finer analysis of the mechanism of imitation, which was quite different from Tarde's. This analysis led to his first theory of "organic selection." His second theory of organic selection, however, led him right back to artfully reconstructing the lineage of his ideas.

The Evolutionary Analysis of Consciousness in the Individual and the Race

Baldwin returned to Princeton in 1893 and remained for ten years. Shortly after arriving, he opened up a psychological laboratory, in which he continued experimental study of infant behavior. His seminars on development in 1893–1894 and 1896–1897 formed the more proximate bases for his *Mental Development in the Child and the Race* and *Social and Ethical Interpretations*. Baldwin spent a good portion of his last five years at Princeton organizing his famous two-volume *Dictionary of Philosophy and Psychology* (1901–1902). The dictionary incorporated the work of over sixty contributors and required of him enormous editorial and persuasive skills.

Baldwin, a man of prickly ego, must have often strained the forbearance of other faculty members at Princeton who also knew their own worth. Certainly his relations with the professor of jurisprudence could not have been cordial. Almost immediately after Baldwin arrived at the university, he took a dislike to Woodrow Wilson. In his autobiography he recalled with particular distaste Wilson's speech at the celebration of the 150th year of the college in 1896: "Clothed in the style of which he was master, impressionistic and grave, rhetorical and unprecise, the address left one thought, if no more, in his hearers' minds—that of the menace of modern science."[57] Undoubtedly the scorn Baldwin recollected in tranquility had a sharper edge because of the later president's war policies. At the outbreak of the Great War, when Baldwin was exiled in France, he wrote bitter tracts decrying Wilson's morally effete neutrality.[58]

While at Princeton, Baldwin enjoyed his most creative period. He gradually worked out his theory of organic selection, the first part of which he incorporated into *Mental Development*. The second part took shape during 1896, achieving final definition by 1902. It was his theory of organic selection that came to bear the eponymous title "the Baldwin Effect." In the construction of the first part of his theory, which differs

57. Baldwin, *Between Two Wars* 1:59.
58. Many of these pieces are included in the second volume of *Between Two Wars*

significantly from the second, he analyzed mental growth as a function
of the adaptations achieved by imitation.

Organic Selection: the Theory of Individual Adaptation

A theory of individual mental development through imitation seems
liable to the objection that repetitions of past behavior (the essential
ingredient of Baldwin's conception of imitation) cannot always accommodate new situations. He, of course, recognized, though with typical
parental amazement, that despite this theoretical obstacle his children
did learn new skills. They might begin by crudely imitating some behaviors (e.g., drawing, pronouncing difficult words), but with persistence would achieve greater accuracy. Baldwin designed his theory of
organic selection to explain his children's ability to learn accommodating behaviors, not only at the level of persistent imitation but also from
the very first stages of biological and cortical imitation up through the
higher reaches of apperception and intelligence. He conceived each
stage as represented by a series of habits imitatively achieved after passing through a prior stage. Accommodation to new environments by
organic selection, then, engineered the progress of an individual
through the various levels of mental development. Though Baldwin
meant his theory of mental development to be quite general, his theory
building began with observations of his children's persistent efforts at
imitation of behaviors they had witnessed.

He analyzed persistent imitation as a process comparable to natural
selection. During attempts to learn a new behavior, two regions in the
child's brain become coordinated. One region is excited by the original
suggestive perception or image (e.g., the mother's tying her child's
shoe), the other by the child's observations of its own imitative behavior. According to Baldwin's theory, these excitations coalesce into a
greater mass, which, obeying the dynamogenic law, produces a more
diffuse reaction. With continued efforts, the exciting mass enlarges, and
consequently more elements of the original copy are reproduced in behavior. Finally success will accidentally be achieved (e.g., the child will
tie its shoe). After initial success, selection will work to eliminate those
elements that do not fit the criterion (e.g., a tied shoe).[59]

59. Baldwin, "Origin of Volition in Childhood," p. 287: "In persistent imitation the
first reaction is not repeated. Hence we must suppose the development, in a new centre,
of a function of co-ordination by which the two regions excited respectively by the original suggestion and the reported reaction coalesce in a common more voluminous and
intense stimulation of the motor centre. A movement is thus produced which, by reason

One needs to be wary of this analysis of learning, which Baldwin first offered in his *Science* article "Origin of Volition." He did not maintain that the original copy and the images of imitative attempts allowed the child to produce increasingly refined behaviors which would progressively approach the model. He rather believed (and presumably 'observed') that the child's reactions simply massed ad libitum behavioral elements, which because of their number would accidentally include more features of the model.[60] After the profusion of elements happened by chance to fit the problem situation, then a natural selection directed against the nonuseful elements would occur. "The useless elements fall away," Baldwin argued, "because they have no emphasis. The desired motor elements are reinforced by their agreement with the 'copy,' by the dwelling of attention upon them, by the pleasure which accompanies success. In short, the law of survival of the fittest by natural, or, in this case, physiological, selection assures the persistence of the reaction thus gained by effort."[61]

(Though Baldwin's concern at this time was the adaptation of the individual through learning, his use of the model of natural selection here foretold what would become an opinion convergent with that of Eric Wasmann, Hans Driesch, T. H. Morgan, and other evolutionists at the turn of the century. They thought of natural selection as a negative mechanism. It removed the unfit but did not positively select the fit. In the passage just quoted, Baldwin gives the positive role of selection to consciousness, which acts through "emphasis" and "attention." In the formulation of the second part of his theory of organic selection, he extended this executive function of consciousness to biological evolution itself.)

The model by which Baldwin articulated his theory of learning was

---

of its greater mass and diffusion, includes more of the elements of the 'copy.' This is again reported by eye or ear, giving a 'remote' excitement, which is again co-ordinated with the original stimulation and with the after effects of the earlier imitations. The result is yet another motor stimulation, or effort, of still greater mass and diffusion, which includes yet more elements of the 'copy.' And so on, until simply by its increased mass—by the greater range and variety of the motor elements enervated—the 'copy' is completely reproduced." This passage is taken over almost verbatim in *Mental Development in the Child and the Race*, p. 453.

60. D. J. Freeman-Moir—in his "The Origin of Intelligence," in *The Cognitive-Developmental Psychology of James Mark Baldwin*, p. 143—says that Baldwin's paradigm for persistent imitation "is the relatively simple case of a copy or standard set up at the beginning of a series of successive approximations." A simple inspection of the quotation in the preceding note will show that Freeman-Moir is mistaken.

61. Baldwin, "Origin of Volition in Childhood," p. 287.

natural selection, and he boasted in his "Origin of Volition" paper that this application of natural selection "has not occurred elsewhere."[62] In *Mental Development,* he pursued the natural selection interpretation of learning, but dropped the claim to priority. Lloyd Morgan and Theodor Ziehen had both published similar proposals for analyzing learning in 1891,[63] and of course William James made precisely this use of the model of natural selection in *Principles of Psychology.* In his original paper though, Baldwin seems not to have relied on these sources, at least not in a self-conscious way. In *Mental Development,* he more candidly revealed the path by which he came to his theory of accommodation by organic selection—it was laid by Herbert Spencer and Alexander Bain.

Spencer's theory (which Bain took over) attempted to explain how an organism might acquire new responses that could be inherited by its descendants.[64] He argued in the second edition (1872) of his *Principles of Psychology* that the motor response of an organism would tend to produce an overflow discharge of energy to the other muscles, resulting in random movements. If these movements happened to have a favorable consequence, then a pleasurable feeling would result. Such a feeling, Spencer went on, would then draw off energy through the nerve passages that had been so opened by chance, facilitating further energy flow along the same channels and increasing the probability of success on subsequent occasions. Spencer believed this functional reorganization of the nerves could be passed on to descendants, thus explaining evolutionary progress. Baldwin had strong reservations about the inheritance of acquired characteristics, but found the rest of Spencer's theory congenial.

Baldwin did insist, however, on an emendation he thought crucial in light of the model of natural selection. He maintained that successful

62. Ibid.

63. Lloyd Morgan had published his natural selection interpretation of individual learning in 1891 (see chapter 8). Theodor Ziehen, whom Baldwin read during this period, also advanced natural selection as a model of learning. Ziehen put it this way: "We could render the general fitness of our actions just as intelligible as the fitness of automatic and reflex acts, or the fitness of a bird's plumage. In both cases the process of selection is the essential factor in the development of this fitness. In the case of the bird's plumage, of reflex action, and to some extent of automatic action this selection is essentially a phylogenetic process; in the case of actions it is an ontogenetic process" (Theodor Ziehen, *Introduction to Physiological Psychology,* trans. C. van Liew and O. Beyer [London: Sonnenschein, 1892], p. 274). Baldwin cited part of this passage in *Mental Development in the Child and the Race,* p. 456 n. 1.

64. Herbert Spencer's theory may be found in his *Principles of Psychology,* 2d ed. (London: Williams & Norgate, 1872), 1: 544–45. Alexander Bain's version is in his *Emotions and the Will,* 3d ed. (New York: D. Appleton, 1875), pp. 318–20.

movements in themselves brought no pleasure nor consequently any repetition of behavior. Darwinian theory taught that animals were naturally selected to enjoy life-enhancing objects and events in the external environment. Baldwin argued it was thus not the actions which secured them, but the vital stimuli themselves that brought pleasure and produced the repetition of behavior which kept such stimuli in contact with the organism.[65] A small change from Spencer, one might think. Yet for Baldwin it was an important one. For being the kind of biological realist he was, he believed that success in the external environment made the difference; only by being plugged into the vital stimuli of nature could dynamogenesis operate.

Baldwin owed a large intellectual debt to Spencer and admitted as much in the preface of his *Mental Development,* where he credited both Spencer and George Romanes with inspiring his thought on the topics covered. The debt, however, extended beyond that theory of learning for which Baldwin gave special acknowledgment. More generally, Spencer provided an example of developmental analysis; he showed how to account for the different stages of mental progress by the reiterative use of a single principle. Moreover Spencer used the mental growth of the individual as a model for mental growth in the evolution of species. Baldwin also thought that the biogenic law, that ontogeny recapitulated phylogeny, when used with extreme caution offered a way of grasping the evolutionary history of human mind—it must have been similar to the evolution of individual mind.[66] Finally Spencer contended, just as Baldwin would, that individual organisms and the race, that is, the species, developed through accommodation to the external environment, the social as well as the natural environment. Spencer (along with Tarde and others) undoubtedly brought Baldwin to appreciate the function of the social environment in individual mental development. But Baldwin's own refinement of the idea into a conception of "social heredity" went beyond the received suggestions. It was this conception that may be his most lasting contribution to social and evolutionary psychology. It certainly demonstrates his peculiar genius.

Social Heredity, an Alternative to Biological Heredity

Baldwin's social theory reverberated with the evolutionary and social conceptions of the period: Wundt's ideas about the development of self-consciousness; James's depiction of the several selves with which we identify; the various features of Spencer's psychology just mentioned;

65. Baldwin, *Mental Development in the Child and the Race,* pp. 190–94.
66. Ibid., pp. 14–31.

and Tarde's notions of social imitation. Baldwin cast his experimental-ist's eye on his own children—not a harsh gaze to be sure—in order to confirm the fundamental assumptions of these theorists, that the developing self was to its very core social, that life in the family and then in the broader society shaped the particular personalities which evolved.

Baldwin was convinced that most of the social elements forming the self did not stem from seeds of physical heredity. Nor were they simply chance accretions. Cultural rules, social norms and expectations, habits of conduct peculiar to certain families—these all endured through generations. Each new generation received its social deposit from the previous. "It is inheritance," Baldwin insisted, "for it shows the attainments of the fathers handed on to the children; but it is not physical heredity, since it is not transmitted physically at birth." Nonetheless it was, in Baldwin's judgment, just as inexorable as the physical determination of Socrates' snub nose and Emperor Charles's jutting jaw.[67]

Organic selection served as the main engine of this heredity. It operated in reciprocating fashion, producing first a *projective,* then a *subjective,* and finally an *ejective* consciousness. Each of these imitative accomplishments fed the growing self with the staples of common social life. During the projective stage, the child begins to discriminate objects from people. The behavior of objects is predictable, while people capriciously provide pleasure and pain. The child then gradually becomes projectively aware of her own body, though with this difference: she feels an inner series of experiences corresponding to the outer representations. This initiates proper subjective awareness. Finally, the child ejects this subjectivity onto the persons constituting her social environment: she recognizes that they have internal experiences just as she. They are also me's. "My sense of myself," Baldwin concluded, "grows by imitation of you, and my sense of yourself grows in terms of my sense of myself. Both ego and alter are thus essentially social; each is a socius and each is an imitative creation."[68]

In *Mental Development in the Child and the Race* and its companion volume *Social and Ethical Interpretations,* Baldwin depicted social heredity as an alternate line of hereditary transmission. Not that the physical and social lines never crossed. Like James, he believed that the social environment, for example, could render an individual biologically unfit, when, say, a society eliminated its miscreants and thus removed a germ line that might have inclined to criminal behavior.[69] But generally, to

---

67. Baldwin, *Social and Ethical Interpretations,* p. 60.
68. Baldwin, *Mental Development in the Child and the Race,* p. 338.
69. Baldwin, *Social and Ethical Interpretations,* pp. 77–78.

account for the transmission of social behaviors, social heredity offered a mechanism superior to its biological rivals (Darwinian and Lamarckian). For Baldwin, the advantages of social heredity in explaining the transmission of knowledge and morality demonstrated its importance both as an authentic evolutionary principle and as a needed complement to theories of physical heredity.

The Social Evolution of Knowledge and Morality

The Darwinian Theory of Knowledge and Truth

William James produced the first thoroughgoing Darwinian epistemology. He proposed that creative ideas were the result of selection of fit thought variations from among the multitude spontaneously generated. But like the Darwin of the *Origin,* he had no theory of heredity. Without such a theory, the vast amount of our more pedestrian knowledge lies unexplained; for most of that knowledge is acquired socially, through listening to others, reading, imitative practice, and the application of learned heuristics and algorithms (e.g., adding a column of figures and discovering their sum). And even in our more creative endeavors, thought trials are not sheer chance occurrences. Idea production must be constrained, else an infinity of worthless thought variations would gush forth, making the selection of fit, that is, relatively true ideas entirely miraculous.[70] In Baldwin's judgment, James's evolutionary epistemology thus failed on two accounts: it ignored the social aspect of knowledge, and it seemed to deny any constraints on the production of mental variations. Baldwin constructed his theory of social heredity to remedy these deficiencies.

The theory of social heredity, particularly as Baldwin developed it in *Social and Ethical Interpretations* and *Development and Evolution* (1902), construed knowledge as social in two senses. First, according to the theory, the very meaning of truth includes the notion of social confirmability. When a thinker asserts to herself or to others that something is truly the case, she concomitantly conceives other people as agreeing. In affirming something, we also implicitly claim that the situation is (or should be) similarly understood by others. In Baldwin's terms, our thought is ejective: it imposes a subjective state on others and then assimilates the construction of others again to our own perception of the situation.[71]

Baldwin regarded knowledge as social also in a second sense. Tradi-

70. Baldwin, *Development and Evolution,* p. 241.
71. Baldwin, *Social and Ethical Interpretations,* pp. 112–15.

tions of knowledge become established in a society and form the hereditary deposit for each generation. A tradition consists of ideas that are fit. Though, according to Baldwin, their fitness "is not in any sense fitness for struggle; it is fitness for imitative reproduction and application."[72] In making this emendation for conceptual evolution, however, he misinterpreted the biological analogue: in organic evolution fitness is also a matter of differential reproduction and only metaphorically a matter of struggle. Conceptual evolution is no different, even as Baldwin proposed it: the differential spread of ideas in an environment evinces fitness. Now as Baldwin insightfully recognized, once ideas have been selected and have infiltrated a society, they become part of the environment against which the new ideas of those living in that society are selected. Ultimately, "the environment of thought can only be thoughts; only processes of thought can influence thoughts and be influenced by them."[73] This is the further sense in which knowledge, which results from the selection of ideas in particular environments, is a social construction.

The second contribution that the theory of social heredity made to evolutionary epistemology was the postulation of levels of organization to mark off the conceptual space of thought variations. Idea generation is not scatterbrained. In attempting to solve a problem, the thinker must generate thoughts within certain limits, lest an infinite variety of completely worthless ideas pass through his head. The "thought-variations by the supply of which selective thinking proceeds," Baldwin supposed, "occur in the processes at the level of organization which the system in question has already reached—a level which is thus the platform for further determinations in the same system."[74] Baldwin maintained that the fundamental level of knowledge organization consisted in the motor adaptations of the child to the physical world. Movement variations thus must precede thought variations. The child accommodates to his world by reconstructing it in movement. The adaptations of attention, wherein a conceptualized world takes shape, will be selected for, then, not immediately against the physical world but against the motor reconstruction of that world. For Baldwin, this motor accommodation constitutes our initial and mute contact with the external world.[75] The child thereafter slowly builds up a fund of knowledge that is in increasingly remote contact with this original but ultimate rock of reality.

72. Ibid., p. 183.
73. Baldwin, *Development and Evolution*, pp. 260–61.
74. Ibid., p. 243.
75. Baldwin, *Social and Ethical Interpretations*, p. 97.

Each stage of already-achieved knowledge provides the main features of the environment against which new ideas are tested. In Baldwin's view, they are tested in two ways. First is the test of habit. Already-achieved knowledge easily accommodates that which is familiar, that which is, as it were, preadapted for its environment, as when we learn, for example, that another Chicago politician has been indicted for graft. More novel ideas, for instance those that may be found as hypotheses in science, will not be readily assimilated to our habitual knowledge (i.e., our lived, thoroughly-taken-for-granted experience); but they must be familiar enough to allow some connections with that foundation, lest they not enter consciousness at all. If, say, a scientific hypothesis cannot be confirmed by our established ideas, then it must be tested against a different standard. It must be more reflectively selected against a refined conceptual environment of acceptable theories and established facts. As new ideas pass this muster, they gradually enlarge the deposit of habitual knowledge. This legacy, then, becomes our guarantee of the real. Echoing James's conception of reality, Baldwin argued that to regard something as real was to make its idea "part of that copy system which hangs together in our memory, as representing a consistent course of conduct and the best adjustment we have been able to effect to our physical and moral environment."[76] In this Darwinian conception of reality, truth still means "correspondence with the world," but it is a world sifted through socially constructed knowledge, that is, it is largely a social world.[77]

Moral Evolution

Under the heading "Emotions of Relation," Baldwin had discussed ethical feeling in the second volume of his *Handbook*. The conception expressed there differed from the moral sense theory of McCosh, who simply proclaimed that just as the physical eye discriminated color, so the moral eye perceived the hues of good and evil.[78] Baldwin proposed, rather, that human conscience judged a given act as good or bad in relation to a moral ideal, whose objective character consisted of three

76. Ibid., p. 324.
77. Baldwin's theory of the adaptation of ideas is remarkably consilient with views of the eminent population geneticist Richard Lewontin, who has recently insisted that animals adapt or construct their environment as much as they become adapted to it. See Richard Lewontin, "How Do We Explain the Major Features of Evolution," Fishbein Center Symposium on Persistent Controversies in Evolutionary Theory, The University of Chicago, March 1982.
78. James McCosh, *Psychology, The Motive Powers* (New York: Scribner's Sons, 1888), pp. 195–200.

elements: the harmony of the act with established interests, its approval by others, and its imperative character.[79] "The ethical ideal, therefore, as far as it is conscious, is the degree of harmony and universality in conduct which I find my emotional nature responding to with imperative urgency."[80] Baldwin's conception had Kantian features, but already a strong social component. The latter was intensified in the evolutionary interpretation of morality he finally produced in *Mental Development* and *Interpretations*. The principal inspiration for his new theory was not Kant, though, but a couple of recalcitrant children.

In Baldwin's revised theory, morality consisted essentially in the construction of an ideal self from socially inherited norms, initially those of the parents. He developed this conception by specifying a particular dimension in the evolution of the self. As in the earlier moral theory, there were three levels. First, the child puts on the "self of habit," the already settled accommodations and acquired practices. Concomitantly active is the "accommodating self," the self that "learns, that imitates, that accommodates to new suggestions from persons in the family and elsewhere."[81] This self continually reconstructs the habitual self. In time the "ethical self" emerges. It is formed through the child's obedient submission to the will of another. The child finds itself constantly stimulated, usually by the stinging hand of authority, to deny impulses coming from the habitual self. But what begins as a projective awareness of authoritative coercion gradually insinuates itself into the subjective role of internal guide. Then, in the ejective mode, the child comes to expect others to adhere to what it understands to be the rules of conduct. Finally, through the continuing dialectic of projection, subjection, and ejection, the child begins to absorb the attitudes of the various other ethical selves of its social environment. The subjective judge of what is law-abiding, correct, and required thus becomes a public self, a truly universal moral authority. The sense of moral obligation, in this interpretation, stems from a lack of harmony between the habitual and the ethical self. "My sense of moral ideal, therefore," Baldwin concluded, "is my sense of a possible perfect, regular will, taken over *in me,* in which the personal and the social self—my habits and my social calls,—are brought completely into harmony; the sense of obligation in me, in each case, is a sense of lack of harmony—a sense of the actual discrepancies in my various thoughts of self, as my actions and tendencies give rise to them."[82] Baldwin's new theory of morality (new, that

79. James Mark Baldwin, *Handbook of Psychology: Feeling and Will,* pp. 212–14.
80. Ibid., p. 226.
81. Baldwin, *Social and Ethical Interpretations,* p. 34.
82. Baldwin, *Mental Development in the Child and the Race,* p. 345.

is, in the way evolved species are new) developed considerations already present in the *Handbook,* but provided a detailed account of the growth of the ideal self, showing it to be a thoroughly social product.

Baldwin believed his analysis escaped the two-pronged criticism Thomas Huxley leveled in his "Romanes Lecture" against evolutionary theories of morals. Huxley argued (in Baldwin's reconstruction), first, that if feelings of obligation arose from lack of assimilation of new elements to old categories of action (e.g., when a given act ignored the pull of instinctive sympathy, as in Darwin's theory), then all such lack of assimilation (e.g., inability to walk steadily on ice) should generate feelings of ethical obligation.[83] Baldwin thought this a fair criticism of Spencer and Darwin. Ethical judgments should be ruled by considerations other than the accidental suggestions of sympathy.[84] But he found that his own theory could comfortably escape Huxley's objection. For the child's imitative growth into a sense of ideal personality set a "higher category of action than either of the two concrete categories recognized by Darwin, Spencer, and the naturalists generally, i.e., those of spontaneous egoism and equally spontaneous generosity or sympathy."[85] (The reader sympathetic to either Darwin's or Spencer's theory of morals might find more to them than Baldwin's jejune sketch suggests.)

Huxley's second criticism hit the ethics of evolutionary naturalism more squarely. He simply pointed out that evolved impulses to perform particular acts had no guarantee of moral appropriateness. One could always morally and reasonably ask: Ought I heed my impulses? And the moral answer would frequently be no. As Huxley urged, we must often fight against the "cosmic process." Whatever is, is not necesssarily what morally ought to be.

Baldwin responded, as every evolutionary moralist must, by deriving an "ought" from an "is". The sense of "ought" reflects, he argued, our anticipation of experience, that is, anticipation of what our ideal personality, which waits in our future, would regard as its experience. Hence what from our present perspective is an "ought," from the perspective of our ideal self, that self toward which we tend and which is involved in the "cosmic process," is an expression of what ideally will be.[86]

83. Baldwin picked out two of the most salient of the several criticisms Huxley brought against an evolutionary construction of ethics. See Thomas Huxley, "Evolution and Ethics" (The Romanes Lecture for 1893) and "Evolution and Ethics: Prolegomena" (1894), in *Collected Essays* (New York: D. Appleton, 1896–1902), 9:1–116. See also chapter 7.

84. Baldwin, *Social and Ethical Interpretations,* p. 43.

85. Ibid., p. 307.

86. I believe I have interpreted Baldwin correctly here. His own statement has the deep clarity of the Chicago River on St. Patrick's Day: "The sense of ought, then, from my

Baldwin failed, I think, to respond adequately to Huxley's objection, since even the anticipations of the ideal self presumably reflected the merely factitious consequences of social heredity. One would be as moral as the nurturing society; but to admit that appears to reintroduce the problem. (Baldwin would likely have been more sensitive to this difficulty had he already experienced what he came to regard as the perfidy of German culture and society during the Great War.) He perhaps believed, however, that his moral theory nonetheless escaped the naturalistic fallacy, since, as his analysis made clear, social evolution unfolded relatively independently of biological evolution. In the ethical sphere, he concluded in *Social and Ethical Interpretations*, "there seems to be *very little natural heredity, and a great deal of plasticity; in short, a great deal of social heredity.*"[87] Social heredity, being the deposit of social norms and moral standards, would be autonomous and thus could run counter to the cosmic process of biological heredity. Yet at approximately the same time he penned this last passage, probably early in 1896, Baldwin was refitting his theory of organic selection to blur the line between biological and social heredity; he was doing so, however, in a way that would allow social heredity to lay down a track to be followed by biological heredity. In short, he began to construct a theory that would postulate mind as a moral force directing the cosmic process. We will next examine his reformulated theory of organic selection and its consequences for evolutionary ethics.

Organic Selection and the Politics of Scientific Discovery

A New Factor in Evolution

By the spring of 1896 Baldwin believed he had discovered a new and extremely important mechanism of evolutionary development. The mechanism conformed to ultra-Darwinian assumptions, but nonetheless allowed consciousness and intelligence a role in directing evolution. By philosophic disposition and conviction, Baldwin was a spiritualistic metaphysician. He felt the beat of consciousness in the universe; it pulsed through all the levels of organic life. Yet he understood the

---

point of view, is the anticipation of more experience, not yet reached under the rubrics of description; but so far as it is identified with any object of desire, so far it is thought to exemplify the canons of description of that object, as being most nearly the sort of experience that expectation is reaching toward. And natural science, the 'cosmic process,' *is the same series read backward.*" See James Mark Baldwin, "The Cosmic and the Moral," in *Fragments in Philosophy and Science*, p. 73. The essay was first published in *International Journal of Ethics* 6 (1895): 93–97.

87. Baldwin, *Social and Ethical Interpretations*, p. 303.

Figure 10.2 James Mark Baldwin (standing, left), E. B. Poulton (seated, left), Robert Bridges (seated, right), photograph from 1900.

power of mechanistic explanations of evolution. Indeed, the mechanisms of variation and natural selection appealed to the metaphysician who understood both sides of the Spinozistic equation *Deus sive Natura*. The Lamarckian principle of the inheritance of acquired habit could not even explain individual learning, much less the way a species learned to adapt to new circumstances. Baldwin compared Lamarckian theory to special creationism, since it required the prescient exhibition of preadapted behavior.[88] Yet some of the objections to ultra-Darwinism voiced by neo-Lamarckians such as Spencer and Cope did leave nagging difficulties. Baldwin thought his new principle could handle these difficulties and thus vindicate Darwinism. The new principle appeared to him, and to many of his contemporaries, just as important as Darwin's natural selection.

88. Baldwin, *Development and Evolution,* p. 88.

Before tracing the twisting path by which Baldwin made his way to the principle and the straighter road he himself described, a general characterization of the new factor will provide some direction. Baldwin termed the new factor "organic selection," thereby forging a link with the ideas previously traveling under that name. He kept the confusingly identical designation for strategic and political reasons, as I will explain in a moment. The new principle of organic selection derived from the hard-won recognition, for which Baldwin could take some credit, of two evolutionary facts we now take for granted: first, that "all characters [of an organism] are partly congenital and partly acquired"—that the hereditary *Anlage* develops only in ways that the environment permits; and second, that natural selection operates immediately only on traits with this dual determination.[89] From these two facts the conclusion seemed to follow that acquired characteristics did influence the operations of natural selection and thus the direction of evolution. Baldwin's principle of organic selection embodied these facts and conclusion. In 1902 he formulated the principle in this way:

> This position is the general one that it is the individual accommodations which set the direction of evolution, that is, which determine it; for if we grant that all mature characters are the result of hereditary impulse plus accommodation, then only those forms can live in which *congenital variation is in some way either 'coincident' with, or correlated with the individual accommodations which serve to bring the creatures to maturity. Variations which aid the creatures in their struggle for existence will, where definite congenital endowment is of utility, be taken up by the accommodation processes, and thus accumulated to the perfection of certain characters and functions.*[90]

Baldwin envisioned as the paradigm for his principle animals' acquiring innate behaviors similar to behaviors they originally had to learn. If a group of animals migrates into a new environment for which they initially lack congenital adaptations, those plastic enough to accommodate themselves through conscious learning will survive. Their ontogenetic behaviors will buffer them against the winnowing hand of natural selection. This safety net, according to Baldwin, will allow natural selection opportunity to accumulate chance variations that follow the path laid down by the acquired behaviors, which indeed have already been favored by selection.

Baldwin announced this principle as a "New Factor in Evolution" in

89. Ibid., p. 35.
90. Ibid., pp. 37–38.

the June 1896 issue of *The American Naturalist,*[91] a journal edited by the neo-Lamarckian E. D. Cope. In the article Baldwin tried to show the power of organic selection to deal with the objections of the Lamarckians. It could explain the toughest difficulties they threw up, those based on coadapted variations, difficulties that compelled even Darwin's disciple George Romanes to admit Lamarckian inheritance. Romanes or Spencer might offer as an instance of such objections the complex of coadaptive instincts displayed by the honeybee in constructing its cells. It appeared unlikely that the crude mechanism of chance variation and selection could coadapt and finely tune the bee's instincts for measuring sixty-degree angles, producing six planar walls, constructing romboidal surfaces, abutting the cells back to back, and stacking them to achieve the least expenditure of wax. Moreover the selective values of these instincts were interdependent, since the ability to measure sixty-degree angles, for example, would be useless unless the bee also had the instinct for completing an interior surface using six sides, and vice versa. Hence even if harmonious variations appeared over long stretches of time, they could not be selected seriatim. Each by itself had no selective value. And it was infinitely improbable that all elements would chance to arise at once. The only solution seemed to be Spencer's. He contended that intelligence would allow an animal to acquire complex habits that would later solidify into instincts. But such transformation required Lamarckian inheritance, or so it seemed. Baldwin offered another answer, one he thought rescued neo-Darwinism. The solution of organic selection supposed that an animal's conscious intelligence might, in response to an environmental need, initially produce a coadaptive behavior system which would stave off extinction. But then natural selection could begin to save up those congenital variations that chanced to appear; the selective value of the system's elements would be maintained, while physical evolution gradually replaced learned traits with instinctive ones.

The theory of organic selection had another dividend, as Baldwin undoubtedly felt satisfaction in demonstrating in the journal of America's leading Lamarckian paleontologist. His theory could help explain both the progressive character of evolution and the gaps in the fossil record. The traits of a species, left to the devices of natural selection alone, would vary randomly around some relatively fixed mean. But imitative learning, he proposed, could provide a definite direction and would accumulate adaptive, socially inherited traits each generation that

91. James Mark Baldwin, "A New Factor in Evolution," *American Naturalist* 30 (1896): 441–51, 536–53.

would move a species away from the previous population mean. Natural selection would then plod steadily behind, nailing the traits down with physical heredity. "So there is," Baldwin urged upon the readers of the *American Naturalist,* "continual phylogenetic progress in the directions set by ontogenetic adaptations."[92] Moreover the theory might also be used to help account for the gaps in the fossil record. If a sweeping change in the environment occurred, then those animals that survived, because their innately determined traits lay at some distance from the population average, might further their cause if they could also intelligently accommodate to the new order. These ontogenetic accommodations, wedded to the sample of deviant traits, would leave abrupt transitions in the fossil record.[93] Lamarckian expedients were no longer necessary.

Baldwin's "New Factor" paper emphasized what we might now regard as his most important insight, the social aspect of evolution. He insisted, and with justice, that social relations and traditions, no less than the rocks and streams of nature's colder side, formed the environment in which natural selection operated. He believed that social progress served as the template for organic selection; it laid out the path for phylogenetic development. But even without the operation of the so-called "Baldwin effect," social evolution might, its author thought, explain a nagging problem that Darwin and his critics had long since recognized: that advanced society might inhibit its own progress, since the less fit would be preserved by reason of social sympathy. Baldwin felt confident that the progress of the human race was insured, since social heredity could fill in for physical heredity "by making the individual learn what the race has learned, thus preventing social retrogression."[94]

Organic Selection and Ethics

Baldwin's principle of organic selection had important consequences for his moral theory. Prior to discovering the principle, he had constructed his theory of moral development in terms of the older principle of organic selection, that is, in terms of social learning transmitted through social heredity, not biological heredity. This meant that ultimately moral norms were vested in individual lineages within a society, but that no guarantee could be provided that the particular strains of norms would have universal moral validity. Moreover the social rules of

92. Ibid., p. 448.
93. Ibid., p. 450.
94. Ibid., p. 539.

a society might be peculiar to it and the contingent result of an idiosyncratic history. Yet moral rules, certainly as Baldwin understood them, needed to be valid for any rational being. They could not be unique and factitious, but had to be universal and apodeictic. His new principle of organic selection, coupled with a study of Darwin's theory of group selection, prompted him to conceive the evolution of a moral sense as having precisely the required universalizing quality.

He first applied the principle of organic selection to the development of the moral faculty in his *Darwin and the Humanities* (1909), a monograph written on the occasion of the fiftieth anniversary of the *Origin of Species*. Baldwin prepared for his study by reexamining *Descent of Man* and focusing especially on Darwin's theory of community selection. In his monograph, Baldwin came to view the origin of the moral sense much as Darwin had. That is, Baldwin held that the altruistic sentiments evolved within human societies because of group selection. Those societies within which individuals adopted altruistic practices would be selected for in competition with other groups of more egoistically disposed members. Natural selection thus worked not to preserve utterly selfish individuals but to maintain socially cohesive groups composed of other-regarding individuals. As argued in *Darwin and the Humanities*: "the fitness of the group for its struggle *requires organization within the group, and this in turn requires a socialized rather than an egoistic individual* . . . Utility for the group *presupposes self-control and altruism in the individual*. It is the extension of the application of natural selection to groups, rather than its direct application to individuals, that has given birth to morals."[95]

The process of organic selection did not, as Baldwin formulated it, dictate the relative ratio of social heredity to biological heredity in the maintenance of any trait. In *Darwin and the Humanities,* he gave only a little weight to biological heredity in determining the sense of altruism. He rather suggested that the social instincts, though indeed biologically inherited products of group selection, formed the bare skeleton upon which hung the real muscle of learned social behaviors. "Sympathy and altruism," he cautioned, "are the socialized and transformed impulses of the growing individual, who is educated into a higher selfhood."[96] Even though social heredity largely produced the specific moral character of the individual, the principle of organic selection, as applied now to the group, nonetheless provided the universalizing form: because

95. James Mark Baldwin, *Darwin and the Humanities* (Baltimore: Review, 1909), p. 88. This monograph was published as volume 2 of "Library of Genetic Science and Philosophy," a *Psychological Review* series.

96. Ibid., p. 65.

organic selection operated to preserve those communities that had a generalized social structure composed of unselfish, sympathetic, and altruistic individuals. Presumably, these faculties would evolve in any society, regardless of its peculiar circumstances, just as would, for example, the general faculty of parental nurture or even of intelligence. Moreover the moral sense was not required to be a purely biological product in order to be a fitness trait of social groups and thus be selected for. The theory of organic selection, of course, implied that the moral sense would gradually become a more biologically fixed faculty, but its moral character did not depend on that consummation. Though initially highly critical of Darwin's moral theory, Baldwin came to adopt a more advanced variation of it.

The view at the turn of the century might easily have persuaded an observer of science that organic selection made an important contribution to the understanding of evolution. It was a universal principle, since even well-fixed innate traits manifested the influence of their ontogenetic environment.[97] The principle certainly seemed to dispatch Lamarckism, while supplying that positive factor in evolution for which even staunch Darwininists like Lloyd Morgan longed. And to those of metaphysical appetite, it revealed that under the clanking, mechanical vesture of Darwinian nature, mind could be found. This suggested that an evolutionary account of ethics need not reduce moral behavior to the blind play of cosmic forces, but that it might rather elevate the forces of evolution, interpreting them as conscious, moral agents. The principle even gave comfort to the profoundest of religious sentiments. "It is natural," Baldwin observed, "to look upon the class of phenomena which show the mind taking part in the determination of natural evolution as being in some way in harmony with, or as furthering, the operation of the larger Purpose which a theory of cosmic teleology postulates."[98] The discovery of the principle shone brightly, especially on Baldwin. Only the shadows from Lloyd Morgan and Osborn dimmed the glory. For they also claimed to have discovered the principle in 1896.

The Politics of a Scientific Discovery

James McKeen Cattell reviewed Baldwin's "New Factor" paper in the *Psychological Review* of September 1896.[99] He began his short review by noting that Morgan, Osborn, and Baldwin all seemed to have arrived

97. Baldwin, *Development and Evolution,* p. 183.
98. Ibid., p. 237.
99. James McKeen Cattell, "Review of 'A New Factor in Evolution by J. Mark Baldwin,'" *Psychological Review* 3 (1896): 571–72.

Figure 10.3 James McKeen Cattell, 1860–1944,
photograph from ca. 1900.

independently at the "new factor" and at about the same time. After a
brief analysis of the principle, he concluded with the complaint that
readers would find "the author's vigorous thinking too often obscure to
an unfortunate degree."[100] The opacity of Baldwin's article was due primarily to the way he constructed it. For the most part he simply pasted
together quotations from the 1895 book *Mental Development in the Child
and the Race* and from a series of six papers he had published from
January 1894 to May 1896.[101] The chaos of quotations from various

100. Ibid., p. 572.
101. The articles from which Baldwin formed the mosaic of his "New Factor" paper
were: "Imitation: A Chapter in the Natural History of Consciousness," *Mind* 19 (1894):
26–55; "Consciousness and Evolution," *Science* 2 (1895): 219–23; "Heredity and Instinct
(I)," *Science* 3 (1896): 438–41; "Heredity and Instinct (II)," *Science* 3 (1896): 558–61;
"Physical and Social Heredity," *American Naturalist* 30 (1896): 422–30; and "Consciousness and Evolution," *Psychological Review* 3 (1896): 300–8. The last article was a report of
the discussion at the American Psychological Association Meeting in December 1895.

sources produced a bewildering effect, but Cattell seems to have caught Baldwin's intention. He wondered whether Baldwin was implying that the principle of organic selection could already be found in *Mental Development*. He protested that his memory of the book would not let him decide the question.

Baldwin quickly responded to the review in Cattell's own journal *Science*. He admitted the obscurity of the presentation, due, he explained, to the need for condensation. And yes, he meant to suggest that the principle of organic selection could be found in his book. He then offered several quotations to demonstrate the fact. The first typifies the rest:

> "It is necessary to consider further how certain reactions of one single organism can be selected so as to adapt the organism better and give it a life history. Let us at the outset call this process 'organic selection,' in contrast with the natural selection of whole organisms." . . . "The facts show that individual organisms do acquire new adaptations in their lifetime, and that is our first problem. If, in solving it, we find a principle which may also serve as a principle of race development, then we may possibly use it against the 'all-sufficiency of natural selection,' or in its support." [102]

An artful quotation, as were the others. In their original contexts, these passages told a different story. In *Mental Development,* Baldwin used the term "organic selection" to refer only to his theory of individual learning, the theory in light of which he proposed that social heredity formed a line of transmission parallel to and usually quite separate from that of biological heredity. The ellipses of the quotations tracked over passages in the book where it was clear that the "principle of race development" referred to social heredity transmitted via social learning. Nowhere in the book did Baldwin even suggest that social acquisitions might prepare the way for and promote biological adaptations, the idea forming the heart of the "new factor." *Mental Development* argued, on the contrary, that the principle of social heredity eliminated the need for any biological theory (Darwinian or Lamarckian) to explain social behavior.

Baldwin never overtly claimed priority for the discovery of organic selection. But in his discussions of the principle in the succeeding years, he staked out his claim for all to see. First, he got hold of the name of

102. James Mark Baldwin, "On Criticisms of Organic Selection," *Science* 4 (1896): 727. The passages in this quotation were from *Mental Development in the Child and the Race,* pp. 174–76.

the concept. In a *Science* article of 1897, he attempted to establish a canon of terms (e.g., "Organic Selection," "Social Heredity," and "Orthoplasy"—the last designating the directive influence of organic selection in evolution) in order "to facilitate the discussion of these problems of organic and mental evolution."[103] A few years later he wrote the article describing the concept for his *Dictionary of Psychology and Philosophy* and persuaded Lloyd Morgan, E. B. Poulton, and G. F. Stout also to sign it; the article bore the title "Organic Selection." In the bibliography appended to the article, Baldwin's own *Mental Development* is listed, with the note: "where the term organic selection was first used." Then came further editions of *Mental Development*. In the French and German editions, he added a chapter describing at some length the new version of organic selection. In the third English edition (the second was a reprint a few months after the first), he did not include a new chapter, but did emend the text and sprinkle in footnotes which urged the reader to find the new factor in the original text. So, for instance, the note to the second part of the long quotation above reads: "This passage anticipates the explicit development of 'organic selection' in later publications—the view, that is, that individual accommodations, by supplementing certain variations, guide evolution in definite lines."[104] In 1902 Baldwin published *Development and Evolution,* a book that incorporated all his earlier articles that touched on organic selection (even those that described only the first version).

Scientists and other scholars make arguments explicitly and implicitly, in the text and through the text, by logic and the appeals to evidence, and by the style and manners of the tribe. For the adepts of the scientific community, Baldwin's claims to priority were clear and persuasive. One only need be reminded, we now refer to the new factor as "the Baldwin effect."

I wish to emphasize, I do not believe Baldwin did anything intentionally deceitful in marshalling his argument in the way he did. When his mind was drawn directly to the question of priority, he frankly acknowledged the work of Morgan and Osborn. He wrote Morgan on 20 November 1896:

> As you have seen in all that I have printed, I have taken pains to refer to you as having reached similar views. . . . So I suggest that we mention each other in our respective books on having reached the position independently (citing references) & thus

103. James Mark Baldwin, "Organic Selection," *Science* 5 (1897): 634–36. Baldwin published a similarly titled essay having a similar purpose in *Nature* 55 (1897): 558.

104. James Mark Baldwin, *Mental Development in the Child and the Race,* 3d ed. (New York: Macmillan, [1906] 1925), p. 167; see also, for example, the note on p. 165 and the emendation of the text on p. 195.

avoid the abominable discussions by outsiders respecting priority, of which I have had enough of from my use quite inno-
cently, following Osborn, of the word "new" in my Naturalist
article.[105]

My argument is that, like many other ambitious, aggressive, and gifted
scientists, Baldwin absorbed into his blood the social norms and customs of his community. He implicitly understood that getting to an
idea first counted the most; independent discovery, if at a later time,
got only honorable mention (a message pounded home by David
Hull).[106] He knew that controlling the terms under which ideas traveled
and making others use your language identified the ideas with you. This
tacit knowledge of the social structure of science enables the scientist to
foster the survival of his ideas. And survival down the road is the surest
sign, in the evolution of organisms as well as ideas, of original fitness.

Evolution of the Principle of Organic Selection

The actual lineage of Baldwin's principle of organic selection was
somewhat different than the one he reconstructed. It shows the slow
and fitful emergence of a creative idea.

The theory of organic selection, as Baldwin finally formulated it, had
components specifying the mechanism of individual learning and characterizing the social heredity that resulted. These parts of the theory
slowly evolved in the years prior to 1894. The books *Mental Development
in the Child and the Race* and *Social and Ethical Interpretations* reflected
Baldwin's refinement of these notions, but they did not contain the
distinctive proposal that social learning prepared the way for physical
inheritance. This last idea flickered into life in August 1895 in a paper,
"Consciousness and Evolution," which was devoted to a criticism of
Edward Cope's insistence that Lamarckian mechanisms were the only
ones allowing for the necessary workings of consciousness in evolution.

Baldwin responded to Cope that the preformist view (i.e., Weismann's ultra-Darwinism) could admit consciousness as a naturally selected variation; the plastic and variable actions of consciousness would
then ultimately be due to preformist factors. Baldwin further suggested,
in vague and groping terms, a consequent effect consciousness might
have on the course of further biological evolution: "these most plastic

105. James Mark Baldwin to Conwy Lloyd Morgan (20 November 1896), Lloyd Morgan Papers 128/81, University Library, Bristol, England.
106. On numerous occasions of personal communication and argument. See also
David Hull, "Exemplars and Scientific Change," *PSA 1982* 2 (1983): 479–503.

individuals will be preserved to do the advantageous things for which their variations show them to be the most fit. And the next generation will show an emphasis of just this direction in its variations. So the fact of Social Heredity—the fact of active use of consciousness in ontogeny—becomes an element in phylogeny, also, even on the Preformist theory." [107] In his "New Factor" paper, Baldwin would cite this passage as evidence that he had the idea in 1895. At this point he may have had it, but he certainly did not know that he did. Like Darwin before reading Malthus, Baldwin required the right circumstances for the central idea to break into awareness, and those circumstances had not yet arrived.

That the significance of his own remarks failed to register with him is evident from his later encounter with Cope, this at the American Psychological Association meeting on 28 December 1895. Baldwin, Cope, and James delivered papers on the role of consciousness in evolution. Baldwin contended that social heredity could perform all the tasks Cope had reserved for a Lamarckian physical heredity. As he remarked, with uncharacteristic clarity: "As soon as there is much development of mind, the gregarious or social life begins; and in it we have a new way of transmitting the acquisitions of one generation to another, which tends to supersede the action—if it exists—of natural heredity in such transmissions." [108] If Baldwin had really employed the principle of organic selection in his previous encounter with Cope, one would think he would roll it out again when the issue and the opponent were the same. But as the quotation shows, Baldwin thrust forward the conclusion of *Mental Development*, which was contrary to the new idea of organic selection. At the end of 1895, Baldwin still did not have the principle, at least not in a self-conscious way.

His first definite expression of organic selection, though not under that name, came in a paper completed on 5 February 1896 and published in the March issue of *Science*. The paper was a revision of a talk he had delivered before the New York Academy of Sciences on 31 January 1896, a talk which immediately followed one by Morgan, who was then visiting the States. Morgan had sketched the principle in his lecture, and Baldwin followed with his own version. Baldwin appended a note to his published talk in which he asserted his independence of Morgan. [109] A note was needed, since it undoubtedly appeared then, as it does now,

107. Baldwin, "Consciousness and Evolution," *Science* 2 (1895): 221.
108. Baldwin, "Consciousness and Evolution," *Psychological Review* 3 (1896): 301.
109. Baldwin, "Heredity and Instinct (I)," p. 441.

an extraordinary coincidence that they should have announced the same, but independently derived, theories precisely on the same day, within a few minutes of each other.

A historian who appeals to evolutionary theory for his model must be prepared for chance occurrences. History simply does not have a Spencerian inevitability. Accident did play a role in these events, I believe. But the particular developmental histories of the ideas of Morgan and Baldwin make intelligible this sort of convergent evolution.

Both Morgan and Baldwin had previously adopted natural selection as a model to explain ontogenetic learning and the transmission of social behavior and culture; they offered these social-cultural mechanisms as a substitute for Lamarckian heredity.[110] Both men accorded biological evolution the role of producing a plastic-enough intelligence to make such transmission possible. In 1895 Morgan had edited and published the posthumous second volume of George Romanes's series *Darwin and After Darwin*. It was this book on which Baldwin chose to comment at the New York meeting. In the book Romanes set out to defend the inheritance of acquired characteristics against neo-Darwinians such as Wallace and Weismann. As a prelude to his defense, he sketched a theory of mental and cultural evolution—directed by "a kind of non-physical natural selection"—almost identical to that of Morgan and Baldwin.[111] He then marshalled his arguments, the most forceful being those based on the evidence of coadaptive traits in reflex behavior and instinct, arguments which, as we have seen, Baldwin would later combat by using the principle of organic selection. Another kind of argument he offered, however, caught Baldwin's eye and, I think, may have provided just the right atmosphere for the emergence of the new idea of organic selection. Romanes proposed that often an organism's intelligent reaction to its environment would protect it from predation by natural selection, so that those reflexes which might be duplicated by intelligent responses (e.g., voluntarily removing one's hand from a hot stove) could have no selective value (since they were redundant), and thus use inheritance, not natural selection, had to be their source. In the published version of his talk, Baldwin dwelt on Romanes's observation that intelligence might buffer an individual from natural selection. This, of course, is central to the concept of organic selection. Further, Romanes suggested that intelligently acquired modifications, if at all heritable, would be extremely important "in furnishing to natural

110. C. Lloyd Morgan, *Animal Life and Intelligence* (Boston: Ginn, 1891), pp. 486–96.
111. George Romanes, *Post-Darwinian Questions, Heredity and Utility,* vol 2 of *Darwin and After Darwin,* ed. C. Lloyd Morgan (Chicago: Open Court, 1895), p. 33.

selection ready-made variations in required directions, as distinguished from promiscuous variations in all directions."[112] And this is the final element in the idea of organic selection, the notion that acquired modifications paved the way for natural selection.

Well, Romanes might have provided the common problem environment against which the theories of Baldwin and Morgan evolved. But evolutionists—whether epistemological or biological—like to tell just-so stories, and perhaps this is one.

The Fate of Organic Selection

During 1896 and for several years thereafter, the scientific literature was replete with discussions of organic selection. Part of the subsequent debate took up the question of priority of discovery. Cattell nettled Baldwin by suggesting that the principle could be found in Darwin. Osborn insisted that Weismann had expressed it in his Romanes Lecture of 1894.[113] And Morgan made substantial use of Weismann in explicating the principle in his own work. Of course reservations were expressed by many. Yet the principle seemed to hold great promise. Alfred Wallace endorsed it as a substantial contribution to evolutionary theory that removed all need for appealing to Lamarckian mechanisms,[114] while Osborn recommended it because it reconciled Lamarckism and Darwinism.[115]

112. Ibid., p. 50.

113. August Weismann, *The Romanes Lecture, 1894: The Effects of External Influences upon Development* (Oxford: Clarendon Press, 1894). In his lecture, Weismann advanced his theory of "intraselection of parts" to answer Spencer's objection that coadaptation required inheritance of individual functional acquisitions. Weismann argued that during ontogenetic development an individual's organic parts would undergo competition and selection (though the outcomes would not be heritable). This would allow internal coadaptation to a part that spontaneously changed due to chance alterations of the germ plasm. "Thus when an advantageous increase in the size of the antlers [i.e., Spencer's example of the coadaptations required for large-antlered deer] has taken place, it does not lead to the destruction of the animal in consequence of the other parts being unable to suit themselves to it" (p. 18). As further phyletic changes gradually occurred, "the secondary adaptations would probably as a rule be able to keep pace with them" (p. 19). As Baldwin pointed out, Weismann's theory entailed that individual modifications followed the phyletic variations; they did not set the prior direction for them. Hence Weismann had not really discovered the principle of organic selection. See also chapter 8.

114. In his review of Lloyd Morgan's *Habit and Instinct,* Alfred Wallace enthusiastically endorsed the principle of organic selection, agreeing that it eviscerated the objections of Spencer and Romanes based on coadaptive variations. As a result of the principle (as well, perhaps, as that of Weismann's "germinal selection"), Wallace concluded, "it now appears as if all the theoretical objections to the 'adequacy of natural selection' have been theoretically answered." See Alfred Wallace, "The Problem of Instinct" (1897), in his *Studies,*

Baldwin responded to much of this literature. He showed that neither Darwin nor Weismann had really had the idea, and he took on those others who attacked the principle. In 1902 he published *Development and Evolution,* which stood as a seemingly impenetrable wall defending the principle and his priority. But one new difficulty, already inherent in Romanes's remarks and subsequently stressed by Delage and Plate, revealed a large crack.[116] It was simply that if individual accommodations were sufficient to preserve an organism, then congenital variations would have little or no utility and thus would not be naturally selected for. Baldwin's treatment of the problem in *Development and Evolution* surely appeared weak even in his own eyes. He appealed to the idea of correlated variations to reestablish the integrity of his theory. He asked his reader to suppose that an individual accommodation might preserve an animal that had the beginnings of an unrelated congenital variation. He further required his reader to allow this congenital variation to be correlated with a biological variation similar to the learned accommodation. Hence, the animal would be preserved by individual accommodation, while natural selection would operate on the unrelated trait.[117] The unrelated trait, in turn, would pull along the relevant replacement character. The objection was powerful, the response anemic. Baldwin never mentioned this objection again. He was no Popperian. Some modern biologists and psychologists, such as C. H. Waddington and Jean Piaget, have yet stood by the Baldwin effect.[118] And the cul-

Scientific and Social (London: Macmillan, 1900), 1:497–508. Wallace, however, became increasingly less convinced of the importance of the principle in evolution. In response to receiving a complementary copy of *Development and Evolution,* he wrote Baldwin: "Your account of Organic Selection, as originated by yourself and Lloyd Morgan, is very clear and I have no doubt is occasionally a real factor in evolution. But I do not think that it is an important or even an essential one." See the exchange of correspondence in Baldwin's *Between Two Wars* 2:246–49.

115. Henry F. Osborn, "The Limits of Organic Selection," *American Naturalist* 31 (1897): 944–951.

116. Baldwin became aware of the criticisms of organic selection by Delage and Plate from the précis given their books *L'hérédité* (1894) and *Selectionsprinzip* (1900) respectively, in the journal *L'année biologique.* See Yves Delage and G. Poirault, "L'origine des espèces," *L'année biologique* 3 (1897): 512; and L. Defrance, "Plate," *L'année biologique* 5 (1899–1900): 388.

117. Baldwin, *Development and Evolution,* pp. 209–12.

118. See, for example, C. H. Waddington, "The 'Baldwin Effect,' 'Genetic Assimilation' and 'Homeostasis,' " *Evolution* 7 (1953): 386–87; and "Evolutionary Adaptation," in *The Evolution of Life,* vol. 1 of *Evolution after Darwin,* ed. Sol Tax (Chicago: University of Chicago Press, 1960), pp. 381–402; and Jean Piaget, *Behavior and Evolution,* trans. D. Nicholson-Smith (New York: Random House, 1978), chap. 2. See also Julian Huxley's earlier endorsement in his *Evolution: the Modern Synthesis* (London: Allen & Unwin,

tural anthropologist Clifford Geertz has used the idea (though without apparent awareness of its origins) to do battle against biological reductionists.[119] But objections such as that just mentioned have persuaded Simpson that, while the principle describes a logically possible mechanism, it probably has played little role in evolution.[120] Other difficulties, revealed in subsequent literature, undoubtedly weakened the appeal of the principle.[121] But a social phenomenon, I believe, also operated to discredit the principle, at least in the eyes of American psychologists, and to remove its author from the center of the scientific community.

Conclusion: Scandal and Professional Extinction

In 1903 Baldwin was invited by President Ira Remsen to reestablish the department of philosophy and psychology at the Johns Hopkins University and to refound the psychological laboratory. (The department and the laboratory had been closed some years before during an economic crisis.) The attractiveness of the position at Hopkins was keen. So was Baldwin's growing discomfort at Princeton. He had considerable duties as an undergraduate teacher, for which he had little heart. Hopkins promised both relief from this and greater numbers of graduate students. And, then, Woodrow Wilson was no friend of the sciences. Baldwin thought Wilson's elevation to the presidency of Princeton "only made him more august!"[122] Finally, Hopkins offered a substantial increase in salary, which bound up all the other reasons for a fairly easy decision.

Shortly after he arrived at Hopkins, he began work on what would become his *Genetic Logic,* the three volumes of which debuted in 1906, 1908, and 1911; the *Genetic Theory of Reality* appeared in 1915.[123] In 1909 Baldwin published *Darwin and the Humanities,* which demonstrated

1942), pp. 17, 114, 296, 304, 523. Jacques Vonèche provides a nice analysis of the intellectual connection between Baldwin and Piaget in "Evolution, Development, and the Growth of Knowledge," in *The Cognitive-Developmental Psychology of James Mark Baldwin,* pp. 51–79. Emily Cahan offers a detailed comparison between the various features of Baldwin's and Piaget's genetic psychology. See her extremely instructive "The Genetic Psychologies of James Mark Baldwin and Jean Piaget," *Developmental Psychology* 20 (1984): 128–35.

119. See Clifford Geertz, "The Growth of Culture and the Evolution of Mind," in his *Interpretation of Cultures* (New York: Basic Books, 1973), pp. 55–86.

120. George G. Simpson, "The Baldwin Effect," *Evolution* 7 (1953): 110–17.

121. These are amply recounted by Ernst Mayr in his *Animal Species and Evolution* (Cambridge: Harvard University Press, 1963), pp. 110–12.

122. Baldwin, *Between Two Wars* 1:99.

123. James Mark Baldwin, *Thought and Things or Genetic Logic* (London: Sonnenschein, 1906–1911); and *Genetic Theory of Reality* (New York: Putnam, 1915).

the value of organic selection for understanding a variety of areas in the social sciences and humanities. These several volumes caused few reverberations in the American intellectual community, since the *Genetic Logic* reeked of the argot of deep idealistic philosophy, particularly the epistemologies of Lotze, Wundt, and Bradley, while the philosophic temper of America had warmed to pragmatism and empiricism. And for his fellow psychologists, who were trying to cut their ties to philosophy and to begin speaking the language of mathematics, Baldwin's metaphysical schemes and uncontrolled style proved quite incomprehensible. But his solid accomplishments in developmental biopsychology were muted for a different reason. Baldwin had left the United States in 1909 finally settling in France. He had become an exile from his intellectual community. And therein lies a tale.

Scandal at Hopkins

Though the main features of the story can be reconstructed from the material in the Presidential Archive at Hopkins, the most important question can be answered only tentatively. President Remsen described the incident this way.[124] In early summer of 1908, Baldwin visited a black house of prostitution in Baltimore. The police raided the establishment and took the inmates to the station, where one man, who originally gave a false name, was discovered to be a professor of the Hopkins. Since the woman involved was over sixteen, the police dropped the criminal complaint, and a station-house lawyer got the remaining charges quashed. On 1 January 1909, the head of the Hopkins board of trustees, Judge Harlan, received the police report along with a rumor indicting Baldwin. Harlan made no move, fearing the possible scandal. In March the mayor of Baltimore, who had nominated Baldwin to serve on the public school board, was furnished a whiff of the summer scandal. He confronted Baldwin with the charge, and Baldwin acknowledged his arrest. By this time Harlan thought the University had to act, since the rumor was spreading. Now the Judge placed the

124. I have reconstructed the story in this paragraph and the one following from the material in the Presidential Archives at Johns Hopkins University. I was guided in this by Philip Pauly who made available to me his unpublished manuscript "Money, Morality, and Psychology at Johns Hopkins University, 1881–1942," which traces the fortunes of psychology at Hopkins through the years of G. Stanley Hall, Baldwin, John Watson, and Knight Dunlap, and into the decline of 1936 to 1942. Pauly has published a shorter version of this study as "Psychology at Hopkins," *Johns Hopkins Magazine* 30 (1979): 36–41. In a personal communication Pauly also set me straight about several matters. Karin Wetmore kindly provided me with copies of relevant Titchener-Cattell correspondence as found in the Cattell Papers at the Library of Congress.

police report before Baldwin. Remsen related that "He [Baldwin] admitted that he was the man arrested, <that the charges were true> that he realized that he would have to resign his Professorship and that he would leave town at once."[125]

A few days after the meeting with Harlan, Baldwin left for Mexico, where he took up an educational project with which he had been previously involved. He wrote Remsen asking for a leave of absence for the remaining part of the year and for the next academic year 1910–1911. Apparently he did not have the understanding about his tenure that Remsen thought he did. The president quickly wrote back saying that Baldwin would have to submit an undated, unconditional resignation immediately, which he did. In early summer, Baldwin traveled with his family to Geneva, where he attended the International Congress of Sociology; and after serving a year as professor in the new National University of Mexico, he and his family took up residence in Paris.

The reason for Baldwin's resignation from Hopkins only slowly leaked out. He was to have presided over the first United States meeting of the International Congress of Psychology in 1913. But when his colleagues Cattell and E. B. Titchener smelled the brewing scandal, they, like the good citizens of Baltimore, felt action was necessary. The letter of Titchener to Cattell is worth quoting in full:

> Dear Cattell,
>   I have just heard, in a letter from a colleague at another university, that Baldwin was caught in a negro dive at Baltimore, and in consequence was summarily dismissed from the Hopkins. A newspaper paragraph, saying the same thing indirectly, has also been sent to me.
>   If this thing is true, we cannot let Baldwin hold the presidency of the Congress. His scientific eminence is not impaired, but he becomes socially impossible, and the Congress is largely a social matter. I am as sorry as I can be to have to act, on many counts; and I have no doubt that you feel in the same way; but as vice-presidents we have to act.
>   I propose that we send a joint letter, officially, to Remsen, asking the straight question whether B. was dismissed for suspicion of immorality. If he says No, well and good. If Yes, then I propose that we ask B. quietly to resign.
>   The whole thing may be a mare's nest; I hope it is.
>   (Signed) E. B. Titchener.[126]

125. Ira Remsen's notes in the Presidential Archives, Johns Hopkins University. The remarks in wedge brackets were scratched out.

126. Edward B. Titchener to James McKeen Cattell (10 January 1910), Cattell Papers (Subject File, APA 1), Library of Congress.

Figure 10.4   E. B. Titchener, 1867–1927, photograph from ca. 1900.

Hugo Münsterberg, at Harvard, undertook the task of inquiry for his colleagues. In his letter to Remsen (8 February 1910), he quoted an earlier correspondence in which Baldwin had given his side of the story. Baldwin had written:

> In the early summer of 1908 I foolishly accepted a suggestion made after a dinner to go to a house of a colored social sort and see what was done there. I did not know that women were harbored there. I was found there entirely by reason of my ignorance and to save themselves the proprietors made a serious charge against me. The charge was dismissed at once by the presiding officer and the people making it were convicted and sent up. The justice called on me at my house afterward and

assured me there was nothing either legal or moral against me.[127]

Baldwin's explanation, which he sent to several of the leaders of American psychology, including William James, had to fight against a rumor that grew ever more ugly. Josiah Royce, who had been Baldwin's long-time friend at Harvard, mentioned (10 February 1910) to James what he had heard:

> apparently authentic report has it, he [Baldwin] admitted, at the time, that frequent and habitual practices of his own, deliberately pursued, had led to the final scene, [and so] if, in the eyes of his University, he had long carried on a mode of life that violated his trust as an officer, and if he had been dismissed for *this* turpitude,—well then it wouldn't do to have him trying by further false statements, to hold himself in the position of a man worthy of the moral support of the general body of his fellow workers in this country.[128]

I have not located Remsen's reply to Münsterberg's inquiry, but apparently he related the story outlined above. Münsterberg's acknowledgment to Remsen is extant, however. It reveals prominently two things: that a hostility to Baldwin had built up over the years, undoubtedly because of his abrasiveness and arrogance; and that Remsen had no further evidence against Baldwin. Münsterberg wrote to the Hopkins president:

> I beg to thank you very sincerely for your confidential letter which closes the whole matter for us Harvard men and practically for the psychologists.
> The essential point for us is that the explanation which Baldwin gives to the facts now is new to you. It seems evident that he would have brought before you everything which might excuse him. As he did not present the matter in this light to you it is obvious that his present excuses are free inventions. That makes it entirely impossible for us to help him. I personally cannot deny that this experience with him is in full harmony with some previous occurrences with reference to money mat-

127. Hugo Münsterberg to Ira Remsen (8 February 1910), the Presidential Archives, Johns Hopkins University. Copy in Munsterberg Papers, MS. 2389, Boston Public Library.

128. Josiah Royce to William James (10 February 1910), bMS AM 1092.9, 560, William James Papers, Houghton Library, Harvard University.

Figure 10.5 Hugo Münsterberg, 1863–1916,
photograph from ca. 1900.

ters concerning the Psychological Review; he behaved at that
time dishonestly without doubt.[129]

Münsterberg's gratuitous remarks about Baldwin's rumored dishonesty
suggest that the community's reaction had something of the vindictively
personal about it. Remsen apparently furnished to Münsterberg, and
he to his colleagues, only the information that Baldwin had not originally defended himself in the terms he later had used with the psychologists. Baldwin never denied the facts of the case, that he was found in
the bordello and was initially charged with a criminal offense. He apparently regarded just those facts, if released to the public, as enough to
embarrass his university severely and thus sufficient cause to resign. He
never admitted anything more than those facts to Remsen and Harlan.
His wife's faith, which must have been tested at the time, gives poignant
witness to the unreliability of the rumor Royce heard. When she visited
the United States in December of 1910, she went to Remsen, without

129. Hugo Münsterberg to Ira Remsen (14 February 1910), the Presidential Archives,
Johns Hopkins University.

her husband's knowledge, to plead his case.[130] Remsen's notes of the meeting indicate that Helen Baldwin knew as much as the president did about the affair, sugesting that Baldwin's explanation might well have been true, or, at least, that Royce's gossip had not been corroborated. But whether Baldwin were 'morally' guilty or not, the reaction of the psychological community discloses much about the regnant scientific and professional norms.

Under pressure from his colleagues, Baldwin resigned the presidency of the Psychological Congress (which, in fact, was not held in the United States). And Titchener reported this to Cattell. He closed his letter with a remark about Baldwin's conduct that tinctures the gentle breezes around his own:

> Baldwin declares his legal and moral innocence of any and everything. He promises me details, but says that he could not send an explanation of his conduct before it had been asked for. In fact I did not ask for it. And in fact it seems poor policy to let yourself be kicked without hollering if you really do not deserve the kicking. However, I need not say that I shall be glad enough if the poor chap can rehabilitate himself.[131]

Professional Extinction

Two years after Baldwin's death in 1934, the social philosopher Charles Ellwood judged him to have been, at the turn of the century, "almost universally regarded as the leading American social psychologist."[132] This evaluation certainly reflected the opinion of his contemporaries. In 1903 Cattell surveyed American psychologists and asked them to rank their colleagues in order of eminence. Baldwin was listed fifth, behind James, Cattell himself, Münsterberg, and G. S. Hall, but ahead of Titchener, Royce, John Dewey, and E. L. Thorndike.[133] Baldwin's score would probably have been one better had Cattell the decency to have eliminated himself from the competition or to have had the responses blindly refereed. James's own ranking excluded himself

130. Three letters from Helen (Mrs. James Mark) Baldwin to Ira Remsen (7 December 1910, 11 December 1910, and 11 January 1911), the Presidential Archives, Johns Hopkins University.

131. Edward B. Titchener to James McKeen Cattell (18 February 1910), Cattell Papers (Subject File, APA 1), Library of Congress.

132. Charles A. Ellwood, "The Social Philosophy of James Mark Baldwin," *Journal of Social Philosophy* 2 (1936): 55.

133. James McKeen Cattell, "Psychology in America," in *Addresses and Formal Papers*, vol. 2 of *James McKeen Cattell, Man of Science* (Lancaster, Penn.: Science Press, 1947), p. 452; the tabulation was originally published in *Science* 70 (1929): 335–47.

and Cattell, and placed Baldwin second, after Münsterberg (whom James had brought to Harvard).[134] Another index also attests to Baldwin's stature. George Howard's analytical reference syllabus for social psychology, published in 1910, gives Baldwin first place among contemporary psychologists (before James, Royce, and Avenarius) on the study of the self.[135] But Baldwin's star faded after his retreat to Paris. Already in 1910, in their *Guide to Reading in Social Ethics and Allied Subjects,* the professors at Harvard University, who knew of the scandal, listed hardly any of his books.[136] By 1924 Baldwin had been eclipsed. He received only passing reference in Floyd Allport's *Social Psychology,* the book most widely used in the early establishment of the field.[137] Though all older scientists gradually vanish from the citations of their successors, Baldwin's disappearance, even while he was still alive, suggests certain accelerating factors at work.

One factor that cannot be ignored was the changing intellectual environment of American psychology. During the second decade of the new century, in the face of the rapid growth of the natural sciences, American psychologists turned in great numbers to behaviorism, which promised a scientifically hard foundation for psychology, one that would pave over the embarrassment of an older, softer-minded and philosophically bloated psychology that appeared to many to have retarded progress in the science. The invasion of behaviorism, however, did not occur as quickly in social psychology.

Two other factors in Baldwin's professional demise must, I think, also be recognized. First, with his exile to France, Baldwin lost his institutional position as a leader of American psychology: he would have no students, he would edit no journals, he would preside over no meetings. These underpinnings of professional stature gave way beneath him. But second, Baldwin had clearly violated certain norms of his discipline. At the turn of the century, standards for personal sexual conduct (and crossing racial boundaries) could hardly be extricated, at least in America, from the understood professional canon of a scientist who specialized in social psychology and ethics. When Titchener learned of Remsen's reply to Münsterberg, he feared that "psychology suffers, and

134. William James to James McKeen Cattell (10 June 1903), Cattell Papers (Subjet File, APA 1), Library of Congress.
135. George E. Howard, *Social Psychology: An Analytical Reference Syllabus* (Lincoln: University of Nebraska Press, 1910).
136. Teachers in Harvard University (chapters individually authored), *A Guide to Reading in Social Ethics and Allied Subjects* (Cambridge: Harvard University, 1910).
137. Floyd Allport, *Social Psychology* (New York: Houghton Mifflin, 1924).

American science suffers, and we are all a bit implicated."[138] Such remarks suggest that Baldwin's colleagues implicitly judged him to have breached that admixture of professional and personal principles. And a scientist perceived to have sinned against norms of professional honesty and integrity (e.g., Cyril Burt's manufacturing IQ test data) must struggle against enormous odds to have his work henceforth taken seriously. Baldwin, I believe, succumbed to those odds in his native land, but fortunately the French, who learned of his Baltimore travails, apparently cared little.

In Paris his intellectual circle included Emile Boutroux, the dean of French philosophers, Henri Bergson, who had come into high intellectual fashion, Théodule Ribot, whose book he had translated, Henri Poincaré, the great mathematician and philosopher, Pierre Janet, who occupied the chair of experimental psychology in the Collège de France, and many other notables. French intellectuals recognized Baldwin as one of their own by electing him in 1910 to succeed John Stuart Mill and William James as correspondent of the Academy of Moral and Political Sciences in the Institute of France. Upon learning of Baldwin's election, Titchener observed to Münsterberg: "They do these things differently in France!"[139]

138. E. B. Titchener to Hugo Munsterberg (5 March 1910), MS. 2191, Münsterberg Papers, Boston Public Library.
139. E. B. Titchener to Hugo Munsterberg (24 June 1910), MS. 2191, Munsterberg Papers, Boston Public Library.

11

Transformation of the Darwinian
Image of Man in
the Twentieth Century

Contemporary Darwinism focuses a stark image of man. Through its lens we have come to perceive man as a completely material being, whose reason traces the narrow paths of fixed brain circuits, whose religious sentiments bespeak the need for conformity rather than a passion after transcendence, and whose moral feeling, driven from below by selfish genes, quickens to secret pleasure for self rather than to the welfare of others. Some evolutionary thinkers take Stoic comfort in this vision, suggesting that those who reject it only indulge in high romance or low liberal politics. Others, while admitting that contemporary Darwinism forms this image, yet think it a chimera, since it cannot represent civilized human beings. Culture, they argue, has emancipated us from the tyranny of our genes and has thus given us a more humane aspect.

The received view of Darwinian man, though a potent icon in modern culture, does not resemble that image shaped by Darwin, Spencer, and the Darwinians writing in the last part of the nineteenth century. The metaphysics that supported the evolutionary conceptions of Romanes, Morgan, James, and Baldwin—as well as the grand theory of Spencer—was quite inimical to blind, mechanistic materialism. They conceived of mind, in its various manifestations, as guiding the evolutionary process. And Darwin, perhaps more forcefully than any of his disciples, attempted to infuse human nature with an authentic moral sense: altruistic behavior did not disguise a more fundamental utilitarian selfishness but instead revealed a divine spark lighting the rest of nature.

But this brighter and, in a more traditional way, nobler view of the human animal did not penetrate deeply into the next century. Rather that original image became refracted and transformed into the specter that hovers over contemporary debates about sociobiology. The transformation came as the result of powerful disciplinary and social forces acting during the early part of this century, forces which initially inhibited the further development of evolutionary theories of mind and behavior, but which in the end served to recast the image of Darwinian man. The strongest of these forces erupted in the social sciences.

Decline of Evolutionary Theory in the Social Sciences

James, Baldwin, and Morgan—three principal contributors to the Anglo-American development in evolutionary biopsychology—had entered the twentieth century with philosophy on their minds. James formally turned over teaching and laboratory duties in psychology at Harvard to Hugo Münsterberg in 1892, and requested of President Eliot a transfer into philosophy.[1] From century's end until his death in 1910, James rushed to work out the implications, especially for religion, of a monism of radical empiricism. Baldwin, exiled in Paris, spent the last twenty-five years of his life also pursuing the higher metaphysics, though the still-vital juices of his evolutionary psychology seeped into the early work of Jean Piaget.[2] And Morgan swung more sharply in his last two decades to the philosophy of emergent evolutionism. As evolutionary theories of mind became elaborated into grand philosophy instead of being stitched into the fabric of advancing science, they lost their attraction for those new scientists of the laboratory craft.

John Watson, the progenitor of modern behaviorism in psychology, was the very model of the laboratory man.[3] He originally came to the University of Chicago to study with John Dewey. Dewey's philosophical naturalism rippled with evolutionary ideas and seems initially to have stimulated Watson to follow a particular research path.[4] But the gauzy character of Dewey's speculations could not hold the intellectual

1. Matthew Hale discusses the transition in psychology at Harvard from the leadership of James to that of Münsterberg. See Matthew Hale, *Human Science and Social Order: Hugo Münsterberg and the Origins of Applied Psychology* (Philadelphia: Temple University Press, 1980), pp. 45–55.

2. For an assessment of Baldwin's influence on Piaget, see Jacques Vonèche's introductory remarks to his interview with Piaget on the subject: Jacques Vonèche, "An Interview Conducted with Piaget: Reflections on Baldwin," in *The Cognitive Developmental Psychology of James Mark Baldwin,* eds. John Broughton and D. Freeman-Moir (Norwood, N. J.: Ablex, 1982), pp. 80–86.

3. Concerning the beginnings of behaviorism, see John O'Donnell, *The Origins of Behaviorism: American Psychology, 1870–1920* (New York: New York University Press, 1985), and John Burnham, "On the Origins of Behaviorism," *Journal of the History of the Behavioral Sciences* 4 (1968): 143–51.

4. O'Donnell argues for the residual influence of Dewey on Watson in his *The Origins of Behaviorism,* pp. 212–14.

enthusiasms of this empirically directed researcher. "I never knew what he was talking about then," Watson recalled in his autobiographical sketch, "and unfortunately for me, I still don't know."[5] Watson retained the democratic and pragmatic part of the Deweyan vision, but the decidedly evolutionary part faded. In 1924, when he had become the recognized spokesman for the new behavioral technology, Watson exclaimed: "Give me a dozen healthy infants, well-formed, and my own specified world to bring them up in and I'll guarantee to take any one at random and train him to become any type of specialist I might select—doctor, lawyer, artist, merchant-chief and, yes, even beggar-man and thief, regardless of his talents, penchants, tendencies, abilities, vocations, and race of his ancestors."[6]

The democratic potential implied by Watson's behaviorism struck a resonant cord in the hearts of most of his American readers. Much of the academic world, which perhaps allowed for Madison Avenue hyperbole, and large portions of the lay public believed his boast.[7] At least they thought every good American boy might grow up to be president, if he worked hard and received the proper training. To set children along the right path, Watson offered advice to the mothers of America in a number of popular articles and in a Spock-like baby book, *Psychological Care of Infant and Child*.[8] The message of these publications was clear: the mind of the child had no inherited groves that determined its station in life; it could be putty in the hands of the behavioral technologist, or the well-informed mother.

The triumph of behaviorism in psychology was one of the signal causes of the decline in theorizing—at least in English-speaking countries—about the evolution of mind and behavior.[9] There were other

5. John Watson, "Autobiography," in *History of Psychology in Autobiography*, vol. 3, ed. Carl Murchison (Worcester, Mass.: Clark University Press, 1936), p. 274.

6. John B. Watson, *Behaviorism*, revised ed. (Chicago: University of Chicago Press, [1924] 1930), p. 104.

7. Franz Samelson traces the uneven acceptance of Watsonian behaviorism in American psychology. He shows that though the letter of Watson's doctrine, especially about the elimination of mind as a subject of psychological concern, was often resisted by later American psychologists, the ethos of experimental analysis yet flooded over the discipline. See Franz Samelson, "Struggle for Scientific Authority: The Reception of Watson's Behaviorism, 1913–1920," *Journal of the History of the Behavioral Sciences*, 17 (1981): 399–425.

8. John B. Watson, *Psychological Care of Infant and Child* (New York: Norton, 1928).

9. Though Watson became the celebrated spokesman for behaviorism, his particular scientific conceptions were not always shared by those rallying to the name "behaviorism." James McKeen Cattell, E. L. Thorndike, Edward Tolman, and Robert Yerkes all described themselves as behaviorists, though they dissented from several important Watsonian assumptions. They shared the belief, however, in an objective psychology founded on precise measurement of and experiment on behavior.

important factors, of course, and these I will mention in a moment. But behaviorism became the most potent internal force that checked further developments in evolutionary biopsychology. There is some irony in this, since in Watson's first book, *Behavior: An Introduction to Comparative Psychology,* he framed his study of sensation, instincts, and learning in terms of Darwinian evolutionary theory.[10] His own earlier field research had been on the instincts of the noddy and sooty terns in the Tortugas.[11] In his first book, he joined these studies with controlled investigations of instinct and learning in many other animals. Watson initially aimed to construct an objective psychology of behavior based on mildly Darwinian assumptions.

During the time Watson was constructing his theories of instinct, that paradigm of evolved behavior, other researchers working in all quarters of psychology were assembling their own ideas about instinct and stockpiling masses of unrefined studies. Speaking disciplinarily incomprehensible languages, they built a tower that swayed and shook on collapsing foundations. L. L. Bernard, a social psychologist who also received his training at the University of Chicago, surveyed the writings on instinct of over three hundred authors publishing between 1900 and 1924. With some amazement, he cataloged 1594 different classes of instinct that had been attributed to men and animals.[12] This alone, he thought, demonstrated the utter confusion and vacuity of most research on the topic. In the theorizing of psychologists instinct had become a *vis dormitationis.* As an example of the Molièrean use of the concept, he chose the eugenicist Charles Davenport's studies of human instinct, which he described in a mordant chapter entitled "A Reductio ad Absurdum." Bernard himself was willing to admit inherited behavior, but only of a very primitive sort, mainly simple reflexes. He argued that on the bases of these elemental inborn traits more complex habits—previously thought of as instincts—had been built up through learning. This was a conclusion at which Watson also arrived, but with more powerful illustration.

Watson elaborated a theory of learning in his 1924 book *Behaviorism* that he formulated largely from Bekhterev's and Pavlov's techniques of conditioning. While this theory—and technology—required the assumption of some unconditioned reflexes, some innate behavior, Wat-

10. John B. Watson, *Behavior: An Introduction to Comparative Psychology* (New York: Henry Holt, [1914] 1967).
11. John B. Watson, "The Behavior of Noddy and Sooty Terns," *Carnegie Publications* no. 103 (1908): 187–255.
12. L. L. Bernard, *Instinct: a Study in Social Psychology* (New York: Henry Holt, 1924), pp. 172–220.

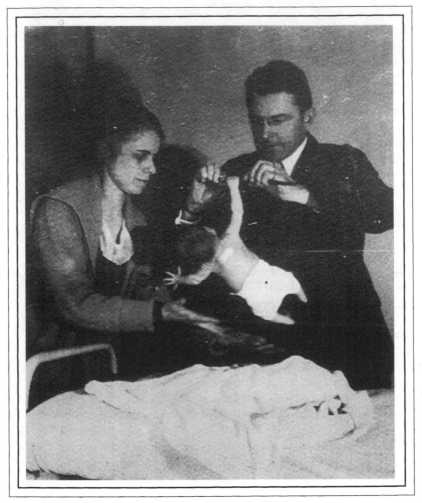

Figure 11.1 John B. Watson, 1878–1958, from film of 1919.

son acknowledged only three inborn emotional responses: love, rage, and fear. He urged that out of these common bricks, very complex edifices of character and behavior could be built up, without recourse to instinct or to anything so insubstantial and useless as conscious mind. He and his graduate student Rosalie Rayner dramatically demonstrated this conception by conditioning emotional response in little Albert, an orphaned infant upon whom they experimented.[13] When

13. John B. Watson and Rosalie Rayner, "Conditioned Emotional Reactions," *Journal*

little Albert moaned in fear of a white rat that he had happily played with before his reconstruction, the reverberating echo in the discipline of psychology drowned out further talk of evolutionary approaches to mind.

Watson's creation, however, turned even on him. Z. Y. Kuo, a militant spokesman for the radical behaviorists, attacked Watson's retention of residual hereditarian ideas. In his article "The Net Result of the Anti-Heredity Movement in Psychology" (1929),[14] Kuo charged that Watson inconsistently admitted hereditary emotions while "denying the inheritance of instincts and 'mental traits.'"[15] Though other behaviorists during the twenties and thirties did not go as far as Kuo—many granted some innate components to behavior—they nevertheless presumed that complex behavior and mental activity were learned, and they set out to establish the laws governing that learning. During the next two decades, through the forties and fifties, B. F. Skinner's radical behaviorism coupled with logical positivism's principle of verification—which held that the meaning of a term was to be fixed by the method of its empirical test—pushed those still referring to mind into the camps of the metaphysicians. Questions about evolution could not arise for what was deemed a "ghost in the machine," as Gilbert Ryle referred to the now evanescing mind.[16] It is not surprising, then, that between 1927 and 1958, listings in *Psychological Abstracts* of entries under the term "instinct," as compared with listings under the learning theory terms "drive," "reinforcement," and "motivation" steadily and rapidly declined from 68 percent of the total to 8 percent.[17]

Evolutionary approaches to human mind and behavior received another check during the early decades of this century. The blow came from anthropology. The newly burgeoning science of human culture turned its back on evolution because of the ethnocentrism and racism that evolutionary theory seemed to imply. Franz Boas, the German-born leader of American anthropology objected, early on, to the as-

of *Experimental Psychology* 3 (1920): 1–14. See also Watson, *Behaviorism,* pp. 158–66. The pitiable case of little Albert is put in the context of Watson's development by Franz Samelson in "J. B. Watson's Little Albert, Cyril Burt's Twins, and the Need for a Critical Science," *American Psychologist* 35 (1980): 619–25. See also Ben Harris, "Whatever Happened to Little Albert?" *American Psychologist* 34 (1979): 151–60.

14. Z. Y. Kuo, "The Net Result of the Anti-Heredity Movement in Psychology," *Psychological Review* 36 (1929): 181–99.

15. Ibid., pp. 188–89.

16. Gilbert Ryle, *The Concept of Mind* (London: Hutchinson, 1949), pp. 15–18.

17. See R. J. Herrnstein, "Nature as Nurture: Behaviorism and the Instinct Doctrine," *Behaviorism* 1 (1974): 23–52.

sumption that certain apparently constant features of mind and character resulted from the universal biological nature of men: "We must . . . consider," he cautioned, "all the ingenious attempts at constructions of a grand system of the evolution of society as of very doubtful value, unless at the same time proof is given that the same phenomena must always have had the same origin."[18] No one could provide for him the proof that similar cultural practices of different societies could be produced only by a common biological cause. Through the later part of his career, Boas became ever more hostile to the idea of evolutionary laws governing the development of mind in different races. While admitting that particular families might transmit mental characters to offspring, he argued that there were no lines of descent binding existing human races into distinct hereditary types. Therefore "any attempt to explain cultural forms on a purely biological basis is doomed to failure."[19] Boasian anthropology came more and more to ascribe the peculiar mental characteristics of different races to cultural differences, which were argued to have derived principally from social diffusion and learning.[20] American and British cultural anthropologists in the years since Boas have developed considerable theory and method by which to appropriate the concept of culture as their special intellectual domain. Invasion of this territory by evolutionary biologists can evoke extreme reaction, as, for example, when Marshall Sahlins rushed into print his *The Use and Abuse of Biology* (1976) a few short months after the publication of Edward Wilson's *Sociobiology* (1975).

Boasian environmentalism, which emphasized immediate and measurable processes of social learning rather than remote and speculative events of biological history, spread quickly into sociology. As Carl Degler has shown, the early sociologists Charles Ellwood, Carl Kelsey, Howard Odum, W. I. Thomas, and Ellsworth Faris came to appreciate Boas's arguments and methods as congenial to the needs of their own discipline, especially its evangelistic goals.[21] Many of the fresh recruits

18. Franz Boas, "Limitations of the Comparative Method of Anthropology" (1896), in his *Race, Language, and Culture* (Chicago: University of Chicago Press, [1940] 1982), p. 276.

19. Boas, "Some Problems of Methodology in the Social Sciences" (1930), in *Race, Language, and Culture*, p 265.

20. George Stocking, "Franz Boas and the Culture Concept in Historical Perspective," *Race, Culture, and Evolution* 2d ed. (Chicago: University of Chicago Press, 1981), pp. 195–233. An extra dimension to the source and influence of Boas's ideas about culture can be found in Stocking's address to the American Historical Association Meeting in New York City, 29 December 1968: "Boasian Anthropology and Popular Racial Thought In America."

21. Carl Degler describes the spread of the Boasian explanation of racial differences to

who came to the new departments of sociology at Chicago, Pennsylvania, and Columbia had recently shucked off religious fundamentalism, but retained a residual missionary zeal to improve the lot of the wretched, especially blacks and recent immigrants. Their social calipers measured class, profession, education, and health care—environmental causes the reorganization of which might immediately alter the lot of the poor. Their techniques, however, could not easily grasp hidden evolutionary factors; nor were they inclined to try to measure what they could not change. So if the Boasian revaluation of other cultures did not persuade all of the new sociologists, their own disciplinary techniques and professional ethos constrained them to ignore biological causes.

Thus in the second and third decades of this century, the weight of the social sciences—pressed by Watsonian behaviorism and Boasian environmentalism, along with immediate empirical methods and a social gospel—shifted against the Darwinian interpretation of mind and behavior. And then, biology itself began to give way.

The Reactions of the Biological Community to Theories of Mental Evolution

The New Genetics

It is easy enough to forget that at the turn of the century, prior to the new synthesis of natural selection theory and genetics in the 1930s and 1940s, the fortunes of Darwinism in biology seemed poor. The theory of natural selection and gradual evolution appeared to be eclipsed by theories based firmly in laboratory work—the new theories in genetics. The field biology of the Darwinians could be consigned to the sweetly decaying speculations of other kinds of romanticism. Erik Nordenskiöld bid it farewell in the last chapters of his masterful *History of Biology* (1920–1924): "With this we can leave the doctrine of descent in the old Darwinistic sense. Modern heredity-research has introduced quite a different and essentially experimental treatment of the problems of evolution, and the old morphological speculation upon the origin of species and genera has proportionately lost ground—as it has always happened in the history of the exact sciences that speculation must give way to facts."[22] Nordenskiöld believed, as did many in the biological

sociology in "The Beast in Humanity: Biological Ideas in American Social Thought, 1900 to the Present," the Samuel Paley Lectures, delivered at The Hebrew University, 26 and 30 April and 3 May 1984.

22. Erik Nordenskiöld, *The History of Biology* (New York: Tudor, [1920–1924] 1936),

community, that Mendelian factors and De Vriesian mutations furnished the real mechanisms of species change. At the turn of the century, Darwin's pangenetic notions about inheritance had succumbed to more careful studies and experiments in heredity. And his idea of natural selection, which the new geneticists believed to exhale the aura of a mystically creative energy, seemed to evaporate when struck by the cold light of the laboratory. Thomas Hunt Morgan, whose establishment of experimental techniques at Columbia University provided the tools for the rapid advance of Mendelian theory, initially objected vigorously to the hypothesis of natural selection. Natural selection simply could not produce novel variations nor determine their continued production to form new species. At best, selection acted negatively to weed out the unfit, but it had no creative power to shape original adaptations. Mutation, Morgan proposed, was the mechanism that drove evolution.[23] But if Darwinism failed to account for the rise of new species, it could hardly provide an adequate theory of human mind and behavior.

The Taint of Eugenics

Prior to the new synthesis, genetics itself momentarily replaced Darwinian "speculation." It promised not only an experimental understanding of mind and behavior, but also a technology for improving even further the stock of the higher races. Eugenics societies in America, Great Britain, and Germany at the turn of the century included very distinguished geneticists in their lists. The Americans Thomas Hunt Morgan, Herbert Spencer Jennings, Raymond Pearl, William Castle, and Edward East concurred in the theory and the social policy of the eugenics movement. During the first few decades of this century that movement came to represent the new scientific approach to evolution of human mind and behavior. Later, however, as its promise calcified into reactionary ideology, its distorted grimace reminded professionals and laymen alike of the dangers of applying evolutionary theory to account for the human state.

---

p.573. Peter Bowler charts the response of the Mendelian geneticists to selection theory at the beginning of this century in his *The Eclipse of Darwinism* (Baltimore: Johns Hopkins University Press, 1983), pp. 182–213.

23. See Thomas H. Morgan, *Evolution and Adaptation* (New York: Macmillan, 1903) and *A Critique of the Theory of Evolution* (Princeton: Princeton University Press, 1916). Garland Allen provides the most comprehensive analysis of Morgan's shifting position on natural selection theory, in his *Thomas Hunt Morgan: the Man and His Science* (Princeton: Princeton University Press, 1978), pp. 106–16, 301–17. William Provine discusses the battles between the biometricians—who defended Darwinism—and the Mendelians, in his *The Origins of Theoretical Population Genetics* (Chicago: University of Chicago Press, 1971), pp. 25–89

The most influential geneticist working out the theoretical foundations of eugenics and relentlessly advancing its social goals was Charles
Davenport. After leaving teaching positions at Harvard University and
the University of Chicago, Davenport became director of the Station
for Experimental Evolution at Cold Spring Harbor as well as of the
Eugenics Records Office there. Davenport and his colleagues, especially
Harry Laughlin (superintendent of the Eugenics Records Office) and
Paul Popenoe (editor of the *Journal of Heredity*), operated under the
initial assumption of the Mendelians that each human trait was due to
one gene—the doctrine of one gene, one trait. They traced out the
genealogical path in family pedigrees of such traits as stature, musical
and artistic talent, general intellectual ability, epilepsy, insanity, pauperism, criminality, and a host of other somatic and mental attributes. Davenport's analysis of the celebrated Jukes family immediately suggests
both the level of empirical security and the perceived therapeutic value
of eugenics doctrine in the first third of the century:

> The case of the "Jukeses" is well known. We are first introduced
> to a man known in literature as Max, living as a backwoodsman
> in New York state and a descendant of the early Dutch settlers;
> a good-natured, lazy sot, without doubt of defective mentality.
> He has two sons who marry two of six sisters. . . . One of these
> sisters is known as "Ada Jukes," also as "Margaret, the mother
> of criminals." She was indolent and a harlot before marriage.
> Besides an illegitimate son she had four legitimate children.
> The first, a son, was indolent, licentious and syphilitic; he mar-
> ried a cousin and had eight children all syphilitic from birth.
> Of the 7 daughters 5 were harlots and of the others one was an
> idiot and one of good reputation. Their descendants show a
> preponderance of harlotry in the females. . . . Ada had an ille-
> gitimate son who was an industrious and honest laborer and
> married a cousin. Two of the three sons were licentious and
> criminalistic in tendency and the third, while capable, drank
> and received out-door relief. All of the three daughters were
> harlots or prostitutes and two married criminals. The third
> generation shows the eruption of criminality. . . . The differ-
> ence in the germ plasm determines the difference in the pre-
> vailing trait. But however varied the forms of non-social behav-
> ior of the progeny of the mother of the Jukes girls the result
> was calculated to cost the State of New York over a million and
> a quarter of dollars in 75 years—up to 1877, and their proto-
> plasm has been multiplied and dispersed during the subsequent
> 34 years and is still marching on.[24]

24. Charles Davenport, *Heredity in Relation to Eugenics* (New York: Henry Holt, 1911),
pp. 233–34.

One way of slowing the tramp of degenerate protoplasm of the Jukes kind would be to prevent immigration into the United States of races having a greater proportion of those who were similarly rotten at the core. In 1924, the United States Congress passed the Immigration Restriction Act, the design of which was to prevent, in the terms of Senator Shields, "the great American type of citizenship [from being] diluted, mongrelized, and destroyed."[25] Members of the Eugenics Committee of the United States of America lobbied particularly with the House Committee on Immigration and Naturalization to restrict immigration of southern and eastern Europeans, whose intellectual inferiority, cultural primitiveness, criminal traits, and Bolshevik tendencies threatened the original Nordic stock that gave rise to the great American civilization. As Davenport put it some time earlier: "Society must protect itself, as it claims the right to deprive the murderer of his life, so also it may annihilate the hideous serpent of hopelessly vicious protoplasm."[26]

The eugenics movement, to sustain these last powerful thrashings, had to be pricked increasingly by scientists of little distinction. By the mid-1920s, most of the geneticists of serious scientific repute—for example, Morgan, Pearl, and Castle—had withdrawn from the movement, and Jennings testified before Congress against the Restriction Act.

Several causes internal to the science of genetics began to operate in the third and fourth decades of the century to turn the professional biological community away from eugenics. First was the recognition of the difficulty of eliminating unwanted Mendelian traits even under ideal conditions. Edward East and D. F. Jones, in their brief for eugenics—*Inbreeding and Outbreeding* (1919)—realized that if feeblemindedness were exhibited in the homozygous condition in one out of two hundred marriages, then one out of fourteen in the general population would be carriers of the defective germ cells but would not themselves be feebleminded. The implications were obvious: "The problem of cutting off defective germ plasm is not the theoretically simple one of preventing the multiplication of the afflicted; it is the almost hopeless task of reducing the birth rate among the personally unaffected transmitters where there is little prudential restraint and consequently a high reproductive rate."[27] They did hold out the hope, however, that those hetero-

25. Senator Shields, *Congressional Record—Senate* 65, part 7 (1924): 6461.
26. Charles Davenport, "Report of the Committee on Eugenics," *American Breeders' Magazine* 1 (1910): 129.
27. Edward East and D. F. Jones, *Inbreeding and Outbreeding* (Philadelphia: Lippincott, 1919), pp. 242–43.

zygous for feeblemindedness might be "more or less dull, stupid, lack-ing in real ability," so that discrimination of carriers might not be impossible.[28] Nonetheless, the problem of silent carriers provided, they recognized, "food for thought."[29] Another difficulty made eugenics technology even more dubious: most human mental traits, geneticists came increasingly to recognize, fell under the control of several genetic factors, not just one. Again, East was instrumental in laying the foun-dation for this important adjustment in Mendelian theory. His contri-bution here helped unhinge the crucial assumption—namely that each mental trait was controlled by one gene—upon which depended the pedigree studies of the eugenicists and their suggestions for therapeutic reform[30] Genetic recombination and assortment of the many genes un-derlying mental and behavioral traits meant, in Pearl's estimation, "that an enormously wide variety of new and different combinations of quali-ties is always possible, and may be expected to appear in some degree in virtually every mating."[31] With the zeal of the shriven sinner, Pearl indicted the eugenicists' theoretical sophistication: "For their public teaching, their legislative enactments, and their moral fervor are plainly based chiefly upon a pre-Mendelian genetics, as outworn and useless as the rind of yesterday's melon."[32]

The scientific floor supporting the eugenicists' program slumped even more precariously as geneticists came to insist that traits were de-termined not simply by heredity, but also by the environment. When wretched people live in wretched conditions, their faults need not lie in their genes. Jennings, a once staunch advocate of the full eugenics pro-gram, put it pointedly in his book *The Biological Basis of Human Nature* (1930):

> Persons may become idle and worthless, insane or criminal or tuberculous—either through bad genes or bad living condi-tions, or through a combination of both. So long as living con-ditions are bad, we do not know what ills are due to poor genes. We must therefore correct the bad living conditions, not only for their directly beneficial effect, but also for the sake of eugenics. When this is done, it will be possible to discover what defects are primarily the result of defective genes, and

28. Ibid., p. 243.

29. Ibid.

30. See, for example, the numerous pedigree studies in Charles Davenport, *Heredity in Relation to Eugenics* (New York: Henry Holt, 1911).

31. Raymond Pearl, "The Biology of Superiority," *American Mercury* 12 (1927): 257–66; quotation from p. 262.

32. Ibid., p. 261.

then to plan measures for getting rid of these genes: measures
for stopping the propagation of their carriers.[33]

By the 1930s, almost all geneticists of distinction had abandoned the
eugenics movement to the sophistic propagandists.[34]

Abstruse genetics may have spoiled the kernel of eugenics theory. But
it took wicked wit, penetrating journalistic analysis, and the horrors of
Nazi racial policy finally to crack the movement's hard shell of socially
and economically generated racial prejudice. Just back from Dayton,
Tennessee, where he defended one version of evolutionary biology,
Clarence Darrow took up the prosecution of another. In *The American
Mercury,* in 1925 and 1926, he examined the "Eugenics Cult."[35] He sym-
pathized with the plight of the eugenicists: "The good old *Mayflower*
stock is suffering the same unhappy fate as the good old pre-Prohibition
liquor. It is being mixed with all sorts of alien and debilitating substances."[36] Nonetheless, he was bound to find their analyses guilty of
both bad logic and worse taste. He compared the line of the Jukeses
with that of Jonathan Edwards, the hell fire preacher of Mayflower stock
that eugenicists liked to contrast with the Jukeses. Darrow thought the
Jukeses' infamous sexual pullulations as a dry stream compared with the
enormous drive gushing from the loins of the Edwardses: Jonathan's
grandfather had thirteen children; his grandmother was put away for
adultery and immorality; and he himself was one of eleven offspring.
But of greatness, the family actually had little. Out of the some forty
thousand estimated descendants of Edwards's grandparents, Darrow
reckoned only six hundred were of any note—about 1.5 percent of the
total. And this despite the Edwardses' social and educational opportunities, which were denied the Jukes family, who lived a poor and squalid
rural life. Darrow confessed that if he had to choose for a neighbor a
man like Edwards—who preached that "the God that holds you over
the pit of hell, much as one holds a spider over the fire, abhors you, and
is dreadfully provoked"—or one who lived the simple life of a dirt
farmer, well, Max Jukes would be his choice.[37]

33. H. S. Jennings, *The Biological Basis of Human Nature* (New York: Norton, 1930),
p. 250.

34. This is Kenneth Ludmerer's argument in his splendid book *Genetics and American
Society* (Baltimore: Johns Hopkins University Press, 1972), p. 77. One major geneticist
who remained faithful to the eugenics movement was Edward East. See Edward East,
Heredity and Human Affairs (New York: Scribner's, 1927).

35. Clarence Darrow, "The Edwardses and the Jukeses," *American Mercury* 6 (1925):
147–57; "The Eugenics Cult," *American Mercury* 7 (1926): 129–37.

36. Darrow, "The Eugenics Cult," p. 129.

37. Darrow, "The Edwardses and the Jukeses."

Walter Lippmann did not have Darrow's blunt sarcastic manner. Yet the cool patrician attitude of his *New Republic* articles on IQ testing (1922–1923) must have stopped the hearts of many blue-bloods (as his editorials certainly did during the Vietnam era).[38] Lippmann arched an eyebrow over the claim of the eugenicist Lothrop Stoddard that the average mental age of Americans stood at a meager fourteen. Stoddard based his conclusion on the mental testing done on Army inductees during the First World War. The tests, when translated into Stanford-Binet scores, showed that the average IQ of some 1,700,000 men tested—certainly an adequate sample of the American public—was about eighty-seven. Stoddard cried a warning over this revelation. He feared that inferior racial types, flowing especially out of southern and eastern Europe, had begun to sully the American mind. But Lippmann was suspicious. He suggested that a bit of legerdemain had been worked to demonstrate that the average American IQ was below average.

Lippmann examined the evidence for Stoddard's conclusion, and in doing so made comprehensible to a literate public the conceptually fragile nature of such intelligence tests as the Stanford-Binet. He pointed out, for instance, that the norms for the test were set by relatively few children and adults living in California, whereas it would have been more reasonable to set new norms based on the Army tests, since the population sample for the tests was quite large: then, of course, the average mental age of Americans would turn out to be average. But with greater sting, Lippmann disdained the social attitudes of eugenicists like Stoddard, William McDougall, and Lewis Terman, who claimed the support of the tests.

Lippmann's series of articles provoked response from Terman, the principal designer of the Stanford-Binet. The several exchanges pitted the assertions of the social technocrat against the incisive objections of an intellectual of humane but increasingly passionate temper. In moving terms Lippmann declared the motivation for his analysis; these same sentiments still bring sharpened intellects to cut away the presumptions of much IQ ideology:

> I hate the impudence of a claim that in fifty minutes you can judge and classify a human being's predestined fitness in life. I

38. Lippmann's series of articles with Lewis Terman's responses are conveniently collected in *The IQ Controversy,* eds. N. Block and G. Dworkin (New York: Pantheon, 1976), pp. 4–44. Terman, in addition to being the author of the "Stanford" revision of the Binet IQ test, was a member of the Eugenic Society of America (1923–1935) and endorsed the "Report of the Committee on Selective Immigration," which members of that society advanced in support of the Immigration Restriction Act of 1924.

hate the pretentiousness of that claim. I hate the abuse of scientific method which it involves. I hate the sense of superiority which it creates and the sense of inferiority which it imposes. And so, while I honestly think that there is a considerable future for mental testing, if it is approached with something like the caution employed by the editors of the army report, I believe also that the whole field is destined to be the happy hunting ground of quacks and snobs if loose-minded men are allowed to occupy positions of leadership much longer.[39]

Lippmann identified the theoretical deficiencies of IQ tests and racial stereotyping, as well as their moral falsity. The moral indictment, more than the abstract difficulties, diminished the attractiveness of eugenics doctrine for American scientists of stature during the 1920s and 1930s. Finally, the moral Götterdämmerung in Germany during the late 1930s—clearly enough reflected in the Nuremberg Laws of 1935—shed a spectral light not only on Nazi racial theories but also back on their eugenical counterparts in Britain and America. J. B. S. Haldane, in 1938, slyly suggested that a fifth column had to exist among the German eugenicists, since their proposals were so morally repugnant as well as scientifically absurd that they could only have been designed to undermine faith in all eugenics doctrine.[40] In the postwar period, scientists and laymen alike perceived an inverse relation between the weight of Nazi propaganda concerning race and the scientific probity of racial theories.

Psychologists, anthropologists, and sociologists in the early part of this century largely rejected what they took to be a Darwinian construction of man. They insisted that human behavior could be understood only through culture and the principles of learning theory. They were impatient to utilize new technologies in order to change conditions—and consequently the nature of men. The social scientists suggested that human nature when hardened in biological theory had no

39. Ibid., p. 42. In his important history of the persistent efforts to scale intelligence, *The Mismeasure of Man* (New York: Norton, 1981), Stephen Jay Gould has undertaken in our day much the same mission as Lippmann. Gould's deeper motivation also recalls Lippmann's. In the introduction to his book, Gould writes (p. 28): "If this subject were merely a scholar's abstract concern, I would approach it in more measured tone. But few biological subjects have had a more direct influence upon millions of lives. Biological determinism is, in its essence, a *theory of limits*. It takes the current status of groups as a measure of where they should and must be. . . . We pass through this world but once. Few tragedies can be more extensive than the stunting of life, few injustices deeper than the denial of an opportunity to strive or even to hope, by a limit imposed from without, but falsely identified as lying within."
40. J. B. S. Haldane, *Heredity and Politics* (New York: Norton, 1938), pp. 13–44.

potential, that for the evolutionists culture hung limply like a loincloth that could be quickly pushed aside for the real business of life. The new scientists of man, particularly the psychologists and sociologists, aimed not to understand the human world in biological terms, but to transform it in social terms. Though the program of the eugenicists looked initially promising as a plan for social transformation, its technology failed and its blood ran coldly reactionary. Finally, the cataclysm of World War II uprooted in Britain and America interest in biological theories of mind and behavior. So as the program of the eugenicists came to stand for the evolutionary approach to human biopsychology, the resulting picture of Darwinian man grew to caricature. From the early decades of this century to the present, the picture painted not only failed to recall the original, it looked harsh, ugly, and sinister to those of humane temper. It was not an image to be prized. Outside the Anglo-American tradition, however, Darwinian biopsychology branched somewhat differently. In Germany it gave birth not only to eugenics but also to the science of ethology, a discipline that reveled in rigorous formulation, that recognized the new advances in genetics, and that would give interesting, if finally suspect, account of cultural life.

The Biology of Mind and Behavior in Germany: From Darwinism to Neo-Darwinism

Biology of Mind and Behavior before the Turn of the Nineteenth Century

In Germany, prior to the reception of Darwinism, inquiries had long been made into the biology of behavior, though the studies had been intermittent. The early investigations, like the later, focused on animal behavior but drew lessons for the conception of human behavior. For instance, Hieronymus Rorarius, nuncio of Pope Clement VII, composed the wonderfully mordant tract *Quod animalia bruta saepe ratione utantur melius homine* (1547) to argue that animals displayed rational abilities that often outstripped those of men.[41] The book, published posthumously in 1654, was pressed into service against Descartes, who insisted that numinous reason guided human behavior, while animals displayed only mechanical instincts. With the relish of a Plutarchian moralist, Rorarius would compare the perfidy of humans with the pro-

41. Hieronymus Rorarius, *Quod animalia bruta saepe ratione utantur melius homine,* ed. Gabriel Naude (Amsterdam: Ravesteinium, 1654).

bity, say, of elephants, "who care for their weak"[42] and who "not only do not know of adultery, but think the sexual act one of turpitude; for after copulating with females they do not return to their herd until they wash themselves in a stream."[43]

In the eighteenth century, Hermann Samuel Reimarus made the first sustained studies of animal behavior that suppressed Rorarius's kind of anthropomorphizing urge. Wilhelm Wundt regarded Reimarus, a professor of Oriental Languages at the Hamburg Academie Gymnasium, as the founder of modern instinct theory.[44] Though a devoted student of animal behavior, Reimarus had more ultimate agenda. Like later natural theologians, he wished to demonstrate that the wonderful instincts of animals indicated the existence of a most wise God, as the title of his early tract made clear: *Instinctum brutorum existentis Dei, eiusdemque sapientissimi, indicem* (1725). His later studies of instinct submerged the theological intention, though not the combative. He set his *Allgemeine Betrachtungen über die Triebe der Thiere* (1760) in stated opposition to the Cartesian mechanists, on the one hand, and the Condillacian sensationalists on the other. Philosophically he inclined toward Aristotle. He proposed that certain "representational drives" (*Vorstellungstriebe*) brought animals to picture to themselves objects which impinged on their sensory organs.[45] These images directed the "intentional drives" (*willkührliche Triebe*) to certain actions that promised pleasure or to others that prevented pain.[46] "Skill drives" (*Kunsttriebe*)—or instincts—formed the most important class of intentional drives. Skill drives had as their aim "the welfare and preservation of each animal according to its mode of life, or the well being and preservation of the species or progeny."[47]

Reimarus undertook a minute analysis of instincts in different species; he wished to demonstrate that neither the mechanists nor the sensationalists could give them a proper account. Against the Cartesians, especially La Mettrie and Buffon, he offered examples of animals whose behavior could not result simply from fixed corporeal structures: for instance, young calves, rams, and goats attempted to butt with horns that had yet to sprout—which showed that the soul, not anatomy,

42. Ibid., p. 21.

43. Ibid., p. 70.

44. Wilhelm Wundt, *Vorlesungen über die Menschen- und Thierseele,* (Leipzig: Voss, 1863), 1:490.

45. Hermann Samuel Reimarus, *Allgemeine Betrachtungen über die Triebe der Thiere,* 3d ed. (Hamburg: Bohn, [1760] 1773), pp. 15–16.

46. Ibid., p. 2.

47. Ibid., p. 102.

guided the animal in the use of its organs.[48] Against Condillac, Guer, and other sensationalists—who believed instincts really to be learned habits—Reimarus produced many instances of behavior stereotyped in species, especially behavior that appeared immediately after birth: "A large number of skill drives," he observed, "are executed from birth on, without any experience, instruction, or example, and without error; and thus certainly they are naturally innate and hereditary."[49] As I have argued in the first chapter of this book, Reimarus produced the challenge that later biological theorists had to meet: the explanation of behavior that was unlearned and uniform in a species. In Germany, Wilhelm Wundt was the first to meet the challenge with an evolutionary theory of instinct.

Wundt appears to have read Bronn's translation (1860) of Darwin's *Origin of Species* while preparing the lecture series that formed the second volume of his *Vorlesungen über die Menschen- und Thierseele* (1863).[50] In the first volume, Wundt surveyed the intellectual talents of beasts, from polyps to beavers, and followed a continuous progression up to man. Under the influence of Hegel, Fichte, and Fechner, he was prepared to discover the glimmerings of consciousness even among the infusoria and to trace its development into the transcendent brilliance of human reason. Wundt's philosophical disposition—especially his acceptance of Fechnerian monism—preadapted his response to Darwin's chief explanatory principle, natural selection. In the second volume of the *Vorlesungen,* he recruited the principle to explain mental development:

> If the principle of natural selection lightens the considerable darkness engulfing the natural history of physical organisms, it serves no less to illuminate puzzles of psychic development. Both of the laws that are essential ground of the principle in the physical connection—namely the law of variability and the law of the inheritance of individual characteristics—are yet today applicable, albeit in a smaller effective range, to the mental realm. If these laws, which have operated for an almost limitless time, have segregated—according to the principle of natural selection—the physical characteristics of creatures, so also will their psychic conditions have been gradually but constantly differentiated in a distinctive manner. The perfection and diver-

48. Ibid., pp. 167–68.

49. Ibid., p. 160.

50. See Robert Richards, "Wundt's Early Theories of Unconscious Inference and Cognitive Evolution in their relation to Darwinian Biopsychology," in *Wundt Studies,* ed. W. Bringmann and R. Tweney (Toronto: Hogrefe, 1980), pp. 42–70.

gence of species in their corporeal and mental relations run in parallel development. Both reciprocally determine each other. Where particular nerves, muscles, and central organs are often stimulated by the functioning of psychic impulses, there the physical structure will be developed; the psychic structure, in turn, will develop through physical adjustments.[51]

Propelled by the idealistic assumption that mind guided all natural processes, Wundt interpreted natural selection—not without some suggestion from the *Origin*, of course—as dependent upon the inheritance of purposefully acquired characteristics. He never shed his belief, exemplified in this quotation, that individual intention and habit were the principal molding instruments of evolution.

Wundt was encouraged in his reading of the *Origin* by Ernst Haeckel, whose own monism and Lamarckism led him to construe natural selection much as Wundt had. When Wundt finally came to recognize that the Darwinians—that is, the ultra-Darwinians led by August Weismann—conceived natural selection as a blind mechanism, he rejected the principle, though not, to be sure, his belief in mental and behavioral evolution. In his later psychology, for instance, Wundt used the supposition of the inheritance of purposeful adaptations to explain the universal and progressive stages of moral evolution in different societies.[52]

At the turn of the century in Germany, as in England and America, evolutionary biopsychologists divided themselves into three fairly distinct classes. Natural selection provided the demarcation criterion. There were those, like Haeckel, Ludwig Büchner, Georg Schneider, and Wilhelm Preyer, who recognized natural selection as a principle of mental development, but who joined its operation with functionally acquired adaptations.[53] Haeckel, for example, conceived instincts as "inherited," but insisted that "experiences and thus new adaptations of the animal mind are also transmitted through inheritance."[54] Others, like Wundt, Georg Eimer, August Pauly, Eduard von Hartmann, Hans

51. Wilhelm Wundt, *Vorlesungen uber die Menschen- und Thierseele* 2:pp. 355–56.

52. Wilhelm Wundt, *Ethik* (Stuttgart: Enke, 1886), pp. 204–33, 402–10.

53. Haeckel regarded natural selection as operating on adaptations that had been acquired through direct impingements of the environment or through habit. See Ernst Haeckel, *Natürliche Schopfungsgeschichte*, 4th ed. (Berlin: Reimer, [1868] 1875), pp. 197, 207–10. Büchner and Schneider made similar assumptions. See Ludwig Buchner, *Aus dem Geistesleben der Thiere* (Berlin: Unbekant, 1879), chaps. 1–3; Georg Schneider, *Der therische Wille* (Leipzig: Abel, 1880), p. 426; and Wilhelm Preyer, *Die Seele des Kindes: Beobachtungen über die geistige Entwicklung des Menschen in den ersten Lebensjahren*, 4th ed. (Leipzig: Grieben, [1882] 1895), pp. 155–84.

54. Haeckel, *Natürliche Schopfungsgeschichte*, p. 636.

Figure 11.2   Ernst Haeckel (seated), 1834–1919, photograph from 1866.

Driesch, and Erich Wasmann, virtually abandoned natural selection for internal, vital principles of psychic evolution.[55] Von Hartmann, representative of this sect, rejected the Darwinists' 'mechanistic' approach to instinct; he rather explained heritable behavior as arising from an "unconscious purposive and formative activity" (*zweckmässige unbewusste*

55. See, for instance, Wilhelm Wundt, "Die Thierpsychologie," *Essays* (Leipzig: Engelmann, 1885) and *Grundriss der Psychologie* (Leipzig: Engelmann, 1896), p. 331; Theodor Eimer, *Die Entstehung der Arten auf Grund von Vererbung erworbener Eigenschaften* (Jena: Fischer, 1888), chap. 6; August Pauly, *Darwinismus und Lamarckismus: Entwurf einer psychophysischen Teleologie* (München: Reinhardt, 1905); Eduard von Hartmann, *Das Problem des Lebens* (Bad Sachsa im Harz: Haacke, 1906), pp. 362–440; Hans Driesch, *The Science and Philosophy of the Organism,* Gifford Lectures for 1907 and 1908 (London: Black, 1907–1908); and Erich Wasmann, *Instinct und Intelligenz im Thierreich* (Freiburg i. B.: Herder, 1897).

Bildungsthätigkeit) that produced "an unconscious knowledge" (*unbe-wussten Erkenntniss*) in the individual.[56] Driesch referred to such a pur-posive principle as an "entelechy," which he supposed might operate on organic chemical and physiological responses by "suspend[ing] move-ment, transforming kinetic energy into potential energy, and . . . [by setting] free suspended movement as circumstances require."[57] The vitalists believed that natural selection and other kinds of mechanical causality simply failed to account for such phenomena as instinct and intelligence, so that inherent vital principles had to be invoked. These thinkers faced off especially against the third group of evolutionists, the ultra-Darwinists, who advanced the natural selection of chance varia-tions as the chief agent of evolutionary transformation. August Weis-mann was the leader of this circle, but in the area of biopsychology its chief representatives were Heinrich Ziegler, Karl Groos, and Otto Zur Strassen.[58] Ziegler identified this group (along with Conwy Lloyd Mor-gan and Charles Otis Whitman) as advocates of the "new animal psy-chology." What united them, according to Ziegler, was their disavowal of vital principles in biology, their rejection of Lamarckism, and their commitment to physiological (as opposed to mentalistic) explanations of behavior.[59]

At the turn of the century, German evolutionary theorists struggled with one another over biopsychological issues on narrower, technical grounds, but often to secure larger, doctrinal territory, especially in re-ligion and politics.

In religion, a bitter but instructive conflict occurred between Ernst Haeckel and Erich Wasmann. In the spring of 1905, Haeckel gave a series of popular lectures on evolution to very large audiences in Berlin. He focused particularly on what he took as the subversion of evolution-ary theory by Erich Wasmann, a Jesuit and perhaps Europe's leading entomologist.[60] Early in his career, Wasmann had dismissed evolution-ary theory, ostensibly because it could not explain certain marvelous instincts of insects (especially those of *der Trichterwickler,* the leaf-rolling

56. Eduard von Hartmann, *Philosophie des Unbewussten,* 6th ed. (Berlin: Dunckers, [1868] 1874), pp. 77 and 84.

57. Driesch, *Science and Philosophy of the Organism* 2:221.

58. See for instance, Heinrich Ziegler, *Der Begriff des Instinktes einst und jetzt,* 3d ed. (Jena: Fischer, [1904] 1920); Karl Groos, *Die Spiele der Thiere* (Jena: Fischer, 1896), pp. v–ix; and Otto Zur Strassen, *Die Neuere Tierpsychologie* (Leipzig: Teubner, 1908).

59. Ziegler, *Der Begriff des Instinktes,* pp. 75–86.

60. Haeckel reacted specifically to Wasmann's newly published book on evolution. See Erich Wasmann, *Die moderne Biologie und die Entwicklungstheorie* (Freiburg i. B.: Herder, 1904).

Figure 11.3 August Weismann (center), photograph from 1880.

beetle).[61] But in the early 1890s, he had converted to a Drieschian version of evolutionary theory—principally for empirical reasons. He had investigated the slave-making instinct of several groups of ants—the same subject that had occupied Darwin in the *Origin*'s chapter on instinct—and determined the instinct's various stages of development among related species and subspecies.[62] He also discovered an order of beetles—the Myrmecophilae, the so-called "guests of ants"—whose instinctive and anatomical mimicry of ants varied across geographically connected species. The appearance and behavior of the beetles ranged

61. Erich Wasmann, *Der Trichterwickler* (Münster: Aschendorf, 1884).

62. Erich Wasmann, "Zur Entwicklung der Instincte," *Verhandlungen zoologisch-botanischen Gesellschaft in Wien* 47 (1897): 168–83; and *Vergleichende Studien über das Seelenleben der Ameisen und der höhern Thiere*, 2d ed. (Freiburg i. B.: Herder, 1900).

from slight similarity to that of their ant hosts to virtual identity.[63] These geospatial representations of the process of descent brought Wasmann to adopt a qualified version of evolutionary theory. Haeckel became incensed over Wasmann's turn of mind. He objected not to the generous strain of vitalism in Wasmann's rendition of evolution, but to his exclusion of the human soul from the evolutionary process. Haeckel regarded Wasmann's conversion to descent theory as a "master piece of Jesuitical perversion and sophistry," since it was obvious, at least to Haeckel, that this was simply a Jesuit plot to subvert science with Papist dogma.[64] Haeckel, though, had no compunctions about enmeshing his own evolutionary ideas in a religion that oozed Weltschmaltz. He installed a "monistic God, an all-embracing World-essence," which he identified with an "eternal, psychic energy that does not stand in fiendish and strange opposition to space-filling matter, but that is bound with it in an eternal and immortal substance."[65] Haeckel's differences with Wasmann really amounted to an *odium theologicum,* sweetly disguised as a scientific conflict.

In Germany as in the United States, biological disputes became instruments of political war by other means. The bloody rhetoric flowed from many different factions, but one of the more interesting contests pitted Marxist socialists against Prussian conservatives for the Darwinian palm. The issue was joined when August Bebel, the leader of the Social Democratic party, performed a nice reverse on Rudolph Virchow's argument against introducing evolutionary theory into the schools of Germany. Bebel contended that if Darwinian theory led to socialism, as Virchow claimed, then that was but an argument in favor of socialism. Bebel thought that capitalism put artificial restraints on the operations of natural selection, so that the idiot son of the factory owner had the advantage over the talented son of the factory worker. Were socialism actually put into practice, then the natural forces of progressive evolution would produce a classless society, in which property would cease to exist and women would no longer suffer political and sexual subjugation. Moreover, with the dissolution of the state and the common possession of property, men would learn war no more. These were, according to Bebel, the socialist lessons of Darwinism.[66]

63. See, for instance, Erich Wasmann, *Die Gastpflege der Ameisen: ihre biologischen und philosophischen Probleme* (Berlin: Borntraeger, 1920).

64. Ernst Haeckel, *Der Kampf um den Entwickelungs-Gedanken* (Berlin: Reimer, 1905), p. 32.

65. Ibid., p. 92. See also Niles Holt, "Ernst Haeckel's Monistic Religion," *Journal of the History of Ideas* 32 (1971): 265–80.

66. August Bebel, *Die Frau unter der Sozialismus,* 16th ed., unchanged (Stuttgart:

Figure 11.4 Ernst Haeckel, photograph from 1914.

Bebel's claim to Darwinian authority for socialism ignited the quick fuse of the Monist League. Ziegler, a founding member along with Haeckel, exploded with counteranalyses ostensibly sanctioned by the principles of natural and sexual selection. Thus against Bebel's theory of feminine liberation, Ziegler objected that "by nature the sexes have different psychological and social tasks . . . [consequently] the demand that women hold the same political and social position as men has no basis in science."[67] Moreover, Darwinism hardly taught cooperation and human fellowship:

Dietz, [1879] 1892), pp. 195–99. See Alfred Kelly's incisive discussion of the impact of Darwinism on German politics in his *The Descent of Darwin: The Popularization of Darwinism in Germany, 1860–1914* (Chapel Hill: University of North Carolina Press, 1981), pp. 123–141.

67. Heinrich Ziegler, *Die Naturwissenschaft und die Socialdemokratische Theorie* (Stuttgart: Enke, 1893), pp. 25–26.

> Bebel does not know that according to Darwin's doctrine war
> has been of the greatest importance for the general progress of
> the human race, since the physically weaker, the less intelligent,
> and the morally degenerate must make way for the stronger
> and better developed people. . . . If one accepts the insights of
> modern science, he must see war between different races or
> people as a form of the struggle for existence in the human
> race.[68]

As even this brief sketch might make clear, evolutionary biopsychology in Germany was extremely fragmented. On narrower issues of biological theory, evolutionists of mind and behavior exhibited a wide
range of views, from various forms of vitalism at the one extreme,
through the varieties of Lamarckism-Darwinism, to the opposite pole
of ultra-Darwinism. Ziegler, a disciple of Weismann, had few scientific
qualms about uniting with his mentor's enemy, Ernst Haeckel, to oppose the Catholic Wasmann and the Marxist Bebel. The history of these
scientific, political, and religious alliances ought, I think, give caution
about any easy identification of "Darwinism" with a particular political
or social philosophy.

Out of this scientific and cultural mix came one of the most influential
of recent thinkers in biopsychology, Konrad Lorenz. Lorenz gave conceptual and empirical shape to the modern science of ethology, the science which has been further elaborated into (and Edward Wilson believes absorbed by) sociobiology. Lorenz concentrated his attention on
the theoretical development and experimental justification of the concept of instinct. But the circumstances of its evolution, particularly in
the intellectual environment of Nazi Germany, have encouraged the association of the evolutionary theory of human mind and behavior with
conservative, or—as many engaged in the contemporary debate would
put it—Fascist ideology.

Lorenz's Ethological Theory and Nazi Ideology

Lorenz grew up among chickens, ducks, dogs, cats, and semitame
jackdaws on his family's estate in Altenberg, Vienna.[69] Though he was
trained in the profession of his father, taking an M.D. at the University
of Vienna in 1928, his soul followed a different call. Even when he was

---

68. Ibid., pp. 168–69.
69. For further biographical details see, O. Koehler, "Konrad Lorenz 60 Jahre," *Zeitschrift für Tierpsychologie* 20 (1963): 385–401; and Alec Nisbett, *Konrad Lorenz: A Biography*
(New York: Harcourt Brace Jovanovich, 1977).

a visiting medical student at Columbia University in the fall of 1922 and had the opportunity to work in T. H. Morgan's laboratory, he spent most of his time on Long Island studying ducks. The year before he received his degree, he published his first paper; but it was on a topic at some distance from medical concerns, the flight-behavior of jackdaws.[70] In composing this and his next several essays, he received encouragement and theoretical direction from his friend and mentor Oskar Heinroth, a fellow ornithologist and the director of the Berlin Zoological Gardens.

Lorenz credits Heinroth with two original and foundational contributions to the evolutionary study of animal behavior. Heinroth (and, Lorenz acknowledges, C. O. Whitman) made the "discovery that there are motor patterns of constant form which are performed in exactly the same manner by every healthy individual of a species." Heinroth also forged the principles of comparative ethological study, particularly the use of hybridization experiments to isolate inherited elements of behavior.[71] Though Lorenz never possessed any sense of the history of biology, there is little doubt of his debt to the ultra-Darwinian Heinroth.[72]

In his third published paper (1932),[73] Lorenz began to refine Heinroth's theory of "drive activities" (*Triebhandlungen*), or what he later would call "instinctive activities" (*Instinkthandlungen*)—using the older, Latinate name to distinguish his conception from psychoanalytic

70. Konrad Lorenz, "Beobachtungen an Dohlen," *Journal für Ornithologie* 75 (1927): 511–19.

71. Konrad Lorenz, "Part and Parcel in Animal and Human Societies" (1950), in *Studies in Animal and Human Behavior,* trans. Robert Martin (Cambridge: Harvard University Press, 1971), 2:131. I have generally cited Martin's translations for those articles included in these two volumes; I have, however, compared them with the German edition: Konrad Lorenz, *Über tierisches und menschliches Verhalten* (München: Piper, 1965). See also, Robert Richards, "The Innate and the Learned: The Evolution of Konrad Lorenz's Theory of Instinct," *Philosophy of the Social Sciences* 4 (1974): 111–33; Theo Kalikow, "History of Konrad Lorenz's Ethological Theory, 1927–1939," *Studies in the History and Philosophy of Science* 6 (1975): 331–41; and "Konrad Lorenz's Ethological Theory, 1939–1943," *Philosophy of the Social Sciences* 6 (1976): 15–34.

72. Three recent and important essays on the history of modern ethology are Richard Burkhardt, "On the Emergence of Ethology as a Scientific Discipline," *Conspectus of History* 1 (1981): 62–81; "Development of an Evolutionary Ethology," in *Evolution from Molecules to Men,* ed. D. Bendall (Cambridge: Cambridge University Press, 1983), pp. 429–44; and John Durrant, "Innate Character in Animals and Man: A Perspective on the Origins of Ethology," in *Biology, Medicine and Society: 1840–1940,* ed. Charles Webster (Cambridge: Cambridge University Press, 1981): 157–92.

73. Konrad Lorenz, "A Consideration of Methods of Identification of Species-Specific Instinctive Behavior Patterns in Birds" (1932), in *Studies in Animal and Human Behavior* 1:57–100.

530 THE TWENTIETH CENTURY

and behaviorist views of drive.[74] In his 1937 paper, "The Establishment of the Instinct Concept," Lorenz discriminated five criteria that would give observational definition to the concept of "inherited drives of fixed behavior" (*Erbtriebe bedingter Handlungen*). He thought the two most obvious criteria were that an instinctive behavioral pattern would appeared in nearly all members of a species and that animals reared in isolation would still exhibit the pattern. A more blurred mark, which he nonetheless assumed the experienced naturalist would intuitively recognize, was the complexity of the behavior in question. If the behavioral pattern exceeded the learning capacities of the animal, then, as he supposed, that pattern would have to be inherited. Actions performed incompletely (*Intentionsbewegungen*) or in nonappropriate circumstances (*Leerlaufreaktionen*) constituted a fourth criterion of innate drives. The greylag goose that made fragmented efforts all year round to build a nest or the kitten that 'stalked' a ball of yarn—these animals could not be pursuing an individual purpose, but were simply releasing fixed behaviors at inopportune times. Finally, the fixity and rigidity of behavioral patterns themselves bespoke innately determined reflex chains rather than individually acquired responses. Lorenz insisted on the reflex character of instinct and (quoting Ziegler) "the histological conceptual framework."[75] He installed these as defenses against insidious vitalisms on the one hand, and incipient Lamarckisms on the other.

Despite Lorenz's adamantine commitment to ultra-Darwinism, his instinct theory bore the sign of an openly anti-Darwinian thinker—Jacob von Uexküll, a Drieschian vitalist. From von Uexküll, an independent scholar of spousal means, Lorenz adapted the notion of a "functional system" (*Functionskreis*). According to von Uexküll's theory, a functional or interactive system constituted the relation between an animal, with its special organs and needs, and its own experienced world (*die Umwelt*), the lived reality of which corresponded to the animal's sensory abilities and requirements.[76] Lorenz transformed von Uexküll's conception of the functional system into that of the "innate releasing schemata" (*angeborenen Auslöse-Schemata*).[77] This innate releasing mechanism (IRM), as he also termed it, was the receptor correlate in the animal that responded with a particular pattern of behavior to specific elicitory cues in the environment. The male three-spined

74. Konrad Lorenz, "The Establishment of the Instinct Concept" (1937), in *Studies in Animal and Human Behavior* 1:260.

75. Ibid., p. 301.

76. Jacob von Uexküll, *Strafzüge durch die Umwelten von Tieren und Menschen* (Berlin: Springer, 1934).

77. Lorenz, "Establishment of the Instinct Concept," p. 273.

stickleback fish, for example, when confronted with another male with a red belly, or even a crude model with a red underside, releases a characteristic threat posture; but when it beholds a female with a nonred, swollen belly, it dances the fixed steps of the mating ritual. Thus a red belly releases one kind of instinctive pattern in the male stickleback, while a swollen belly releases another.

Following Ziegler's lead, Lorenz originally conceived instinct as a chain of reflexes that had been forged through natural selection. Though under this interpretation instinct had the advantages of a mechanical and naturally heritable system, Lorenz gradually grew dissatisfied with the construction. Even in his early papers he recognized that certain instincts seemed to display an emotional component, a force driving an animal to find release through instinctive acts. But did anyone enjoy self-stimulating the patellar reflex? Lorenz also observed that instincts might go off in vacuo, as if dammed energy burst through containing valves—again, a condition uncharacteristic of reflexes. Undoubtedly Wallace Craig, a student of Whitman with whom Lorenz carried on an extensive correspondence between 1935 and 1937, sensitized him to the motivational features of instinctive behavior. But Lorenz formally abandoned the reflex conception of instinct only after discussions with Erich von Holst, who discovered endogenous neural centers of rhythmic discharge and coordination.[78] Von Holst provided the physiological basis for Lorenz's hypothesis of response-specific energy gradually accumulating within an animal and motivating the creature to seek energy release through the IRM.

After the war, Lorenz's work in ethology was introduced to American science through an essay review by the biological psychologist Daniel Lehrman.[79] The introduction was hostile. Lehrman indicted Lorenz's conception ostensibly for theoretical reasons; but he apparently was also angered by the brown shirt which the concept of instinct had recently worn.[80] Lehrman complained, as did others in American psychology during the 1950s, that Lorenz's simplistic division of behavior into the instinctive and the learned forced researchers to homogenize many kinds and levels of activity.[81] The two fixed categories frustrated

78. Lorenz first mentions von Holst in his "Uber den Begriff der Instinkthandlung," *Folia Biotheoretica* 2 (1937): 17–50.

79. Daniel Lehrman, "A Critique of Konrad Lorenz's Theory of Instinctive Behavior," *Quarterly Review of Biology* 28 (1953): 424–34.

80. Personal communication from a friend of Lehrman and also an early critic of Lorenz.

81. Other important critics were, T. C. Schneirla, "A Consideration of Some Conceptual Trends in Comparative Psychology," *Psychological Bulletin* 49 (1952): 559–97; Donald

precise analysis of behavior. In Lehrman's view, Lorenz simply ignored the genetic principle that phenotypic structures depended not only on innate factors but also on internal and external environmental factors. Lorenz thus failed to recognize that the behavior an animal displayed could never be simply categorized as either innate or learned. So-called instinctive behavior required environmental contributions for its development. It was simply to be blinded by dogma to think that experiments could be devised, as Lorenz had tried, to isolate the innate from the learned in behavior. To illustrate these objections concretely, Lehrman offered the studies of Z. Y. Kuo, who argued against the notion of the perfect isolation experiment. The chick, which 'instinctively' pecks with coordinated movements shortly after emerging from its egg, might still have been conditioned even while ensconced: Kuo pointed out that the head of the embryonic chick, which was bent over the thorax, would bob up and down with its heartbeat, so the chick could 'learn' to peck while still in the egg. It was an elementary mistake to think of phenotypic behavior as itself innate or exclusively determined by innate factors.[82]

The special passion with which Lehrman rejected Lorenz's theory may have been motivated by its incriminating past. The tainted political and social circumstances surrounding certain aspects of its evolution have recently been explored by Kalikow.[83] She focuses on Lorenz's hypothesis of human biological degeneration. The hypothesis holds that the city-dweller's instinctual behavior, which originally evolved in rustic and pastoral settings, has become, like that of the domesticated animal, increasingly pathological. Lorenz insisted that the protection from natural selection afforded by advanced society had its crippling consequences: behavioral, intellectual, and moral degeneration. Kalikow argues that Lorenz's hypothesis stemmed from the eugenically contami-

---

Hebb, "Heredity and Environment in Mammalian Behaviour," *British Journal of Animal Behaviour* 1 (1953): 43–47; and Frank Beach, "The Descent of Instinct," *Psychological Review* 62 (1955): 401–10.

82. Lorenz did grudgingly take account of this criticism. He resolved to refer only to the information stored in the genes as innate. See Konrad Lorenz, *Evolution and the Modification of Behavior* (Chicago: University of Chicago Press, 1965), p. 40. Lehrman responded to this new proposal in "Semantic and Conceptual Issues in the Nature-Nurture Problem," in *Evolution and Development of Behavior*, ed. L. Aronson et al. (San Francisco: Freeman, 1970), pp. 17–51. See also my "The Innate and the Learned: the Evolution of Konrad Lorenz's Theory of Instinct," pp. 124–29.

83. Theo Kalikow, "Konrad Lorenz's Ethological Theory: Explanation and Ideology, 1938–1943," *Journal of the History of Biology* 16 (1983): 39–73. See also Theo Kalikow, "Konrad Lorenz's 'Brown Past': A Reply to Alec Nisbett," *Journal of the History of the Behavioral Sciences* 14 (1978): 173–79.

nated theories of Haeckel and was further elaborated to support Nazi racism. In forging the Haeckel connection, she relies largely on the work of Gasman, who maintains that Haeckel's social Darwinist ideas fostered Nazi biology.[84] Gasman has orchestrated a strong case. Nonetheless, the influence of Haeckel and the Monist League on Nazi ideology was hardly straightforward. The Nazi elite resisted evolutionary theory, despite its scientific charms.[85] After all, could the Aryan race have descended from a tribe of baboons? Moreover, as Holt has emphasized, the Monist League had a pacifist and socially liberal orientation and had explicitly rejected the "new religion" of Nazism.[86] The League disbanded in 1933 rather than accept "coordination" into the policies of National Socialism. Nonetheless, some Nazi propagandists did eulogize Haeckel for having supplied scientific support for central ideas of the new regime.[87] Haeckel and some of the Monists undoubtedly created an intellectual environment congenial to the growth of Nazi pseudoscience.[88] But did Lorenz also swallow and then regurgitate the doctrines upon which the Nazis fed?

Kalikow tries to establish points of similarity among the views of Lorenz, Haeckel, and the Nazis in order to demonstrate the Haeckelian (and thus effectively Nazi) sources of central Lorenzian ideas. She discriminates four shared theses:[89] (1) the "biologistic view that the laws of nature are the laws of society"; (2) "the view that phylogeny has followed a steady course until now when it is suddenly surrounded by a number of perils: over-population, softened attitudes, weakened pulsions"; (3) the idea that human external form mirrors internal moral conditions; and (4) the belief "that the ancient Greeks were forebears of the Aryan race." But the first two proposals can be found generally in evolutionary literature;[90] and the latter two, as Kalikow admits, were ubiquitous at the turn of the century. If such vague similarities suffice here, we should all be hustled to the gallows. There are, I believe, two

84. Daniel Gasman, *The Scientific Origins of National Socialism* (New York: Science History Publications, 1971).

85. Something Kalikow herself recognizes ("Konrad Lorenz's Ethological Theory," p. 62).

86. Niles Holt, "Monists & Nazis: a Question of Scientific Responsibility," *Hastings Center Report* 5 (1975): 37–43.

87. See, for example, Heinz Brücher, "Ernst Haeckel, ein Wegbereiter biologischen Staatsdenkens," *Nationalsozialistische Monatshefte* 6 (1935): 1087–98.

88. See Gasman, *The Scientific Origins of National Socialism,* pp. 147–82.

89. Kalikow, "Konrad Lorenz's Ethological Theory," pp. 47–48.

90. As we have already seen in some detail, chapter 5 of Darwin's *Descent of Man* examines the possible deleterious consequences to society of the abridgment of evolutionary principles.

Figure 11.5   Konrad Lorenz, 1903–, photograph from 1973.

strong objections to the argument that Lorenzian biology has Haeckelian roots: (1) Lorenz never cited Haeckel's work in support of his views,
but rather the studies of biologists like Heinroth, Whitman, Craig,
and von Uexküll; and (2) he regarded any biologist proposing the inheritance of acquired characters—the kernel of Haeckel's hereditary
theory—as scientifically senile.

In 1938 Lorenz, like many other German scientists—and particularly
physicians—joined the NSDAP (*Nationalsozialistische Demokratische Ar-
beiter Partei*).[91] Prior to the war he had spoken at a respected scientific

91. Kalikow, "Konrad Lorenz's Ethological Theory," p. 56. Robert Proctor describes
the high incidence of physicans among the Nazis and their various roles in the party; see
his comprehensive treatment *Racial Hygiene: Medicine under the Nazis* (Oxford: Oxford
University Press, forthcoming).

congress (a meeting of the German Psychological Association), which was, however, sponsored by the Nazis. He had also published two articles (one commemorating the birth of Heinroth) in a journal having explicit Nazi connections (*Der Biologe*). On these few occasions of public Nazi association, he had touched on a theme that Kalikow identifies as the principal evidence of endorsement of National Socialist ideology: that like domesticated animals, civilized, urban men and women were in peril of biological degeneration. Passages such as the following, she believes, reveal the Haeckel-Nazi source of the main elements of Lorenz's biology of behavior:

> Whether we share the fate of the dinosaurs or whether we raise ourselves to a higher level of development, scarcely imaginable by the current organization of our brains, is exclusively a question of biological survival power and the life-will of our people. Today especially the great difference depends very much on the question whether or not we can learn to combat the decay phenomena in our people and in humanity which arise from the lack of natural selection. In this very contest for survival or extinction, we Germans are far ahead of all other cultural peoples.[92]

But this is a gossamer thread by which to tie Lorenzian biology to the Nazis. The vast bulk of Lorenz's work on instinct rests squarely in the Darwinian evolutionary tradition. Even his concern about behavioral and mental degeneration has deep roots in that tradition. The above passage, mutatis mutandis, could have been written by any number of British or American evolutionary biologists in the last century or the early part of this one. And again, recall that Darwin too warned of the dangers to human progress consequent on the disengagement of natural selection in civilized societies.

Lorenz undoubtedly descended to accommodate some of his biological views to the ideology of his time and place. He may have reached his nadir in 1940, when in an article on domestication, he wrote:

> If it should turn out . . . the mere removal of natural selection causes the increase in the number of existing mutants and the imbalance of the race, then race-care must consider an even more stringent elimination of the ethically less valuable than is done today, because it would, in this case, literally have to

92. Konrad Lorenz, "Systematik und Entwicklungsgedanke im Unterricht," *Der Biologe* 9 (1940), p. 29 (quoted by Kalikow, and translation adapted from hers in "Konrad Lorenz's Ethological Theory," p. 63).

replace all selection factors that operate in the natural environment.[93]

In view of the history we have traversed in the preceding chapters, even this stain of ideology is hardly surprising. At this point in Lorenz's career, certain well-entrenched evolutionary ideas happened to intersect with despicable Nazi dogma. Certainly he fostered the union of biology and propaganda, but I doubt that his main concerns would have been markedly different had the Weimar Republic survived. Nonetheless, just this sort of public association of Nazism with human evolutionary bio-psychology froze any enthusiasm for the discipline immediately after the war, and continues to chill its development within contemporary biology of behavior as well as within the social sciences.

The Rise of Sociobiology

The principles and techniques of behaviorism dominated research in American psychology from the late 1920s to the early 1960s. The domi-nation of this brand of science, along with the after-war distaste for theories of biological hierarchy, effectively barred ethological ap-proaches to mind and behavior. Behaviorists usually did not deny the influence of evolution upon behavior, but they conceived the influence as uniform. Natural selection may have produced organisms of wildly different anatomical structures but, perhaps exhausted of imagination, made them one and all into drive-reducing machines. This assumption allowed behavioral technologists to discover the generic principles of learning by studying only two species of animal: the white rat and the college sophomore. Even those few comparative psychologists who had retained a more sophisticated biological perspective were yet shaded in their thinking by the eclipse of Darwinism during the early part of the century. N. R. Maier and T. C. Schneirla, in their influential book *Prin-ciples of Animal Psychology* (1935 and 1964), reflected the nervousness of biologists of the period: they shrank back from the slightest hint of "purposiveness" as a biological construct and even from the Darwinian presumption that structures evolved because of utility. "No matter how the structure first appeared in the race," they cautioned, "there is no justification for saying that utility was a *primary* factor in the matter."[94] The instincts an animal might display, they insisted, would often be the

93. Konrad Lorenz, "Durch Domestikation verursachte Störungen arteigenen Verhal-tens," *Zeitschrift für angewandte Psychologie und Charakterkunde* 59 (1940): 66 (quoted and translated by Theo Kalikow, "Konrad Lorenz's 'Brown Past,'" p. 176).

94. N. R. Maier and T. C. Schneirla, *Principles of Animal Psychology*, 2d ed. (New York: Dover, [1935] 1964), p. 122.

mechanical result of inherited anatomical structures; but one should not suspect that either the morphological form or the consequent behavior had been naturally selected. In any case, they wished to prescind from such phylogenetic problems, so to concentrate on the ontogenetic, "the nature and causes of behavior in the individual organism."[95] Even in the second edition of their book, in 1964, which included several essays in response to the new ethological theory coming out of Germany, they resisted a more evolutionary approach to behavior.[96]

Some early biologists of behavior, especially field researchers trained in the first two decades of this century, did not share all the phobias of the psychologists and their biological colleagues. They continued to nurture conceptions that had come to birth with the Darwinians of the previous century. Chief among these paleo-Darwinians was Warder C. Allee, who advanced the principle of cooperation as crucial to explaining the evolution of social organization among animals and men. Allee took his Ph.D. at the University of Chicago in 1912 and returned nine years later to the zoology department, where he would number among his colleagues Frank Lillie, Alfred Emerson, and Sewall Wright. Through his graduate study of freshwater crustaceans, Allee became interested in the phenomenon of animal aggregation, the tendency of members of a species of simple organisms to clump together. His theoretical understanding of this observation slowly evolved through reading in the works of Wheeler, Espinas, Deegener, Kropotkin, and Spencer.[97] He seems to have been especially struck by the views of the Earl of Shaftesbury, the eighteenth-century British philosopher who recognized, in Allee's biological paraphrase, that "there are racial drives that go beyond personal advantage, and can only be explained by their advantage to the group."[98]

In his first book, *Animal Aggregations: A Study in General Sociology* (1931), Allee explained the evolution of social organization, from its primitive stages in animal aggregation to its altruistic basis in human groups, by appeal to the selective principle of cooperation.[99] He realized that if altruistic behavior were to be explained, selection had to occur at the level of the community or group rather than at the level of the individual. In defending the principle of cooperation, he relied on the

95. Ibid.

96. Ibid., pp. 511–652.

97. W. C. Allee, *Social Life of Animals* (New York: Norton, 1938), pp. 20–26. See the conclusion to chapter 7, above.

98. Ibid., p. 24.

99. W. C. Allee, *Animal Aggregations: A Study in General Sociology* (Chicago: University of Chicago Press, 1931).

population-genetic analyses of his colleague Sewall Wright. Wright's mathematical demonstrations, whose intricacies Allee confessed not to understand, supported the conclusion that "the random differentiation of local populations furnishes material for the action of selection on types as wholes, rather than on the mere average adaptive effects of individual genes."[100]

In the 1930s, appeal to group selection hardly disturbed any geneticists who might have been listening.[101] After all, it seemed to have been sanctioned by Wright's formidable mathematics, which few of his colleagues (save Fisher and Haldane) could penetrate, but all respected.[102] In the 1960s, however, assertions of animal 'altruism' and its explanation by group selection fell upon the ears of neo-Darwinians of considerable genetic and mathematical acumen. What initially captured their attention was the account of altruism constructed by V. C. Wynne-Edwards, whose massive volume *Animal Dispersion in Relation to Social Behavior* (1962) obliquely referred back to Allee's *Animal Aggregations: a Study in General Sociology*. Wynne-Edwards, a Scots ecologist and fisheries expert, focused on the evolutionary principles governing "the placement of individuals and groups of individuals within the habitats they occupy."[103] He was particularly interested in the way social communities of organisms controlled their own population densities. Such control seemed dictated by evolutionary exigencies; for if group numbers outstripped the carrying capacity of the niche, then the whole community might suddenly tip over into extinction. Wynne-Edwards found, however, that despite the expected tendency of each individual to increase its representation in the next generation, community numbers kept within optimum limits. It was as if individuals altruistically gave up a selfish advantage in order to contribute to the welfare of their society. Wynne-Edwards argued that nature provided species with homeostatic

100. Allee, *Social Life of Animals,* p. 127.

101. J. B. S. Haldane—in his *Causes of Evolution* (London: Longmans, Green, 1932), pp. 207–10—did consider the likelihood of altruism emerging as the result of group selection. Like later genic selectionists, he had his mathematical doubts about the possibility of such evolution. He did note—perhaps optimistically, given his Marxist politics—that "altruism is commonly rewarded by poverty, and in most modern societies the poor breed quicker than the rich" (p. 210).

102. For an extensive and felicitous examination of the question of group selection, see Elliott Sober, *The Nature of Selection* (Cambridge: M.I.T. Press, 1984), pp. 215–76. William Provine describes the rather ambiguous position Sewell Wright took in the early 1930s on group selection. See Provine's carefully wrought *Sewall Wright and Evolutionary Biology* (Chicago: University of Chicago Press, 1986), pp. 277–91.

103. V. C. Wynne-Edwards, *Animal Dispersion in Relation to Social Behavior* (New York: Hafner, 1962), p. 1.

regulatory mechanisms, most of which involved social behaviors of great variety (e.g., flocking, schooling, forming dominance hierarchies), in order to control reproduction. And nature did it through the agency of group selection, as Wynne-Edwards explained:

> Evolution at this level can be ascribed, therefore, to what is here termed group-selection, . . . and for everything concerning population dynamics, much more important than selection at the individual level. The latter is concerned with the physiology and attainments of the individual as such, the former with the viability and survival of the stock or the race as a whole. Where the two conflict, as they do when the short-term advantage of the individual undermines the future safety of the race, group-selection is bound to win, because the race will suffer and decline, and be supplanted by another in which antisocial advancement of the individual is more rigidly inhibited. In our own lives, of course, we recognize the conflict as a moral issue, and the counterpart of this must exist in all social animals.[104]

In the mid-1960s the minimalist movement in the social and biological sciences reached a peak. Behavioristic analysis reduced apparently intentional activities of animals and men to chained sequences of atomic bits of behavior, with each atom decreasing drive by a fraction. In biology, what Mayr has called "bean-bag genetics" came into style: the success of population genetics led to the mathematical disassembly of a species into a disembodied and unstructured pool of genes.[105] Natural selection a la mode became genic selection, and evolution became merely the change in allelic frequencies over time. But the heavy highland structure Wynne-Edwards erected simply could not be supported by spare, lone genes.

George Williams, in his very influential book *Adaptation and Natural Selection* (1966), sought to topple the unstable idea of group selection, as well as other "distractions that impede the progress of evolutionary theory."[106] Principally he tried to show that the conditions of population structure which Wright had prescribed for group selection would rarely obtain. So while group selection might be theoretically possible, it could play no significant role in evolution.[107] He further suggested

104. Ibid., p. 20.

105. Ernst Mayr, "Where Are We?" (1959), in his *Evolution and the Diversity of Life: Selected Essays* (Cambridge: Harvard University Press, 1976), pp. 307–28.

106. George Williams, *Adaptation and Natural Selection: A Critique of some Current Evolutionary Thought* (Princeton: Princeton University Press, 1966), p. 4.

107. Ibid., pp. 92–124.

that many of the assumed group adaptations described by Wynne-Edwards were better construed either as individual adaptations or as the accidental consequences of individual adaptations. His most striking example of an apparent group trait that could be explained by individual selection concerned human altruism. Williams observed that our small acts of apparent selflessness might actually redound to our larger benefit, so that individual rather than group selection could give apparently altruistic traits satisfactory explanation.

> Simply stated, an individual who maximizes his friendships and minimizes his antagonisms will have an evolutionary advantage, and selection should favor those characters that promote the optimization of personal relationships. I imagine that this evolutionary factor has increased man's capacity for altruism and compassion and has tempered his ethically less acceptable heritage of sexual and predatory aggressiveness.[108]

William's proposal received a more precise formulation a few years later, in 1971, when Robert Trivers advanced his theory of reciprocal altruism. The theory supposed that if a certain behavior evolved whose future reciprocation made it a probabilistically good investment (e.g., when your rescue of a drowning stockbroker elicits a reciprocal tendency to aid you on a future occasion), then genes promoting such an apparently altruistic trait would increase in the population.[109]

Not everyone might believe, though, that virtue was indeed its own reward. But Williams had confidence that the more patent cases of altruism, in which the action generally did reduce individual fitness, might be explained through employment of the theory of inclusive fitness. The bare idea behind this theory was hardly virgin. Darwin's conception of community selection—when the community was composed of close relatives—was a remote progenitor. J. B. S. Haldane, in his book *Causes of Evolution* (1932), more explicitly formulated the idea that "in so far as

108. Ibid., p. 94. In one of those wonderfully rhetorical moves, congenial to many modern evolutionary thinkers, Williams recruited Darwin to support this account of human altruism. He cited chapter 5 of the *Descent of Man* (London: Murray, 1871), where Darwin indeed had mentioned that "as the reasoning powers and foresight of the members [of a tribe] became improved, each man would soon learn from experience that if he aided his fellow-men, he would commonly receive aid in return" (1:163). Darwin had suggested that this experience could have become hereditary; and Williams made the same suggestion, though, of course, with a different mechanism in mind. Unblinkingly Williams neglected to mention, however, that Darwin thought the principal cause of moral behavior would be group selection.

109. See Robert Trivers, "The Evolution of Reciprocal Altruism," *Quarterly Review of Biology* 46 (1971): 35–57.

it makes for the survival of one's descendants and near relations, altruistic behaviour is a kind of Darwinian fitness, and may be expected to spread as the result of natural selection";[110] or as he more convivially put it over a pint, he would be willing to risk his life for two brothers or eight cousins (since on average they would represent his own genetic endowment). The theory of inclusive fitness, which gives mathematical weight to Haldane's burbling jest, was formulated in 1964, a short time before William's own book appeared. In two elegant papers, W. D. Hamilton showed that an altruistic trait would be inclusively fit if the detriment to the individual were outweighed by the product of the recipient's degree of relation and the amount of benefit conferred. For instance, any genetically determined behavior of an individual that greatly helped close relatives but involved little risk to self would tend to increase the representation of the controlling genes in the relatives and their offspring. Thus the generous impulse of a mother to feed and care for an orphaned niece along with her own children, would likely preserve similar generosity genes (supposing their initial existence) in the niece and allow them to be multiplied in the niece's children. Altruistic traits such as generosity would then evolve in a society if the organisms bearing them were more inclusively fit than those without them.[111]

The analyses of Williams, Trivers, and Hamilton together implied that, at the level of the gene, altruistic traits were actually "selfish," that is, they promoted their own representation in the next generation. But this conclusion had slight cultural impact. The minimalist genetics of these biologists was expressed in austere mathematics and directed to apparently technical questions concerning the units of selection. Like Hume's *Treatise*, their work hardly excited even the zealots. This changed in 1975, when a book appeared that accepted the conclusion of the minimalist geneticists, but supported it in a lushly fin de siècle way.

Edward Wilson began his large volume *Sociobiology* with a Mahleresque rhetorical flourish that was unprecedented in recent scientific monographs. The reader could immediately hear romantic themes transformed into the dissonant economics of modern science. His introductory chapter bore the title "The Morality of the Gene," and the opening paragraph reset Camus's singly important philosophical question. It was no longer the question of suicide but of the evolution of the physiological mechanisms determining ethical judgment. Wilson

110. Haldane, *The Causes of Evolution*, p. 131.
111. W. D. Hamilton, "The Genetical Evolution of Social Behavior," *Journal of Theoretical Biology* 7 (1964): 1–16; 17–51.

declared that the question could be answered only through the discipline of sociobiology, which sought systematically to "study the biological basis of all social behavior."[112] After this initial challenge, Wilson, with anything but a minimalist inclination, synthesized a huge volume of literature on population genetics, ethology (especially of the kind associated with Lorenz and Niko Tinbergen), and the evolution of social behaviors in taxons from slime molds through primates.

Wilson employed a Darwinian strategy. Like Darwin in the *Origin of Species* and *Descent of Man*, Wilson foliated the many branches of his argument with numerous examples drawn mainly from studies by field biologists and ethologists, studies accessible to a literate audience. And like Darwin, Wilson had his eye on man. After twenty-six long and profusely illustrated chapters, he reached the citadel itself. In chapter 27, he sketched out the implications—supported by a few suggestive studies—for human social behavior. Our various cultural institutions—marriage, trade, religion, esthetics, habits of labor, and ethics—he argued, all carried an evolutionary legacy. He proclaimed, with undoubted intention to unplug the emotional reservoirs, that our religious aspirations were an artifact of "conformer genes," of the same variety that also prompted obedience to Hitler. He cited IQ theorists to suggest that individuals were pushed by an "upward-mobile gene" into higher socioeconomic classes. And he proposed that "scientists and humanists should consider together the possibility that the time has come for ethics to be removed temporarily from the hands of the philosophers and biologicized."[113] Wilson seized high rhetorical ground by urging that centuries-old questions, which hitherto had wonderfully perplexed intellects in the humanities, could now be given a simple solution, one that had the authority of modern biology and mathematical genetics. From that vantage he fought the Darwinian fight. That is, he attempted to explain the very defining traits of human nature in evolutionary terms. And his efforts evoked from the intellectual community a response similar to that heard a century ago. In our day, as in Darwin's, what has so incited not only the zealots, but almost every person of scientific and humane curiosity, has not been simply Wilson's bold conclusions about the roles of biological factors in molding mind and behavior, but the staggering evidence, so theoretically articulated

112. Edward Wilson, *Sociobiology: the New Synthesis* (Cambridge: Harvard University Press, 1975), p. 4. The term "sociobiology" was already in use in the late 1940s. In 1956, C. Judson Herrick had recruited the term to describe the "search for the biological origins and nature of human patterns of social organization." See C. Judson Herrick, *The Evolution of Human Nature* (New York: Harper Torchbooks, [1956] 1961), p. 189.

113. Wilson, *Sociobiology*, p. 562.

and clearly presented in the first twenty-six chapters of his book, that the conclusions might well be true.

Conclusion: Transformation in the Darwinian Image of Man and the Schism in Contemporary Evolutionary Theory

There is a Darwin who lives in the collective memory of intellectuals and scientists. Through his discovery of evolution by natural selection, he instituted a pervasive materialism and mechanism in the interpretation of life. Darwinian theory when turned on man seems to project a bleak image. As depicted, human beings possess intellectual faculties that betray simian origins. Their social life conforms to rules governing turkey brotherhoods. Their religious rites mask the superstitions of pigeons and lead to the dominance of dictators. And their moral judgment, the mark of their humanity, springs to the urgings of selfish chemicals. Human mind and behavior lie on the same plane as animal mind and behavior, and that plane rumbles and tosses in response to mechanical forces deep within the genes.

I have argued that this image hardly epitomizes the actual accomplishments of Darwinian theorists during the last century. If "materialism" means that only matter exists, that what we call mind is simply a fixed function of matter, that ethical judgments are inescapably subjective and determined by selfish pleasure, then neither Darwin, nor his colleagues and disciples—Wallace, Haeckel, Romanes, Morgan, James, Baldwin—nor even Herbert Spencer constructed materialistic theories. Darwin and Spencer found objective grounds for authentic altruism; our very biology, they believed, rejected solitary pleasure as a motive for moral acts. And Spencer, Haeckel, Romanes, Morgan, James, and Baldwin spun out a monistic metaphysics that made matter epiphenomenal. If "mechanism" implies that in the evolutionary process mind must be derivative and phantasmal, rather than directive and real, then the leading Darwinists of the later nineteenth century were the very opposite of mechanists: each found a role for mind in guiding evolution, from Wallace's higher intelligences, through Romanes's ejective mind of the cosmos and James's effective mind in human evolution, to Morgan's and Baldwin's proposals for organic selection.

The Darwinian image of man underwent significant transformation during the twentieth century. Human nature remained the same, but like the picture of Dorian Gray, its representation altered in striking and for those of humane sensibility rather sinister ways. In the social sciences, several different movements conspired to eliminate evolutionary theories of mind and behavior from general discussion and serious investigation. In psychology, the Watsonian behaviorists not only rejected the concept of mind, they banished hereditarian theories and replaced them with empirical laws of environmental control. At the same time, Boas and his disciples in anthropology and sociology shunned evolutionary approaches because of their apparent ethnocentric and racist implications. Further, the seemingly speculative methods of evolutionary biology promised little by way of immediate social intervention and improvement, while the social sciences had developed techniques that permitted direct experimental control and encouraged the progressive aspirations of social technocrats. In biology, Darwinism itself was abandoned in the rush to perfect the experimental science of genetics. The pragmatic social arm of the new genetics, the eugenics movement, initially gathered in those geneticists who wished to engineer human progress. But during the 1920s and early 1930s, the more sensitive and intellectually acute dropped from the movement as its scientific supports were knocked out by advances in Mendelian theory, and as its darker, more morally repulsive side became exposed. Nazi racial hygienists left little doubt about the possible uses to which a biological approach to human nature might be put. Eugenicists in America and Britain in the early part of this century and the Nazi racial theorists later left a dark stain on evolutionary readings of human nature. The blotch is still quite visible.

In the 1930s and 1940s, Darwinian theory was revitalized and merged with Mendelian genetics in the synthesis that remains dominant today. But the new life given Darwinism left dormant, in the English speaking world, one branch of previous concern—human mental and moral faculties. Social scientists had already reserved these subjects for their province, and any efforts to reintroduce evolutionary considerations—for example, those of Lorenzian ethology—evoked both professional territorialism from social scientists and recollections from everyone of the moral dangers of a biological interpretation of distinctively human faculties. The strongly negative reactions to sociobiology by a good portion of the social scientific community can perhaps be explained by their entrenched commitments to experimental methods and proprietary assumptions, as well as by their moral suspicions. But these causes seem inadequate to account fully for the large fraction within the biological community who resist going the whole orang.

Stephen Jay Gould and Richard Lewontin have been the most persuasive and searching critics of sociobiology. They have marshalled a large camp of biologists and other scholars against those who believe that modern evolutionary theory might give account of human social relationships, cognitive abilities, and moral evaluations. Though differing on some issues, Gould, a paleontologist, and Lewontin, a geneticist, mount similar objections to perceived deficiencies of sociobiology. Their real target, though, is the image of Darwinian man as transformed during the first half of this century.

Their objections, like those of the early critics of eugenics, fall into two intersecting classes, the biologically substantive and the moral. Gould and Lewontin charge that sociobiological theory is ultra-Darwinian, reductive, and deterministic, which properties deliver it up to exploitation by reactionary politicians and neo-conservative thinkers. Sociobiologists, they maintain, assume generic human traits to be adaptive and thus naturally selected for specific purposes. This strong adaptationist assumption decomposes human behavior into specific bits, each having a determined genetic cause and evolutionary design. This suggests that human beings lack potential, that they are shackled by their genes to specific social and psychological roles. Sociobiologists pretend to interpret these roles by studying their analogues among the higher (and sometimes lower) animals: sexual relations, aggressive behavior, mating rituals, dominance hierarchies, and altruistic acts of lowland baboons thus declare the structure of human society and the causes of human behavior.[114]

As against the ultra-Darwinian adaptationist presumption, both Gould and Lewontin suggest other mechanisms for the origin of human behavioral traits: allometry (which partly explains why human beings cannot fly, save by United), cultural learning (which moves with Lamarckian swiftness), and utilization of excess potential—thus we may have been selected to be numerate, but culture rather than genes explains the origins of the calculus.[115] When natural selection does operate, it moves, they believe, at various levels of organization, from the lonely gene up to larger units, such as kin-groups and perhaps beyond. Moreover the complex organization of persons and the interactions of their parts—from DNA through enzymes, cells, organs, personal history, and social entanglements—require explanations at different levels, each with a relative autonomy. The autonomy is only relative, however, since, as Lewontin insists, individuals and their environments maintain a dialectical tension, such that organisms transform environments,

114. See, for example, Stephen Jay Gould, *Ever Since Darwin* (New York: Norton, 1977), pp. 251–59; and Richard Lewontin, Steven Rose, and Leon Kamin, *Not in Our Genes* (New York: Pantheon, 1984), pp. 243–64.

115. Stephen Jay Gould and Richard Lewontin, "The Spandrels of San Marco and the Panglossian Paradigm: A Critique of the Adaptationist Programme," *Proceedings of the Royal Society of London* 205 (1978): 581–98.

which in turn form new naturally selective and culturally influential pressures for development. This means, according to Lewontin, that all "causes of the behavior of organisms . . . are simultaneously both social and biological, as they are all amenable to analysis at many levels."[116] In the view of both biologists, it is the internally interactive and externally dialectical relations constituting human beings that supply the biological potential that grounds social mobility and human freedom.[117] This more hopeful image of man can only be compromised, they believe, by the practices of sociobiologists, who anthropomorphically read constricted human traits into animals and then reductively read them out again, providing justification for politically reactionary theories of human behavior.

Contemporary sociobiologists would not, I suspect, completely embrace as their own the transformed image of Darwinian man against which Gould and Lewontin contend. Many, for example, would undoubtedly agree that organisms are complex and in dialectical relation with their environments. And surely they would admit the liabilities of hasty generalization from animal behavior to human. They certainly believe, though, that such generalizations, when exercised with due caution, are quite legitimate. They would, moreover, locate the wellspring of generic primate traits, those that appear common to most humans and the higher animals, in the genes, and would attempt to discriminate the causal channels gouged out by the genetic flow. This is particularly true of that family of behavioral traits falling under the rubric of altruism. There is thus little doubt that Wilson, Dawkins, and Alexander—to name the more prominent contemporary sociobiologists—are legatees of the transformed image of Darwinian man.[118]

In their hostile reaction to the transformed image, Gould and Lewontin have elaborated another image of human nature that inchoately resembles the older nineteenth-century version. While they claim to be materialists, their vision begins to melt into the monistic view of Spencer, Romanes, Morgan, and Baldwin. According to Lewontin, "the property of being a mind—of 'minding'—must be seen as the activity of the brain as a whole; the product of the interaction of all of its cellular processes with the external world."[119] This is but short logical distance

116. Lewontin, Rose, and Kamin, *Not in Our Genes*, p. 282.

117. Ibid., pp. 265–90; and Stephen Jay Gould, *Hen's Teeth and Horse's Toes* (New York: Norton, 1984), pp. 241–50.

118. In addition to Edward Wilson's *Sociobiology*, see also Richard Dawkins, *The Selfish Gene* (Oxford: Oxford University Press, 1976); and Richard Alexander, *Darwinism and Human Affairs* (Seattle: University of Washington Press, 1979).

119. Lewontin, Rose, and Kamin, *Not in Our Genes*, p. 284.

from the admission that both "minding" and "cellular processing" are distinct properties of a common substrate—the basic proposal of monism. Further, they insist that it is the whole organ which reveals potentials not foretold by the selective forces that molded it. "Our large brain," remarks Gould, "may have originated 'for' some set of necessary skills in gathering food, socializing, or whatever; but these skills do not exhaust the limits of what such a complex machine can do."[120] From this complexity and flexibility comes human freedom. As Gould and Lewontin begin to shift into a moral appreciation, they declaim a freedom that genetic determinism would deny. "Our biology," exults Lewontin, "has made us into creatures who are constantly re-creating our own psychic and material environments, and whose individual lives are the outcomes of an extraordinary multiplicity of intersecting causal pathways. Thus, it is our biology that makes us free."[121]

Yet when reflecting more steadily about what human freedom can mean in a world fixed by material causality, Gould and Lewontin themselves accept a substantial part of the transformed image. As Lewontin reluctantly recognizes:

> What characterizes human development and actions is that they are the consequence of an immense array of interacting and intersecting causes. Our actions are not at random or independent with respect to the totality of those causes as an intersecting system, for we are material beings in a causal world. But to the extent that they are free, our actions are independent of any one or even a small subset of those multiple paths of causation: that is the precise meaning of freedom in a causal world.[122]

Probably no sociobiologist would deny that events are multiply caused. Indeed, the most rigid biological determinist—of requisite philosophical sophistication—could well admit this meaning for freedom. Nothing is won for human freedom or social responsibility in this conception. Thus while decrying the transformed image of Darwinian man, Gould and Lewontin have nevertheless accepted some of its nastier features. By contrast, in their more rhapsodical moments, they depict a human freedom that longs to be really free.

They might have found resources for justifying their sentiments in the original image of Darwinian man. It has virtues that accord well with their own aspirations for a socially concerned biological science.

120. Stephen Jay Gould, *The Panda's Thumb* (New York: Norton, 1982), p. 56.
121. Lewontin, Rose, and Kamin, *Not in Our Genes*, p. 290.
122. Ibid., p. 289.

For instance, William James argued, as we have seen (chapter 9), that choice of first principles need not be coerced by the current state of science (or even by the Marxist metaphysics Gould and Lewontin profess). James suggested that the social and moral experiences of responsibility and choice have as much claim to have their justifying principles recognized as do the experiences of the material-causal framework of a constricted science. The neutral monism elaborated by late nineteenth-century evolutionists—a metaphysics that held mind and matter to be two logically and descriptively distinct properties of a fundamental stuff (which was conceived yet more after the manner of mind)—this monistic metaphysics comports better with the sort of biology Gould and Lewontin would like to establish. Further, Morgan and Baldwin's organic selection bears familial resemblance to Lewontin's conception of the dialectical relation of organism and environment. Both theories suggest that animals—and human beings—alter their physical and social environment, so that it might become a new selective force on their own evolution. Finally, that original image depicted man as authentically moral; it showed altruism to have seeped deeply into human hereditary stock. In the second appendix, I have tried to restore that older image in order to bring out its bright moral features, to show that if our morality has profound roots in our animal past and has evolved by natural selection, this conviction hardly demeans our humanity, rather it elevates our biology, our evolutionarily human and moral biology.

Conclusion: Darwinism Is Evolutionary

The Oxford Debate Revisited

In the late 1890s, Leonard Huxley began collecting his father's correspondence for inclusion in the condign monument erected for every
great intellectual of the nineteenth century, a two-volume life and letters. He was especially interested in the event that made T. H. Huxley
"a personal force in the world of science"—the confrontation with the
Bishop of Oxford, Samuel Wilberforce, over Darwin's theory.[1] The
younger Huxley sought the recollections of those who were present at
the celebrated meeting of the British Association that late June weekend
in 1860. He gathered several varying accounts, but prominently featured
the story related almost forty years afterward that I have quoted in my
introduction. Some of his correspondents, though, could not endorse
the "Grandmother's Tale" in *Macmillan's*.[2] And the memories of others
were like the worn fragments of a mosaic, which when reset seemed to
show an outline, but may not have come from the same depiction. Evolutionary partisans at the meeting remembered that the bishop, after
citing the contrary scientific evidence—force-fed, as they believed, by
Richard Owen—directly attacked Huxley, and indirectly Darwin, by
sneering the question whether it was through his grandfather's side or
his grandmother's that he claimed his descent from a monkey. Some
others at the meeting, however, recalled it differently. The canon of
Durham, Professor Farrar, thought the bishop had not personally
boxed Huxley, but had simply uttered the flippancy: "If any one were
to be willing to trace his descent through an ape as his grandfather,
would he be willing to trace his descent similarly on the side of his

---

1. Leonard Huxley, *Life and Letters of Thomas H. Huxley* (New York: D. Appleton, 1900), 1: 193.
2. See my "Introduction," note 1.

grandmother?"[3] The Bishop may thus have only issued a scientific challenge and then hid behind Victorian skirts.

Nevertheless, the fortuitous coincidence of bishop and ape undoubtedly allowed Huxley to make a retort that signaled the coming defeat
of both religious and biological orthodoxy. But did Huxley really remark meanly on the bishop's "equivocal success in his own sphere of
activity," as a young student in the audience recalled?[4] Huxley himself
was sure he would not have said that, though he could not remember
exactly how he did respond. Perhaps he returned in kind, decrying the
bishop's "aimless rhetoric" and "appeals to religious prejudice," as the
student maintained; though another auditor remembered Huxley's
"self-restraint, that gave dignity to his crushing rejoinder."[5]

A mangled bishop makes a poignant, though mildly ludicrous sight.
Did Huxley's retort merely kick over an unsteady prelate intoxicated
with vintage science and ancient religion? Or did the bishop fall from
grace by tripping over high Victorian propriety? One auditor believed
that Wilberforce was defeated not by science, but because he "had forgotten to behave like a perfect gentleman."[6]

Leonard Huxley wished to reconstruct as exactly as he could the
bishop's comments and his father's reply, since from the vantage of the
late 1890s the encounter appeared to be the pivot point of Huxley's
career and the first public victory in the ultimately successful war of
science over religion, of Darwinian theory over biological orthodoxy.
Seen from the end of the century—Darwin's century—the events of
that Saturday achieved a significance that only a later period in science
could bestow. A contemporary account would, of course, miss the preg-
nancy of the moment. The reporter for the weekly *Athenaeum* described
the meetings in fair detail just afterward. He gave an account of Wilberforce's and Huxley's remarks, along with those of several others during the session. But he did not record either the bishop's famous question or the young zoologist's rejoinder. He did mention the witticism
of Darwin's eminent friend Dr. Joseph Hooker, who feigned no authority on evolution, because Wilberforce had already declared that "all
men of science were hostile to Mr. Darwin's hypothesis."[7] The reporter,
who betrayed sympathy for the new views, actually gave the older

3. L. Huxley, *Life and Letters of Thomas H. Huxley* 1: 197.
4. Ibid., p. 199.
5. Francis Darwin, *Life and Letters of Charles Darwin* (New York: D. Appleton, 1891), 2: 115.
6. L. Huxley, *Life and Letters of Thomas H. Huxley* 1: 197n.
7. "Meeting of The British Association, Section D., Saturday," *The Athenaeum* (July 14, 1860): 65.

Hooker more than three times the coverage of the younger Huxley. He lacked the retrospective advantage of the 1890s.

Though history has no argument, historians do. Their arguments require distance, enough that consequences might indicate the events of importance. The consequences, though, will always be caught up in the intellectual currents of the later period, which inevitably will wash back upon the earlier events. When Leonard Huxley reconstructed those times of his father, he found the triumph of Darwinism the signal beacon by which the exchange at Oxford was not simply revealed, but by a kind of light metaphysics came historically to be. My own narrative in this volume bears witness to the teleological transformation of the past. I surely have perceived the last century through a lens ground by contemporary debates in evolutionary theory. But equally important have been the consequential stages of historiographic analysis of Darwinism since the nineteenth century.

The Generations of Darwinian Historiography

Historians of science, as other peoples, have their generations. Leonard Huxley and Francis Darwin, who also published his father's life and letters (shorn of agnostic sentiments), might be considered historians of the first generation. Their scientific attitudes, if not their filial piety, were formed at the height of the Darwinian victory, and their reconstructions must reflect that. Others of this generation came together, fittingly, in *Darwin and Modern Science* (1909), a volume wrought to celebrate the fiftieth anniversary of the *Origin of Species* and the hundredth of Darwin's birth.[8] Weismann writing on selection theory, De Vries on heredity, Haeckel on anthropology, and Morgan on mental evolution—all considered the histories of their various disciplines in light of the science that had triumphed during those fifty years. Even the neo-Lamarckian Henry F. Osborn read the significance of Darwin's forerunners, in his history *From the Greeks to Darwin,* in respect of the English victory in 1859 and the biology confirming it in the 1890s.[9]

The second generation of historians of evolutionary theory presided over the decline of Darwinism. The new genetics of the first two decades of this century, heralded by the recovery of Mendel's work, cast a deep shadow on the previous period (see chapter 11). Nordenskiöld, in his still useful *History of Biology,* recited the obsequies for Darwinian theory. It had succumbed, in his estimation, to real science, which was

8. A. C. Seward, ed., *Darwin and Modern Science* (Cambridge: Cambridge University Press, 1909).

9. Henry F. Osborn, *From the Greeks to Darwin* (New York: Scribner's, 1894).

for him the new laboratory genetics.[10] Since romantic field biology and the hypothetical force of natural selection had failed as real science, Darwin's theory must have been supported by hidden wires, unobserved by biologists of the time. Nordenskiöld suggested that the political and social sentiments of liberalism had held aloft what empirical evidence and scientific reasoning could not. Similar considerations, fostered by the success of controlled experiments in the early part of this century, seem to have led Karl Popper to dismiss Darwinian theory, along with Freudian psychology and Marxism, as metaphysical proposals not amenable to the hard falsification procedures of the laboratory.[11]

After the new synthesis of genetics and natural selection theory, in the 1930s, historians of biology reexamined Darwin's ideas, though they were now convinced that natural selection could operate with Mendelian genes to produce the kind of evolutionary gradualism that Darwin had described. One of the architects of the new synthesis later became a powerful molder of historical opinion, turning as he often did to reconstruct the biology of the nineteenth century. Ernst Mayr, though publishing his principal historical works only in the last two decades, might be considered a historian of the third generation. In his book *The Growth of Biological Thought,* he synthesized a vast amount of recent literature in the history of biology and ordered it in a straightforward and systematic way, reflecting the virtues of his own previous scientific efforts. Mayr adopted a revolutionary model of scientific development (the features of which I have detailed in the first appendix). He constructed his history to demonstrate the progress of evolutionary thought from Darwin through the new synthesis. Curiously, he conceived that progress in pre-Darwinian terms, as an internally self-generating series of biological ideas that suffered no alterations from encroaching conceptual systems. As Mayr characterized his book: "The emphasis is on the background and the development of the ideas dominating modern biology; in other words, it is a developmental, not a purely descriptive, history. Such a treatment justifies, indeed necessitates, the neglect of certain temporary developments in biology that left no impact on the subsequent history of ideas."[12] Since the kernel of

---

10. Erik Nordenskiöld, *The History of Biology,* 2d ed. (New York: Tudor, [1920–1924] 1936), p. 574.

11. See, for instance, Karl Popper, "Darwinism as a Metaphysical Research Programme," in *The Philosophy of Karl Popper,* ed. Paul Schilpp (LaSalle, Ill.: Open Court, 1974). Michael Ruse discusses Popper's view of Darwinian theory in "Karl Popper's Philosophy of Biology," *Philosophy of Science* 44 (1977): 638–61.

12. Ernst Mayr, *The Growth of Biological Thought* (Cambridge: Harvard University Press, 1982), p. vii.

biological thought grew in isolation from philosophical, theological, moral, cultural, and social influences, Mayr could safely exclude such matters from historical consideration. Nor did he need to travel those evolutionary paths that appeared, from the perspective of the new synthesis, as a blind alley. Hence, no counterevidence could surface against his elevation of features he supposed to be essential to the Darwinian revolution: "the replacement of essentialism by population thinking"; "the refutation of cosmic teleology"; and "the explanation of 'design' in the world by the purely materialistic process of natural selection."[13] By ignoring the epistemological tradition of sensationalism (see above, chapters 1 and 2), he could believe that Darwin's rendering species as populations sprang from the founder's brain alone. And by neglecting such thinkers as Romanes, Morgan, James, and Baldwin (discussed in chapters 8–10) he could believe that Darwinism implied essentially a materialistic, mechanistic, and nonteleological approach to nature.

There are two ways to assess the implications of a conceptual system. One can adopt the procedure of Imre Lakatos, as Mayr has implicitly done. Lakatos admonished historians to discriminate the core or essence of a research tradition, such as the Darwinian, and extract its logical implications—even those which the practitioners of the tradition themselves failed to recognize. But this procedure assumes that a heritable tradition retains an essence over time and that the historian can choose the real deductive consequences of that essence from the infinite number possible. The other way, which does not make these unwarranted assumptions (as I have argued in chapter 8 and in the first appendix), traces out the actual evolutionary lineages of a particular conceptual system. I believe it is those descendent systems—Romanes's, Morgan's, James's and Baldwin's—that determine the real implications of Darwinism in the nineteenth century.

The disappearance of Spencer from the history of science stands as perhaps the most damaging evidence against those historians of biology, like Mayr, who employ the revolutionary model of science. This model leads to the presumption that the historian should only include thinkers whose ideas resemble those found in contemporary science. No evolutionary ecologist, however, would dream of attempting to account for a lineage without assessing the selective pressures of powerful competing species lines—though a biological systematist might. Spencer's ideas occupied the same intellectual environment as Darwin's. Their conceptual systems exerted mutually selective pressures on the further evolution of their respective systems, and both left a rich legacy for the

13. Ibid., p. 501.

subsequent development of Darwinism.[14] Resemblance, then, should not be the ultimate criterion for historians; they must also attend to the causal pressures and selective forces of other impinging conceptual systems.

Michael Ghiselin, another biologist who has turned his attention to history, also falls into the third generation. His *Triumph of the Darwinian Method* had the great merit of attempting to see Darwin whole. He considered not only the *Origin of Species,* but *The Structure and Distribution of Coral Reefs, Geological Observations on Volcanic Islands, The Descent of Man, Various Contrivances by Which Orchids Are Fertilised,* and several other books of the corpus. Ghiselin read sensitively; yet guided by a revolutionary model, he was encouraged, just as Mayr, to discover in Darwin an Urprogenitor of the defining features of recent evolutionary theory. Darwin, he presumed, had overturned philosophical approaches to biology, introduced population thinking, and employed a triumphant method—which on Ghiselin's account turns out to be the hypothetical deductive method as the later logical empiricists understood it. Further, and particularly pertinent to my study (see chapter 5), Ghiselin happily attributed to Darwin a moral theory of genetic selfishness that was actually formulated in the twentieth century.

Third-generation historians, such as Mayr and Ghiselin, move us smartly back from the present to the Darwin that rooted modern biology, and they force us to appreciate an extraordinary accomplishment. But often enough it is the accomplishment of modern science they appraise, and not the evolutionary consequences of conceptual lineages descending from pre-Darwinian history and shaped by mid-nineteenth century science and culture.

The hundredth anniversary of the *Origin of Species* in 1959 opened a gate for historical studies of Darwinism, the flood of which shows no signs of cresting. The fourth generation of historians, writing at that time, turned to examine more carefully the intellectual antecedents and conceptual environment of Darwinian ideas. Loren Eiseley, John Greene, and Gertrude Himmelfarb, prominent within this generation, approached their subject in a decidedly more evolutionary way.[15] They

14. John Greene also argues that Darwin and Spencer shared a number of fundamental ideas, which have historical claim on the name "Darwinism." See his dispute with Ernst Mayr on this issue in John Greene, *Science, Ideology and World View* (Berkeley: University of California Press, 1981), pp. 151–55.

15. See in particular Loren Eiseley, *Darwin's Century* (Garden City: Doubleday, 1958) and *Darwin and the Mysterious Mr. X* (New York: Dutton, 1979); John Greene, *The Death of Adam* (Ames: Iowa State University Press, 1959) and *Science, Ideology, and World View: Essays in the History of Evolutionary Ideas* (Berkeley: University of California Press, 1981);

stressed the continuities of Darwin's thought with that of his predecessors, while showing how he constantly accommodated his changing ideas to the objections of contemporaries. But it was a kind of Spencerian evolutionary perspective they adopted. The progress toward Darwin's ideas was smooth and inevitable; had Darwin not existed, Eiseley or Greene would surely have invented him: "If Wells [an obscure predecessor] had been a zoologist as well as a physician," Greene insisted, "Charles Darwin's theory of the origin of species might have been anticipated by almost fifty years. All the elements of the theory were present in the scientific world by 1818." [16] The distinctive features of Darwin's thought vanished in the anticipations of his forerunners. Himmelfarb also stressed Darwinism's premodern origins and, with Eiseley, Darwin's perplexed responses to contemporaries. She portrayed Darwin's conception as being confused and contradictory, on the one hand, and inhumane on the other. Darwin, who simply absorbed the laissez-faire liberalism and utilitarianism of his time, left human nature drained of moral significance. Recently Himmelfarb has joined Darwin's views with those of Edward Wilson to pull the Darwinian rabbit yet again out of the Spencerian hat: she assumes that the sociobiological end of one branch of the Darwinian heritage must be exactly comparable to its beginnings.[17] Himmelfarb simply failed to analyze closely the *Descent of Man* and passed quickly over the views of such Darwinians as have been treated in this volume.

Social Darwinism vs Evolutionary Darwinism

Robert Young, in a series of perceptive essays in the late 1960s and early 1970s, reoriented examination of nineteenth-century evolutionary theory.[18] Along with the Edinburgh sociologists of science, then just beginning to be heard, Young employed with considerable effect a strong social-psychological model (described in appendix 1). These fifth-generation studies focused on the intellectual background, espe-

---

and Gertrude Himmelfarb, *Darwin and the Darwinian Revolution* (New York: Norton, [1959] 1968) and *Marriage and Morals among the Victorians* (New York: Knopf, 1986). Antonello La Vergata discusses developments in Darwinian scholarship from the late 1950s with considerable clarity and insight. See his "Images of Darwin: A Historiographic Overview," in *The Darwinian Heritage,* ed. David Kohn (Princeton: Princeton University Press, 1985).

16. Greene, *Death of Adam,* p. 245.

17. See Himmelfarb's "Social Darwinism, Sociobiology, and the Two Cultures," in her *Marriage and Morals among the Victorians.*

18. Most of these essays have been collected in Robert Young, *Darwin's Metaphor* (Cambridge: Cambridge University Press, 1985).

cially the social philosophy, that made Darwinian thought possible. There had been a long tradition of regarding Darwin's accomplishment as directly dependent upon Malthusian and Smithian political economy and Benthemite utilitarianism. Young, however, gave that approach gritty reality in his careful investigations of great batches of nineteenth-century monographic and journal literature. With a Marxian broom he beat the books of Paley, Malthus, Smith, Lyell, Chambers, Galton, and articles from the *Westminster, Edinburgh, Fortnightly,* and *Quarterly* reviews to raise liberal clouds of unhealthy dust, which then resettled over the works of Darwin, Spencer, and their successors.

Young's guiding premise, a corollary of the strong social-psychological model, was that all categories fundamental to Darwin's theory, as well as those imbedded in the opposed theological views, were social. The debates over evolutionary theory in the nineteenth century—and the descendent fights over sociobiology—amounted thus to clashes of social ideology and interests. "Once it is granted," Young asserted,

> that natural and theological conceptions are, in significant ways, projections of social ones, then important aspects of all of the Darwinian debates are social ones, and the distinction between Darwinism and Social Darwinism is one of level and scope, not of what is social and what is asocial. . . . The point I'm making is that biological ideas have to be seen as consti-tuted by, evoked by, and following an agenda set by, larger social forces that determine the tempo, the mode, the mood, and the meaning of nature.[19]

To claim that Darwinism is social—or with those of the Edinburgh persuasion, that all science is social—is maddeningly vague. The proposition has a liquid meaning in the work of recent sociological historians of science. At times, especially when challenged, these historians appear to mean only what Aristotle meant: man is a social animal, so all his works are social. In this sense, perfectly true and perfectly useless. But in the model they employ—and in their often disarming admissions—the claim has more substance, and becomes interesting, provocative, surely instructive, and ultimately refutable. It means that scientific descriptions and theories that appear to be about one thing, nature and her lawful operations, are really about another, man and his ideological manipulations. The Darwinian scientist, then, is only, to use Young's term, an "ideologue." Young thinks that this thesis is easily sustained by casual examination of the works of present-day ethologists and sociobiologists, who exude from every pore a repellent ideology.

19. Robert Young, "Darwinism is Social," in *The Darwinian Heritage,* pp. 610 and 622.

The quick refutation of this global sociologism is, of course, the tu quoque argument: what criterion shall we employ to choose between the ideology of these sociological historians and that of the sociobiologists they condemn? The only standard when ignorant ideologies clash by night must be the vulgar Darwinian one: the strongest shall survive. This ultimately ethical objection can be spiked down by an ontological-epistemological observation. It seems obvious (though it should be argued out in more philosophically appropriate terms) that biological organisms—and the rest of nature—are not made of play dough. They cannot be construed in just any way; nature will not bend without breaking just to meet the cut of today's political interests; not every social philosophy will fit. If nature has a structure and if evolution has provided us sense organs and nervous systems to grasp that structure, though inadequately, then there are ways other than ideological contest to decide among competing scientific conceptions. The Darwinian scientist, therefore, need not be supposed to advance biological propositions from purely ideological grounds.

The natural selection model that I have adopted in this history does not deny the social causation of scientific ideas. But in imitation of its biological counterpart, the model recognizes that many different kinds of pressure can be exerted on an evolving conceptual system, of which the distinctively social constitutes just one kind. There is no justification for historians to assume a priori that only social forces (i.e., political and ideological interests) finally determine ideas. Historians must rather empirically analyze both the conceptual system in question and its intellectual environment to discriminate the several kinds of force that are operative and to estimate their respective strengths. In this history, I have tried to show, for instance, that Spencer's particular ethical principles, ultimately stemming from the nonconformist society he kept, gave logical shape to his evolutionary theory. While this was a major factor in the development of Spencer's biological ideas, it certainly was not the only one: others derived from Carpenter's physiology, Lamarck's theory via Lyell, Hamilton's neo-Kantianism, and Darwin's theory of community selection—just to mention a few of the shaping causes discussed in chapters 6 and 7. I certainly believe, as I argued in chapters 2 and 5, that Darwin constructed his biological theory of morality around the philosophical (and I guess social) theories of James Mackintosh. But the story would only be partly told if I were to omit Darwin's analysis of the instincts of social insects, which provided him the biological mechanism for human altruism. The progressive social program of the philosophes makes comprehensible the orientation of Cabanis and Lamarck, as well as that of Darwin's grand-

father Erasmus; but it would be a mistake to credit Enlightenment social philosophy for the details of their views. In the case of James Mark
Baldwin, social factors did not so much mold his evolutionary biopsychology as bring it to an end, in the shattering personal and professional tragedy he suffered. To claim, as Young and others do, that all is
social admits that nothing is: the social becomes the music of the
spheres, constantly sounding and thus never heard.

We have entered the present generation of Darwinian scholars. The
cri de coeur of Young's essay "Darwinism is Social" laments this generation because of its obsession with minutiae of the Darwin archive at
Cambridge.[20] Since the early 1970s, scholars have made increasing use
of Darwin's unpublished manuscripts, notes, correspondence, and even
shop bills, to puzzle out the development of his thought. Their micro-
studies—citations of which accumulate at the bottom of my pages—
have focused, for the most part, on the origins of the young Darwin's
theory. They exhibit the intellectual environment of his theorizing and
gauge the smaller and larger pressures that curved his ideas along their
trajectory. These scholars do not neglect the social features of Darwin's
thought, but do not apotheosize them either. As sensitive historians,
they have implicitly constructed their narratives according to those assumptions I have tried to formalize in the natural selection model that
frames the history told in this volume.

My own effort has been to extend in three ways the methods and
resources used by this generation of Darwinian scholars: first, to be
explicit about the historiographic principles employed; second, to examine several subjects that have thus far received little attention; and
finally to trace the development of these subjects from the Enlightenment, through Darwin and Spencer, and down to their successors. I
hope the results of this study will help undermine the received view of
Darwinism—that it formed man in the image of a materialistic, mechanistic, and amoral being. I also hope that we may now come to understand that Darwinism is evolutionary.

20. Ibid., pp. 609–38.

Appendix 1
The Natural Selection Model and Other Models in the Historiography of Science

The writing of science history may itself be regarded as a scientific enterprise, involving evidence, hypotheses, theories, and models. I wish here to investigate several historiographic models and their variants. While these undoubtedly do not exhaust the store available to imaginative historians of science, they nonetheless represent, I believe, those that have played the significant roles in the development of the discipline, either as models that have long functioned in historical writing or as models more recently proposed in metahistorical works.

The models described in the first part of this appendix represent major assumptions that have guided the construction of histories of science since the Renaissance. They thus embody directive ideas concerning the character of science, its advance, and the nature of scientific knowing. Since the models are idealizations, they do not always precisely reflect the structures of particular written histories. Yet they can serve to elucidate those controlling assumptions that have shaped our understanding of science and its history.

The second part of this appendix will attend to the class of models that appears the most powerful for capturing the actual movement of science: evolutionary models. I will briefly examine two instances of this class, the models of Popper and Toulmin, and consider their deficiencies. I will then develop a natural selection variant that, I believe, escapes their liabilities. Finally, I will test the resources of this variant against Lakatos's model of scientific research programs, perhaps its strongest competitor.

I have four reasons, accelerating in importance, for offering a dissection of historiographic models: first, to show that historians of science have typically constructed their narratives in light of distinct sets of assumptions; second, to display the major forms these assumptions have taken during the development of the discipline; third, to explore the advantages and disadvantages of the models; and, fourth, to provide comparative standards by which to judge the virtues of the natural selection model detailed in the second part of this appendix and employed in the construction of the history related in this volume.

Five Models in the Historiography of Science

The Static Model

Many historians and scientists of the late Renaissance and early Enlightenment shared R. Bostocke's conviction, as expressed in his *The difference Betwene the Auncient Phisicke and the Latter Phisicke* (1585), that God had infused certain men (such as Adam or Moses) with scientific knowledge, which was passed on to successive generations intact.[1] Even Newton, in his historical musings, employed a static model, maintaining that his *Principia* was a recovery of wisdom known to the ancients.[2]

Use of a static model in history of science accorded with the Renaissance presumption that ancient thought embodied the highest standards of knowledge and style. But another consideration also promoted the acceptance of the model. This may be found in Olaus Borrichius's *De ortu et progressu chemiae dissertatio* (1668), a standard textbook of the history of chemistry during the late seventeenth and early eighteenth centuries.[3] In accord with the tradition, Borrichius credited Tubalcain, a descendant of Cain and a figure he identified with Vulcan, as having received from God the divine knowledge of chemistry. The Cartesian argument he used to fortify his baroque sentiments displays an important justification for use of the static model. He reasoned that "the priests of Tubalcain would have been unable to discover, shape, and form the metals of iron and copper except that their *ratio* was prior known; that the natures of these minerals might be investigated and that they might be cooked, purged, and segregated could not occur except that knowledge of this were divinely inborn. Once this knowledge is had, however, these techniques follow for any skillful people."[4] Borrichius, tinctured with the Cartesian spirit, knew that chemical knowledge and science in general must be innate, at least in their fun-

1. R. Bostocke, *The Difference Between the Auncient Phisicke and the Latter Phisicke* (1585), in Allen Debus, "An Elizabethan History of Medical Chemistry," *Annals of Science* 18 (1962): 1–29.

2. J. E. McGuire and P. Rattansi, "Newton and the 'Pipes of Pan,'" *Notes and Records of the Royal Society of London* 21 (1966): 108–43.

3. Olaus Borrichius, *De ortu et progressu chemiae dissertatio* (1668), in *Bibliotheca chemica curiosa*, ed. J. Manget (Geneva: Chouet, 1702).

4. Ibid., p. 1.

damentals; for unilluminated natural induction could never of itself lead to such scientific achievements as his age had witnessed. And if the essential features of a science had this kind of origin, then from its first discoverer such knowledge could only be passed on or rediscovered again by succeeding generations. This model of the origin and course of science can be detected in transmogrified form in Thomas Kuhn's Gestalt model (described below), which assumes that in a moment of insight the transformed vision of an inspired genius may establish the framework and fundamental premises of a science, the details of which may be left to the normal plodding of disciples.

The Growth Model

After the late Renaissance, historians of science began to discard the static model, replacing it with one still in use today. By the eighteenth century, the growth model clearly prevailed, as Freind's *The History of Physick from the Time of Galen to the Beginning of the Sixteenth Century* (1725)[5] and Watson's essay "On the Rise and Progress of Chemistry" (1793)[6] testify. Indeed, Freind's history may be read as a sustained argument against the Renaissance tendency to overprize the ancients and to suppose that the essential concepts and principles of science lay with them, only to be ornamented by succeeding generations. Freind proposed to show that the knowledge of medicine did not begin and end with Hippocrates and Galen. Instead, as a careful study of the writings of subsequent physicians demonstrated, "Physick was still making progress 'till the Year 600".[7] (He charted the gradual advance of the science since the beginning of the medieval period in the second volume of his history.) As a consequence of the particular model he had chosen, that of gradual, cumulative growth, Freind could recommend reading in the history of medicine as "the surest way to fit a man for the Practice of this Art."[8] This is a piece of advice annulled by historians advocating other models.

Watson's essay highlights an assumption of the growth model that was to have particular importance in later controversies, namely, that science in its conceptual development is relatively isolated from other human occupations, even from the technology that fostered it. Watson

5. John Freind, *The History of Physick from the Time of Galen to the Beginning of the Sixteenth Century* (London: Walthe, 1725).

6. R. Watson, "On the Rise and Progress of Chemistry," in vol. 1 of his *Chemical Essays,* 6th ed. (London: Evans, 1793).

7. Freind, *History of Physick,* 1: 298.

8. Ibid., p. 9.

felt assured of this independence, since he understood science to have a rational integrity not found in the less "liberal and philosophical" pursuits.[9] This presumption also bound together the various parts of that monument to the growth model, the *Encyclopédie* of Diderot and d'Alembert. In the *Discours Préliminaire* to the *Encyclopédie*, d'Alembert projected the prescriptions of the growth model back even to prerecorded thought, suggesting that primitive sensory awareness might gradually have established the foundational principles of scientific advance.[10]

In the nineteenth century, William Whewell supplied the most elaborate employment and justification of the growth model in his *History of the Inductive Sciences* (1837).[11] He rejected the idea that discontinuous intellectual upheaval marked the development of the various sciences: "On the contrary, they consist in a long-continued advance; a series of changes; a repeated progress from one principle to another, different and often apparently contradictory."[12] If the progress of science occured by contradictory ideas replacing one another, there would not, of course, be organic growth, but revolutionary saltation. That is why Whewell urged his reader to remember that the contradictions were only apparent:

> The principles which constituted the triumph of the preceding stages of the science may appear to be subverted and ejected by the later discoveries, but in fact they are (so far as they were true) taken up in the subsequent doctrines and included in them. They continue to be an essential part of the science. The earlier truths are not expelled but absorbed, not contradicted but extended; and the history of each science, which may thus appear like a succession of revolutions, is, in reality, a series of developments.[13]

The central assumptions embodied in the growth model are compendiously present in the work of George Sarton, the doyen of historians of science in the middle of this century. His several observations on the nature of science therefore afford a convenient summary of the implications of the model. The primary feature of the model is its affirmation of the unalterable and clearly discernible progress of science toward the

9. Watson, "On the Rise and Progress of Chemistry," p. 30.

10. Jean d'Alembert, *Discours Préliminaire,* in vol. 1 of *Encyclopédie ou dictionnaire raisonné des sciences, des arts et des métiers,* 2d ed., ed. Dennis Diderot and Jean D'Alembert (Paris: Lucques, 1758–1771).

11. William Whewell, *History of the Inductive Sciences* (London: Parker, 1837).

12. Ibid. 1:9.

13. Ibid., p. 10.

fullness of truth, progress that can be only momentarily delayed by retarding forces. "The history of science," Sarton declared, "is an account of definite progress, the only progress clearly and unmistakably discernible in human evolution. Of course, this does not mean that scientific progress is never interrupted; there are moments of stagnation and even regression here or there; but the general sweep across the times and across the countries is progressive and measurable."[14]

The steady advance of science, accomplished by the rationally exact methods of quantification and experimentation, and "its astounding consistency (in spite of occasional, partial, temporary contradictions due to our ignorance) prove at one and the same time the unity of knowledge and the unity of nature."[15] Since the unity and continuity of knowledge, which are grounded in the unity and intelligibility of nature, are not, in Sarton's estimation, enjoyed in other human pursuits, these latter are unable conceptually to affect the course of science. Moreover, the clear evidence of history gives no support, he thought, to attempts at the sociologizing of scientific knowledge. The internal progress of science has a force beyond the vicissitudes of men's passions and the subtle pressures of social life. To be sure, science does not grow in a social vacuum: men need food, they are called to war; money for equipment is required. But, Sarton avowed, the man of science remains ultimately untouched in his theoretical endeavor by the ideologies or conditions of society: "Nobody can completely control his spirit; he may be helped or inhibited, but his scientific ideas are not determined by social factors."[16] Insofar as the history of science is independent of the cultural life of the larger community, it can serve as a standard of truth and error in those other domains: "the history of science describes man's exploration of the universe, his discovery of existing relations in time and space, his defense of whatever truth has been attained, his fight against errors and superstitions. Hence, it is full of lessons which one could not expect from political history, wherein human passions have introduced too much arbitrariness."[17]

The Revolutionary Model

A brief examination of the history of the term *revolution* suggests that its application to scientific thought is not necessarily derived from

14. George Sarton, *Sarton on the History of Science: Essays by George Sarton* (Cambridge: Harvard University Press, 1962).

15. Ibid., p. 15.

16. Ibid., p. 13.

17. Ibid., p. 21.

analogies with political overthrow. The *Oxford English Dictionary* indicates that its use to describe dramatic changes in thought antedates by a considerable period its use to designate political upheavals. By the late eighteenth century, the term was widely employed to denote important transformations in the course of science. When Kant referred to particular "revolutionary" events in the history of science, he employed the word in the manner of contemporary historians: to describe a profound shift in thinking, after which there is relatively smooth scientific progress to the present time. For Kant, as well as for most recent historians using the model, revolution in a science is a one-time affair. In the preface to the second edition of the *Kritik der reinen Vernunft*, Kant depicted the intellectual revolution undergone by the mathematical and physical sciences, before which we had no science proper and after which we had unimpeded advance into the modern period. Mathematics had to grope during the Egyptian era, but with the Greeks came the revolution that set it on its present course. Natural science had to wait a bit longer for its revolution, as Kant explained: "It took natural science much longer before it entered on to the road of science; for it is only about a century and a half since the proposal of the ingenious Bacon of Verulam partly fostered its discovery and, since some were already on its trail, partly gave encouragement. But this can be explained only as a suddenly occurring revolution in the mode of thought (*eine schnell vorgegangene Revolution der Denkart*)."[18]

While the use of the term *revolution* to describe radical changes in thought is older than its use in the specifically political context, the political analogy is often implied and does seem justified. Political revolutionaries have particular enemies with whom they wage their ideological and bloody battles; the scientific revolutionaries of the sixteenth and seventeenth centuries also had their foes: Aristotle, Ptolemy, Galen, and the Scholastics. Political revolutionaries aim at overturning an undesirable system and replacing it with one that will perdure and serve as a base for further progress; their scientific counterparts harbor similar goals. Significant political revolutions are not usually spontaneous; they have their doctrinal basis formed in the work of men who may be long dead before the revolution. Historiographers of scientific revolutions also acknowledge necessary foundations: the groundwork of modern physics laid, for example, by the Merton school of mathematical physics or the Paduan Aristotelians of the early Renaissance. The historical im-

18. Immanuel Kant, *Kritik der reinen Vernunft* (1787), vol. 2 of *Immanuel Kant Werke in sechs Bänden*, ed. W. Weischedel (Wiesbaden: Insel, 1956), p. 23 (B xii).

portance of a political revolution lies more in the fruit of the new ideas and systems that the revolution inaugurates—fruit that may take time in ripening. Those writing of the scientific revolution fomented by Copernicus, Kepler, Galileo, Harvey, and Descartes construe the ideas of these scientists as establishing the foundations for thoroughly modern science, even though their specific conceptions may no longer be acceptable.

The most influential historian to employ the revolutionary model was Alexandre Koyré, whose views set out its essential features. To the history of science Koyré brought the philosopher's eye for metaphysical assumptions and the intellectual historian's concern for doctrinal context. In his view, the scientific revolution of the sixteenth and seventeenth centuries outwardly expressed a more fundamental turn of mind, a "spiritual revolution" having two basic features. There was, first of all, the Platonically motivated dismissal of the qualitative space of Aristotle and the Scholastics and its replacement with abstract geometrical space. Galileo's contribution to the scientific revolution was precisely his insistence on mathematical reasoning rather than sense experience as the foundation for scientific success.[19] But this alteration in thought about the universe was only a phase, though the crucial one, of a more pervasive revolution, one that brought about, according to Koyré,

> the destruction of the Cosmos, that is, the disappearance, from philosophically and scientifically valid concepts, of the conception of the world as a finite, closed, and hierarchically ordered whole . . . and its replacement by an indefinite and even infinite universe which is bound together by the identity of its fundamental components on the same level of being. This, in turn, implies the discarding by scientific thought of all the considerations based upon value concepts, such as perfection, harmony, meaning, and aim, and finally the utter devalorization of being, the divorce of the world of value and the world of facts.[20]

Insofar as the revolution had banished from explanatory rule such concepts as "perfection, harmony, meaning, and aim," historians of science under the banner of Koyré have felt justified in dismissing from serious consideration neo-Platonic mysteries and Paracelsian occultism, which were contemporary with what has become known as the new

19. Alexandre Koyré, *Metaphysics and Measurement: Essays in Scientific Revolution* (Cambridge: Harvard University Press, 1968).
20. Alexandre Koyré, *From the Closed World to the Infinite Universe* (Baltimore: Johns Hopkins University Press, 1957), p. 2.

science. Perhaps paradoxically, Koyré himself was quite willing to consider the influence of such spirits on the new science, though he denied the influence was specifically scientific.[21]

Models in historiography, as well as in science, provide more than a mere heuristic for investigation. They focus attention, exclude possibilities, and reveal hidden connections. Whether as covert assumptions or as consciously accepted devices, models intervene (inevitably, I believe) between the historian and his or her subject. Yet the sensitive historian is not often led far astray by the odd magnifications a model might produce; a distorted perspective can be corrected by the feel of hard facts that he or she continues to accumulate. Moreover, the use of a model and the application of its embedded hypotheses require an artful intelligence, one that individualizes the crafted product. Thus historians who generally employ the revolutionary model may offer different perspectives 'on the same issues. Alistair Crombie, Rupert Hall, and Charles Gillispie, for example, in some contrast to Koyré, locate the revolution in scientific thought in the application of mathematics to mechanics and in the resultant construction of formal systems for the construal of nature.[22] Hall believes that the instruments and techniques developed by craftsmen have provided a stimulus and auxiliary to the new sciences.[23] 'But Koyré virtually ignores the crafts, since the science of Galileo and Descartes "is made not by engineers or craftsmen, but by men who seldom built or made anything more real than a theory."[24] Hall regards pre-seventeenth-century investigations of nature as essentially discontinuous with science after that period.[25] Crombie, who devotes considerable attention to the medieval development of the foundations of modern science, believes that "a more accurate view of

21. The role of occult influences on the development of science is highly controverted. The dispute may be followed in the following discussions: Francis Yates, *Giordano Bruno and the Hermetic Tradition* (Chicago: University of Chicago Press, 1964); P. Rattansi, "Some Evaluations of Reason in Sixteenth- and Seventeenth-Century Natural Philosophy," in *Changing Perspectives in the History of Science*, ed. M. Teich and R. Young (London: Heinemann, 1973); and Mary Hesse "Reasons and Evaluation in the History of Science," in *Changing Perspectives in the History of Science*. The various parties are brought together in Roger Stuewer, ed., *Historical and Philosophical Perspectives on Science*, vol. 5 of *Minnesota Studies in the Philosophy of Science* (Minneapolis: University of Minnesota Press, 1970).

22. Alistair Crombie, *Medieval and Early Modern Science*, 2d ed. (Cambridge: Harvard University Press, 1961), 2:125; A. Rupert Hall, *The Scientific Revolution, 1500–1800* (Boston: Beacon, 1966), pp. 370–71; Charles Gillispie, *The Edge of Objectivity* (Princeton: Princeton University Press, 1960), pp. 8–16.

23. Hall, *The Scientific Revolution*, pp. 217–43.

24. Koyré, *Metaphysics and Measurement*, p. 17.

25. Hall, *The Scientific Revolution*, p. 370.

seventeenth-century science is to regard it as the second phase of an intellectual movement in the West that began when philosophers of the thirteenth century read and digested in Latin translation the great scientific authors of classical Greece and Islam."[26] Gillispie, too, acknowledges the debt of Renaissance science to Greek mathematical rationalism.[27]

Yet those who generally employ the revolutionary model agree—and this constitutes the essential feature of the model—that a revolution in thought, a decisive overthrow of distinctly ancient modes of conception, is necessary to set a discipline on the smooth course of modern science. Hall clearly highlights the core of the model. For him the medieval period did have its quasi-science; and though that enterprise set the stage for the appearance of modern science, yet the mathematical methods of the latter were radically different from the methods of its predecessor: "Rational science, then, by whose methods alone the phenomena of nature may be rightly understood, and by whose application alone they may be controlled, is the creation of the seventeenth and eighteenth centuries."[28] It is the method of rational science that guarantees its further progress—without fear of taking fundamentally wrong paths. Whatever revisions in science have come since the revolution are revisions in content only, not in structure.[29]

The Gestalt Model

In recent years ideas from particular currents within the social and psychological sciences have joined those springing from conceptual studies in the history of science, especially those studies whose epistemological channels run to neo-Kantianism. From this confluence has emerged what might be called a Gestalt model of science. Among those most influential in employing this model are Norwood Russell Hanson, Thomas Kuhn, and Michel Foucault.

Both Hanson and Kuhn explicitly use devices drawn from Gestalt psychology and the psychology of perception. The Necker cube, the goblet-faces display, pictures of creatures looking alternately like birds or antelope, and similar puzzles illustrate for them the ways in which context, past experience, and familiar assumptions control our perceptual and conceptual experiences of things. In the scientific domain, as construed by Hanson, it is the well-entrenched theory that determines

26. Crombie, *Medieval and Early Modern Science* 2:110.
27. Gillispie, *The Edge of Objectivity,* pp. 8–16.
28. Hall, *The Scientific Revolution,* p. xii.
29. Ibid., p. xiii.

the perception of facts: "Physical theories provide patterns within which data appear intelligible. They constitute a 'conceptual Gestalt.' A theory is not pieced together from observed phenomena, it is rather what makes it possible to observe phenomena as being of a certain sort, and as related to other phenomena."[30] Likewise in Kuhn's judgment: "Assimilating a new sort of fact demands more than additive adjustment of theory, and until that adjustment is completed—until the scientist has learned to see nature in a different way—the new fact is not quite a scientific fact at all."[31]

If facts and their organizing theories are mutually implicative and constitute a perceptual-conceptual whole—a "paradigm," to use the by-now debased coin—and if "the switch of Gestalt . . . is a useful elementary prototype for what occurs in full-scale paradigm shift," then the model of scientific advance through the gradual increment of new facts and ideas proves inadequate for the historian's needs. "The transition from a paradigm in crisis to a new one from which a new tradition of normal science can emerge," argues Kuhn, "is far from a cumulative process, one achieved by an articulation or extension of the old paradigm. Rather, it is a reconstruction that changes some of the field's most elementary theoretical generalizations as well as many of its paradigm methods and applications."[32]

The lens of the Gestalt model can be focused more narrowly on the immediate scientific community, or dilated to situate the scientific community within a broader cultural context, as Foucault attempted in *The Order of Things: An Archaeology of the Human Sciences*. He intended in this work to explore the "*positive unconscious* of knowledge: a level that eludes the consciousness of the scientist and yet is part of scientific discourse."[33] He held that there were different epochs in Western history in which the sciences and related disciplines were bound together in the general cultural reticulum by unconscious principles of order. These principles yielded an "entire system of grids which analysed the sequences of representations (a thin temporal series unfolding in men's minds), arresting movement, fragmenting it, spreading it out and redistributing it in a permanent table."[34] Such ordering structures func-

30. Norwood Russell Hanson, *Patterns of Discovery* (Cambridge: Cambridge University Press, 1970), p. 90.

31. Thomas Kuhn, *The Structure of Scientific Revolutions*, 2d ed. (Chicago: University of Chicago Press, 1970), p. 53.

32. Ibid., pp. 84–85.

33. Michel Foucault, *The Order of Things: An Archaeology of the Human Sciences* (New York: Vintage, [1966] 1973), p. xi.

34. Ibid., pp. 303–4.

tioned, in Foucault's estimation, to determine both the domain of problems existing for the sciences and the methods of their resolution. His inquiry revealed to him three distinct epistemological ages—the Renaissance, the classical period (seventeenth and eighteenth centuries), and the modern period—each radically discontinuous from the previous one, so that terms of description used by one (e.g., "man," "society," "language," "nation") would have fundamentally different meanings when used by others. In the transition from one epoch to the next, it was "not that reason made any progress: it was simply that the mode of being of things, of the order that divided them up before presenting them to understanding, was profoundly altered." [35] As a consequence of the shift in the patterns of representation—the switch of the Gestalt—man as we now construe him in the human sciences came into existence only at the beginning of the nineteenth century. This is the paradoxical thesis of Foucault's work.

The Gestalt model makes two principal demands: first, that the historian should attempt sympathetically to assimilate and reconstruct the context of scientific discourse of a given period, and in this way to determine the social, psychological, and historical influences that controlled the ways scientists patterned their theoretical concepts and perceived through them the facts constituting the domain of scientific inquiry; and second, that the historian should regard history of science not as an internal and smooth flow of observations and theoretical generalizations across the ages but as the sudden shift of different world views, linked only by the extrinsic contingencies of time and place.

The Gestalt model as employed by Hanson, Kuhn, and Foucault bears similarities to the revolutionary model—and, of course, Kuhn's express aim is to describe the structure of revolutions in science. But the differences between the revolutionary model as commonly used and the Gestalt model are marked. Those employing a revolutionary model discover in the course of a particular science a signal awakening of thought, an overturning of what the model characterizes as a decidedly archaic mode of thinking, and the establishment of a lasting foundation for future progress by, as Hall puts it, "accretion." [36] Since the logic of the revolutionary model hinges on the dichotomy between ancient and modern methods of scientific thought, historians of this persuasion usually assume that revolution in a science is a one-time affair. The Gestaltists, however, emphasize multiple "scientific revolutions," no one of which secures a position that is any more scientific or more stable than

35. Ibid., p. xii.
36. Hall, *The Scientific Revolution,* pp. xiii–xiv.

others that have preceeded it. The revolutionists believe that revolutions happen for good reasons, reasons that sustain the future growth of a science. The Gestaltists, as is consonant with the source of their model, tend to stress psychological and sociological factors in scientific change. In their view, scientific change is rarely the result of good reasons; indeed, reasons have weight only against a background of commonly accepted theory. The revolutionists view science as a search for truth about the world. The Gestaltists argue that there is no truth about the world; truth is a function of the coherence of the theoretical arrangement which holds at any one time; there are no independent, theory-free standards against which a hypothesis might be measured to assess its truth.[37] The revolutionists are apt to regard postrevolutionary science as better than or more true than prerevolutionary science. The Gestaltists believe that the perceptual-conceptual paradigm adopted by a given community of scientists is incommensurable with those assumed by their predecessors: in the Gestalt switch of the goblet-faces display, the goblet is no better or truer than the faces.[38]

The Gestalt model encourages the historian to interpret scientific ideas as parts of a larger complex of meanings; it emphasizes the mutual determination of these elements. The hermeneuticist of the scientific Gestalt begins with a node of experience or a paradigmatic idea and moves laterally, interpreting one symbol of the pattern in terms of the others, ultimately including socially and culturally entwined meanings. Another recent model, however, suggests that the interpretive relation is vertical and unidirectional, and that scientific patterns of thought merely reflect deeper and more covert social or psychological structures.

Social-Psychological Model

From the ancient through the modern periods, scientists have frequently justified their theories by appeal to the more general doctrines—metaphysical, religious, or social—to which those theories have been related. Samuel Clarke defended Newton's science, since it would "confirm, establish, and vindicate against all objections those great and fundamental truths of natural religion."[39] But it was only at the beginning of our century, after transformations in the social and psychological sciences (by Marxism, Durkheimian social anthropology, Freudianism, and similar conceptual movements) that historians seriously

37. Hanson, *Patterns of Discovery*, p. 15.
38. Kuhn, *Structure of Scientific Revolutions*, pp. 170–71.
39. Samuel Clarke, *The Leibniz-Clarke Correspondence*, ed. H. Alexander (Manchester: Manchester University Press, [1717] 1956), p. 6.

attempted to organize their narratives under the assumption that scientific programs might be fueled by social interests and psychological needs. What united both socially oriented and psychologically disposed historians was the conviction that apparently extrinsic conceptual structures, whether embedded in social relationships or in psychological complexes, might covertly determine the generation, formulation, and acceptance of scientific ideas. Moreover, though Freudians have insisted on the primacy of sedimented attitudes, they usually have admitted that these originated in certain real or imagined social situations. Similarly, Marxist historians have recognized that the effects of class stratification are mediated by subtle patterns of individual belief. Because of these common features, social and psychological models may be considered as forming one class of historiographic models.

Social-psychological models can be divided into those prescribing weak determination and those prescribing strong determination of scientific development. The weak version of the model is the central organizing device of J. D. Bernal's four-volume *Science in History*. The model guided Bernal in mapping an enlarged field of investigation. "Science," he proposed, "may be taken as an institution; as a method; as a cumulative tradition of knowledge; as a major factor in the maintenance and development of production; and as one of the most powerful influences molding beliefs and attitudes to the universe and man."[40] Such a generous conception compelled him to trace the social and psychological patterns in the terrain of science. For example, he initially explained Darwin's hypothesis of natural selection as a reformulation of Malthusian economics in other terms—that is, as a biological construction of a "theory built to justify capitalist exploitation."[41] For Bernal, the source of scientific thought, the institutions of science, its methods, the economic forces driving it, and its impact on society were all fit subjects for social-psychological analysis. Yet science as a "cumulative tradition of knowledge" was not.

Bernal could not bring himself to extend his Marxist vision to the heart of science. He confessed that the cumulative nature of science distinguished it from such other human pursuits as law, religion, and art. Though science, like these other enterprises, grows in a field of social relations and class interests, its claims, unlike theirs, can be checked directly "by reference to verifiable and repeatable observations in the material world."[42] The weak model thus protects the internal

40. J. D. Bernal, *Science in History*, 3d ed. (Cambridge, Mass.: M. I. T. Press, 1971), 1:31
41. Ibid. 2:644.
42. Ibid. 1:43−44.

logical and justificatory structure of science from the hands of the sociologist and the psychologist. That is, it does so when the science goes right.

When it goes wrong, the historian has a sure sign that extrinsic social or psychological factors have intruded. For instance, Erik Nordenskiöld, in his influential *History of Biology,* felt constrained to invoke the weak model in his account of the unwarranted (as he believed) acceptance of Darwinian theory by scientists in the latter half of the nineteenth century:

> From the beginning Darwin's theory was an obvious ally to liberalism; it was at once a means of elevating the doctrine of free competition, which had been one of the most vital cornerstones of the movement of progress, to the rank of natural law, and similarly the leading principle of liberalism, progress, was confirmed by the new theory. . . . It was no wonder, then, that the liberal-minded were enthusiastic; Darwinism must be true, nothing else was possible.[43]

It is ironic that one powerful tradition in the sociology of science, led by Robert Merton[44] and Joseph Ben-David[45], endorses the weak model as the only one appropriate for respecting the cognitive content of science. Ben-David, for example, admits that socially conditioned biases and ideology "might have played some role in the blind alleys entered by science." In those darkened corners, sociology can prove illuminating. But the main scientific roads are "determined by the conceptual state of science and by individual creativity—and these follow their own laws, accepting neither command nor bribe."[46] Sociologists in this tradition confine their empirical analyses to questions of institutional organization, the spread of scientific knowledge, social controls on the focus of scientific interest, and public attitudes toward science. They regard the cognitive content of science, however, as the reserve, not of the sociologists, but of those intellectual historians whose concerns are principally logical and methodological.

Yet even with the support of the dominant tradition in the sociology of science, can the weak model be justified? Those employing it usually fail to supply a convincing reason why social or psychological analyses might explain error in science but not truth. The persuasiveness of this

43. Erik Nordenskiöld, *The History of Biology,* 2d ed. (New York: Tudor, 1936), p. 477.

44. Robert Merton, *The Sociology of Science* (Chicago: University of Chicago Press, 1973).

45. Joseph Ben-David, *The Scientist's Role in Society* (Englewood, N.J.: Prentice-Hall, 1971).

46. Ibid., pp. 11–12.

model is further diminished when one considers that, in a strict sense, most past science is "erroneous," at least by contemporary standards. Hence if the logicist assumptions of the weak model are consistently heeded, the content of virtually all past science ought to be amenable to social and psychological interpretation. Nor should contemporary science be exempt, since there is no reason to suspect that it has achieved final truth.

The logic of the preceding line of reasoning appears to have persuaded, implicitly at least, those using a strong version of the model. For example, Margaret Jacob has detected social interests at the root of seventeenth-century mechanical philospohy. In her *The Newtonians and the English Revolution*, she argues that the Newtonians, those traditional harbingers of contemporary science, constructed matter as passive (i.e., having no occult powers) not because reason and evidence required it but because their latitudinarian and religious ideology demanded it.[47] The Edinburgh sociologist of science David Bloor supports Jacob's use of the strong version of the model. In similar fashion, he maintains that the seventeenth-century scientist Robert Boyle, his colleagues, and his opponents "were arranging the fundamental laws and classifications of their natural knowledge in a way that artfully aligned them with their social goals." The lesson Bloor drew from this drama of Restoration science was that quite generally in the history of science, "the classification of things reproduces the classification of men."[48]

The strong version of the model, then, asserts that the structure of scientific knowledge is determined not by nature but by social patterns or psychological complexes. The model stipulates that logic and the appeal to natural facts are on the surface and that what really matters in comprehending the work of scientists are dominance struggles with the father—as in the case of Mitzman's reconstruction of Weber's social science[49]—or the social practices of a society—as in the case of Bloor's account of Greek mathematics.[50]

The strong version of the social-psychological model, despite initial implausibility, does focus the historian's sight on a cardinal feature of scientific development: that science depends on norms—norms sug-

47. Margaret Jacob, *The Newtonians and the English Revolution* (Ithaca, N.Y.: Cornell University Press, 1976).

48. David Bloor, "Klassifikation und Wissenssoziologie: Durkheim und Mauss neu betrachtet," *Kölner Zeitschrift für Soziologie und Sozialpsychologie* 22 (1980): 20–51.

49. Arthur Mitzman, *The Iron Cage: An Historical Interpretation of Max Weber* (New York: Grosset & Dunlap, 1971).

50. David Bloor, *Knowledge and Social Imagery* (London: Routledge and Kegan Paul, 1976), pp. 95–116.

gesting what is appropriate both to investigate and to accept. Norms, however, are dictated not directly by nature but by the decisions of men. The logic of scientific argument cannot coerce, except insofar as men feel moved to abide by its rules and adopt its premises. Ultimately, the acceptance of metarules and first premises appears to be a function of social enculturation, of psychological conditioning, and perhaps of biological disposition. For as Aristotle pointed out, only the fool tries to demonstrate the principles upon which all his arguments are based.

Nonetheless, the strong version of the social-psychological model seems too strong. It is liable to a tu quoque response. Why, after all, should we be convinced by the account of a historian who uses the strong version, if that account itself merely reflects his inferiority complex or his Calvinistic upbringing? The destruction of scientific rationality also undermines the plausibility of historical argument. To restrain the destructive relativism of both the social-psychological model and the Gestalt model, while preserving the edge of their insights, is one of the chief tasks for which evolutionary models have been constructed.

Evolutionary Models of Scientific Development

The use of evolutionary theory in explanations of cultural phenomena can easily be traced back to the mid-nineteenth century. John Lubbock, Walter Bagehot, Lewis Henry Morgan, Edward Tylor, Herbert Spencer, and a host of others applied evolutionary concepts to societal institutions in an effort to account for the descent from primitive culture.[51] More recently, the specialized use of evolutionary notions, aping its biological counterpart, has proceeded from the macroconsideration of culture to the microconsideration of the development of ideas, particularly scientific ideas. Gerald Holton, for instance, makes detailed use of the evolution analogy in his *Thematic Origins of Modern Science;*[52] and in reconsidering his theory of paradigms, Kuhn has suggested that the appropriate approach to science history is evolutionary.[53] But Holton and Kuhn wield evolutionary constructs only as

51. See John Burrow, *Evolution and Society* (Cambridge: Cambridge University Press, 1966); and George Stocking, *Race, Culture, and Evolution,* 2d ed. (Chicago: University of Chicago Press, 1981).

52. Gerald Holton, *Thematic Origins of Scientific Thought: Kepler to Einstein* (Cambridge: Harvard University Press, 1973).

53. Thomas Kuhn, "Reflections on My Critics," in *Criticism and the Growth of Knowledge,* ed. Imre Lakatos and Alan Musgrave (Cambridge: Cambridge University Press, 1970), p. 264.

vague analogies. Others believe they promise more. Evolutionary theory, duly generalized, provides, it is argued, the very explanation of scientific growth. Not only are ideas conceived, but like Darwin's finches they also evolve. In the nineteenth century, George Romanes, Conwy Lloyd Morgan, William James, and James Mark Baldwin all proposed that Darwinian theory explained the development of ideas. More recently Karl Popper, Stephen Toulmin, and Donald Campbell have advanced a strict epistemological Darwinism. In what follows, I will briefly examine the proposals of Popper and Toulmin, indicate the deficiencies of their models, and then elaborate a natural selection version that comes close in spirit to the fertile ideas of Campbell. My analysis of the natural selection model, which has guided the construction of the history portrayed in this book, will include an evaluative comparison with Lakatos's model of scientific research programs and will conclude with a consideration of the natural selection model's historiographic advantages over other models.

The Models of Popper and Toulmin

In *The Logic of Scientific Discovery,* Popper describes the scientific community's selection of theories not as a process by which a given theory is justified by the evidence but as one by which a theory survives because its competitors are less fit. Thus he argues that the preference for one theory over another "is certainly not due to anything like an experimental justification of the statements composing the theory; it is not due to a logical reduction of the theory to experience. We choose the theory which best holds its own in competition with other theories; the one which, by natural selection, proves itself the fittest to survive."[54]

In Popper's judgment, our scientific and pedestrian quests for knowledge always begin not with pure observation but with a problem that has arisen because some expectation has not been met. In confronting the problem, the cognizer makes unrestrained conjectures about possible solutions, much as nature makes chance attempts at solving particular survival problems.[55] These conjectures are then tested against empirical evidence and rational criticism. The rational progress of science, therefore, consists in replacing unfit theories with those that have solved more problems. These latter, according to Popper, should imply more empirical statements that have been confirmed than their prede-

54. Karl Popper, *The Logic of Scientific Discovery,* 2d ed. (New York: Harper & Row, 1968), p. 108.
55. Karl Popper, *Objective Knowledge* (Oxford: Oxford University Press, 1972), p. 145.

cessors.[56] This condition enables us to describe successor theories as closer to the truth and consequently more progressive. I will not expand further on Popper's conception, since Lakatos has already done this with concision, fashioning from it a model of scientific research programs, which I will discuss below.

The evolutionary model permits Popper to avoid the presumption that theories are demonstrated by experience; it also allows him to dismiss the view that theories and creative ideas arise from any sort of logical induction from observation. Thus the older and newer problems of induction are skirted. Popper believes that the model directs one to interpret scientific discovery as fundamentally an accidental occurrence, a chance mutation of ideas. He consequently fails to emphasize that the intellectual environment not only selects ideas but restricts the kinds of ideas that may be initially entertained by a scientist. Attention to the environment of scientific ideas, however, is precisely what Toulmin requires for an adequate account of scientific growth.

Toulmin's thesis is that scientific disciplines are like evolving biological populations, that is, like species. Each discipline has certain methods, general aims, and explanatory ideas that provide its coherence over time, its specific identity, while its more rapidly changing content is constituted of loosely related conceptions and theories, "each with its own separate history, structure, and implications."[57] To comprehend the evolution of a science so structured requires that one attend to the cultural environment promoting the introduction of new ideas, as well as to the selection processes by which some few of these ideas are perpetuated.

The content of a discipline, according to Toulmin's scheme, adapts to two different (though merging) environmental circumstances: the intellectual problems the discipline confronts and the social situations of its practitioners. Novel ideas emerge as scientists attempt rationally to resolve the conceptual difficulties with which their science deals; but often those new sports will also be influenced by institutional demands and social interests. Therefore, in explaining the appearance of innovative ideas within an evolving science, one must consider both *reasons* and *causes*. After such variations are generated, however, one must turn to the processes, rationally and socially causal, by which the variations are selected and preserved.

The processes that shape the growth of a discipline—selection pro-

56. Karl Popper, *Conjectures and Refutations: The Growth of Scientific Knowledge,* 2d ed. (New York: Harper & Row, 1968), pp. 215–50.

57. Stephen Toulmin, *Human Understanding* (Oxford: Oxford University Press, 1972), p. 130.

cesses—also occur within particular intellectual and social settings. The intellectual milieu consists of the immediate problems and entrenched concepts of the science and its neighbors. Within this environment, rational appraisal by the scientific community tests the mettle of new ideas. The survivors are incorporated into the advancing discipline. The social and professional conditions of the discipline also work to cull ideas, sanctioning some and eliminating others. Both of these selection processes—selection against intellectual standards and against social demands—may act either in complementary fashion or in opposition. But both must be heeded, in Toulmin's judgment, if one is to understand the actual history of a science. The historian will look only to intellectual conditions, however, when pursing a rational account of the devel opment of a particular science. When investigating, say, the causes accelerating or retarding scientific growth, he or she will turn to the social and professional institutions of that science.[58]

It is the mark of the recent past in the historiography of science that it is the rational continuity of science that appears to require explanation. Toulmin has proposed his evolutionary model to meet this need. In his view, the continuity of disciplines, like the continuity of biological species, involves transmission of previously selected traits to new generations. In science this process is, according to Toulmin, one of enculturation: junior members of a discipline serve an apprenticeship in which they learn by tutored doing, by exercising certain "intellectual techniques, procedures, skills, and methods of representation, which are employed in 'giving explanations' of events and phenomena within the scope of the science concerned."[59] What principally gets inherited, he believes, is not a disembodied set of mental concepts but particular constellations of explanatory procedures, techniques, and practices that give muscle to the explicating representations and methodological goals of the science. Through the active participation in an ongoing scientific community, the novice inherits two kinds of instantiated concepts. The first comprises the specific substantive ideas and theories, the special explanations and techniques that solve recognized problems at any one period in the evolution of a discipline. The second kind of inheritance remains continuous over much longer periods and changes only slowly. It consists of the explanatory ideals, the general aims, and the ultimate goals that distinguish the disciplines from one another. It is within this more general inherited tradition that large-scale conceptual changes in substantive theory occur "by the accumulation of smaller modifications,

58. Ibid., pp. 307–13.
59. Ibid., p. 159.

each of which has been selectively perpetuated in some local and im-mediate problem situation."[60] But such changes should not suggest, as they do for those adopting the Gestalt model, that there are not good reasons for the shifts. The basic structure against which reasons can be measured is the continuity of explanatory aims and ideals which a dis-cipline manifests through long periods of its history.

Toulmin further attempts to ensure that his model will allow rational criteria to operate in science by adjusting it with the postulate of "coupled evolution." The neo-Darwinian theory of organic evolution requires that variability within a species be independent ("decoupled" as Toulmin puts it) of natural selection. According to his interpretation of the modern synthesis, there is no preselection or direction given classes of variations. But coupled evolution, which he regards as an-other species of the larger genus of evolutionary processes, postulates that variation and selection "may involve related sets of factors, so that the novel variants entering the relevant pool are already preselected for characteristics bearing directly on the requirements for selective perpetuation."[61]

What Toulmin has suggested by his postulate of coupled evolution is not, however, a Darwinian sort of mechanism, in which production of variations is blind or random, but a Lamarckian one, in which con-scious acts preshape the material in anticipation of the exigencies of survival. Accordingly the cardinal feature of the Darwinian perspective, competitive struggle against environmental demands, is largely obvi-ated. Natural selection has no pivotal role in Toulmin's scheme.

But Toulmin need not have abandoned the device of natural selection so quickly. For the neo-Darwinian synthesis does recommend clearly acceptable senses in which individuals within a species might be de-scribed as preselected or preadapted to an altered environment: when, for example, heterozygote superiority leads to the retention of alleles that would be fit in different circumstances; or when linkage holds in a population certain alleles that would enhance adaptation to changed surroundings; or when alleles at certain loci have fixed rates of muta-tion. Such mechanisms for storing variation act as constraints on selec-tion, making specific kinds of adaptive responses to a given situation more likely. It is of course true that the variations stored and the meth-ods of their preservation are products of previous selection over many generations. In any case, classes of variations characteristic of elephants are not likely to occur in the species *Rattus rattus*. The genetic back-

60. Ibid., p. 130.
61. Ibid., p. 337.

ground of a species will restrict, and in that sense preselect, the kinds of variations that are immediately possible. In a moment I will indicate what this feature of biological evolution suggests for understanding conceptual evolution.

The Natural Selection Model

Popper's version of the evolutionary model of science emphasizes that theories succeed one another something like species: that theory is selected which solves more problems than its competitors. Toulmin's version complements Popper's by focusing on the cultural environments in terms of which new ideas appear and are incorporated into an evolving discipline. But Toulmin relinquishes a formal Darwinian device in an attempt to capture the way problem solutions originally emerge. As just indicated, abandonment of a natural selection mechanism is not necessary in order to model the birth of new scientific theories. In this section I want to build upon the Popperian and Toulminian variations and thereby refine a natural selection model for historiographic use. I will do this in two stages: first, by further specifying exactly what it is that evolves in scientific change; and second, by adding a psychosocial theory of idea production and selection, one similar to that proposed by Campbell.

According to Toulmin's model, the specieslike entity that evolves is the intellectual discipline. But this, I think, is the wrong analogue. Intellectual disciplines are, after all, composed of heterogeneous theories, methods, and techniques, while a species is a population of interbreeding individuals that bear genetic and phenotypic resemblance. Disciplines, moreover, are organized formally into subdisciplines and overlapping and competing specialties and are interlaced with invisible networks of communication.[62] Disciplines seem more like evolving ecological niches, consisting of symbiotic, parasitic, and competing species. The proper analogue of a species is, I believe, the conceptual system, which may be a system of theoretical concepts, methodological prescriptions, or general aims. The gene pool constituting such a species is, as it were, the theory's individual ideas, which are united into genotypes or genomic individuals by the bonds of logical compatibility and implication and the ties of empirical relevance. These connecting principles may themselves, of course, be functions of higher-order regulatory ideas. Biological genotypes vary by reason of their components, the genes, and the specific linkage relations organizing them; these genotypes display different phenotypes according, both as they have

62. See Diana Crane, *Invisible College* (Chicago: University of Chicago Press, 1972).

slightly different components and componential relationships and as they react to altered environments. Analogously, the cognitive representation of a scientific theory—its phenotypic expression in terms of the model here proposed—will vary from scientist to scientist by reason of the slightly different ideas constituting it, their relations, and the changing intellectual and social environment that supports it. So, for instance, Darwin and Wallace both advanced *specifically* the same evolutionary theory, though the components of their respective representations were not exactly the same, and the intellectual problems to which they applied their views and for which they sought resolutions also differed in some respects. Yet we still want to say that Darwin and Wallace developed the "same"—specifically the same—theory of evolution by natural selection. Constructing the model in this way also allows us to appreciate that, like the boundaries between species, the boundaries separating theories may be indefinite and shifting.

If a historiographic model of scientific development proposes that conceptual systems, like biological species, evolve against a problem environment, then that model, to tighten the Darwinian analogy, should include a mechanism accounting for adaptive change in scientific thought. During the last quarter century, Donald Campbell has worked out a psychological theory of idea production and selection that meets this demand.[63] His mechanism of "blind variation and selective retention" not only illuminates a fundamental feature of creative thinking in science (and in other cognitive pursuits) but, as an unintended consequence, also explains why some ideas seem to come (as Toulmin believes) preadapted to their intellectual tasks. Let me first sketch the essential aspects of Campbell's natural selection mechanism and then add some refinements.

In the Darwinian scheme, species become adapted to solve the problems of their environment through chance variations and selective perpetuation. Campbell supposes that the creative thinker exhibits counterpart cognitive mechanisms; these mechanisms blindly gener-

63. Donald Campbell has developed his theory in a series of papers: "Methodological Suggestions from a Comparative Psychology of Knowledge Processes," *Inquiry* 2 (1959): 152–82; "Blind Variation and Selective Retention in Creative Thought as in Other Knowledge Processes," *Psychological Review* 67 (1960): 380–400; "Blind Variation and Selective Retention in Socio-Cultural Evolution," in *Social Change in Developing Areas,* ed. H. Barringer, G. Blanksten, and R. Mack (Cambridge, Mass.: Schenkman, 1965); "Evolutionary Epistemology," in *The Philosophy of Karl Popper,* ed. Paul Schilpp (La Salle, Ill.: Open Court, 1974); "Unjustified Variation and Selective Retention in Scientific Discovery," in *Studies in the Philosophy of Biology,* ed. Francisco Ayala and Theodosius Dobzhansky (London: Macmillan, 1974); "Discussion Comment on 'The Natural Selection Model of Conceptual Evolution,'" *Philosophy of Science* 44 (1977): 502–507.

ate possible solutions to intellectual problems, select the best-adapted thought trials, and reproduce consequently acquired knowledge on the appropriate occasions. A distinctive postulate of this model is that cognitive variations are produced blindly, which is to say that initial thought trials are not justified by induction from the environment, or by previous trials, or by "the eventual fit or structured order that is to be explained."[64] The production of thought variations by the scientist—or the creative thinker in any realm—is therefore precisely analogous to chance mutations and recombinations in organic evolution.

The bones of Campbell's conception can be fleshed out in ways that make it fit for the historiographic model I have in mind. The following additional postulates serve this function.

1. The generation and selection of scientific ideas, both as the hypotheses that guide a scientist's work and as the relatively sedimented doctrine of the scientific community, should be understood as the result of a feedback mechanism. Such a mechanism, which we may consider only formally without worrying about its physiological realization, will generate ideas in a biased rather than in a purely random fashion. For without some restraints on generation, a scientist might produce an infinity of ideas with virtually no probability of hitting on a solution to even the simplest problem. But of course even mutations and recombinations of genes do not occur completely at random. The constraints on idea production are determined by the vagaries of education and intellectual connections, the social milieu, psychological dispositions, previously settled theory, and recently selected ideas. This postulate therefore suggests that, though ideas may come serendipitously, their generation is not unregulated but can be comprehended by the historian. Thus, for example, when Darwin began musing on the nature of a mechanism to explain species change, he did so in a conceptual environment formed partly of ideas stimulated by his *Beagle* voyage and partly of ideas acquired from his grandfather, from Lamarck, and from a host of authors he read between 1836 and 1838. These ideas not only determined the various problems against which successful hypotheses were selected, but they also initially fixed the restraints on the generation of trial solutions. It is within a certain (albeit vaguely defined and shifting) conceptual space that chance variations are displayed. And it is because of such constraints that even a scientist's rejected hypotheses can make sense to the historian.

2. To think scientifically is to direct the mind to the solution of prob-

---

64. Campbell, "Unjustified Variation and Selective Retention in Scientific Discovery," p. 150.

lems posed by the intellectual environment. Novel ideas are not produced in an environment where perceptual or theoretical situations are
settled. As Popper (and Dewey before him) has argued, for thinking to
occur there must be a troubled, unsettled cognitional matrix; the perceived environment must be changing. Conversely, alterations in the
intellectual situation that are not perceived or that are ignored must lead
to the arrest of scientific thought and the eventual extinction of a scientific system.

3. Ideas and ultimately well articulated theories are originally generated and selected within the conceptual domain of the individual scientist. Only after an idea system has been introduced to the scientific
community (or communities, since scientists usually belong to several
interlocking social networks) does public scrutiny result. The broader
conceptual environment established by the community may present
somewhat different problem situations and standards of competitive
survival. To the extent, however, that the problem environments of the
individual and the community coincide, individually selected ideas or
theories will be fit for life in the community. If the historian neglects
(as Toulmin does) to consider the processes of idea generation and
evaluation at the individual level, then scientific ideas will appear to
come mysteriously preadapted to their public environment.

4. Finally, if this model is to be used in construing the acquisition of
knowledge in science, then one must suppose that selection components operate in accord with certain essential criteria: logical consistency, semantic coherence, standards of verifiability and falsifiability,
and observational relevance. These criteria may function only implicitly,
but they form a necessary subset of criteria governing the development
of scientific thought throughout its history. Without such norms, we
would not be dealing with the selection of *scientific* ideas. The criteria
thus aid historians of science in distinguishing their subject from other
cognitive occupations. It should be stressed, however, that these selection criteria are themselves the result of previous idea generation and
continuous selection, processes by means of which science has descended from protoscience—just as the mammals have descended from
the reptiles. The complete set of selection criteria define what in a given
historical period constitutes the standard of scientific acceptability.
The above-specified criteria are only elements of this more comprehensive set.

Natural Selection Model vs. Scientific Research Programs

Since the natural selection model of science (NSM for short) is a
model, it implicitly represents a theory about science, about its struc-

ture, growth, and rationality. The model and its imbedded theory portray scientific conceptual systems as quasi-organisms that compete for survival; and it proposes that the system which best solves the problems of its cultural environment will survive, gradually displacing its competitors. In order to assess the model's viability, we might compare it with another powerful model, which appears to offer it the keenest competition—Lakatos's model of scientific research programs (SRP).[65] Lakatos designed SRP to serve both as a standard for appraising the scientific and rational status of contemporary conceptual systems and as a historiographic device for constructing explanatory accounts of science's growth. Because of this explicit intention and the model's rigorous formulation, SRP furnishes an exceptional standard by which to evaluate NSM.

Lakatos formulated his model expressly for the purpose of interpreting the history of science as rationally progressive. He contrasts his conception with the Kuhnian model, which he regards, correctly I believe, as forbidding judgments of general scientific progress across problem shifts and as supplying no criteria for distinguishing scientific rationality from doctrinaire opinion.[66] Yet like Kuhn, Lakatos chooses a larger unit of analysis than the solitary idea or theory, since he recognizes the historical and epistemological fact that ideas cannot be evaluated in isolation from the auxiliary concepts which specify normal conditions, relevant evidence, and theoretical pertinence. He takes this larger conceptual scheme, the SRP, as the entity to be judged as progressing (or degenerating), as competitive with other programs, and as the basis for estimating the rationality of a particular scientific enterprise.

As Lakatos characterizes its structure, SRP has a "hard core" of central principles and a "belt of surrounding auxiliary hypotheses" that continues to change during the life of a program.[67] Newton's program, for example, had a stable center consisting of his three laws of dynamics and principle of attraction; it also had a belt of hypotheses composed of assumptions about the gravitational center of large bodies, the viscosity of different resisting media, the paths of planets, the distance of the fixed stars, and a host of other boundary conditions. If a program is to be pursued, the hard core embodying its defining ideas must be

65. See Imre Lakatos, "Falsification and the Methodology of Scientific Research Programmes" and "History of Science and Its Rational Reconstructions," in *The Methodology of Scientific Research Programmes: Philosophical Papers of Imre Lakatos,* vol. 1, ed. John Worrall and Gregory Currie (Cambridge: Cambridge University Press, 1978).

66. Lakatos, "Falsification and the Methodology of Scientific Research Programmes," pp. 8–10.

67. Ibid., pp. 48–52.

protected, especially in the early stages of growth, from noxious facts and the harmful competition of rival programs. The program's "negative heuristic," then, bids falsification attempts be deflected to the auxiliary hypotheses. It is the protective girdle of hypotheses that is challenged by the facts and adjusted to escape the force of contrary evidence. The "positive heuristic" of the program complements the negative imperative by proposing means of advancing the empirical content of the program through development of the auxiliary hypotheses. The positive heuristic discharges this function principally by suggesting replacement hypotheses when evidence or internal logic require that and by setting the plan which the program will stubbornly follow in the face of anomalies and the claims of rival programs.

Lakatos offers his model as a refinement of Popper's. It nevertheless differs from Popper's selection model on an important point. Popper sometimes suggests that theories can be falsified directly, by infection from toxic facts, and that such falsified theories are (or scientific honor demands that they should be) rendered immediately extinct.[68] Lakatos, in contrast, recognizes that theories may accumulate anomalies but that scientists properly adjust their auxiliary hypotheses to avoid them, even, if possible, to turn them into dramatic corroborations of the program. Darwin, for example, was initially stumped by the seemingly inexplicable adaptations of neuter insects—they left no progeny to inherit favorable variations. But when after several years he finally developed his mechanism of community selection, what originally threatened to falsify his theory became the strongest evidence for it. Such manipulations, however, can be abused.

To prevent ad hoc alterations from turning rational science into empirically immune pseudoscience, Lakatos stipulates that such adjustments should be capacious enough to extend the empirical content of the theory beyond the refuting cases, so that such extension yields the prediction of new facts: "A given fact is explained scientifically only if a new fact is also explained with it."[69] Predictive extension is for Lakatos the mark of a scientifically authentic research program. If a program accounts for the empirical content of rivals but also generates further predictions, then the program is "theoretically progressive" and thus scientific. If the predictive excess is corroborated, then the program is also "empirically progressive"; otherwise, it is "degenerating." But if a program confronts incompatible facts that its constituent theories can-

68. Popper, *The Logic of Scientific Discovery*, pp. 86–87.
69. Lakatos, "Falsification and the Methodology of Scientific Research Programmes," p. 34.

not neutralize, while theories of a rival program can give them account, and if that rival also generates further predictions, some of which are corroborated, then the original program is "falsified." Programs that persist in the face of a rival's success can only be judged "pseudo-scientific."[70]

Lakatos has constructed his model as a device for appraising research programs, both recent and historically remote ones. He believes that appropriate standards are required if the historian is to do his job properly. SRP allows the historian to distinguish the internal history of science, which expresses the rational growth of objective knowledge, from "empirical (sociopsychological) 'external history.'"[71] Since the principal meaning for "science" is "accomplished objective knowledge," the internal history of science captures all that is essential to it. SRP allows investigators to select out of the morass of historical clutter precisely their special subject, the internal logical development of theories. After all,

> most theories of the growth of knowledge are theories of the growth of disembodied knowledge: whether an experiment is crucial or not, whether a hypothesis is highly probable in the light of the available evidence or not, whether a problem shift is progressive or not, is not dependent in the slightest on the scientists' beliefs, personalities, or authority. These subjective factors are of no interest for any internal history.[72]

Indeed, with the help of SRP, the historian of science should construct an ideal history, the normative fabula that the logic of a given research program demands. Lakatos offers Niels Bohr's program as illustrative:

> Bohr, in 1913, may not have even thought of the possibility of electron spin. He had more than enough on his hands without the spin. Nevertheless, the historian, describing with hindsight the Bohrian program, should include electron spin in it, since electron spin fits naturally in the original outline of the program. Bohr might have referred to it in 1913. Why Bohr did not do so is an interesting problem which deserves to be indicated in a footnote.[73]

Despite his comic exaggeration, Lakatos does not intend that the historian should literally write a bilevel history, one narrative in the text and another in the footnotes. But his model does require a history

70. Ibid, pp. 32–35.
71. Lakatos, "History of Science and Its Rational Reconstructions," p. 102.
72. Ibid., p. 118.
73. Ibid., p. 119.

structured with a logically distinctive internal core which is to control the significance assigned to external social and psychological events. This core, in his view, should express not only the research program that some scientist actually established but also an enlarged program that contains features implicitly derivable from the original.

In Lakatos's application, SRP exudes a peculiar Platonic odor, which I think most historians would find offensive. For the model appears to demand not a historian but a Laplacean demon who could extract from a program all that was logically or compatibly contained therein. Insofar as SRP is used by human historians, it seems to urge them to read history backwards, to find in earlier, inchoate concepts the results of more recent research. In Lakatos's hands, SRP would obscure the vision of historians wishing to detect the emergence of scientific ideas from previously developed ideas, community expectations, and personal aims.

Because an instrument is badly used does not mean, of course, that it is defective. SRP could conceivably be given a historically justified employment. But the model itself is, I believe, radically deficient for historical work. A comparison with NSM should make clear the advantages of the latter. In the following analysis, I will use episodes in the development of Darwin's conceptual system as the test base. Lakatos's selective use of examples from the history of physics has prejudiced his case.

1. As Lakatos structures it, a research program is essentially immutable; its hard core remains stable and defines the period of the program's existence. For Darwin's program, the mechanism of natural selection must certainly be regarded as a core principle. Yet when he initially attacked the problem of species change, he developed and used several other mechanisms before hitting upon natural selection. And even after he formulated that key principle, he continued to modify both its logic and its scope of application. Despite changes and reorganization of core principles, Darwin's conceptual system retains a historical identity.

NSM allows for this kind of alteration. Evolving conceptual systems may undergo fundamental changes, changes more basic than simple adjustment of peripheral principles. A system will be regarded as forsaken only when historical continuity has been broken and the problem situation vacated. Barring this, NSM encourages the expectation that the introduction of fundamental ideas will alter a developing system's more remote principles and that changes in these latter—the adaptations by which a conceptual system more immediately meets the requirements of its environment—will in turn affect the central principles. Expectation of reciprocity, not unilateral alteration, is the methodological rule.

2. SRP evaluates a program as nondegenerating only if it continues to make novel predictions that are empirically confirmed. What success rate must be maintained in order to keep a favorable evaluation is, however, unspecified. More seriously, if this criterion of progress were actually operative in the mid-nineteenth century, it would have counseled the immediate rejection of Darwin's conceptual system; for his theory made no real predictions (certainly not comparable to Lakatos's favorite—Einstein's forecast of starlight bending near the sun). Darwin simply did not use his system as a predictive instrument in the conventional way. He fairly estimated that the advantage of his theory was that it made sense of a medley of facts. Its cogency lay in uniting what had before seemed disparate.

NSM interprets a conceptual system as progressive for much the same reasons we want to regard biological systems as progressive—if they continue to solve the problems of their environment. In the case of conceptual systems, we must judge such progress by using the measuring standards provided by the intellectual environment. For example, astrology continues to exist within a rather specialized cultural niche, but its central environment is not that of contemporary science. The requirements for the survival of scientific ideas have changed over the course of ages, making new demands that astrology cannot meet. The conceptual system that at one time existed within the same intellectual milieu as ancient astronomy has now migrated to a more logically tolerant climate.

3. SRP assumes that competitive programs vie head-to-head, each claiming the same explanatory ground, with one inching out the other by a few more predictions. This is hardly ever the case. Rival theories usually have preferred evidentiary bases, which at best only partly overlap. Thus Darwin's conceptual system could explain the evolution of neuter insects, but Spencer's could more easily account for coadaptive evolution of organs. Appraisal in this situation is impossible if guided by SRP alone.

NSM recognizes, by contrast, that competing conceptual systems may occupy partly coincident but not identical problem spaces and that in such cases the preferred evidentiary ground of each—that which offers the strongest support for their particular claims—will likely differ. For the historian using NSM, this is ordinarily expected. But it does not mean that comparative evaluations are precluded. On common ground arbitration comes more easily. But when there is no overlap,

NSM directs the historian to assay the central intellectual environment of the scientific community to determine what problems it regarded as significant at the time. That Darwin could explain the wonderful instincts of neuter insects was perceived as particularly dramatic, since the complex instincts of animals constituted the province many prominent natural theologians had reserved as the most supportive of their account. NSM, in this respect, does not evaluate all explanations according to the same scale. It weighs the significance of particular explanations or predictions by their importance to the scientific community. SRP permits no such discriminative evaluation.

4. SRP should require that every emerging program be judged as falsified, since immature programs cannot usually compete with rivals in empirical coverage or predictive success. SRP fails to recognize that the existence of a conceptual system depends on the character of several intellectual and cultural environments. The continued viability of a conceptual system is, first of all, a function of the set of problems individual scientists have determined for themselves, which set of problems may only partly coincide in scope and significance with the problem set of the larger scientific community. Emerging conceptual systems, then, are rationally pursued when they solve those immediate problems. As scientists publicize their efforts, they thereby introduce their systems into a different environment. In that larger conceptual space, a particular system may compete with rivals and, depending on the terrain, will survive or perish (or mutate or hybridize or migrate).

5. SRP extirpates conceptual systems from their historical situations. The only relation that can exist between systems is that of fundamental opposition (for if their cores were logically similar, they would constitute the same program). There is no sense that conceptual systems may evolve into different systems, or branch off from a parent system, or merge with close relatives to form a hybrid system, or exist as part of the intellectual environment of other systems. These vital historical relationships are obscured by an essentialistic model of the kind SRP represents, but they are highlighted by NSM.

6. Since SRP is designed only to provide a standard of appraisal, it cannot direct the historian in an attempt to capture ideas as they are born, or to explore the immediate environment that shapes their content. Appraisal is a procedure of justification—as Lakatos construes it, "public" justification—which for the historian is only part of the story: the historian wants also to record the birth of new ideas and chronicle their growth. NSM, by contrast, functions both for appraisal and for guiding the historical reconstruction of the environment of discovery. It suggests appraisal of conceptual systems from three perspectives: the

problems of the individual scientist; the problems of the scientific community (or communities, since the scientist may be a member of more than one); and the problems of subsequent communities. Insofar as a system continues to solve the problems that the individual scientist recognizes as important, it is *rational* to pursue development of the system. (This supposes, of course, that among the particular standards an individual scientist sets for a successful solution are those few necessary criteria of consistency, empirical pertinence, and the like.) If the problems he or she resolves are also those recognized by the scientific community or subsequent communities, we can describe the system as *scientific*. And if the system adapts more effectively than rivals to newly uncovered problems in the community or subsequent communities, it is to that extent *progressive*. These evaluations can be made with some confidence, of course, only in retrospect: they are distinctively historical as opposed to merely philosophical judgments.

In regard to the context of discovery, NSM urges a reconstitution of the scientist's own beliefs, which form the most intimate environment out of which his or her ideas are generated and against which they are selected. It is that environment that shapes those perduring features of a system as it adjusts to the demands of the wider community. Close scrutiny of the private space out of which theories arise indicates to the historian, therefore, the logical structure of the ideas in question. To understand the core of a conceptual system, then, necessitates precisely the kind of investigation of an individual's beliefs that SRP neglects and that Lakatos's own instincts deny.

7. SRP stipulates that conceptual systems be judged as resolving problems only if they meet certain contemporary criteria of scientific acceptability (for example, dramatically confirmed predictions): SRP is insistently presentistic. Though there are, I believe, some few 'eternal principles' of rational discourse in empirical science (for example, logical consistency, observational pertinence, and procedures of verification and falsification), yet it is clear that different ages also invoked special standards (for example, compatibility with theological doctrine, inductive support, and axiomatic formulation), which must be considered if we are to appreciate the rational objective of scientific ideas and the conceptual forces molding them.[74] SRP ethnocentrically requires that all criteria of scientific reason conform to our own. NSM, on the other hand, establishes several standards of evaluation: those appropriate to the scientist's own conception of his or her problems; those operating

74. See Larry Laudan, *Progress and Its Problems: Toward a Theory of Scientific Growth* (Berkeley: University of California Press, 1977), pp. 128–33.

in the scientific community of the time; and those utilized by subsequent communities, including our own. In this way, scientific appraisal is contextualized and thereby made into a truly historical evaluation.

8. Finally, SRP directs the historian to separate the scientific domain from the social-psychological domain (a dislocation, incidentally, that Toulmin also requires). It treats scientists of past ages as if they were prototypes of the neo-Popperian philosopher. NSM, however, recognizes that within the private conceptual milieu of the individual scientist such fixed boundaries cannot be observed. The relations governing his or her thought constructions may be logical or psychological, sanctioned by the scientific community of the time or derived from social and religious concerns. But which of these evaluational descriptions and distinctions are to be used will depend on the standards of comparison the historian chooses. The scientist likely could not make these distinctions, nor could the scientific community. NSM, unlike SRP, directs the historian to make such discriminations against several criterial environments: the private milieu, the larger social and scientific communities, and succeeding communities. The muses of a Popperian third world have no final say here.

Conclusion: The Natural Selection Model as a Historiographic Model

Every model bears both similarities and dissimilarities to its primary analogue. I have stressed the analogies of NSM to biological selection and evolution. Of course, one may as a purely logical exercise construct a theory of scientific cognition and development which simulates the theory of biological evolution. But even if an exact fit between the formal structures of the two domains were possible, only the metaphysical mind would find this frightfully compelling. What the historian of science wants to know is how a model of this kind can be of use. Through the chapters of this volume, I have tried to demonstrate the utility of NSM, and in the previous sections, to suggest its analytic advantages. I will conclude by indicating more generally what I see as the principal historiographic merits of NSM.

1. NSM is an articulated model having definite implications, though—as with any model—craft is required for its application. That historians typically use models, I have attempted to show in the first part of this appendix. That they must use models is perhaps less obvious, but, I think, epistemologically demonstrable: without some guide to the past, we would not know what to count as significant science, as rational reflection and hypothesis building, or as the tradition of em-

pirical investigation. A model directs inquiry to the proper domains, suggests ways of analyzing the historical evidence, and leads to the construction of explanations. If a model is therefore necessary for historical work, then one that is articulated and that forces explicit and conscious application will have greater value for the historian.

2. NSM, like its biological counterpart, is flexible enough to serve as a higher-order model for more specialized theories of scientific advance. Darwin's scheme subsumed particular theories of, for instance, embryogenesis, while at the same time it became more securely anchored to its empirical bases by ties with these lower-level theories. In a similar way, NSM receives instantiation by its relation to well-founded epistemological, psychological, or social theories. For example, Darden and Maul's conception of the role of interfield theories (i.e., theories formed from parent theories in closely connected scientific fields)[75] can be interpreted in terms of NSM as a more particularized characterization of the general phenomenon of hybridization in the evolution of theoretical systems. Several such well-founded, lower-level theories can be systematized under NSM and can thereby provide a unified but highly resolved vision for the historian.

3. NSM preserves the traditional distinction between the process of discovery, when ideas are generated and generational criteria are continually adjusted, and the process of justification, when ideas are selected. The model recognizes that at times the environment in which a discovery is made will differ extensively from the environment of its justification, but that both must be scrutinized by the historian. Yet it also recognizes that the environments will often largely coincide, that the considerations promoting discovery will be similar to those serving as justificatory norms. That is to say, in the actual movement of science, the same set of criterial ideas and situations may be both generational and selective, just as in the biological sphere epistatic relations and ultimately a given environment may control the kinds of allelic alternatives available, and consequently may select a set of alleles for perpetuation once they are generated.

4. NSM guides the careful survey of central environments in which ideas have been generated and selected—those environments constituted by the specific problems of the individual scientist and his or her community. It also directs the exploration of intersecting and neighboring niches formed by other kinds of cultural concerns. The model encourages the historian to attend not only to the logic and particular

75. Lindley Darden and Nancy Maul, "Interfield Theories," *Philosophy of Science* 44 (1977): 43–64.

content of scientific theory development, but also to its psychology, sociology, economics, and politics. NSM thus protects the historian against the self-refuting reductionism of the strong version of the social-psychological model by focusing interpretive efforts on central environments. At the same time, NSM avoids the insularity of the traditional growth and revolutionary models, and opposes head-on the newer isolationism of Lakatos's SRP by building bridges to recent studies in the psychological, sociological, and wider intellectual aspects of scientific activity.

5. NSM suggests, in conformity to the data of science's history, that a science advances neither by reason of a fixed set of universal standards, as implied by the growth and revolutionary models, nor through irrational macrosaltations of radically distinct paradigms, as required by the Gestalt model. It does lead the historian to recognize that standards of scientific acceptability themselves evolve, while yet retaining some stable features; to regard the raw material of scientific evolution—its ideas—as discrete but genetically related to prior conceptual states; to hold the usual source of variability to be recombinations of ideas rather than novel mutations; and generally to assume that conceptual systems will change more slowly in some climates, more rapidly in others, but never in radically discontinuous fashion.

6. NSM makes intelligible the nonprogressive character of some conceptual systems in the history of science. The growth and revolutionary models must regard such systems as actually nonscientific, since they did not lead to modern science. The Gestalt model and the strong version of the social-psychological model must consider them as logically and scientifically indistinguishable from other competing systems. But NSM is able to construe them either as systems which failed to respond adequately to a changing intellectual environment, though they were inertially perpetuated for a period, or as systems which continued to develop, but in an environment different from that of the main stem of the scientific community. Thus, NSM does not compel the historian to ignore such systems as not being authentically scientific, or to treat them as if they were conceptually equivalent to the more direct forebears of contemporary science.

7. NSM allows the historian to achieve both a diachronic and a synchronic perspective on his or her subject. That is, at each stage in the development of a science, the historian will be prompted to isolate the selection criteria for scientific ideas, so that the value of a given conceptual system can be judged by relevant standards, and so that one is not constrained to describe something as pseudo-science, or crank science, or mysticism, simply because it does not conform to all present-day-

norms of scientific acceptability. From a diachronic point of view, the historian can discriminate patterns of early science in a way that its practitioners themselves could not, distinguishing it from religion, superstition, myth, and other human pursuits. Thus the synchronic approach permits one to understand science on its own terms, while the diachronic approach aids in determining the progenitors of more recent science.

In sum, NSM has decided advantages over the other historiographic models we have examined: it renders normative what sensitive historians do instinctively.

Appendix 2
A Defense of Evolutionary Ethics

"The most obvious, and most immediate, and most important result of the *Origin of Species* was to effect a separation between truth in moral science and truth in natural science," so concluded the historian of science Susan Cannon.[1] In Cannon's view, Darwin had demolished the truth complex that joined natural science, religion, and morality in the nineteenth century. He had shown, in Cannon's terms, "whatever it is, 'nature' isn't any good."[2] Those who attempt to rivet ethics and science together again must therefore produce a structure that can bear no critical weight. Indeed, most contemporary philosophers suspect that the original complex cracked decisively because of intrinsic logical flaws, so that any effort at reconstruction must necessarily fail. G. E. Moore believed those making such an attempt would perpetrate the "naturalistic fallacy," and he judged Herbert Spencer the most egregious offender. Spencer uncritically transformed scientific assertions of fact into moral imperatives. He and his tribe, according to Moore, fallaciously maintained that evolution, "while it shews us the direction in which we *are* developing, thereby and for that reason shews us the direction in which we *ought* to develop."[3]

Those who commit the fallacy must, it is often assumed, subvert morality altogether. Consider the self-justificatory rapacity of the Rockefellers and Morgans at the beginning of this century, men who read Spencer as the prophet of profit and preached the moral commandments of social Darwinism. Marshall Sahlins, in his *The Use and Abuse of Biology,* warns us against the most recent consequence of the fallacy,

1. Susan Cannon, *Science in Culture: The Early Victorian Period* (New York: Science History Publications, 1978), p. 276.
2. Ibid.
3. G. E. Moore, *Principia Ethica* (Cambridge: Cambridge University Press, [1903] 1929), p. 46.

the ethical and social preachments of sociobiology. This evolutionary theory of society, he finds, illegitimately perpetuates Western moral and cultural hegemony. Its parentage betrays it. It came aborning through the narrow gates of nineteenth-century laissez-faire economics: "Conceived in the image of the market system, the nature thus culturally figured has been in turn used to explain the human social order, and vice versa, in an endless reciprocal interchange between social Darwinism and natural capitalism. Sociobiology . . . is only the latest phase in this cycle."[4] An immaculately conceived nature would remain silent, but a Malthusian nature urges us to easy virtue.

The fallacy might even be thought to have a more sinister outcome. Ernst Haeckel, Darwin's champion in Germany, produced out of evolutionary theory moral criteria for evaluating human "Lebenswerth." In his book *Die Lebenswunder* (1904), he seems to have prepared instruments for Teutonic horror:

> Although the significant differences in mental life and cultural conditions between the higher and lower races of men is generally well known, nonetheless their respective *Lebenswerth* is usually misunderstood. That which raises men so high over the animals—including those to which they are closely related—and that which gives their life infinite worth is culture and the higher evolution of reason that makes men capable of culture. This, however, is for the most part only the property of the higher races of men; among the lower races it is only imperfectly developed—or not at all. Natural men (e.g., Indian Vedas or Australian negroes) are closer in respect of psychology to the higher vertebrates (e.g., apes and dogs) than to highly civilized Europeans. Thus their individual *Lebenswerth* must be judged completely differently.[5]

Here is science brought to justify the ideology and racism of German culture in the early part of this century: sinning against logic appears to have terrible moral consequences.

But was the fault of the American industrialists and German mandarins in their logic or in themselves? Must an evolutionary ethics commit the naturalistic fallacy? And is it a fallacy after all? These are questions I wish to consider in this appendix.

4. Marshall Sahlins, *The Use and Abuse of Biology* (Ann Arbor: University of Michigan Press, 1976), p. xv.

5. Ernst Haeckel, *Die Lebenswunder: Gemeinverständliche Studien öder Biologische Philosophie* (Stuttgart: Kröner, 1904), pp. 449–50.

Social Indeterminacy of Evolutionary Theory

Historians such as Richard Hofstadter have documented the efforts of the great capitalists at the turn of the century to justify their practices by appeal to popular evolutionary ideas. John D. Rockefeller, for instance, declared in a Sunday sermon that "the growth of a large business is merely a survival of the fittest." Warming to his subject, he went on: "The American Beauty rose can be produced in the splendor and fragrance which bring cheer to its beholder only by sacrificing the early buds which grow up around it. This is not an evil tendency in business. It is merely the working-out of a law of nature and a law of God."[6] More recently, however, other historians have shown how American progressives[7] and European socialists[8] made use of evolutionary conceptions to advance their political and moral programs. For instance, Enrico Ferri, an Italian Marxist writing at about the same time as Rockefeller, sought to demonstrate that "Marxian socialism . . . is only the practical and fruitful complement in social life of that modern scientific revolution, which . . . has triumphed in our days, thanks to the labours of Charles Darwin and Herbert Spencer."[9] Several important German socialists also found support for their political agenda in Darwin: Eduard Bernstein argued that biological evolution had socialism as a natural consequence;[10] and August Bebel's *Die Frau und der Sozialismus* (1879) derived the doctrine of women's liberation from Darwin's conception.[11] Rudolf Virchow had forecast such political uses of evolutionary theory when he warned the Association of German Scientists in 1877 that Darwinism led logically to socialism.[12]

6. Quoted by Richard Hofstadter in his *Social Darwinism in American Thought*, rev. ed. (Boston: Beacon, 1955), p. 45.

7. See, for instance, Robert Banister, *Social Darwinism: Science and Myth in Anglo-American Social Thought* (Philadelphia: Temple University Press, 1979). Banister traces the knotted skein of American social thought fostered by Darwin, in order to show that progressive reform was as much a consequence of Darwinism as was conservative brutality. This constitutes a major conclusion of his study (p. 11).

8. See, for example, Greta Jones, *Social Darwinism and English Thought* (Atlanta Highlands, N. J.: Humanities Press, 1980).

9. Enrico Ferri, *Socialism and Positive Science (Darwin-Spencer-Marx,* Socialist Library—1, ed. Ramsay McDonald, M.P., trans. E. Harvey (London: Independent Labour Party, 1909), p. xi.

10. Eduard Bernstein, "Ein Schüler Darwin's als Vertheidiger des Sozialismus," *Die Neue Zeit* 9 (1890–1891): 171–177.

11. August Bebel, *Die Frau und der Sozialismus,* 16th ed. (Stuttgart: Dietz, [1879] 1892).

12. Alfred Kelly, *The Descent of Darwin: the Popularization of Darwinism in Germany, 1860–1914* (Chapel Hill: University of North Carolina Press, 1981), pp. 59–69.

While Virchow might have been a brilliant medical scientist, and even a shrewd politician, his sight dimmed when he inspected the finer lines of logical relationship: he failed to recognize that the presumed logical consequence of evolutionary theory required special tacit premises imported from Marxist ideology. Add different social postulates, of the kind Rockefeller dispensed along with his dimes, and evolutionary theory would demonstrate the natural virtues of big business. Though, as I will maintain, evolutionary theory is not compatible with every social and moral philosophy, it can accommodate a broad range of historically representative doctrines. Thus in order for evolutionary theory to yield determinate conclusions about appropriate practice, it requires a mediating social theory to specify the units and relationships of concern. It is therefore impossible to examine the 'real' social implications of evolutionary theory without the staining fluids of political and social values. The historical facts thus stand forth: an evolutionary approach to the moral and social environment does not inevitably support a particular ideology.

Those apprehensive about the dangers of the naturalistic fallacy may object, of course, that just this level of indeterminacy—the apparent ability to give witness to opposed moral and social convictions—shows the liability of any wedding of morals and evolutionary theory. But such an objection ignores two historical facts: first, that moral barbarians have frequently defended heinous behavior by claiming that it was enjoined by holy writ and saintly example, so no judgment about the viability of an ethical system can be made simply on the basis of the policies that it has been called upon to support; and second, that several logically different systems have traveled under the name "evolutionary ethics," so one cannot condemn all such systems simply because of the liabilities of one or another of them. In other words, we must examine particular systems of evolutionary ethics to determine whether they embody any fallacies and to discover what kinds of acts they sanction.

In the preceding chapters I have described the moral systems of several evolutionary theorists and have attempted to assess the logic of those systems, so I need not repeat that in this appendix. Rather I will draw on those systems to develop the outline of an evolutionary conception of morals, one that I believe escapes the usual objections to this approach. In what follows, I will first summarize Darwin's theory of morals, which provides the essential structure for the system I wish to advance, and then compare it to a recent and vigorously decried descendant, the ethical ideas formulated by Edward Wilson in his books *Sociobiology* (1975) and *On Human Nature* (1978). Next I will describe my own revised version of an evolutionary ethics. Then I will consider the most pressing objections brought against an ethics based on evolutionary theory. Finally, I will show how the proposal I have in mind escapes these objections.

Constructing an Evolutionary Ethics

Darwin's Moral Theory

In the *Descent of Man,* Darwin insisted that the moral sense—the motive feeling which fueled intentions to perform altruistic acts and which caused pain when duty was ignored—be considered a species of social instinct. He conceived social instincts as the bonds forming animal groups into social wholes. Social instincts comprised behaviors that nurtured offspring, secured their welfare, produced cooperation among kin, and organized the group into a functional unit. The principal mechanism of their evolution, in Darwin's view, was community selection: that kind of natural selection operating at levels of organization higher than the individual. The degree to which social instincts welded together a society out of its striving members depended on the species and its special conditions. Community selection worked most effectively among the social insects, but Darwin thought its power was in evidence among all socially dependent animals, including that most socially advanced creature, man.

In the *Descent,* Darwin elaborated a conception of morals that he first outlined in the late 1830s. He erected a model depicting four overlapping stages in the evolution of the moral sense. In the first, well-developed social instincts would evolve to bind protomen into social groups, that is, into units that might continue to undergo community selection. During the second stage, creatures would develop sufficient intelligence to recall past instances of unsatisfied social instincts. The primitive anthropoid that abandoned its young because of a momentarily stronger urge to migrate might, upon brutish recollection of its hungry offspring, feel again the sting of unfulfilled social instinct. This, Darwin contended, would be the beginning of conscience. The third stage in the evolution of the moral sense would arrive when social groups became linguistically competent, so that the needs of individuals and their societies could be codified in language and easily communicated. In the fourth stage, individuals would acquire habits of socially approved behavior that would direct the moral instincts into appropriate channels—they would learn how to help their neighbors and advance the welfare of their group. So what began as crude instinct in our predecessors, responding to obvious perceptual cues, would become, in Darwin's construction, a moral motive under the guidance of social custom and intelligent decision. As the moral sense evolved, so did a distinctively human creature.

Under prodding from his cousin Hensleigh Wedgwood, Darwin expanded certain features of his theory in the second edition of the *Descent*. He made it clear that during the ontogenesis of conscience, individuals learned to avoid the nagging persistence of unfulfilled social instinct by implicitly formulating rules about appropriate conduct. These rules would take into account not only the general urgings of instinct, but also the particular ways a given society might sanction their satisfaction. Such rules, Darwin thought, would put a rational edge on conscience and in time would become the publicly expressed canons of morality. With the training of each generation's young, these moral rules would recede into the very bones of social habits and customs. Darwin, a child of his scientific time, also believed that such rational principles, first induced from instinctive reactions, might be transformed into habits, and then infiltrate the hereditary substance to augment and reform the biological legacy of succeeding generations.

Darwin's theory of moral sense was taken by some of his reviewers to be merely a species of utilitarianism, one that gave scientific approbation to the morality of selfishness. Darwin took exception to such judgments. He thought his theory was completely distinct from that of Bentham and Mill. Individuals, he emphasized, acted instinctively to avoid vice and seek virtue without any rational calculations of benefit. Pleasure may be our sovereign mistress, as Bentham painted her, but some human actions, Darwin insisted, were indifferent to her allure. Pleasure was neither the usual motive nor the end of moral acts. Rather moral behavior, arising from community selection, was ultimately directed to the vigor and health of the group, not to the pleasures of its individual members. This meant, according to Darwin, that the criterion of morality—that highest principle by which we judged our behavior in a cool hour—was not the general happiness, but the general good, which he interpreted as the welfare and survival of the group. This was no crude utilitarian theory of morality dressed in biological guise. It cast moral acts as intrinsically altruistic.

Darwin noticed, of course, that people sometimes adopted the moral patterns of their culture for somewhat lower motives: implicitly they formed contracts to respect the person and property of others, provided they received the same consideration themselves; they acted, in our terms, as reciprocal altruists. Darwin also observed that his fellow creatures glowed or smarted under the judgments of their peers; accordingly they might at times practice virtue in response to public praise rather than to the inner voice of austere duty. Yet individuals did harken to that voice, which they understood to be authoritative, if not always coercive.

From the beginning of his formulation of a moral theory in the late 1830s, Darwin recognized a chief competitive advantage of his approach. He could explain what other moralists merely assumed: he could explain how the moral criterion and the moral sense were linked. Sir James Mackintosh, from whom Darwin borrowed the basic framework of his moral conception, declared that the *moral sense* for right conduct had to be distinguished from the *criterion* of moral behavior. We instinctively perceive murder as vile, but in a cool moment of rational evaluation, we can also weigh the disutility of murder. When a man jumps into the river to save a drowning child, he acts impulsively and without deliberation, while those safely on shore may rationally evaluate his deed according to the criterion of virtuous behavior. Mackintosh had no satisfactory account of the usual coincidence between motive and criterion. He could not easily explain why impulsive actions might nevertheless be what moral deliberation would recommend. Darwin believed he could succeed where Mackintosh failed; he could provide a perfectly natural explanation of the linkage between the moral motive and the moral criterion. Under the aegis of community selection, men in social groups evolved sets of instinctive responses to preserve the welfare of the community. This common feature of acting for the community welfare would then become, for intelligent creatures who reacted favorably to the display of such moral impulses, an inductively derived but dispositionally encouraged general principle of appropriate behavior. What served nature as the criterion for selecting behavior became the standard of choice for her creatures as well.

Wilson's Moral Theory

In his book *On Human Nature,* Edward Wilson elaborated a moral theory that he had earlier sketched in the concluding chapter of his massive *Sociobiology.* Though Wilson's proposals bear strong resemblance to Darwin's own, the similarity appears to stem more from the logic of the interaction of evolutionary theory and morals than from Wilson's intimate knowledge of his predecessor's ethical views. Wilson, like Darwin, portrays the moral sense as the product of natural selection operating on the group. In light of subsequent developments in evolutionary theory, however, he more carefully specifies the unit of selection as the kin, the immediate and the more remote. The altruism evinced by lower animals for their offspring and immediate relatives can be explained, then, by employing the Hamiltonian version—that is, the kin-selection version—of Darwin's original concept of community selection. Like Darwin, Wilson also suggests that the forms of altruistic behavior are constrained by the cultural traditions of particular societies. But unlike Darwin, Wilson regards this "hard-core" altruism, as he calls it, to be insufficient, even detrimental, to the organization of societies larger than kin groups, since such altruism does not reach beyond blood relatives. As a necessary compromise between individual and group welfare, men have adopted implicit social contracts; they have become reciprocal altruists.

Wilson calls this latter kind of altruism, which Darwin also recognized, "soft-core," since it is both genetically and psychologically selfish: individuals agree mutually to adhere to moral rules in order that they may secure the greatest amount of happiness possible. Though Wilson deems soft-core altruism to be basically a learned pattern of behavior, he conceives it as "shaped by powerful emotional controls of the kind intuitively expected to occur in its hardest forms."[13] He appears to believe that the "deep structure" of moral rules, whether hard-core or soft-core, expresses a genetically determined disposition to employ rules of the moral form. In any case, the existence of such rules ultimately can have only a biological explanation, for "morality has no other demonstrable ultimate function" than "to keep human genetic material intact."[14]

Wilson's theory has recently received vigorous defense from Michael Ruse. Ruse endorses Wilson's evaluation of the ethical as well as the biological merits of soft-core altruism: "Humans help relatives without hope or expectation of the ethical return. Humans help nonrelatives insofar as and only insofar as they anticipate some return. This may not be an anticipation of immediate return, but only a fool or a saint (categories often linked) would do something absolutely for nothing."[15] Ruse argues that principles of reciprocal altruism have become inbred in the human species and manifest themselves to our consciousness in the form of feelings. The common conditions of human evolution mean that most human beings share feelings of right and wrong. Nonetheless ethical standards, according to Ruse, are relative to our evolutionary history. He believes we cannot justify moral norms through other means: "All the justification that can be given for ethics lies in our evolution."[16]

13. Edward Wilson, *On Human Nature* (Cambridge: Harvard University Press, 1978), p. 162.
14. Ibid., p. 167.
15. Michael Ruse, "The Morality of the Gene," *Monist* 67 (1984): 171.
16. Ibid., p. 177.

A Revised Theory of Evolutionary Ethics

The theory of evolutionary ethics I wish to advocate is based on Darwin's original conception and has some similarities to Wilson's proposal. It is a theory, however, which augments Darwin's and differs in certain respects from Wilson's. For convenience I will refer to it as the revised version (RV). RV has two distinguishable parts, a speculative theory of human evolution and a more distinctively moral theory based on it. Evolutionary thinkers attempting to account for human mental, behavioral, and, indeed, anatomical traits usually spin just-so stories, projective accounts that have more or less theoretical and empirical support. Some will judge the evidence I suggest for my own tale too insubstantial to bear much critical weight. My concern, however, will not be to argue the truth of the empirical assertions but to show that if those assertions are true they adequately justify the second part of RV, the moral theory. My aim, then, is fundamentally logical and conceptual: to demonstrate that an ethics based on presumed facts of biological evolution need commit no sin of moral logic but rather can be justified by using those facts and the theory articulating them.

RV supposes that a moral sense has evolved in the human group. "Moral sense" names a set of innate dispositions that in appropriate circumstances move the individual to act in specific ways for the good of the community. The human animal has been selected to provide for the welfare of its own offspring (e.g., by specific acts of nurture and protection); to defend the weak; to aid others in distress; and generally to respond to the needs of community members. The individual must learn to recognize, for instance, what constitutes more subtle forms of need and what specific responses might alleviate distress. But, so RV proposes, once different needs are recognized, feelings of sympathy and urges to remedial action will naturally follow. These specific sympathetic responses and pricks to action together constitute the core of the altruistic attitude. The mechanism of the initial evolution of this attitude I take to be kin selection, aided perhaps by group selection on small communities.[17] Accordingly, altruistic motives will be strongest when be-

17. The usual models of group selection assume that individual selection and group selection work at cross-purposes, that, for instance, the individual must pay a high price for altruistic behavior (e.g., bees' disemboweling themselves by stinging enemies; risking one's life to save a drowning child, etc.). But in most familiar cases, individuals perform altruistic acts at little practical cost. In a hostile environment, those small tribal groups populated by altruists and cooperators would have a decided advantage. Cheating would not likely become widespread, since the gain would be quite small and the cost quite high (e.g., ostracism of the individual or death of the tribe). Under such circumstances, group selection, especially on tribes laced with relatives, might well become a force to install

havior is directed toward immediate relatives. (Parents, after all, are apt
to sacrifice considerably more for the welfare of their children than for
complete strangers.) Since natural selection has imparted no way for
human beings or animals to perceive blood kin straight off, a variety of
perceptual cues have become indicators of kin. In animals it might be
smells, sounds, or coloring that serve as the imprintable signs of one's
relatives. With human beings, extended association during childhood
seems to be a strong sign. Maynard Smith, who has taken some exception to the evolutionary interpretation of ethics, nevertheless admits he
changed his mind about the incest taboo.[18] The reasons he offers are:
(1) the deleterious consequences of inbreeding; (2) the evidence that
even higher animals avoid inbreeding; and (3) the phenomenon of kibbutz children not forming sexual relations. Children of the kibbutz appear to recognize each other as 'kin,' and so are disposed to act for the
common good by shunning sex with each other.

On the basis of such considerations, RV supposes that early human
societies consisted principally of extended kin groups, of clans. Such
clans would be in competition with others in the geographical area, and
so natural selection might operate on them to promote a great variety
of altruistic impulses, all having the ultimate purpose of serving the
community good.

Men are cultural animals. Their perceptions of the meaning of behaviors, their recognition of 'brothers' or 'sisters,' their judgments of what
acts would be beneficial in a situation—all of these are interpreted according to the traditions established in the history of particular groups.
Hence, it is no objection to an evolutionary ethics that in certain
tribes—whose kin systems only loosely recapitulate biological relations—the natives may treat with extreme altruism those who are cultural but not biological kin.[19] In a biological sense, this may be a mistake; but on average the cultural representation of kin will serve nature's
ends.

---

virtuous behavior. For an analysis of the problematic assumptions of most group selection
models, see Michael Wade, "A Critical Review of the Models of Group Selection," *Quarterly Review of Biology* 53 (1978): 101–114. For some suggestions as to other evolutionary
strategies that might have installed altruistic impulses, see Donald Campbell, "Social Morality Norms as Evidence of Conflict Between Biological Human Nature and Social System Requirements," in *Morality as a Biological Phenomenon,* ed. Gunther Stent (Berkeley:
University of California Press, 1980), pp. 76–79.

18. John Maynard Smith, "The Concepts of Sociobiology," in *Morality as a Biological
Phenomenon,* pp. 21–30.

19. This is largely the objection of Sahlins (*Use and Abuse of Biology*) to the sociobiology
of human behavior.

RV supposes that the moral attitude will be informed by an evolving intelligence and cultural tradition. Nature demands we protect our brother and sister, but we must learn who they are. During human history, evolving cultural traditions may translate "community member" as "red Sioux," "black Mau Mau," or "white Englishman," and the "community good" as "sacrificing to the gods," "killing usurping colonials," or "constructing factories." But as men become wiser and old fears and superstitions fade, they may come to see their brothers and sisters in every human being and to discover what really does foster the good of all people.

RV departs from Wilson's sociobiological ethics and Ruse's defense of it, since they both regard reciprocal altruism as the chief sort, and "keeping the genetic material intact" as the ultimate justification.[20] Reciprocal altruism, as a matter of fact, may operate more widely than the authentic kind; it may even be more beneficial to the long-term survival of human groups. But this does not elevate it to the status of the highest kind of morality, though Wilson and Ruse suggest it does. And while the evolution of authentically altruistic motives may serve to perpetuate genetic stock, that justifies altruistic behavior only in an empirical sense, not in a moral sense. That is, the biological function of altruism may be understood (and thus justified) as a consequence of natural selection, but so may aggressive and murderous impulses. Authentic altruism requires a moral justification. Such a justification, which I will undertake below, will show it to be morally superior to contract altruism.

The general character of RV may now be a little clearer. Its further features can be elaborated in a consideration of the principal objections to evolutionary ethics.

Objections to Evolutionary Ethics

Systems of evolutionary ethics, of both the Darwinian and the Wilsonian varieties, have attracted objections of two distinct kinds: those challenging their adequacy as biological theories and those their adequacy as moral theories. Critics focusing on the biological part have complained that complex social behavior does not obviously fall under the direction of any genetic program, and indeed, that the conceptual structure of evolutionary biology prohibits the assignment of any behavioral pattern exclusively to the genes. Critics certainly reject the belief that behaviors which must be responsive to complex and often highly abstract circumstances are biologically determined.[21] A present-

20. Wilson, *On Human Nature*, p. 167.
21. See, for example Stephen Jay Gould, "Biological Potential vs. Biological Determin-

day opponent of Darwin's particular account might also insist that the kind of group selection his theory requires has been denied by many recent evolutionary theorists,[22] and that even among those convinced of group selection (e.g., Wilson), a number doubt that it has played a significant role in human evolution. And if kin selection, instead of group selection of unrelated individuals, is proposed as the source of altruism in humans, a persistent critic might contend that human altruistic behavior is often extended to nonrelatives. Hence kin selection cannot be the source of the ethical attitude.[23]

Within the biological community, the issues raised by these objections continue to be strenuously debated. So, for instance, some ethologists and sociobiologists would point to very intricate animal behaviors that are nonetheless highly heritable.[24] And Ernst Mayr has proposed that complex instincts can be classified as exhibiting a relatively more open or a more closed genetic program: the latter remains fairly impervious to shifting environments, while the former responds more sensitively to changing circumstances.[25] Further, various animal species show social hierarchies of amazing complexity (e.g., societies of lowland baboons) and display repertoires of instinctive behaviors whose values are highly context dependent (e.g., the waggle dance of the honeybee, which specifies the direction and distance of food.) This suggests the likelihood that instinctual and emotional responses in humans can be triggered by subtle interpretive perceptions (e.g., the survival responses of fear and flight can be activated by a stranger who points a gun at you in a Chicago back alley). Anthropological studies, moreover, have discovered similar patterns of moral development across diverse cultures. The similarity of patterns could be explained, at least in part, as the result of a biologically based program determining the sequence

---

ism," in his *Ever Since Darwin* (New York: Norton, 1977), pp. 251–59; Richard Burian, "A Methodological Critique of Sociobiology," in *The Sociobiology Debate*, ed. Arthur Caplan (New York: Harper, 1978); and Richard Lewontin, Stephen Rose, and Leon Kamin, *Not in Our Genes: Biology, Ideology, and Human Nature* (New York: Pantheon, 1984), pp. 265–90.

22. George Williams, *Adaptation and Natural Selection* (Princeton: Princeton University Press, 1966), pp. 92–124.

23. See, for example Ruth Mattern, "Altruism, Ethics, and Sociobiology," in *The Sociobiology Debate*; and Lewontin, Rose, and Kamin, *Not in Our Genes,* p. 261.

24. Edward Wilson, *Sociobiology: the New Synthesis* (Cambridge: Harvard University Press, 1975); and Irenaus Eibl-Eibesfeldt, *Ethology: the Biology of Behavior* (New York: Holt, Rinehart and Winston, 1970).

25. Ernst Mayr, "Behavior Programs and Evolutionary Strategies," in his *Evolution and the Diversity of Life* (Cambridge: Harvard University Press, 1976).

of moral stages that individuals in conventional environments follow.[26] Further, recent impressive experiments have shown that group selection may well be a potent force in evolution.[27] Finally, some anthropologists have found kin selection to be a powerful explanation of social behavior in primitive tribes.[28] How these issues will eventually fall out, however, is not my immediate concern, since only developing evolutionary theory can properly arbitrate them. At this time we can say, I believe, that the objections based on a particular construal of evolution seem not to be fatal to an evolutionary ethics—and this concession suffices for my purposes.

Concerning the other class of objections, those directed to the distinctively moral character of evolutionary ethics, resolution does not have to wait, for the issues are factually mundane, though conceptually tangled. Against the moral objections, I will attempt to show that the evolutionary approach to ethics need abrogate no fundamental meta-ethical principles. For the sake of getting at the conceptual difficulties, I will assume that the biological objections concerning group and kin selection and an evolutionary account of complex social behavior have been eliminated. With this assumption, I can then focus on the question of the moral adequacy of an evolutionary ethics.

The objections to the adequacy of the distinctively moral component of evolutionary ethics themselves fall into two classes: objections to the entire framework of evolutionary ethics and objections based on the logic or semantics of the conceptual relations internal to the framework. For convenience I will refer to these as *framework questions* and *internal questions*. Questions concerning the framework and the internal field overlap, since some problems will be transitive—that is, a faulty key principle may indict a whole framework. The interests of clarity may, however, be served by this distinction. Another helpful distinction is that between ethics as a descriptive discipline and ethics as an imperative discipline. The first tries to give an accurate account of what ethical principles people actually use and the origin of these principles: this may be regarded as a part of social anthropology. The second, ethics as an imperative discipline, urges that its principles be considered the ethi-

26. Wilson, *Sociobiology,* pp. 562–63.

27. Michael Wade, "An Experimental Study of Group Selection," *Evolution* 31 (1977): 134–53; and "Group Selection among Laboratory Populations of Tribolium," *Proceedings of the National Academy of Sciences* 73 (1976): 4604–7.

28. See the collection of studies in Napoleon Chagnon and William Irons, eds., *Evolutionary Biology and Human Social Behavior: An Anthropological Perspective* (North Scituate, Mass.: Duxbury, 1979).

cally adequate principles; it enjoins their adoption. The former kind of theory will require *empirical justification,* the latter *moral justification.*

Let me first consider some important internal challenges to both the empirical and the moral justification of evolutionary ethics. It has been charged, for instance, that the concept of altruism when used to describe a soldier bee sacrificing its life for the nest has a different meaning than the nominally similar concept that describes the action of a human soldier who sacrifices his life for his community.[29] It would be illegitimate, therefore, to base conclusions about human altruism on the evolutionary principles governing animal altruism. Some critics further maintain that the logic of the concept's role in sociobiology must differ from its role in any adequate moral system, since the biological use implies genetic selfishness, while the moral use implies unselfishness.[30] I do not believe these are lethal objections. First, the term "altruism" does not retain a univocal meaning even when used to describe various human actions. Its semantic role in a description of parents' saving money for their children's education surely differs from its role in a description of a stranger's jumping into a river to save a drowning child. Nonetheless the many different applications to human behavior and the several applications to animal behavior intend to pick out a common feature, namely, that the action is directed to the welfare of the recipient and costs the agent some good for which reciprocation would not normally be expected. Let us call this "action altruism." We might then wish to extend, as sociobiologists are wont to do, the description "altruistic" to the genes that prompt such action, but such extension would be based on causal analogy only (as when we call Tabasco sauce "hot"). Hence the explanation of human or animal action altruism by reference to "selfish genes" involves no contradiction; for the concept of genetic selfishness is antithetic neither to action altruism—since it is not applied to the same category of object—nor to genetic altruism, for these concepts are implicitly defined to be compatible by sociobiologists. The real issue in applying the concepts of (action) altruism and (action) selfishness is whether the agent is motivated principally to act

29. See Burian, "A Methodological Critique of Sociobiology"; Mattern, "Altruism, Ethics, and Sociobiology"; and Joseph Alper, "Ethical and Social Implications," in *Sociobiology and Human Nature,* ed. M. Gregory, A. Silvers, and D. Sutch (San Francisco: Jossey-Bass, 1978).

30. Playing on the apparent reduction of altruistic behavior to genetic selfishness and then to selfishness simply, Lewontin, Rose, and Kamin (*Not in Our Genes,* p. 264) complain: "by emphasizing that even altruism is the consequence of selection for reproductive selfishness, the general validity of individual selfishness in behaviors is supported. . . . Sociobiology is yet another attempt to put a natural scientific foundation under Adam Smith."

for the good of another or for the good of self. Of course one could, as a matter of linguistic punctiliousness, refrain from describing any animal behavior or its genetic substrate as "altruistic." The problem would then cease to be semantic and would again become one of the empirical adequacy of evolutionary biology to account for similar patterns of behavior in men and animals.

Though some varieties of utilitarianism define behavior as morally good if it has certain consequences, the evolutionary ethics that I am advocating regards an action as good only if it is intentionally performed from a certain kind of motive and can be justified by that motive. I will assume as an empirical postulate that the motive has been established by community or kin selection. The altruistic motive encourages the agent to attend to the needs of others, such needs as either biology or culture (or both) interpret for the agent. Aristotelian-Thomistic ethics, as well as the very different Kantian moral philosophy, holds that action from appropriate motives, not action having desirable consequences, is necessary to render an act moral. The commonsense moral tradition sanctions the same distinction. That tradition prompts us, for example, to judge those Hippocratic physicians who risked their lives during the Athenian plague as moral heros—even though their therapies as often as not hastened the deaths of their patients. The Hippocratics acted from altruistic motives—ultimately to advance the community good (i.e., the health and welfare of the group), proximately to do so through certain actions directed, unfortunately, by invincibly defective medical knowledge.

This nonconsequentialist feature of RV leads, however, to another important internal objection. The nonconsequentialism suggests that either animal altruism does not stem from altruistic motives, or that animals are moral creatures (since moral creatures are those who act from moral motives).[31] Yet if animal altruism does not arise from altruistic motives and thus is only nominally similar to human altruism, then there is no reason to postulate community selection as the source of both, and we therefore cannot use evidence from animal behavior to help establish RV. Thus either the evolutionary explanation of morals is deficient or animals are moral creatures. But no system that renders animals moral creatures is acceptable. Hence the evolutionary explanation is logically deficient.

To answer this objection we must distinguish between altruistic motives and altruistic intentions. Though my intention is to write a book about evolutionary theories of mind, my motive may be either money,

31. Mattern, "Altruism, Ethics, and Sociobiology."

prestige, professional advancement, or something of a higher nature. Human beings form intentions to act for reasons (i.e., motives), but animals presumably do not. We may then say that though animals may act from altruistic motives, they can neither form the intention of doing so, nor can they justify their behavior in terms of its motive. Hence they are not moral creatures. Two conditions, then, are necessary and sufficient for denominating an action moral: the agent performs the action from an altruistic (or moral) motive; and the agent intends to act from that motive. These two conditions imply the agent could justify his action by appeal to the motive.

The distinction between motives and intentions, while it has the utility of overcoming the objection mentioned, seems warranted for other reasons as well. Motives consist of cognitive representations of goals or goal-directed actions coupled with positive attitudes about the goal (e.g., the Hippocratic physician wanted to reinstate a humoral balance so as to effect a cure). Appeal to the agent's motives and his beliefs about the means to attain desired goals (e.g., the physician believed continued purging would produce a balance) provides an explanation of action (e.g., the physician killed his patient by producing a severe anemia). Intentions, on the other hand, should not be identified with motives or beliefs, though they operate on both. 'To intend' is to perform a conscious act that recruits motives and beliefs to guide behavior (e.g., the physician, motivated by commitment to the Oath, intended to cure his patient through purging). Intentions alone may not adequately explain action (e.g., the physician killed his patient because he intended to cure him!). Intentions, however, confer moral responsibility, while mere motives only furnish a necessary condition for the ascription of responsibility. To see this, consider Sam, a man who killed his mistress by feeding her spoiled pâté. Did he murder her? Before the court, Sam planned to plead that yes, he had the motive (revenge for her infidelity) and yes, he knew spoiled pâté would do it, but that in giving her the pâté he nonetheless did not intend to kill her. He thought he could explain it by claiming that his wife put him in an hypnotic trance that suppressed his moral scruples. Thus, though he acted on his desire for revenge, he still did not intend to kill his erstwhile lover. Sam's lawyer suggested a better defense. He should plead that though he had the motive and knew that spoiled pâté would do her in, yet he did not intend to kill her since he did not know this particular pâté was spoiled. The moral of this sordid little example is threefold. First, simply that motives differ from intentions. Second, that for moral responsibility to be attributed, motives must be not only marshalled (as emphasized by the second defense), but consciously marshalled (as emphasized by the first defense). And finally, that conscious marshalling of motives and beliefs allows a justification of action (or in their absence, an excuse) by the agent (as emphasized by both defenses).

The charge that RV would make animals moral creatures is thus overturned. For we assume that animals, though they may act from altruistic motives, cannot intend to do so. Nor can they justify or defend their behavior by appeal to such motives. Generally we take a moral creature to be one who can intend action and justify it.

In addition to these several objections to specific features of the internal logic and coherence of RV (and other similar systems of evolutionary ethics), one important objection attempts to indict the whole framework by pointing out that the logic of moral discourse implies the agent can act freely. But if evolutionary processes have stamped higher organisms with the need to serve the community good, this suggests that ethical decisions are coerced by irrational forces—that men, like helpless puppets, are jerked about by strands of their DNA. There are, however, four considerations that should defuse the charge that an evolutionary construction of behavior implies the denial of authentic moral choice. First, we may simply observe that the problem of compatibility of moral discourse and scientific discourse (which presumes, generally, that every event has a cause, at least at the macroscopic level) is hardly unique to evolutionary ethics. Almost every ethical system explicitly or implicitly recognizes the validity of causal explanations of human behavior (which explanatory efforts imply the principle that every event has a cause). Hence, this charge is really a challenge not to an evolutionary ethics but to the possibility of meaningful ethical discourse quite generally. Nonetheless, let us accept the challenge and move to a second consideration. Though evolutionary processes may have resulted in sets of instinctual urges (e.g., to nurture children, to alleviate obvious distress, etc.) that promote the welfare of the community, is this not a goal at which careful ethical deliberation might also arrive? Certainly many moral philosophers have thought so. Moreover an evolutionary account of why men generally act for the community good does not invalidate a logically autonomous argument that concludes this same standard is the ultimate moral standard. The similar case of mathematical reasoning is instructive. Undoubtedly we have been naturally selected for an ability to recognize the quantitative aspects of our environment. Those protomen who failed to perform simple quantitative computations (such as determining the closest tree when the sabertooth charged) have founded lines of extinct descendants. A mathematician who concedes that her brain has been designed, in part at least, to make quantitative evaluations need not discard her mathematical proofs as invalid, based on a judgment coerced by an irrational force. Nor need the moralist.[32] Third, the standard of community good must be intelligently applied. Rational deliberation must discover what actions in contingent circumstances lead to enhancing the community welfare. Such choices are not automatic but the result of improvable reason. Finally, the evolutionary perspective indicates that external forces do not conspire to wrench moral acts from a person. Rather, man is ineluctably a moral being. Aristotle believed that men were by nature moral creatures. Darwin demonstrated it.

I wish now to consider one final kind of objection to an evolutionary ethics. It requires special and somewhat more extended treatment, since its force and incision have been thought to deliver the coup de grace to all Darwinizing in morals.

The "Naturalistic Fallacy" Describes No Fallacy

RV Escapes the Usual Form of the Naturalist Fallacy

G. E. Moore was the first formally to charge evolutionary ethicians—particularly Herbert Spencer—with committing the naturalistic fallacy.[33] The substance of the charge had been previously leveled against Spencer by both his old friend Thomas Huxley and his later antagonist Henry Sidgwick.[34] Many philosophers subsequently have endorsed the complaint against those who would make the Spencerian turn. Bertrand Russell, for instance, thumped it with characteristic élan:

> If evolutionary ethics were sound, we ought to be entirely indifferent as to what the course of evolution may be, since whatever it is is thereby proved to be the best. Yet if it should turn out that the Negro or the Chinaman was able to oust the European, we should cease to have any admiration for evolution; for as a matter of fact our preference of the European to the Negro is wholly independent of the European's greater prowess with the Maxim gun.[35]

Anthony Flew glosses this passage with the observation that "Russell's argument is decisive against any attempt to define the ideas of right and wrong, good and evil, in terms of a neutrally scientific notion of evolu-

---

32. C. Fried, "Biology and Ethics: Normative Implications," in *Morality as a Biological Phenomenon*.

33. Moore, *Principia Ethica*, pp. 46–58.

34. Thomas Huxley, "Evolution and Ethics," in *Collected Essays* (New York: D. Appleton, 1896–1902), pp. 46–116; and Henry Sidgwick, *Lectures on the Ethics of T. H. Green, Mr. Herbert Spencer, and J. Martineau* (London: Macmillan, 1902), p. 219. See also chapter 7.

35. Quoted by Anthony Flew, *Evolutionary Ethics* (London: Macmillan, 1967), p. 44.

tion."[36] He continues in his tract *Evolutionary Ethics* to pinpoint the alleged fallacy: "For any such move to be sound [i.e., 'deducing ethical conclusions directly from premises supplied by evolutionary biology'] the prescription in the conclusion must be somehow incapsulated in the premises; for, by definition, a valid deduction is one in which you could not assert the premises and deny the conclusion without thereby contradicting yourself."[37] Flew's objection is, of course, that one could jolly well admit all the declared facts of evolution, but still logically deny any prescriptive statement purportedly drawn from them.

This objection raises two questions for RV: Does it commit the fallacy as here expressed? And is it a fallacy after all? I will endeavor to show that RV does not commit this supposed fallacy, but that even if at some level it derives norms from facts, it would nevertheless escape unscathed, since the "naturalist fallacy" describes no fallacy.

There are two ways in which evolutionary ethics has been thought to commit the naturalist fallacy.[38] Some versions of evolutionary ethics have represented the current state of our society as ethically sanctioned, since whatever has evolved is right. Haeckel believed, for instance, that evolution had produced a higher German culture which could serve as a norm for judging the moral worth of men of inferior cultures. Other versions of evolutionary ethics have identified certain long-term trends in evolution, which they deem ipso facto good. Julian Huxley, for example, held that efforts at greater social organization were morally sanctioned by the fact that a progressive integration has characterized social evolution.[39] But RV (and its parent, Darwin's original moral theory) prescribes neither of these alternatives. It does not specify a particular social arrangement as being best; rather it supposes that men will seek the arrangement that appears best to enhance the community good. The conception of what constitutes such an ideal pattern will change through time and over different cultures. Nor does this theory isolate a particular historical trend and enshrine that. During long periods in our

36. Ibid., p. 45.
37. Ibid., p. 47.
38. Michael Ruse, in *Sociobiology: Sense or Nonsense?* (Dordrecht: Reidel, 1979), agrees that any evolutionary ethics must commit the naturalist fallacy, and admits that the two characteristics mentioned in the following text produce the most potent objections to evolutionary ethics. I am not sure whether this troubles Ruse or not. For a list of the other ways in which fallacious arguments from evolutionary facts to ethical imperatives might be made, see Donald Campbell, "Social Morality Norms as Evidence of Conflict Between Biological Human Nature and Social System Requirements," in *Morality as a Biological Phenomenon*, pp. 70–72.
39. Julian Huxley, "Evolutionary Ethics," in *Touchstone for Ethics, 1893–1943* (New York: Harper, 1947), p. 136.

prehistory, for instance, it might have been deemed in the community interest to sacrifice virgins, and this ritual might in fact have contributed to community cohesiveness and thus have been of continuing evolutionary advantage. But RV does not therefore sanction the sacrifice of virgins, but only acts that, on balance, appear to be conducive to the community good. As the rational capacities of human beings have evolved, the ineffectiveness of such superstitious behavior has become obvious. The theory maintains that the criterion of morally approved behavior will remain constant, while the conception of what particular acts fall under the criterion will continue to change. RV, therefore, does not derive ethical imperatives from evolutionary facts in the usual way.

But does RV derive ethical norms from evolutionary facts in some way? Unequivocally, yes. But to see that this involves no logically or morally fallacious move requires that we first consider more generally the roles of factual propositions in ethics.

Empirical Hypotheses in Ethics

Empirical considerations impinge upon ethical systems both as *framework* assumptions and as *internal* assumptions. In analyzing ethical systems, therefore, framework questions or internal questions may arise. Framework questions, as indicated above, concern the relationships of the ethical system to other conceptual systems and, via those other systems, to the worlds of men and nature. They stimulate such worries as: Can the ethical system be adopted by men in our society? How can such a moral code be justified? Must ethical systems require rational deliberation before an act can be regarded as moral? Internal questions concern the logic of the moral principles and the terms of discourse of a given ethical system. They involve such questions as: Is abortion immoral in this system? What are the principles of a just war in this system? Some apparently internal questions—such as, What is the justification for fostering the community good?—are really framework questions—to wit, How can this system, whose highest principle is "Foster the community good," be justified? The empirical ties an ethical framework has to the worlds of men and nature are transitive: they render the internal principles of the system ultimately dependent upon empirical hypotheses and assumptions.

Every ethical system fit for human beings includes at least three kinds of empirical assumption (or explicit empirical hypothesis) regarding frameworks and, transitively, internal elements. First, every ethical system recommended for human adoption makes certain framework assumptions about man's nature, that is, about the kind of creature man is such that he can heed the commands of the system. Even the austere ethics of Kant supposes human nature to be such, for instance, that intellectual intuitions into the noumenal realm are foreclosed; that behavior is guided by maxims; that human life is finite; and that men desire immortality. An evolutionary ethics also forms empirical suppositions about human nature, suppositions extracted from evolutionary theory and its supporting evidentiary base. Consequently, no objection to RV (or any evolutionary ethics) can be made on the grounds that it requires empirical assumptions—all ethical systems do.

A second level of empirical assumption is required of a system designed for culture-bound human nature: connections must be forged between the moral terms of the system—"goods," "the highest good," and the like—and the objects, events, and conditions realized in various human societies. What are goods (relative and ultimate) in one society (e.g., secular Western society) may not be in another (e.g., a community of Buddhist monks). In one sense these are internal questions of how individual terms of the system are semantically related to characterizations of a given society's attitudes, observations, and theoretical knowledge (e.g., the 'virtue' of sacrificing virgins, since that act produces life-giving crops; the 'evil' of thermonuclear war, since it will likely destroy all human life; etc.). But these quickly become framework questions. So the question of what a society deems the highest good may become the question of justifying a system whose ultimate moral principle is, for example, "Seek the sensual pleasure of the greatest number of people." Since the interpretation of moral terms will occur during a particular stage of development, it may be that certain acts sanctioned by one society's moral system might be forbidden by ours, yet still be, as far as we are concerned, moral. That is, we may be ready not only to make the analytic statement that "The sacrifice of virgins was moral in Inca society," but also to judge the Inca high priest as a good and moral man for sacrificing virgins. Such judgments, of course, would not relieve us of the obligation to stay, if we could, the priest's hand from plunging in the knife.

A third way in which empirical assumptions enter into framework questions regards the methods of justifying the system and its highest principles. Consider an ethical system that has several moral axioms of the kind we might find adopted in our own society: for example, lying is always wrong; abortion is immoral; adultery is bad. If asked to justify these precepts, someone might attempt to show that they conformed to a yet more general moral canon, such as the Golden Rule, the Ten

Commandments, or the Greatest Happiness Principle. But another common sort of justification might be offered. Appeal might be made to the fact that moral authorities within our society have condemned or praised certain actions. Such an appeal, of course, would be empirical. Yet the justifying argument would meet the usual criterion of validity, if the contending parties implicitly or explicitly agreed on a metaethical inference principle such as "Conclude as sound ethical injunctions what moral leaders preach." Principles of this kind—comparable to Carnap's meaning postulates—implicitly regulate the entailment of propositions within a particular community of discourse.[40] They would include rules that govern use of the standard logical elements (e.g., "and," "or," "if . . . then," etc.) as well as the other terms of discourse. Thus in a community of analytic philosophers, the rule "From 'a knows x,' conclude 'x'" authorizes arguments of the kind: "Hilary knows we are not brains in vats, so we are not brains in vats." In a particular community, the moral discourse of its members could well be governed by a metaethical inference principle of the sort mentioned. Such an inference rule would justify the argument from moral authority, because the interlocutors could not assert the premise (e.g., "Moral leaders believe abortion is wrong") and deny the conclusion (e.g., "Abortion is wrong") without contradiction. In this case, then, one would have a perfectly valid argument that derived morally normative conclusions from factual propositions.

The cautious critic, however, might object that this argument does not draw a moral conclusion (e.g., "Abortion is wrong") solely from factual premises (e.g., "Moral leaders believe abortion is wrong"), but also from the metaethical inference principle itself, which is not a factual proposition—hence, that I have not shown a moral imperative can be derived from factual premises alone. Moreover, so the critic might continue, the inference principle actually endorses a certain moral action (e.g., shunning abortion) and thus incorporates a moral imperative— consequently that I have assumed a moral injunction rather than deriving it from factual premises. This two-pronged objection requires a double defense, one part that examines the role of inference principles and the other that analyzes what such principles enjoin.

The logical structure of every argument has, implicitly at least, three distinguishable parts: (1) one or more premises; (2) a conclusion; and

40. For a consideration of inference principles of the kind mentioned, see Rudolph Carnap, *Meaning and Necessity* (Chicago: University of Chicago Press, 1956), pp. 222–32; Wilfrid Sellars, "Concepts as Involving Laws and Inconceivable without Them," *Philosophy of Science* 15 (1948): 287–315; James McCawley, *Everything that Linguists have Always Wanted to Know about Logic* (Chicago: University of Chicago Press, 1981), p. 46.

(3) a rule or rules that permit the assertion of the conclusion on the basis of the premises. The inference rule, however, is not a rule 'from which' a conclusion is drawn, but one 'by which' it is drawn. If rules were rather to be regarded as among the premises from which the conclusion was drawn, there would be no principle authorizing the move from premises to conclusion, and the argument would grind to a halt (as Lewis Carroll's tortoise knew). Hence, the first prong of the objection may be bent aside.

The second prong may also be diverted. An inference principle logically only endorses a conclusion formed on the basis of the premises—that is, it enjoins not a moral act (e.g., shunning abortion) but an epistemological act (e.g., accepting the proposition "Abortion should be shunned"). Once we are convinced of the truth of a proposition, we might, of course, act in light of it; but that is an entirely different matter—at least logically. These two considerations, I believe, take the bite out of the objection.

We have just seen how normative conclusions may be drawn from factual premises. This would be an internal justification if the contending parties initially agreed about inference principles. However, they may not agree, and then the problem of justification would become the framework issue of what justifies the inference rule. It would also turn out to be a framework question if the original challenge were not to an inference rule, but to a cardinal principle (e.g., the Greatest Happiness, the Golden Rule, etc.) that was used as the axiom whence the moral theorems of the system were derived. To meet a framework challenge, one must move outside the system in order to avoid a viciously circular justification. When philosophers take this step, they typically begin to appeal (and ultimately must) to commonsense moral judgments. They produce test cases to determine whether a given principle will yield the same moral conclusions as would commonly be reached by individuals in their society. In short, frameworks, their inference rules, and their principles are usually justified in terms of intuitively clear cases—that is, in terms of matters of fact. Such justifying arguments, then, proceed from what people as a matter of fact believe to conclusions about what principles would yield these matters of fact.

This method of justifying norms is not confined to ethics. It is also used, for example, in establishing modus ponens as the chief principle of modern logic: that is, modus ponens renders the same arguments valid that rational men consider valid. But this strategy for justifying norms utilizes empirical evidence, albeit of a very general sort. Quite simply the strategy recognizes what William James liked to pound home: that no system can validate its own first principles. The first prin-

ciples of an ethical system can be justified only by appeal to another
kind of discourse, an appeal in which factual evidence about common
sentiments and beliefs is adduced. (It is at this level of empirical appeal,
I believe, that we can dismiss Wilson's suggestion that contract altruism—i.e., "I'll scratch your back, if and only if you'll scratch mine"—is
the highest kind. For most men would declare an action nonmoral if it
were done only for personal gain.)

The contention that the inference principles or cardinal imperatives
of a moral system can ultimately be justified only by referring to common beliefs and practices seems degenerately relativistic. To what beliefs, to what practices, to what men shall we appeal? Should we look
to the KKK for enlightenment about race relations? Further, even if the
argument is correct about the justification of logical rules by appeal
to the practices of rational men, the same seems not to hold for moral
rules, because persons differ far less in their criteria of logical soundness
than in their criteria for moral correctness. The analogy between logical
imperatives and moral imperatives thus appears to wither. These objections are potent, though I believe they infect all attempts to justify
moral principles.[41] In the case of evolutionary ethics, however, I think
the prognosis is good. I will take up the last objection now and then
turn to the first to sketch an answer that will be completed in the final
section of this appendix.

The last objection actually grants my contention that logical-infer-
ence rules or principles are justified by appeal to beliefs and practices;
presumably the objection would then be deflated if a larger consensus
were likely in the case of moral justification. The second objection, then,
either accepts my analysis of justificatory procedures or it amounts to
the first objection, that appeal to the beliefs and practices of men fails
to determine the reference class and becomes stuck in the moral muck
of relativism. My sketchy answer to the second objection, which will be
filled in below, is simply that the reference class is moral men (just as in
logical justification it is the class of rational men) and that we can count
on this being a rather large class because evolution has produced it so
(just as it has produced a large class of rational creatures). Indeed, one
who cannot comprehend the soundness of basic moral principles, along
with one who cannot comprehend the soundness of basic logical principles, we regard as hardly a man. Moreover, we have evolved, so I
maintain (and ask the reader to accept), to recognize and approve of
moral behavior when we encounter it (just as we have evolved to rec-

41. Alan Gewirth, *Human Rights: Essays on Justification and Applications* (Chicago:
University of Chicago Press, 1982), pp. 43–45.

ognize and approve of logical behavior). Those protohuman lineages that have not had these traits selected for, have not been selected at all. This does not mean, of course, that every infant slipping fresh out of the womb will respond to others in altruistic ways or be able to formulate maxims of ethical behavior. Cognitive maturity must be reached before the individual can become aware of the signs of human need and bring different kinds of response under the common description of altruistic or morally good behavior. Likewise, maturity and cultural transmission must complement the urges for logical consistency that nature has instilled: even the baby Russell could not be expected to understand right off that 'if . . . then' propositions are logically equivalent to 'not . . . or' propositions. We should not, therefore, be misled by the KKK example. Most Klansmen are probably quite moral people. They simply have unsound beliefs about, among other things, different races, international conspiracies, and so forth. Our chief disagreement with them will not be with their convictions about heeding the community good, but with their beliefs about what leads to that good.

This brief discussion of justification of ethical principles indicates, I believe, how the concept of justification must be employed. "To justify" means "to demonstrate that a proposition or system of propositions conforms to a set of acceptable rules, a set of acceptable factual propositions, or a set of acceptable practices." The order of justification is from rules to empirical propositions about beliefs and practices. That is, if rules serving as inference principles or the imperatives serving as premises (e.g., the Golden Rule) of a justifying argument are themselves put to the test, then they must be shown to conform either to still more general rules or to empirical propositions about common beliefs and practices. Barring an infinite regress, this procedure must end in what are regarded as acceptable beliefs or practices. Aristotle, for instance, justified the forms of syllogistic reasoning by showing that they made explicit the patterns employed in argument by rational men. Kant justified the categorical imperative and the postulates of practical reason by demonstrating that they were the necessary conditions of common moral experience: that is, he justified normative principles by showing that their application to particular cases reproduced the common moral conclusions of eighteenth-century German burghers and Pietists.

If this is an accurate rendering of the concept of justification, then the justification of first moral principles and inference rules must ultimately lead to an appeal to the beliefs and practices of men, which of course is an empirical appeal. So moral principles ultimately can be justified only by facts. The rebuttal, then, to the charge that at some level evolutionary ethics must attempt to derive its norms from facts is simply that every ethical system must. Consequently, either the naturalistic fallacy is no fallacy, or no ethical system can be justified. But to assert that no ethical system can be justified is just to say that ultimately no reasons can be given for or against an ethical position, that all ethical judgments are nonrational. Such a view sanctions the canonization of Hitler along with Saint Francis. Utilizing, therefore, the general rational strategy of appealing to common beliefs and practices to justify philosophical positions, we must reject the idea that the "naturalistic fallacy" is a fallacy.

Justification of Evolutionary Ethics

RV stipulates that community welfare is the highest moral good. It supposes that evolution has equipped human beings with a number of social instincts, such as the need to protect offspring, to provide for the general wellbeing of members of the community (including oneself), to defend the helpless against aggression, and other dispositions that constitute a moral creature. These constitutionally imbedded directives are instances of the supreme principle of heeding the community welfare. Particular moral maxims, which translate these injunctions into the language and values of a given society, would be justified by an individual's showing that, all things considered, following such maxims would contribute to the community welfare.

To justify the supreme principle, and thus the system, requires a different kind of argument. I wish to remind the reader, however, that I will attempt to justify RV as a moral system *under the supposition that it correctly accounts for all the relevant biological facts*. I will adopt the forensic strategy that several good arguments make a better case than one. I have three justifying arguments.

First Justifying Argument

The first argument to justify RV morally is adapted from Alan Gewirth who, I believe, has offered a very compelling approach to deriving an 'ought' from an 'is.' He first specifies what the concept of 'ought' means (i.e., he implicitly indicates the rule governing its deployment in arguments). He suggests that it typically means "necessitated or required by reasons stemming from some structured context."[42] Thus in the inference "It is lightning, therefore it ought to thunder," the "ought" means, he suggests, "given the occurrence of lightning, it is

42. Ibid., p. 108.

required or necessary that thunder also occur, this necessity stemming from the law-governed context of physical nature."[43] Here descriptive causal laws provide the major (unexpressed) premise of the derivation of 'ought' from 'is.' The rule governing inferences of this sort in our scientifically modern community (a topic to which Gewirth does not attend) would be: "From 'x causes y' infer 'since x, y ought to occur.'" The practical sphere of action also presents structured contexts. So, for example, as a member of the university, I ought to prepare my classes adequately. Now Gewirth observes that derivation of a practical 'ought,' such as the one incumbent on a university professor, requires first that one accept the structured context. But then, he contends, only hypothetical 'ought's are produced: for example, "*If* I am a member of the university, then I ought to prepare classes adequately." Since nothing compels me to become a member of the university, I can never be categorically enjoined: "Prepare classes adequately." Gewirth further argues that if one decides to commit oneself to the context, for instance, to university membership, then the derivation of 'ought' will really be from an obligation assumed, that is, from one 'ought' to another 'ought.' He attempts to overcome these obstacles by deriving 'ought's from a context that the person cannot avoid, cannot choose to accept or reject. He claims that the generic features of human action impose a context that cannot be escaped and that such a context requires the agent regard as good his freedom and wellbeing. From the recognition that freedom and wellbeing are necessary conditions of all action, the agent can logically derive, according to Gewirth, the proposition "I have a right to freedom and basic wellbeing." This 'rights' claim, which indeed implies "I ought to have freedom and wellbeing," can only be made if the agent must grant the same right to others. Since the claim depends only on what is required for human agency and not on more particular circumstances, Gewirth concludes that everyone must logically concede the right to any other human agent.

Gewirth's derivation of 'ought' from 'is' has been criticized by Alistair MacIntyre among others. MacIntyre simply objects that because I have a need for certain goods does not entail that I have a right to them, that is, that others are obliged to help me secure them.[44] This, I believe, is a sound objection to Gewirth's formulation. Gewirth's core position, however, can be preserved if we recognize that a generally accepted moral inference principle sanctions the derivation of rights claims from

43. Ibid.
44. Alistair MacIntyre, *After Virtue* (Notre Dame, Ind.: University of Notre Dame Press, 1981), pp. 64–65.

empirical claims about needs common to all human beings. Anyone who doubts the validity of such an inference principle need only perform the empirical test mentioned above (i.e., consult the kind of inferences most people actually draw). Yet even if we granted the force of MacIntyre's objection to Gewirth, the evolutionary perspective permits a similar derivation, though without the objectionable detour through human needs. Evolution provides the structured context of moral action: it has constituted human beings not only to be moved to act for the community good, but also to approve, endorse, and encourage others to do so. This particular formation of human nature does not impose an individual need, not a requirement that will be directly harmful if not satisfied; hence the question of a logical transition from an individual (or generic) need to a right does not arise. Rather the constructive forces of evolution impose a practical necessity on each person to promote the community good. We must, we are obliged, to heed this imperative. We might attempt to ignore the demand of our nature by refusing to act altruistically, but this does not diminish the reality of the demand. The inability of individuals completely to harden their consciences to basic principles of morality means that sinners can be redeemed. Hence just as the context of physical nature allows us to argue "Since lightning has struck, thunder ought to follow," so the structured context of human evolution allows us to argue "Since each person has evolved to advance the community good, each ought to act altruistically."

Two important objections might be lodged at this juncture. First, just because evolution has outfitted men with a moral sense of commitment to the community welfare, this fact ipso sole does not impose any obligation. After all, evolution has installed aggressive urges in men, but they are not morally obliged to act upon them. A careful advocate of RV will respond as follows. An inborn commitment to the community welfare, on the one hand, and an aggressive instinct, on the other, are two greatly different traits. The first trait—that is, the particular complex of dispositions and attitudes produced by evolution (i.e., through kin and group selection in my version)—leads an individual to behave in ways that we can generally characterize as acting for the community good. The second sort of trait does not usually lead to action for the community good. Moral 'ought' propositions are not sanctioned by the mere fact of the evolutionary formation of human nature but by the fact of the peculiar formation of human nature we call "moral," which has been accomplished by evolution. The evolutionary formation of human nature according to other familiar biological relations might well sanction such propositions as "Since he has been constituted an aggressive being by evolution, he ought to react hostilely when I punch him in the nose." In this latter case, a moral 'ought' has not been derived, because the structured context does not comprise the complex traits which produce altruistic behavior.

The second objection points out what appears to be a logical gap between the structured context of the evolutionary constitution of man and an 'ought' proposition. Even if it is granted that evolution has formed human nature in a particular way, call it the "moral way" (the exact meaning of which must yet be explored), yet what justifies concluding that one 'ought' to act altruistically? What justifies the move, of course, is an inference principle to the effect: "From a particular sort of structured context, conclude that the activity appropriate to the context ought to occur." Now, evolution has produced the structured context of a human being with altruistic impulses and dispositions. Thus, from a factual proposition of the sort "This person has evolved to act altruistically" and from the rule specifying the deployment of 'ought' (i.e., "From 'x is y kind of thing' conclude 'x ought to act in y-fashion' "), we are entitled to conclude "This person ought to act altruistically." Gewirth, in his attempt to show that moral 'ought's can be derived from 'is's, depends on such a rule; and significantly, MacIntyre's response does not challenge it. Indeed, MacIntyre employs another inference rule of this kind, which Gewirth would likely endorse: "From 'needs' propositions alone one may not conclude to 'claims' propositions." All meta-level discussions, all attempts to justify ethical frameworks depend on such inference rules, whose ultimate justification can only be their acceptance by rational and moral creatures.

Second Justifying Argument

The second argument justifying RV amplifies the first. It recognizes that evolution has formed a part of human nature according to the criterion of the community good (i.e., according to the principles of kin and group selection). This we call the moral part. The justification for the imperative advice to a fellow creature, "Act for the community good," is therefore: "Since you are a moral being, constituted so by evolution, you ought to act for the community good." To bring a further justification for the imperative would require the premise of this inference to be justified, which would entail furnishing factual evidence as to the validity of evolutionary theory (including RV). And this, of course, would be ultimately to justify the moral imperative by appeal to empirical evidence. The justifying argument, then, amounts to: *the evidence shows that evolution has, as a matter of fact, constructed human beings*

APPENDIX 2

to act for the community good; but to act for the community good is what we mean by being moral. Since, therefore, human beings are moral beings—an unavoidable condition produced by evolution—each ought to act for the community good.

This second justifying argument differs from the first only in stressing: (1) that ultimate justification will require securing the evidentiary base for evolutionary theory and the operations of kin and group selection in forming human nature; and (2) that the logical movement of the justification is from (a) the empirical evidence and theory of evolution, to (b) man's constitution as an altruist, to (c) identifying being an altruist with being moral, to (d) concluding that since men so constituted are moral, they morally ought to promote the community good.

Three points need to be made about this second justifying argument in light of these last remarks, especially those under (2). To begin with, the general conclusion reached—"Since each human being is a moral being, each ought act for the community good"—does not beg the question of deriving moral imperatives from evolutionary facts. The connection between being a rational animal and being moral is contingent, due to the creative hand of evolution: it is because, so I allege, that creatures having a human frame and rational mind have also undergone the peculiar processes of kin and group selection that they have been formed to regard and advance the community good and approve of altruism in others. (There is a sense, of course, in which a completely amoral person will be regarded as something less than human.) Having such a set of attitudes and acting on them is what we mean by being moral.[45] Further, given our notion of what it is to be moral, it is a factual question whether certain activity should be described as "moral behavior."

The second point is an evolutionary Kantian one and refers back to the previous discussion on the nature of justification. If challenged to justify altruism as being a moral act in reference to which 'ought' propositions can be derived, a defender of RV will respond that the objector should consult her own intuitions and those commonly of the run of human beings. If the evolutionary scenario of RV is basically correct, then the challenger will admit her own intuitions confirm that she especially values altruistic acts, that she spontaneously recognizes the authority of the urge to perform them, and that she would encourage

45. Gewirth (*Human Rights,* pp. 82–83) endorses the following criteria as establishing a motive as moral: the agent takes it as prescriptive; he universalizes it; he regards it as over-riding and authoritative; and it is formed of principles that denominate actions right simply because of their effect on other persons. These criteria are certainly met in altruistic behavior described by RV.

them in others—all of which identify altruistic behavior with moral behavior. But if she yet questions the reliability of her own intuitions or if she fails to make the identification (because her own development has been devastatingly warped by a wicked aunt), then evidence for evolutionary theory and kin and group selection must be adduced to show that human beings generally (with few exceptions) have been formed to approve, endorse, and encourage altruistic behavior.

The third point glosses the meaning of "ought." In reference to structured contexts, "ought to occur," "ought to be," "ought to act," typically mean "must occur," "must be," "must act, *provided there is no interference.*" Structured contexts involve causal processes. Typically "ought" adds to "must" the idea that perchance some other cause might disrupt the process (e.g., "Lightning has flashed, so it ought to thunder, that is, it must thunder, provided that no sudden vacuum in the intervening space is created, that there is an ear around to transduce movement of air molecules into nerve potentials, etc."). In the context of the evolutionary constitution of human moral behavior, "ought" means that people must act altruistically, provide they have assessed the situation correctly and a surge of jealousy, hatred, or greed does not interfere. The "must" here is a causal "must"; it means that in ideal conditions—that is, perfectly formed attitudes resulting from evolutionary processes, complete knowledge of situations, absolute control of the passions, and so forth—altruistic behavior would necessarily occur in the appropriate conditions. When conditions are less than ideal—when, for example, the severe stress of war causes an individual to murder innocent civilians—then we might be warranted in expressing another kind of 'ought' proposition: "Under conditions of brutalizing war, some soldiers ought to murder noncombatants." In such cases, of course, the 'ought' is not a moral ought; it is not a moral ought because the 'ought' judgment is not formed in recognition of altruism as the motive for behavior. In moral discourse, expressions of 'ought' propositions have the additional function of encouraging the agent to avoid or reject anything that might interfere with the act. The 'ought' derived from the structured context of man's evolutionary formation, then, will be a moral 'ought' precisely because the activities of heeding the community good and approving of altruistic behavior constitute what we mean and (if RV is correct) must mean by "being moral."

This second justifying argument recognizes that there are three kinds of instances in which moral imperatives will not be heeded. First, when a person misconstrues the situation (e.g., when a person, without warrant, takes the life of another, because she didn't know the gun was loaded and therefore could not have formed the relevant intention). But

here, since the person has misunderstood the situation, no moral obligation or fault can be ascribed. The second case occurs when a person
does understand the moral requirements of the situation but refuses to
act accordingly. This is analogous to the case when we say thunder
ought to have followed lightning, though it did not (because of some
intervening cause). The person who so refuses to act on a moral obligation will not be able logically to justify his action and will be called a
sinner. Finally, there is the case of the person born morally deficient,
the sociopath who robs, rapes, and murders without a shadow of guilt.
Like the creature born without cerebral hemispheres, the sociopath has
been deprived of what we have come to regard as an organ of humanity.
We do not think of him as a human being in the full sense. RV implies
that such an individual, strange as it seems, cannot be held responsible
for his actions. He cannot be held morally guilty for his crimes, since
he, through no fault of his own, has not been provided the equipment
to make moral decisions. This does not mean, of course, that the community should not be protected from him, nor that it should permit his
behavior to go unpunished; indeed, community members have an obligation to defend against the sociopath and inflict the kind of punishment that might restrain unacceptable behavior.

Third Justifying Argument

The final argument justifying RV is second order. This argument
shows RV to be warranted because RV grounds other key strategies for
justifying moral principles. Consider how moral philosophers have attempted to justify the cardinal principles of their systems. Usually they
have adopted one of three methods. They might, with G. E. Moore,
proclaim that certain activities or principles of behavior are intuitively
good, that their moral character is self-evident. But such moralists have
no ready answer to the person who might truthfully say, "I just don't
see it, sorry." Nor do they have any way of excluding the possibility that
a large number of such people exist or will exist. Another strategy is
akin to that of Kant, which is to assert that men have some authentic
moral experiences, and from these an argument can be made to a general principle in whose light their moral character is intelligible. But
this tactic too suffers from the liability that men may differ in their
judgments of what actions are moral. Finally, there is the method employed by Herbert Spencer. He asks someone proposing another principle—Spencer's was that of greatest happiness—to reason with him.
The outcome should be—if Spencer's principle is correct—that the interlocutor will find either that actions he regards as authentically moral

do not conform to his own principle but to Spencer's, or that his principle reduces to or is another version of Spencer's principle. But here again it is quite possible that the interlocutor's principle will cover all the cases of action he describes as moral but will not be reducible to Spencer's principle. No reason is offered for expecting ultimate agreement in any of these cases.

All three strategies suppose that one can find near-universal consent among men concerning what actions are moral and what principles sanction them. Yet no way of conceptually securing such agreement is provided. And here is where RV obliges: it shows that the pith of every person's nature, the core by which he or she is constituted a social and moral being, has been created according to the same standard. Each heart must resound to the same moral cord: acting for the common good. It may, of course, occur that some individuals are born deformed in spirit. There are psychopaths among us. But these, the theory suggests, are to be regarded as less than moral creatures, just as those born severely retarded are thought to be less than rational creatures. But for the vast community of men, they have been stamped by nature as moral beings. RV therefore shows that the several strategies used to support an ultimate ethical principle will, in fact, be successful, successful in showing, of course, that the community good is the highest ethical standard. But for RV to render successful several strategies for demonstrating the validity of the highest ethical principle is itself a justification.

In this defense of evolutionary ethics, I have tried to do three things, to demonstrate that, if we grant certain empirical propositions, then my revised version (RV) of evolutionary ethics: (1) does not commit the naturalistic fallacy as it is usually formulated; (2) does, admittedly, derive values from facts; but (3) does not commit any fallacy in doing so. The ultimate justification of evolutionary ethics can, however, be accomplished only in the light of advancing evolutionary theory.[46]

46. Though this essay was originally prepared as an appendix to the present volume, it was solicited for earlier publication. It appeared as "A Defense of Evolutionary Ethics," in *Biology and Philosophy* 1, no. 3 (1986): 265–93. Camilo Cela-Conde, Alan Gewirth, William Hughes, Laurence Thomas, and Roger Trigg wrote critical essays in response, and I rejoined with "Justification Through Biological Faith." These replies and rejoinder can be found in the same number of *Biology and Philosophy*.

Bibliography

Unpublished Manuscripts and Correspondence

Baldwin, James Mark. Presidential Archives, Johns Hopkins University.

Cattell, James McKean. Cattell Papers, Library of Congress.

Cuvier, Frédéric. Fonds Cuvier, Institut de France, Paris; Correspondence of Frédéric Cuvier, Wellcome Institute Library, London.

Cuvier, Georges. Fonds Cuvier, Institut de France, Paris.

Darwin, Charles. The Darwin Papers, Cambridge University Library; Correspondence of Charles Darwin, American Philosophical Society Library, Philadelphia; Correspondence of Charles Darwin, The British Library.

Galton, Francis. Galton Papers, University College Library, University of London.

Gulick, John. Correspondence of John Gulick, Academy of Natural Sciences of Philadelphia (film in American Philosophical Society Library, Philadelphia).

Haeckel, Ernst. Haeckel Papers, Haeckel-Haus, Jena.

Huxley, Thomas. Huxley Papers, Imperial College Library, University of London.

James, William. James Papers, Houghton Library, Harvard Univesity.

Lyell, Charles. Correspondence of Charles Lyell, American Philosophical Society Library, Philadelphia.

Morgan, C. Lloyd. Lloyd Morgan Papers, University of Bristol Library.

Münsterberg, Hugo. Münsterberg Papers, Boston Public Library.

Romanes, George. Correspondence of George Romanes, American Philosophical Society Library.

Spencer, Herbert. Athenaeum Collection of Spencer's Papers, Senate House Library, University of London.

Wallace, Alfred Russel. Papers of Alfred Russel Wallace, British Museum, London; Correspondence of Alfred Russel Wallace, American Philosophical Society Library.

Printed Material

Abercrombie, John. *Inquiries Concerning the Intellectual Powers.* 8th ed. London: Murray, 1838.

Alexander, Richard. *Darwinism and Human Affairs.* Seattle: University of Washington Press, 1979.

Allee, W. C. *Animal Aggregations: A Study in General Sociology.* Chicago: University of Chicago Press, 1931.

————. *Cooperation Among Animals with Human Implications.* New York: Henry Schuman, 1951.

————. *Social Life of Animals.* New York: Norton, 1938.

Allen, Garland. "The Several Faces of Darwin: Materialism in Nineteenth and Twentieth Century Evolutionary Theory." In *Evolution from Molecules to Men,* edited by D. Bendall. Cambridge: Cambridge University Press, 1983.

————. *Thomas Hunt Morgan: The Man and His Science.* Princeton: Princeton University Press, 1978.

Allen, Gay W. *William James.* New York: Viking, 1967.

Allen, Grant. "Hellas and Civilization." *Popular Science Monthly—Supplement* 13–20 (1878): 398–406.

————. "Nation-Making." *Popular Science Monthly—Supplement* 13–20 (1878): 121–27.

Allport, Floyd. *Social Psychology.* New York: Houghton Mifflin, 1924.

Alper, Joseph. "Ethical and Social Implications." In *Sociobiology and Human Nature,* edited by M. Gregory, A. Silvers, and D. Sutch. San Francisco: Jossey-Bass, 1978.

Anderson, James. "William James's Depressive Period (1867–1872) and the Origins of his Creativity: A Psychobiographical Study." Ph.D dissertation. University of Chicago, 1979.

Appel, Toby. "The Cuvier-Geoffroy Debate and the Structure of Nineteenth-Century-French Zoology." Ph.D. dissertation. Princeton University, 1975.

————. "Henri de Blainville and the Animal Series: A Nineteenth-Century Chain of Being." *Journal of the History of Biology* 13 (1980): 291–319.

Aquinas, Thomas. *Opera omnia.* 41 vols. Romae: Ex Typographia Polyglotta, 1891.

Argyle, 8th Duke of (G. D. Campbell). "The Struggle of Parts in the Organism." *Nature* 24 (1881): 581.

Avicenna. *Avicenna Latinus, Liber de anima.* Edited by S. Van Riet. 2 vols. Leiden: Brill, 1968–1972.

Bain, Alexander. *Mental and Moral Science.* London: Longmans, Green, 1868.

Baker, Keith. *Condorcet: From Natural Philosophy to Social Mathematics.* Chicago: University of Chicago Press, 1975.

Baldwin, James Mark. *Between Two Wars, being Memories, Opinions and Letters Received.* 2 vols. Boston: Stratford, 1926.

————. "Consciousness and Evolution." *Psychological Review* 3 (1896): 300–8.

————. "Consciousness and Evolution." *Science* 2 (1895): 219–23.

————. "Contemporary Philosophy in France." *New Princeton Review* 3 (1887): 137–44.

———. "The Cosmic and the Moral." *International Journal of Ethics* 6 (1895): 93–97.

———. "On Criticisms of Organic Selection." *Science* 4 (1896): 727.

———. *Darwin and the Humanities*. Baltimore: Review, 1909.

———. *Development and Evolution*. New York: Macmillan, 1902.

———. "Distance and Color Perception by Infants." *Science* 21 (1893): 231–32.

———. *Fragments in Philosophy and Science*. New York: Scribner's Sons, 1902.

———. *Genetic Theory of Reality*. New York: Putnam, 1915.

———. *Handbook of Psychology: Feeling and Will*. New York: Holt, 1891.

———. *Handbook of Psychology: Senses and Intellect*. New York: Holt, 1889.

———. "Heredity and Instinct (I)." *Science* 3 (1896): 438–41.

———. "Heredity and Instinct (II)." *Science* 3 (1896): 558–61.

———. "The Idealism of Spinoza." *Presbyterian Review* 10 (1889): 65–76.

———. "Imitation: A Chapter in the Natural History of Consciousness." *Mind* 19 (1894): 26–55.

———. "Infant Psychology." *Science* 16 (1890): 351–53.

———. "Infants' Movements." *Science* 19 (1892): 15–16.

———. *Mental Development in the Child and the Race*. New York: Macmillan, 1895.

———. *Mental Development in the Child and the Race*. 3d ed. New York: Macmillan, 1906.

———. "A New Factor in Evolution." *American Naturalist* 30 (1896): 441–51, 536–53.

———. "A New Method of Child Study." *Science* 21 (1893): 213–14.

———. "Organic Selection." *Nature* 55 (1897): 558.

———. "Organic Selection." *Science* 5 (1897): 634–36.

———. "Origin of Right or Left Handedness." *Science* 16 (1890): 247–48.

———. "Origin of Volition in Childhood." *Science* 20 (1892): 286–87.

———. "Physical and Social Heredity." *American Naturalist* 30 (1896): 422–30.

———. "The Postulates of Physiological Psychology." *Presbyterian Review* 8 (1887): 427–40.

———. *Social and Ethical Interpretations in Mental Development: A Study in Social Psychology*. New York: Macmillan, 1897.

———. "Suggestion in Infancy." *Science* 17 (1891): 113–17.

———. *Thought and Things or Genetic Logic*. 3 vols. London: Sonnenschein, 1906–1911.

———. "With Bernheim at Nancy." *Nation* 55 (1892): 101–3.

Banister, Robert. *Social Darwinism: Science and Myth in Anglo-American Social Thought*. Philadelphia: Temple University Press, 1979.

Barnes, Barry. "On the Conventional Character of Knowledge and Cognition." *Philosophy of the Social Sciences* 11 (1981): 303–33.

———. *Interest and the Growth of Knowledge*. London: Routledge and Kegan Paul, 1977.

Barzun, Jacques. *Darwin, Marx, Wagner*. 2d ed. New York: Doubleday Anchor, 1958.

Beach, Frank. "The Descent of Instinct." *Psychological Review* 62 (1955): 401–10.

Bebel, August. *Die Frau und der Sozialismus*. 16th ed. Stuttgart: Dietz, 1892; 1st ed. in 1879.

Becker, Morton. "Darwinism." In vol. 2 of *The Encyclopedia of Philosophy*, edited by Paul Edwards. 8 vols. New York: Macmillan, 1967.

Bell, Charles. *Expression: Its Anatomy & Philosophy*. 3d ed. New York: Wells, 1873; reprint of 3rd ed. of 1844.

Ben-David, Joseph. *The Scientist's Role in Society*. Englewood, N.J.: Prentice-Hall, 1971.

Bentham, Jeremy. *Introduction to the Principles of Morals and Legislation*. Corrected ed. New York: Hafner, 1973; reprint of final ed. of 1823.

Bernal, J. D. *Science in History*. 4 vols. 3d ed. Cambridge, Mass.: M. I. T. Press, 1971.

Bernard, L. L. *Instinct: A Study in Social Psychology*. New York: Henry Holt, 1924.

Bernstein, Eduard. "Ein Schüler Darwin's als Vertheidiger des Sozialismus." *Die Neue Zeit* 9 (1890–1891): 171–77.

Bjork, Daniel. *The Compromised Scientist*. New York: Columbia University Press, 1983.

Blackwall, John. *Researches in Zoology*. London: Simplin and Marshall, 1834.

Block, N., and G. Dworkin, eds. *The IQ Controversy*. New York: Pantheon, 1976.

Bloor, David. "Klassifikation und Wissenssoziologie: Durkheim und Mauss neu betrachtet." *Kölner Zeitschrift für Soziologie und Sozialpsychologie* 22 (1980): 20–51.

——. *Knowledge and Social Imagery*. London: Routledge & Kegan Paul, 1976.

Blyth, Edward. "On the Psychological Distinctions between Man and All Other Animals." *Magazine of Natural History*, n.s. 1 (1837): 1–9, 77–85, 131–41.

Boakes, Robert. *From Darwin to Behaviourism*. Cambridge: Cambridge University Press, 1985.

Boas, Franz. *Race, Language, and Culture*. Chicago: University of Chicago Press, 1982; reprint of 1st ed. of 1940.

Boring, Edwin. *A History of Experimental Psychology*. 2d ed. New York: Appleton-Century-Crofts, 1950.

Borrichius, Olaus. *De ortu et progressu chemiae dissertatio* (1668). In *Bibliotheca chemica curiosa*, edited by J. Manget. Geneva: Chouet, 1702.

Bostocke, R. *The Difference Betwene the Auncient Phisicke and the Latter Phisicke* (1585). In "An Elizabethan History of Medical Chemistry," by Allen Debus. *Annals of Science* (1962): 1–29.

Bowler, Peter. *The Eclipse of Darwinism*. Baltimore: Johns Hopkins University Press, 1983.

——. *Evolution: The History of an Idea*. Berkeley: University of California Press, 1984.

Brent, Peter. *Charles Darwin, a Man of Enlarged Curiosity*. New York: Harper & Row, 1981.

Brougham, Henry Lord. *Dissertations on Subjects of Science concerned with Natural Theology: Being the Concluding Volumes of the New Edition of Paley's Work*. 2 vols. London: Knight, 1839.

Browne, Janet. "Darwin and the Expression of the Emotions." In *The Darwinian Heritage,* edited by David Kohn. Princeton: Princeton University Press, 1985.

———. "Darwin's Botanical Arithmetic and the 'Principle of Divergence.'" *Journal of History of Biology* 13 (1980): 53–89.

Browning, Don. *Pluralism and Personality: William James and Some Contemporary Cultures of Psychology.* Lewisburg, Pa.: Bucknell University Press, 1980.

Brücher, Heinz. "Ernst Haeckel, ein Wegbereiter biologischen Staatsdenkens." *Nationalsozialistische Monatshefte* 6 (1935): 1087–98.

Büchner, Ludwig. *Aus dem Geistesleben der Thiere.* Berlin: Unbekant, 1879.

Buffon, Georges Louis Leclerc, Comte de. *Oeuvres complètes de Buffon.* Edited by Pierre Flourens. 12 vols. Paris: Garnier, 1853–1855.

Burian, Richard. "A Methodological Critique of Sociobiology." In *The Sociobiology Debate,* edited by Arthur Caplan. New York: Harper, 1978.

Burkhardt, Richard. "Darwin on Animal Behavior and Evolution." In *The Darwinian Heritage,* edited by David Kohn. Princeton: Princeton University Press, 1985.

———. "Development of an Evolutionary Ethology." In *Evolution from Molecules to Men,* edited by D. Bendall. Cambridge: Cambridge University Press, 1983.

———. "On the Emergence of Ethology as a Scientific Discipline." *Conspectus of History* 1 (1981): 62–81.

———. "The Inspiration of Lamarck's Belief in Evolution." *Journal of the History of Biology* 5 (1972): 413–38.

———. *The Spirit of the System.* Cambridge: Harvard University Press, 1977.

Burnham, John. "On the Origins of Behaviorism." *Journal of the History of the Behavioral Sciences* 4 (1968): 143–51.

Burrow, J. W. *Evolution and Society.* Cambridge: Cambridge University Press, 1966.

Butler, Samuel. *Evolution, Old and New.* London: Hardwicke and Bogue, 1879.

———. "Mental Evolution." *Athenaeum* (January–June 1884): 349.

Butts, Robert. *William Whewell's Theory of Scientific Method.* Pittsburgh: University of Pittsburgh Press, 1968.

Cabanis, Pierre-Jean. *Oeuvres complètes de Cabanis.* 5 vols. Paris: Bossange Frères, 1823–1825.

Cahan, Emily. "The Genetic Psychologies of James Mark Baldwin and Jean Piaget." *Developmental Psychology* 20 (1984): 128–35.

Campbell, Donald. "Blind Variation and Selective Retention in Creative Thought as in Other Knowledge Processes." *Psychological Review* 67 (1960): 380–400.

———. "Blind Variation and Selective Retention in Socio-Cultural Evolution." In *Social Change in Developing Areas,* edited by H. Barringer, G. Blanksten, and R. Mack. Cambridge, Mass.: Schenkman, 1965.

———. "Discussion Comment on 'The Natural Selection Model' of Conceptual Evolution." *Philosophy of Science* 44 (1977): 502–7.

———. "Evolutionary Epistemology." In *The Philosophy of Karl Popper*, edited by Paul Schilpp. LaSalle, Ill.: Open Court, 1974.

———. "Methodological Suggestions from a Comparative Psychology of Knowledge Processes." *Inquiry* 2 (1959): 152–82.

———. "Unjustified Variation and Selective Retention in Scientific Discovery." In *Studies in the Philosophy of Biology*, edited by Francisco Ayala and Theodosius Dobzhansky. London: Macmillan, 1974.

Cannon, Susan. *Science in Culture: the Early Victorian Period*. New York: Science History Publications, 1978.

Caplan, Arthur, ed. *The Sociobiology Debate*. New York: Harper, 1978.

Carnap, Rudolph. *Meaning and Necessity*. Chicago: University of Chicago Press, 1956.

Carpenter, William. *Principles of General and Comparative Physiology*. 2d ed. London: John Churchill, 1841.

———. *Principles of Human Physiology*. 5th ed. Philadelphia: Blanchard and Lea, 1853; reprint of 4th London ed. of 1852.

———. *The Principles of Mental Physiology*. New York: D. Appleton, 1874.

Carus, Paul. "The Late Professor Romanes's Thoughts on Religion." *The Monist* 5 (1894–1895): 385–400.

Cattell, James McKeen. "Psychology in America." In *James McKeen Cattell, Man of Science*. Vol. 2 of *Addresses and Formal Papers*. Lancaster, Penn.: Science Press, 1947.

———. "Psychology in America." *Science* 70 (1929): 335–47.

———. "Review of 'A New Factor in Evolution by J. Mark Baldwin.'" *Psychological Review* 3 (1896): 571–72.

Chagnon, Napoleon, and William Irons, eds. *Evolutionary Biology and Human Social Behavior: An Anthropological Perspective*. North Scituate, Mass.: Duxbury, 1979.

[Chambers, Robert.] *The Vestiges of the Natural History of Creation*. 6th ed. London: Churchill, 1847.

Chanet, Pierre. *Considerations sur la sagesse de Charron*. Paris: Le Groult, 1643.

Clarke, Samuel. *The Leibniz-Clarke Correspondence*. Edited by H. Alexander. Manchester: Manchester University Press, 1956; originally published in 1717.

Clifford, William. *Lectures and Essays*. Edited by L. Stephen and F. Pollock. 2 vols. London: Macmillan, 1901; first published in 1879.

Cobb, Francis. "Darwinism in Morals." *Theological Review* (1872): 167–92.

Coleman, William. *Georges Cuvier, Zoologist*. Cambridge: Harvard University Press, 1964.

Collingwood, R. G. *The Idea of History*. Oxford: Oxford University Press, 1956.

Colp, Ralph. *To Be an Invalid*. Chicago: University of Chicago Press, 1977.

———. "'Confessing a Murder,' Darwin's First Revelations about Transmutation." *Isis* 77 (1986): 9–32.

Combe, George. *Essay on the Constitution of Man and Its Relation to External Objects*. 3d ed. Boston: Phillips & Sampson, 1845; reprinted from 3d Edinburgh ed. of 1835.

Conder, Eustace. "Natural Selection and Natural Theology: a Criticism." *Hum-*

boldt Library of Science, no. 40. New York: Humboldt, 1882.

Condillac, Etienne-Bonnot de. *Oeuvres complètes de Condillac.* 23 vols. Paris: Houel, 1798.

Condorcet, Marie Jean, Marquis de. *Esquisse d'un tableau historique des progrès de l'esprit humain.* Paris: Vrin, 1970.

Cornell, John. "Analogy and Technology in Darwin's Vision of Nature." *Journal of the History of Biology* 17 (1984): 303–44.

Crane, Diana. *Invisible Colleges.* Chicago: University of Chicago Press, 1972.

Crombie, Alistair. *Medieval and Early Modern Science.* 2d ed. 2 vols. Cambridge: Harvard University Press, 1961.

Cuvier, Frédéric. "Description d'un orangoutang, et observations sur ses facultés intellectuelles." *Annales du Muséum d'histoire naturelle* 16 (1810): 46–65.

———. "Essay on the Domestication of Mammiferous Animals." *Edinburgh New Philosophy Journal* 3 (1827): 303–18; 4 (1828): 45–60, 292–98.

———. "Examinen de quelques observations de M. Dugal-Stewart, qui tendent à détruire l'analogie des phénomènes de l'instinct avec ceux de l'habitude. *Mémoires du Muséum d'histoire naturelle* 12 (1823): 241–60.

———. "Instinct." In vol. 23 of *Dictionnaire des sciences naturelles.* 61 vols. Strasbourg: Levrault, 1816–1843.

———. "Observations sur le chien des habitants de la Nouvelle-Hollande." *Annales du Muséum d'histoire naturelle* 11 (1808): 458–76.

———. "Observations préliminaires." In *Recherches sur les ossements fossiles,* by Georges Cuvier. 4th ed. 4 vols. Paris: D'Ocagne, 1834.

———. "De la sociabilité des animaux." *Mémoires de Muséum d'histoire naturelle* 13 (1825): 1–27.

Cuvier, Georges. "Eloge de M. de Lamarck." *Mémoires de l'Académie des sciences,* 2d series, 13 (1835): i–xxxi.

———. *Recherches sur les ossements fossiles.* 4th ed. 4 vols. Paris: D'Ocagne, 1834; 1st ed. in 1812.

———. *Le règne animal.* 2d ed. 5 vols. Vols. 4 and 5 by P. A. Latreille. Paris: Deterville, 1829–1830.

D'Alembert, Jean. *Discours Préliminaire.* In *Encyclopédie ou dictionnaire raisonné des sciences, des arts et des métiers,* edited by Dennis Diderot and Jean d'Alembert. 2d ed. 17 vols. Paris: Lucques, 1758–1771.

Danziger, Kurt. "Mid-Nineteenth Century British Psycho-Physiology." In *The Problematic Science: Psychology in Nineteenth-Century Thought,* edited by William Woodward and Mitchell Ash. New York: Praeger, 1982.

Darden, Lindley, and Nancy Maul. "Interfield Theories." *Philosophy of Science* 44 (1977): 43–64.

Darrow, Clarence. "The Edwardses and the Jukeses." *American Mercury* 6 (1925): 147–57.

———. "The Eugenics Cult." *American Mercury* 7 (1926): 129–37.

Darwin, Charles. *The Autobiography of Charles Darwin,* edited by Nora Barlow. New York: Norton, 1969.

———. *Charles Darwin's Diary of the Voyage of H. M. S. Beagle.* Edited by Nora Barlow. Cambridge: Cambridge University Press, 1934.

————. *Charles Darwin's Natural Selection, being the Second Part of his Big Species Book Written from 1856–1858.* Edited by R. Stauffer. Cambridge: Cambridge University Press, 1975.

————. *The Collected Papers of Charles Darwin.* Edited by Paul Barrett. 2 vols. Chicago: University of Chicago Press, 1977.

————. *The Correspondence of Charles Darwin, Vol. 1: 1821–1836.* Edited by Frederick Burkhardt, Sydney Smith, David Kohn and William Montgomery. Cambridge: Cambridge University Press, 1985.

————. *Darwin's Notebooks on Transmutation of Species.* Edited by Gavin De Beer. *Bulletin of the British Museum (Natural History),* historical series 2, nos. 2–6 (1960–1961): 23–183.

————. *Darwin's Notebooks on Transmutation of Species: Pages Excised by Darwin.* Edited by Gavin De Beer. *Bulletin of the British Museum (Natural History),* historical series 3, no. 5 (1967): 133–75.

————. *The Descent of Man, and Selection in Relation to Sex.* 2 vols. London: Murray, 1871.

————. *The Expression of the Emotions in Man and Animals.* Chicago: University of Chicago Press, 1965; first published in 1872.

————. *The Foundations of the Origins of Species: Two Essays Written in 1842 and 1844 by Charles Darwin.* Edited by Francis Darwin. Cambridge: Cambridge University Press, 1909.

————. "Journal." Edited by Gavin de Beer. *Bulletin of the British Museum (Natural History),* historical series 2, no. 1 (1959): 4–21.

————. *Life and Letters of Charles Darwin.* Edited by Francis Darwin. 2 vols. New York: D. Appleton, 1891.

————. *M Notebook.* Transcribed by Paul Barrett. In *Darwin on Man,* by Howard Gruber. New York: Dutton, 1974.

————. *A Monograph of the Fossil Balanidae and Verrucidae of Great Britain.* London: Palaeontographical Society, 1854.

————. *A Monograph of the Fossil Lepadidae or, Pedunculated Cirripedes of Great Britain.* London: Ray Society, 1851.

————. *A Monograph of the Sub-Class Cirripedia. The Balanidae (or Sessile Cirripedes), the Verrucidae, &c.* London: Ray Society, 1854.

————. *A Monograph of the Sub-Class Cirripedia, with Figures of all the Species. The Lepadidae or, Pedunculated Cirripedes.* London: Ray Society, 1851.

————. *More Letters of Charles Darwin.* Edited by Francis Darwin. 2 vols. New York: D. Appleton, 1903.

————. *The Movements and Habits of Climbing Plants.* London: Murray, 1880.

————. *N Notebook.* Transcribed by Paul Barrett. In *Darwin on Man,* by Howard Gruber. New York: Dutton, 1974.

————. *Old and Useless Notes.* Transcribed by Paul Barrett. In *Darwin on Man,* by Howard Gruber. New York: Dutton, 1974.

————. *On the Origin of Species.* London: Murray, 1859.

————. *The Origin of Species and the Descent of Man.* 6th and 2d eds. respectively. New York: Modern Library, 1936.

————. *The Origin of Species by Charles Darwin: A Variorum Text.* Edited by

Morse Peckham. Philadelphia: University of Pennsylvania Press, 1959.

———. "The Darwin Reading Notebooks (1838–1860)." Transcribed and edited by Peter Vorzimmer. *Journal of the History of Biology* 10 (1977): 106–53.

———. *The Red Notebook of Charles Darwin.* Edited by Sandra Herbert. Ithaca: Cornell University Press, 1980.

———. *The Variation of Animals and Plants under Domestication.* 2 vols. London: Murray, 1868.

———. *The Variation of Animals and Plants under Domestication.* 2d ed. 2 vols. New York: D. Appleton, 1899.

———. *On the Various Contrivances by which Orchids are Fertilised by Insects.* London: Murray, 1862.

Darwin, Erasmus. *The Letters of Erasmus Darwin,* edited by Desmond King-Hele. Cambridge: Cambridge University Press, 1981.

———. *Zoonomia or the Laws of Organic Life.* 2d ed. 2 vols. London: Johnson, 1796; 1st ed. in 1794.

Darwin, Francis. "Physiological Selection and the Origin of Species." *Nature* 34 (1886): 407.

"Darwin on the Descent of Man." *Edinburgh Review* 134 (1871): 195–235.

"Mr. Darwin on the Descent of Man." *Times of London* (7 April 1871).

"Mr. Darwin's Descent of Man." *The Spectator* 44 (11 and 18 March 1871): 288–89, 319–20.

"The Darwinian Jeremiad." *The Spectator* 41 (1868): 1215.

Daston, Lorraine. "The Theory of Will versus the Science of Mind." In *The Problematic Science: Psychology in Nineteenth-Century Thought,* edited by William Woodward and Mitchell Ash. New York: Praeger, 1982.

Davenport, Charles. *Heredity in Relation to Eugenics.* New York: Henry Holt, 1911.

———. "Report of the Committee on Eugenics." *American Breeders' Magazine* 1 (1910): 126–29.

Dawkins, Richard. *The Selfish Gene.* Oxford: Oxford University Press, 1976.

Defrance, L. "Plate." *L'année biologique* 5 (1899–1900): 388.

Degler, Carl. "The Beast in Humanity: Biological Ideas in American Social Thought, 1900 to the Present." The Samuel Paley Lectures, delivered at the Hebrew University, 26 and 30 April and 3 May 1984.

Delage, Yves, and G. Poirault. "L'origine des espèces." *L'année biologique* 3 (1897): 512.

Denton, George. "Early Psychological Theories of Herbert Spencer." *American Journal of Psychology* 32 (1921): 5–15.

Descartes, René. *Oeuvres de Descartes.* Edited by C. Adam and P. Tannery. 13 vols. Paris: Cerf, 1897–1913.

"The Descent of Man." *Athenaeum* (4 March 1871): 275–77.

Desmond, Adrian. *Archetypes and Ancestors: Palaeontology in Victorian London, 1850–1875.* Chicago: University of Chicago Press, 1984.

———. "Robert E. Grant: The Social Predicament of a Pre-Darwinian Evolutionist." *Journal of the History of Biology* 17 (1984): 189–224.

Dilly, Antoine. *Traitté de l'ame et de la connoissance des bêtes*. Amsterdam: Gallet, 1676.

Draper, John William. *History of the Conflict between Religion and Science*. 4th ed. International Scientific Series. New York: D. Appleton, 1875; originally published in 1874.

Driesch, Hans. *The Science and Philosophy of the Organism*. Gifford Lectures for 1907 and 1908. 2 vols. London: Black, 1907–1908.

Duncan, David. *Life and Letters of Herbert Spencer*. 2 vols. New York: D. Appleton, 1909.

Durrant, John. "Innate Character in Animals and Man: A Perspective on the Origins of Ethology." In *Biology, Medicine and Society: 1840–1940*, edited by Charles Webster. Cambridge: Cambridge University Press, 1981.

Earle, William. "William James." In vol. 4 of *The Encyclopedia of Philosophy*, edited by Paul Edwards. 8 vols. New York: Macmillan, 1967.

East, Edward. *Heredity and Human Affairs*. New York: Scribner's, 1927.

East, Edward, and D. F. Jones. *Inbreeding and Outbreeding*. Philadelphia: Lippincott, 1919.

Edel, Leon. *Bloomsbury: A House of Lions*. New York: Lippincott, 1979.

Egerton, Frank. "Darwin's Early Reading of Lamarck." *Isis* 67 (1976): 452–56.

Eibl-Eibesfeldt, Irenäus. *Ethology: the Biology of Behavior*. New York: Holt, Rinehart and Winston, 1970.

Eimer, Theodor. *Die Entstehung der Arten auf Grund von Vererbung erworbener Eigenschaften*. Jena: Fischer, 1888.

Eiseley, Loren. *Darwin and the Mysterious Mr. X*. New York: Dutton, 1979.

———. *Darwin's Century*. New York: Doubleday, 1961.

Ellwood, Charles A. "The Social Philosophy of James Mark Baldwin." *Journal of Social Philosophy* 2 (1936): 55–68.

Espinas, Alfred. *Des sociétés animales*. 2d ed. Paris: Bailliere, 1878.

Evans, L. T. "Darwin's Use of the Analogy between Artificial and Natural Selection." *Journal of the History of Biology* 17 (1984): 113–40.

F. B. *A Letter Concerning the Soul and Knowledge of Brutes Wherein is shewn They are Void of One, and Incapable of the Other*. London: Roberts, 1721.

Feinstein, Howard. *Becoming William James*. Ithaca, N. Y.: Cornell University Press, 1984.

Ferri, Enrico. *Socialism and Positive Science (Darwin-Spencer-Marx)*. Socialist Library—1. Edited by Ramsay MacDonald, M.P. Translated by E. Harvey. London: Independent Labour Party, 1905; originally published in French, 1894.

Fleming, John. *Philosophy of Zoology*. 2 vols. Edinburgh: Constable, 1822.

Flew, Anthony. *Evolutionary Ethics*. London: Macmillan, 1967.

Flourens, Pierre. *Résumé analytique des observations de Frédéric Cuvier sur l'instinct et l'intelligence des animaux*. Paris: Langloiset Leclercq, 1841.

Ford, Marcus. *William James's Philosophy*. Amherst: University of Massachusetts Press, 1982.

Foucault, Michel. *The Order of Things: An Archaeology of the Human Sciences*. New York: Vintage, 1973; originally published as *Les mots et les choses*, 1966.

Freeman, Derek. "The Evolutionary Theories of Charles Darwin and Herbert Spencer." *Current Anthropology* 15 (1974): 211–21.

Freind, John. *The History of Physick from the Time of Galen to the Beginning of the Sixteenth Century.* 2 vols. London: Walthe, 1725.

French, John. "An Inquiry Respecting the True Nature of Instincts, and of the Mental Distinction between Brute Animals and Man." *Zoological Journal* 1 (1824): 1–32, 153–73, 346–67.

Fried, C. "Biology and Ethics: Normative Implications." In *Morality as a Biological Phenomenon,* edited by Gunther Stent. Berkeley: University of California Press, 1978.

Galen. *Opera Omnia.* Edited by C. Kühn. 22 vols. Lipsiae: in Libraria Cnoblochii, 1821–1833.

Galton, Francis. "Experiments in Pangenesis." *Proceedings of the Royal Society of London* 19 (1870–1871): 393–410.

———. "Hereditary Improvement." *Fraser's Magazine,* n.s. 7 (1873): 116–30.

———. "Hereditary Talent and Character." *Macmillan's Magazine* 12 (1865): 157–66, 318–27.

———. *Memories of My Life.* 3d ed. London: Methuen, 1909.

———. "Statistical Inquiries into the Efficacy of Prayer." *Fortnightly Review* 18 (1872): 125–33.

Gasman, Daniel. *The Scientific Origins of National Socialism.* New York: Science History Publications, 1971.

Gassendi, Pierre. *Opera omnia.* 6 vols. Lugduni: Anisson and Devenet, 1658.

Geertz, Clifford. "The Growth of Culture and the Evolution of Mind." *Interpretation of Cultures.* New York: Basic Books, 1973.

Geison, Gerald. *Michael Foster and the Cambridge School of Physiology.* Princeton: Princeton University Press, 1978.

Geoffroy Saint-Hilaire, Etienne. "Sur une colonne vertébrale et ses côtes dans les insectes apiropodes." *Isis* 2 (1820): 527–52.

———. "Le degré d'influence du monde ambiant pour modifier les formes animales." *Mémoires de l'Académie royale des sciences* 12 (1833): 63–92.

———. *Principes de philosophie zoologique.* Paris: Didier, 1830.

———. "Rapport fait à l'Académie royale des sciences sur un mémoire de M. Roulin." *Mémoires du Muséum d'histoire naturelle* 17 (1828): 210–29.

———. "Recherches sur l'organization des Gavials." *Mémoires du Muséum d'histoire naturelle* 21 (1825): 95–155.

Gewirth, Alan. *Human Rights: Essays on Justification and Applications.* Chicago: University of Chicago Press, 1982.

Ghiselin, Michael. "Darwin and Evolutionary Psychology." *Science* 179 (1973): 964–68.

———. *The Triumph of the Darwinian Method.* Berkeley: University of California Press, 1969.

Giles, P. "Evolution and the Science of Language." In *Darwin and Modern Science,* edited by A. C. Seward. Cambridge: Cambridge University Press, 1909.

Gillespie, Neil. *Charles Darwin and the Problem of Creation.* Chicago: University of Chicago Press, 1979.

Gillispie, Charles. *The Edge of Objectivity.* Princeton: Princeton University Press, 1959.

————. "The Formation of Lamarck's Evolutionary Theory." *Archives internationales d'histoire des sciences* 9 (1956): 323–38.

Glass, Bentley, ed. *Forerunners of Darwin.* Baltimore: Johns Hopkins University Press, 1968.

Goldschmidt, Richard. "Evolution, as Viewed by One Geneticist." *American Scientist* 40 (1952): 84–135.

Good, Rankin. "Life of the Shawl." *The Lancet* (9 January 1954): 106–7.

Gordon, Scott. "The London *Economist* and the High Tide of Laissez Faire." *Journal of Political Economy* 43 (1955): 461–88.

Gould, Stephen Jay. *Ever Since Darwin.* New York: Norton, 1977.

————. *Hen's Teeth and Horse's Toes.* New York: Norton, 1984.

————. "Introduction." *The Material Basis of Evolution,* by Richard Goldschmidt. New Haven: Yale University Press, 1981.

————. *The Mismeasure of Man.* New York: Norton, 1981.

————. *Ontogeny and Phylogeny.* Cambridge: Harvard University Press, 1977.

————. *The Panda's Thumb.* New York: Norton, 1982.

Gould, Stephen Jay, and Richard Lewontin. "The Spandrels of San Marco and the Panglossian Paradigm: A Critique of the Adaptationist Programme." *Proceedings of the Royal Society of London* 205 (1978): 581–98.

"A Grandmother's Tale." *Macmillan's* 78 (1898): 425–35.

Gray, Asa. "Natural Selection and Natural Theology." *Nature* 27 (1883): 291–92, 527–28; 28 (1883): 78.

Greene, John M. "Biology and Social Theory in the Nineteenth Century: August Comte and Herbert Spencer." In *Critical Problems in the History of Science,* edited by M. Claggett. Madison: University of Wisconsin Press, 1959.

————. "Darwin as a Social Evolutionist." *Journal of the History of Biology* 10 (1977): 11–27.

————. *The Death of Adam.* Ames: Iowa State University Press, 1959.

————. "The Kuhnian Paradigm and the Darwinian Revolution in Natural History." In *Perspectives in the History of Science and Technology,* edited by D. Roller. Norman: University of Oklahoma Press, 1971.

————. *Science, Ideology, and World View: Essays in the History of Evolutionary Ideas.* Berkeley: University of California Press, 1981.

[Greg, William R.] "On the Failure of 'Natural Selection' in the Case of Man." *Fraser's Magazine* 78 (1868): 353–62.

Gregory, Michael, Anita Silvers, and Diane Such, eds. *Sociobiology and Human Nature.* San Francisco: Jossey-Bass, 1987.

Griffin, Donald. *The Question of Animal Awareness: Evolutionary Continuity of Mental Experience.* New York: Rockefeller University Press, 1976.

Grinnell, George. "The Rise and Fall of Darwin's First Theory of Transmutation." *Journal of the History of Biology* 7 (1974): 259–73.

Groos, Karl. *Die Spiele der Thiere.* Jena: Fischer, 1896.

Gruber, Howard. *Darwin on Man.* New York: Dutton, 1974.

Gruber, Jacob. *A Conscience in Conflict: The Life of St. George Jackson Mivart.* New York: Columbia University Press, 1960.

Guer, Jean-Antoine. *Histoire critique de l'ame des bêtes.* 2 vols. Amsterdam: Changuion, 1749.

Haeckel, Ernst. *Anthropogenie oder Entwicklungsgeschichte des Menschen.* 4th ed. 2 vols. Leipzig: Engelmann, 1891.

———. *Generelle Morphologie.* 2 vols. Berlin: Reimer, 1866.

———. *Die Lebenswunder: Gemeinverständliche Studien oder Biologische Philosophie.* Stuttgart: Kröner, 1904.

———. *Natürliche Schöpfungsgeschichte.* 4th ed. Berlin: Reimer, 1875; originally published in 1868.

Hahn, Roger. *The Anatomy of a Scientific Institution: The Paris Academy of Sciences, 1666–1803.* Berkeley: University of California Press, 1971.

Haldane, J. B. S. *Causes of Evolution.* London: Longmans, Green, 1932.

———. *Heredity and Politics.* New York: Norton, 1938.

Hale, Matthew. *Human Science and Social Order: Hugo Münsterberg and the Origins of Applied Psychology.* Philadelphia: Temple University Press, 1980.

Hall, A. Rupert. *The Scientific Revolution, 1500–1800.* Boston: Beacon, 1966.

Haller, Albrecht von. *First Lines of Physiology.* Notes by H. A. Wrisberg. Translation of 4th German ed. Edinburgh: Elliot, 1786.

Hamilton, W. D. "The Genetical Evolution of Social Behavior." *Journal of Theoretical Biology* 7 (1964): 1–16, 17–51.

Hamilton, Sir William. "Philosophy of the Unconditioned." (1829). In *Discussions on Philosophy and Literature.* New York: Harper, 1853.

Hanson, Norwood Russell. *Patterns of Discovery.* Cambridge: Cambridge University Press, 1970.

Harris, Ben. "Whatever Happened to Little Albert?" *American Psychologist* 34 (1979): 151–60.

Harris, Marvin. *The Rise of Anthropological Theory.* London: Routledge and Kegan Paul, 1968.

Hartley, David. *Observations on Man.* 2 vols. London: Leake, Frederick, Hitch, and Austen, 1749.

Hartmann, Eduard von. *Philosophie des Unbewussten.* 6th ed. Berlin: Dunckers, 1874; originally published in 1868.

———. *Das Problem des Lebens.* Bad Sachsa im Harz: Haacke, 1906.

Harvard University, Teachers of. *A Guide to Reading in Social Ethics and Allied Subjects.* Cambridge, Mass.: Harvard University, 1910.

Hebb, Donald. "Heredity and Environment in Mammalian Behaviour." *British Journal of Animal Behaviour* 1 (1953): 43–47.

Herbert, Sandra. "The Place of Man in the Development of Darwin's Theory of Transmutation." *Journal of the History of Biology* 10 (1977): 155–227.

Herrick, C. Judson. *The Evolution of Human Nature.* New York: Harper Torchbooks, 1961; originally published in 1956.

Herrnstein, R. J. "Nature as Nurture: Behaviorism and the Instinct Doctrine." *Behaviorism* 1 (1974): 23–52.

Hesse, Mary. "Reasons and Evaluation in the History of Science." In *Changing*

Perspectives in the History of Science, edited by M. Teich and R. Young. London: Heinemann, 1973.

High, Richard. "Shadworth Hodgson and William James's Formulation of Space Perception." *Journal of History of the Behavioral Sciences* 17 (1981): 466–85.

High, Richard, and William Woodward. "William James and Gordon Allport: Parallels in their Maturing Conceptions of Self and Personality." In *Psychology: Theoretical-Historical Perspectives,* edited by R. Rieber and K. Salzinger. New York: Academic Press, 1980.

Hill, David Jayne. "Introduction." In *The Man versus the State: a Collection of Essays by Herbert Spencer,* edited by Truxton Beale. New York: Kennerley, 1916.

Himmelfarb, Gertrude. *Darwin and the Darwinian Revolution.* New York: Norton, 1968.

———. *The Idea of Poverty.* New York: Knopf, 1984.

———. *Marriage and Morals among the Victorians.* New York: Knopf, 1986.

Hodge, M. J. S., and David Kohn. "The Immediate Origins of Natural Selection." In *The Darwinian Heritage,* edited by David Kohn. Princeton: Princeton University Press, 1985.

Hodgskin, Thomas. *The Natural and Artificial Right of Property Contrasted.* London: Steil, 1832.

Hoeveler, J. David, Jr. *James McCosh and the Scottish Intellectual Tradition.* Princeton: Princeton University Press, 1981.

Hofstadter, Richard. *Social Darwinism in American Thought.* Rev. ed. Boston: Beacon, 1955.

Hollinger, David. "William James and the Culture of Inquiry." *Michigan Quarterly Review* 20 (1981): 264–83.

Holt, Niles. "Ernst Haeckel's Monistic Religion." *Journal of the History of Ideas* 32 (1971): 265–80.

———. "Monists & Nazis: a Question of Scientific Responsibility." *Hastings Center Report* 5 (1975): 37–43.

Holton, Gerald. *Thematic Origins of Scientific Thought: Kepler to Einstein.* Cambridge: Harvard University Press, 1973.

Hopkins, William. "Physical Theories of the Phenomena of Life." *Fraser's Magazine* 61 (1860): 739–53; 62 (1860): 74–90.

Howard, George E. *Social Psychology: An Analytical Reference Syllabus.* Lincoln: University of Nebraska Press, 1910.

Hull, David. "Altruism in Science: A Sociobiological Model of Co-operative Behavior among Scientists." *Animal Behavior* 26 (1978): 685–97.

———. "Central Subjects and Historical Narratives." *History and Theory* 14 (1975): 253–74.

———. *Darwin and His Critics.* Cambridge: Harvard University Press, 1973.

———. "Darwinism as a Historical Entity: A Historiographic Proposal." In *The Darwinian Heritage,* edited by David Kohn. Princeton: Princeton University Press, 1985.

———. "Exemplars and Scientific Change." *PSA 1982* 2 (1983): 479–503.

————. "The Naked Meme." In *Learning, Development, and Culture,* edited by H. C. Plotkin. New York: John Wiley, 1982.

Hume, David. *Treatise of Human Nature.* Edited by L. A. Selby-Bigge. Oxford: Clarendon, 1888; a reprint of the original edition of 1739.

Huxley, Julian. *Evolution: the Modern Synthesis.* London: Allen & Unwin, 1942.

Huxley, Leonard. *Life and Letters of Thomas H. Huxley.* 2 vols. New York: D. Appleton, 1900.

Huxley, Thomas. *Collected Essays.* 9 vols. New York:D. Appleton, 1896–1902.

————. "Mr. Darwin's Critics." *The Contemporary Review* 18 (1871): 443–76.

————. *Evidence as to Man's Place in Nature.* London: Williams & Norgate, 1863.

————. "On the Hypothesis that Animals Are Automata and Its History." *Fortnightly Review* 22 (1874): 555–89.

Huxley, Thomas, and Julian Huxley. *Touchstone for Ethics, 1893–1943.* New York: Harper, 1947.

Inglis, Brian. *Natural and Supernatural: A History of the Paranormal from Earliest Times to 1914.* London: Hodder & Stoughton, 1977.

"Instinct." In vol. 3 of the *Supplement à l'Encyclopédie.* 4 vols. Amsterdam: Rey, 1776–1777.

"Introduction." *Zoological Journal* 1 (1824): vi–vii.

Jacob, Margaret. *The Newtonians and the English Revolution.* Ithaca, N.Y.: Cornell University Press, 1976.

Jaeger, Siegfried. "Origins of Child Psychology: William Preyer." In *The Problematic Science: Psychology in Nineteenth-Century Thought,* edited by William Woodward and Mitchell Ash. New York: Praeger, 1982.

James, Henry. *The Portrait of a Lady.* New York: New American Library, 1963; first published in 1881.

James, William. "Address at the Emerson Centenary in Concord" (1903). In *The James Family: A Group Biography,* by F. O. Mathiessen. New York: Vintage, 1980.

————. "Are We Automata?" *Mind* 4 (1879): 1–22.

————. "Bain and Renouvier." *Nation* 22 (1876): 367–69.

————. "The Feeling of Effort." *Anniversary Memoirs of the Boston Society of Natural History.* Boston: Boston Society of Natural History, 1880.

————. "Great Men, Great Thoughts, and the Environment." *Atlantic Monthly* 46 (1880): 441–59.

————. "Herbert Spencer." *Atlantic Monthly* 94 (1904): 99–108.

————. "Herbert Spencer's Autobiography." (1904). In *Memories and Studies.* New York: Greenwood, 1968; originally published in 1911.

————. "Introduction." *The Literary Remains of the Late Henry James,* edited by William James. Boston: Osgood, 1885.

————. *The Letters of William James.* Edited by Henry James. 2d ed. 2 vols. Boston: Little, Brown, Brown, 1926.

————. *The Principles of Psychology.* 2 vols. New York: Henry Holt, 1890.

————. "Quelques considérations sur la méthode subjective." *Critique philosophique* 2 (1878): 407–13.

————. "Rationality, Activity and Faith." *Princeton Review* 2 (1882): 58–86.

————. "Remarks on Spencer's Definition of Mind as Correspondence." *Journal of Speculative Philosophy* 12 (1878): 1–18.

————. "Review of *Grundzüge der physiologischen Psychologie* by William Wundt." *North American Review* 121 (1875): 195–201.

————. "Review of *Variation of Animals and Plants Under Domestication.*" *Atlantic Monthly* 22 (1868): 122–24.

————. "Review of *Variation of Animals and Plants Under Domestication.*" *North American Review* 107 (1868): 362–68.

————. "Some Human Instincts." *Popular Science Monthly* 21 (1887): 160–70, 666–81.

————. "On Some Omissions of Introspective Psychology." *Mind* 9 (1884): 1–26.

————. "What is an Emotion?" *Mind* 9 (1884): 188–205.

————. "What Is an Instinct?" *Scribner's Magazine* 1 (1887): 355–65.

Jaynes, Julian, and William Woodward. "In the Shadow of the Enlightenment." *Journal of the History of the Behavioral Sciences* 10 (1974): 3–15, 144–59.

Jennings, H. S. *The Biological Basis of Human Nature*. New York: Norton, 1930.

Jesuit Fathers at Coimbra. *In octo libros Physicorum Aristotelis Stagirita,* prima pars. Coloniae: Zetznerius, 1602.

Jones, Greta. *Social Darwinism and English Thought*. Atlanta Highlands, N. J.: Humanities Press, 1980.

Kalikow, Theo. "History of Konrad Lorenz's Ethological Theory, 1927–1939." *Studies in the History and Philosophy of Science* 6 (1975): 331–41.

————. "Konrad Lorenz's 'Brown Past': A Reply to Alec Nesbitt." *Journal of the History of the Behavioral Sciences* 14 (1978): 173–79.

————. "Konrad Lorenz's Ethological Theory: Explanation and Ideology, 1938–1943." *Journal of the History of Biology* 16 (1983): 39–73.

————. "Konrad Lorenz's Ethological Theory, 1939–1943." *Philosophy of the Social Sciences* 6 (1976): 15–34.

Kant, Immanuel. *Critick of Pure Reason*. Translated by Francis Haywood. 2d ed. London: Pickering, 1848.

————. *Kritik der reinen Vernunft* (1787). Vol. 2 of *Immanuel Kant: Werke in sechs Bänden,* edited by W. Weischedel. Wiesbaden: Insel, 1956.

Kelly, Alfred. *The Descent of Darwin: The Popularization of Darwinism in Germany, 1860–1914.* Chapel Hill: University of North Carolina Press, 1981.

Kennedy, James. *Herbert Spencer*. Boston: Twayne, 1978.

King-Hele, Desmond. *Doctor of Revolution: the Life and Genius of Erasmus Darwin*. London: Faber & Faber, 1977.

Kirby, William. *On the Power, Wisdom, and Goodness of God as Manifested in the Creation of Animals and in Their History Habits and Instincts*. 2 vols. Seventh Bridgewater Treatise. London: Pickering, 1835.

Kirby, William and William Spence. *Introduction to Entomology*. 2d ed. 4 vols. London: Longman, Hurst, Rees, Orme, and Brown, 1818.

Kirkman, Thomas. *Philosophy without Assumptions*. London: Longmans, Green, 1876.

Knight, Thomas Andrew. "On the Hereditary Instinctive Propensities of Animals." *Philosophical Transactions of the Royal Society of London* (1837): 365–69.

Knoll, Elizabeth. "The Science of Language and the Evolution of Mind: Max Müller's Quarrel with Darwinism." *Journal of the History of Behavioral Sciences* 22 (1986): 3–22.

Koehler, O. "Konrad Lorenz 60 Jahre." *Zeitschrift für Tierpsychologie* 20 (1963): 385–401.

Kohn, David, ed. *The Darwinian Heritage*. Princeton: Princeton University Press, 1985.

———. "Theories to Work By: Rejected Theories, Reproduction, and Darwin's Path to Natural Selection." *Studies in the History of Biology* 4 (1980): 67–170.

———. "Darwin's Principle of Divergence as Internal Dialogue." In *The Darwinian Heritage*, edited by David Kohn. Princeton: Princeton University Press, 1985.

Kohn, David, Sydney Smith, and Robert Stauffer. "New Light on *The Foundations of the Origin of Species*: A Reconstruction of the Archival Record." *Journal of the History of Biology* 15 (1982): 419–42.

Kottler, Malcolm. "Alfred Russel Wallace, the Origin of Man, and Spiritualism." *Isis* 65 (1974): 145–92.

———. "Charles Darwin and Alfred Russel Wallace: Two Decades of Debate over Natural Selection." In *The Darwinian Heritage*, edited by David Kohn. Princeton: Princeton University Press, 1985.

———. "Charles Darwin's Biological Species Concept." *Annals of Science* 35 (1978): 275–97.

———. "Darwin, Wallace, and the Origin of Sexual Dimorphism." *Proceedings of the American Philosophical Society* 124 (1980): 203–26.

Koyré, Alexandre. *From the Closed World to the Infinite Universe*. Baltimore: Johns Hopkins University Press, 1957.

———. *Metaphysics and Measurement: Essays in Scientific Revolution*. Cambridge: Harvard University Press, 1968.

———. *Newtonian Studies*. Chicago: University of Chicago Press, 1969.

Kropotkin, Prince Peter. *Ethics*. New York: Benjamin Blom, 1968; first published posthumously in 1924.

———. *Mutual Aid*. Boston: Extending Horizons, 1955; reprint of the original edition of 1902.

Kuhn, Thomas. "Reflections on My Critics." In *Criticism and the Growth of Knowledge*, edited by Imre Lakatos and Alan Musgrave. Cambridge: Cambridge University Press, 1970.

———. *The Structure of Scientific Revolutions*. 2d ed. Chicago: University of Chicago Press, 1970.

Kuklick, Bruce. *The Rise of American Philosophy*. New Haven: Yale University Press, 1977.

Kuo, Z. Y. "The Net Result of the Anti-Heredity Movement in Psychology." *Psychological Review* 36 (1929): 181–99.

Lacépède, Bernard-Germain-Etienne, Comte de. *Histoire naturelle des poissons*. 5 vols. Paris: Plassan, 1798–1803.

La Chambre, Cureau de. *Les characters des passions.* 2d ed. Amsterdam: Michel, 1685; first published in 1640.

Lakatos, Imre. *The Methodology of Scientific Research Programmes: Philosophical Papers of Imre Lakatos.* Vol. 1. Cambridge: Cambridge University Press, 1978.

Lamarck, Jean-Baptiste de. "Habitudes" (1817). In vol. 14 of *Nouveau dictionnaire d'histoire naturelle.* 36 vols. Paris: Deterville, 1803–1819.

———. *Histoire naturelle des animaux san vertèbres.* 7 vols. Paris: Verdiere, 1815–1822.

———. *Inédits de Lamarck.* Edited by M. Vachon, G. Rousseau, and Y. Laissus. Paris: Masson et Cie Editeurs, 1972.

———. "Instinct" (1817). In vol. 16 of *Nouveau dictionnaire d'histoire naturelle.* 36 vols. Paris: Deterville, 1803–1819.

———. *Philosophie zoologique.* 2 vols. Paris: Dentu, 1809.

———. *Recherches sur l'organisation des corps vivans.* Paris: Maillard, 1802.

———. *Système des animaux sans vertèbres.* Paris: Lamarck et Deterville, 1801.

La Mettrie, Julien Offray de. *L'homme Machine.* Edited by Aram Vartanian. Princeton: Princeton University Press, 1960.

Laudan, Larry. *Progress and Its Problems: Toward a Theory of Scientific Growth.* Berkeley: University of California Press, 1977.

La Vergata, Antonello. "Images of Darwin: A Historiographic Overview." In *The Darwinian Heritage,* edited by David Kohn. Princeton: Princeton University Press, 1985.

Lehrman, Daniel. "A Critique of Konrad Lorenz's Theory of Instinctive Behavior." *Quarterly Review of Biology* 28 (1953): 424–34.

———. "Semantic and Conceptual Issues in the Nature-Nurture Problem." In *Evolution and Development of Behavior,* edited by L. Aronson et al. San Francisco: Freeman, 1970.

[Le Roy, Charles-Georges.] "Instinct." In vol. 8 of *Encyclopédie ou dictionnaire raisonné des sciences, des arts et des métiers,* edited by Denis Diderot et Jean d'Alembert. 17 vols. Paris: Faulche, 1751–1765.

———. *Lettres sur les animaux.* New ed. Nuremburg: Saugrain, 1781; first published in 1768.

Lesch, John. "The Role of Isolation in Evolution: George J. Romanes and John T. Gulick." *Isis* 66 (1975): 483–503.

Lewontin, Richard. "Adaptation." *Scientific American* 239 (September 1978): 213–30.

———. "How Do We Explain the Major Features of Evolution." Fishbein Center Symposium on Persistent Controversies in Evolutionary Theory, the University of Chicago, March 1982.

———. "The Units of Selection." *Annual Review of Ecology and Systematics* 1 (1970): 1–23.

Lewontin, Richard, Steven Rose, and Leon Kamin. *Not in Our Genes: Biology, Ideology, and Human Nature.* New York: Pantheon, 1984.

Limoges, Camille. "The Development of the Muséum d'Histoire Naturelle of Paris." In *The Organization of Science and Technology in France, 1808–1914,* edited by Robert Fox and George Weisz. Cambridge: Cambridge University Press, 1980.

Locke, John. *An Essay Concerning Human Understanding.* Edited by John Yolton. 2 vols. 5th ed. New York: Everyman, 1965.

Lorenz, Konrad. "Beobachtungen an Dohlen." *Journal für Ornithologie* 75 (1927): 511–19.

———. "Durch Domestikation verursachte Störungen arteigenen Verhaltens." *Zeitschrift für angewandte Psychologie und Charakterkunde* 59 (1940): 1–81.

———. *Evolution and the Modification of Behavior.* Chicago: University of Chicago Press, 1965.

———. "Introduction." In *The Expression of the Emotions in Man And Animals,* by Charles Darwin. Chicago: University of Chicago Press, 1965; reprint of original edition of 1872.

———. *Studies in Animal and Human Behavior.* Translated by Robert Martin. 2 vols. Cambridge, Mass.: Harvard University Press, 1971.

———. "Systematik und Entwicklungsgedanke im Unterricht." *Der Biologe* 9 (1940): 24–36.

———. "Über den Begriff der Instinkthandlung." *Folia Biotheoretica* 2 (1937): 17–50.

———. *Über tierisches und menschliches Verhalten.* 2 vols. München: Piper, 1965.

Lubbock, John. *The Origin of Civilization.* London: Williams & Norgate, 1870.

———. *Prehistoric Times.* London: Williams & Norgate, 1865.

Lubeck, Ian. "Histoire de psychologies sociales perdues: le cas de Gabriel Tarde." *Revue française de sociologie* 22 (1981): 361–95.

Ludmerer, Kenneth. *Genetics and American Society.* Baltimore: Johns Hopkins University Press, 1972.

Lyell, Charles. *Sir Charles Lyell's Scientific Journals on the Species Question.* Edited by Leonard Wilson. New Haven: Yale University Press, 1970.

———. *The Geological Evidences of the Antiquity of Man.* London: Murray, 1863.

———. *Principles of Geology.* 3 vols. London: Murray, 1830–1833.

McCawley, James. *Everything that Linguists have Always Wanted to Know about Logic.* Chicago: University of Chicago Press, 1981.

McCosh, James. "Development: What It Can Do and What It Cannot Do." *Realistic Philosophy.* 2 vols. New York: Scribner's, 1887.

———. *Psychology, The Motive Powers.* New York: Scribner's Sons, 1888.

———. *The Religious Aspects of Evolution.* 2d ed. New York: Scribner's Sons, 1890.

McDougall, William. *Body and Mind.* New York: Macmillan, 1911.

———. "Instinct and Intelligence." *British Journal of Psychology* 3 (1909–1910): 250–66.

McGuire, J. E., and P. Rattansi. "Newton and the 'Pipes of Pan.'" *Notes and Records of the Royal Society of London* 21 (1966): 108–43.

Macintosh, James. *Dissertation on the Progress of Ethical Philosophy.* Edinburgh: Adam and Charles Black, 1836.

MacIntyre, Alisdair. *After Virtue.* Notre Dame, Ind.: University of Notre Dame Press, 1981.

McLennan, John. "The Worship of Animals and Plants." *Fortnightly Review* 12 (1869): 407–27, 562–82; 13 (1870): 194–216.

MacLeod, Roy. "The X-Club, A Social Network of Science in Late-Victorian England." *Notes and Records of the Royal Society of London* 24 (1970): 305–22.

Maier, N. R., and T. C. Schneirla. *Principles of Animal Psychology.* 2d ed. New York: Dover, 1964; 1st ed. in 1935.

Mandelbaum, Maurice. *History, Man & Reason: A Study in Nineteenth-Century Thought.* Baltimore: Johns Hopkins University Press, 1974.

Manier, Edward. *The Young Darwin and His Cultural Circle.* Dordrecht: Reidel, 1978.

Mansel, *Prolegomena Logica: an Inquiry into the Psychological Character of Logical Processes.* Oxford: Graham, 1851.

Maravia, Sergio. "From *Homme Machine* to *Homme Sensible.*" *Journal of the History of Ideas* 39 (1978): 45–60.

Marchant, James. *Alfred Russel Wallace: Letters and Reminiscences.* 2 vols. London: Cassell, 1916.

Marshall, Marilyn, and Russel Wendt. "Wilhelm Wundt, Spiritism, and the Assumptions of Science." In *Wundt Studies,* edited by W. Bringmann and R. Tweney. Toronto: Hogrefe, 1980.

Martineau, Harriet. *Cousin Marshall.* Vol. 7 of *Illustrations of Political Economy.* Boston: Bowles, 1833.

———. *How to Observe: Manners and Morals.* New York: Harper, 1838.

———. *Life in the Wilds.* Vol. 1 of *Illustrations of Political Economy.* Boston: Bowles, 1833.

———. *Weal and Woe in Garveloch.* Vol. 6 of *Illustrations of Political Economy.* Boston: Bowles, 1833.

Marx, Karl, and Friedrich Engels. *The Manifesto of the Communist Party.* In *Karl Marx, The Revolution of 1848: Political Writings, Volume 1,* edited by David Fernbach. New York: Vintage, 1974.

Mattern, Ruth. "Altruism, Ethics, and Sociobiology." In *The Sociobiology Debate,* edited by Arthur Caplan. New York: Harper, 1978.

Mayr, Ernst. *Animal Species and Evolution.* Cambridge: Harvard University Press, 1963.

———. *Evolution and the Diversity of Life.* Cambridge: Harvard University Press, 1976.

———. *The Growth of Biological Thought.* Cambridge: Harvard University Press, 1982.

Mazlish, Bruce. *James and John Stuart Mill.* New York: Basic Books, 1975.

"Meeting of The British Association, Section D., Saturday," *Athenaeum* (14 July 1860), pp. 64–5.

"A Member of the Class of 1878." *Harvard Graduates' Magazine* 39 (1920): 324.

Merz, John Theodore. *A History of European Scientific Thought in the Nineteenth Century.* 2 vols. New York: Dover, 1965; first published in 1904–1912.

Mill, John Stuart. *Autobiography of John Stuart Mill.* New York: New American Library, 1964; originally published in 1873.

————. *The Collected Works of John Stuart Mill.* Edited by J. M. Robson. 20 vols. to date. Toronto: University of Toronto Press, 1963————.

————. *Dissertations and Discussions.* 5 vols. New York: Holt, 1882.

————. *An Examination of Sir William Hamilton's Philosophy.* 4th ed. London: Longmans, Green, Reader, and Dyer, 1872.

————. *A System of Logic.* New York: Harper & Brothers, 1848; reprint of 1st ed. of 1843.

Milne-Edwards, H. *Outlines of Anatomy and Physiology.* Translated by J. Lane. Boston: Little and Brown, 1841.

Mitzman, Arthur. *The Iron Cage: An Historical Interpretation of Max Weber.* New York: Grosset & Dunlap, 1971.

Mivart, St. George. *Collected Essays and Criticism.* 2 vols. London: Osgood, McIlvaine, 1892.

————. "The Continuity of Catholicism." *Nineteenth Century* 47 (1900): 51–72.

————. "Darwin's *Descent of Man*." *Quarterly Review* 131 (1871): 47–90.

————. "Difficulties of the Theory of Natural Selection." *The Month* 11 (1869): 35–53, 134–53, 274–89.

————. "Evolution in Professor Huxley." *Nineteenth Century* 34 (1893): 198–211.

————. "An Explanation." *Fortnightly Review* 46 (1886): 525–27.

————. *On the Genesis of Species.* New York: D. Appleton, 1871.

————. "Happiness in Hell." *Nineteenth Century* 32 (1892): 899–919.

————. "The Happiness in Hell: a Rejoinder." *Nineteenth Century* 33 (1893): 320–38.

————. "Last Words on the Happiness in Hell." *Nineteenth Century* 33 (1893): 635–51.

————. "On Lepilemur and Cheirogaleus and on the Zoological Rank of the Lemuroidea." *Proceedings of the Zoological Society of London* 41 (1873): 485–510.

————. "Letter to the Editor." *Contemporary Review* 44 (1883): 156.

————. "Modern Catholics and Scientific Freedom." *Nineteenth Century* 18 (1885): 30–47.

————. *Nature and Thought: An Introduction to a Natural Philosophy.* London: Kegan Paul, Trench, 1882.

————. "Organic Nature's Riddle." *Fortnightly Review* 43 (1885): 323–37, 519–31.

————. *The Origin of Human Reason.* London: Kegan Paul, Trench, 1889.

————. "The Rights of Reason." *Fortnightly Review* 45 (1886): 61–68.

————. "Scripture and Roman Catholicism." *Nineteenth Century* 47 (1900): 425–42.

————. "Some Recent Catholic Apologists." *Fortnightly Review* 67 (1900): 24–44.

————. "Some Reminiscences of Thomas Henry Huxley." *Nineteenth Century* 42 (1897): 985–98.

[Morley, John]. "Mr. Darwin on Conscience." *Pall Mall Gazette* (12 April 1871): 10–11.

————. "The Descent of Man." *Pall Mall Gazette* (21 March 1871): 11–12.

Montague, Ashley. *Darwin, Competition and Cooperation*. New York: Schuman, 1952.

Montgomery, William. "Charles Darwin's Thought on Expressive Mechanisms in Evolution." In *The Development of Expressive Behavior: Biology-Environment Interactions*. New York: Academic Press, 1985.

Moore, G. E. *Principia Ethica*. Cambridge: Cambridge University Press, 1929; reprint of 1st ed. of 1903.

Moore, James. *The Post-Darwinian Controversies: A Study of the Protestant Struggle to Come to Terms with Darwin in Great Britain and America, 1870–1900*. Cambridge: Cambridge University Press, 1979.

Morgan, C. Lloyd. "Animal Automatism and Consciousness." *Monist* 7 (1896–1897): 1–18.

———. *Animal Behaviour*. London: Arnold, 1900.

———. *Animal Life and Intelligence*. Boston: Ginn, 1890–1891.

———. "Autobiography." In *A History of Psychology in Autobiography*, edited by Carl Murchison. Worcester, Mass.: Clark University Press, 1932.

———. "Biology and Metaphysics." *Monist* 9 (1898–1899): 538–62.

———. "Causation, Physical and Metaphysical." *Monist* 8 (1897–1898): 230–49.

———. "The Doctrine of Auta." *Monist* 3 (1892–1893): 161–75.

———. *Emergent Evolution*. London: Williams & Norgate, 1923.

———. *Instinct and Experience*. London: Methuen, 1912.

———. *Interpretation of Nature*. New York: Putnam's Sons, 1906.

———. *An Introduction to Comparative Psychology*. London: Walter Scott, 1895.

———. "The Law of Psychogenesis." *Mind*, n.s. 1 (1892): 72–93.

———. *Life, Mind, and Spirit*. London: Williams & Norgate, 1926.

———. "The Limits of Animal Intelligence." *Fortnightly Review* 60 (1893): 223–39.

———. "Limits of Animal Intelligence." *International Congress of Experimental Psychology*. Second Session. London: Williams & Norgate, 1892.

———. "Mental Evolution." *Monist* 2 (1891–1892): 161–77.

———. "Of Modification and Variation." *Science* 4 (1896): 733–40.

———. "Naturalism." *Monist* 6 (1895–1896): 76–90.

———. "Natural Selection and Elimination." *Nature* 38 (1888): 370.

———. "The Philosophy of Evolution." *Monist* 8 (1897–1898): 481–501.

———. "The Scope of Psycho-Physiology." *Nature* 49 (1894): 504–5.

———. *The Springs of Conduct: An Essay in Evolution*. London: Kegan Paul, Trench, 1885.

———. "Three Aspects of Monism." *Monist* 4 (1894): 321–32.

Morgan, Thomas H. *A Critique of the Theory of Evolution*. Princeton: Princeton University Press, 1916.

———. *Evolution and Adaptation*. New York: Macmillan, 1903.

Morrell, Jack, and Arnold Thackray. *Gentlemen of Science: Early Years of the British Association for the Advancement of Science*. Oxford: Oxford University Press, 1981.

Müller, Friedrich Max. *Lectures on the Science of Language delivered at the Royal Institution of Great Britain in April, May, & June, 1861.* 4th ed. London: Longman, Green, Longman, Roberts, & Green, 1864.

―――. *Lectures on the Science of Language delivered at the Royal Institution of Great Britain in February, March, April, & May, 1863.* London: Longman, Green, Longman, Roberts, & Green, 1864.

―――. "The Science of Language." *Nature* 1 (1870): 256–59.

―――. *The Science of Thought.* 2 vols. New York: Charles Scribner's Sons, 1887.

Müller, Johannes. *Elements of Physiology.* Translated by W. Baly. 2 vols. London: Taylor and Walton, 1838–1842.

Myers, Charles. "Instinct and Intelligence." *British Journal of Psychology* 3 (1909–1920): 209–18.

Myers, Gerald. "William James' Theory of Emotion." *Transactions of the Charles S. Peirce Society* 2 (1969): 67–89.

Nagel, Thomas. "What Is It Like to be a Bat?" *Philosophical Review* 83 (1974): 435–50.

"Natural and Supernatural Selection." *Spectator* 41 (1868): 1155.

Nisbett, Alec. *Konrad Lorenz: A Biography.* New York: Harcourt Brace Jovanovich, 1977.

Nordenskiöld, Erik. *The History of Biology.* 2d ed. New York: Tudor, 1936; originally published in 1920–1924.

O'Donnell, John. *The Origins of Behaviorism: American Psychology, 1870–1920.* New York: New York University Press, 1985.

Oppenheim, Janet. *The Other World: Spiritualism and Psychical Research in England, 1850–1914.* Cambridge: Cambridge University Press, 1985.

Osborn, Henry F. *From the Greeks to Darwin.* New York: Scribner's, 1894.

―――. "The Limits of Organic Selection." *American Naturalist* 31 (1897): 944–951.

―――. "Organic Selection." *Science* 4 (1897): 583–87.

Ospovat, Dov. *The Development of Darwin's Theory.* Cambridge: Cambridge University Press, 1981.

―――. "The Influence of Karl Ernst von Baer's Embryology, 1829–1859." *Journal of the History of Biology* 9 (1976): 1–28.

Paley, William. *Paley's Natural Theology Illustrated.* Introductory and concluding volumes by Henry Lord Brougham. Edited by Henry Lord Brougham and Sir Charles Bell. 5 vols. London: Knight, 1835–1839.

―――. *The Works of William Paley.* Philadelphia: Woodward, n. d.

Pardies, Gaston. *Discours de la connoissance des bestes.* Paris: Delaulne, 1672.

Parshall, Karen. "Varieties as Incipient Species: Darwin's Numerical Analysis." *Journal of History of Biology* 15 (1982): 191–214.

Pauly, August. *Darwinismus und Lamarckismus: Entwurf einer psychophysischen Teleologie.* München: Reinhardt, 1905.

Pauly, Philip. "Psychology at Hopkins." *Johns Hopkins Magazine* 30 (1979): 36–41.

Pearl, Raymond. "The Biology of Superiority." *American Mercury* 12 (1927): 257–66.

Pearson, Karl. *Life, Letters, and Labours of Francis Galton.* 3 vols. Cambridge: Cambridge University Press, 1914–1930.

Peel, J. D. Y. *Herbert Spencer: The Evolution of a Sociologist.* New York: Basic Books, 1971.

Penfield, Wilder. *Mystery of the Mind.* Princeton: Princeton University Press, 1974.

Perry, Ralph Barton. *In the Spirit of William James.* New Haven: Yale University Press, 1938.

——. *The Thought and Character of William James.* 2 vols. Boston: Little, Brown, 1935.

Piaget, Jean. *Behavior and Evolution.* Translated by D. Nicholson-Smith. New York: Random House, 1978.

Pickering, Andrew. *Constructing Quarks: A Sociological History of Particle Physics.* Chicago: University of Chicago Press, 1984.

Popper, Karl. *Conjectures and Refutations: The Growth of Scientific Knowledge.* 2d ed. New York: Harper & Row, 1968.

——. *The Logic of Scientific Discovery.* 2d ed. New York: Harper & Row, 1968.

——. "Natural Selection and the Emergence of Mind." *Dialectica* 32 (1978): 339–55.

——. *Objective Knowledge: An Evolutionary Approach.* Oxford: Oxford University Press. 1972.

Preyer, Wilhelm. *The Mind of the Child.* Translated by H. Brown. Introduction by G. Stanley Hall. 2 vols. New York: D. Appleton, 1888–1889.

——. *Die Seele des Kindes: Beobachtungen über die geistige Entwicklung des Menschen in den ersten Lebensjahren.* 4th ed. Leipzig: Grieben, 1895; originally published in 1882.

Proctor, Robert. *Racial Hygiene: Medicine under the Nazis.* Oxford: Oxford University Press, forthcoming.

"Proudhon on Government," *Leader* 2 (1851): 997.

Provine, William. *The Origins of Theoretical Population Genetics.* Chicago: University of Chicago Press, 1971.

——. *Sewall Wright and Evolutionary Biology.* Chicago: University of Chicago Press, 1986.

Quinton, Anthony. "Ethics and the Theory of Evolution." In *Biology and Personality,* edited by I. T. Ramsey. Oxford: Blackwell, 1966.

Rattansi, P. "Some Evaluations of Reason in Sixteenth- and Seventeenth-Century Natural Philosophy." In *Changing Perspectives in the History of Science,* edited by M. Teich and R. Young. London: Heinemann, 1973.

Réamur, René-Antoine de. *Mémoires pour servir à l'histoire de insectes.* 6 vols. Paris: L'Imprimerie Royale, 1734–1742.

Reck, Andrew. "The Philosophical Psychology of William James." *Southern Journal of Philosophy* 9 (1971): 293–312.

Reimarus, Hermann Samuel. *Abhandlungen von den vornehmsten Wahrheiten der natürlichen Religion.* 5th ed. Tübingen: Frank und Schram, 1782; first published in 1754.

————. *Allgemeine Betrachtungen uber die Triebe der Thiere.* 3d ed. Hamburg: Bohn, 1773; first published in 1760.

Rennie, James. *Insect Architecture.* London: Knight, 1830.

Renouvier, Charles. *Essais de critique générale, Deuxième essai: Traité de psychologie rationnelle.* 2d ed. 2 vols. Paris: Librarie Armand Colin, 1912; a reprint of ed. of 1875.

"Review of Brougham's *Dissertations on Subjects of Science.*" *Athenaeum* (2 February 1839): 91.

"Review of Michelet, Comte, Mill, Fichte, Spencer, and Newman." *North British Review* 15 (1851): 291–330.

"Review of *The Vestiges of the Natural History of Creation.*" *Blackwood's Magazine* 57 (1845): 448–60.

Ribot, Théodule. *German Psychology of To-day.* Translated by James Mark Baldwin. Introduction by James McCosh. New York: Scribner's, 1886.

Richards, Robert. "Defense of Evolutionary Ethics." *Biology and Philosophy* 1 (1986): 265–93.

————. "Influence of Sensationalist Tradition on Early Theories of the Evolution of Behavior." *Journal of the History of Ideas* 40 (1979): 85–105.

————. "The Innate and the Learned: The Evolution of Konrad Lorenz's Theory of Instinct." *Philosophy of the Social Sciences* 4 (1974): 111–33.

————. "Justification Through Biological Faith: A Rejoinder." *Biology and Philosophy* 1 (1986): 337–54.

————. "The Natural Selection Model of Conceptual Evolution." *Philosophy of Science* 44 (1977): 494–501.

————. "Why Darwin Delayed, or Interesting Problems and Models in the History of Science." *Journal of the History of the Behavioral Sciences* 19 (1983): 45–53.

————. "Wundt's Early Theories of Unconscious Inference and Cognitive Evolution in their relation to Darwinian Biopsychology." In *Wundt Studies,* edited by W. Bringmann and R. Tweney. Toronto: Hogrefe, 1980.

Richerand, Anthelm. *Nouveaux élémens de physiologie.* 5th ed. 2 vols. Paris: Caille et Ravier, 1811; first published in 1802.

Ridely, Mark. "Coadaptation and the Inadequacy of Natural Selection." *British Journal for the History of Science* 15 (1982): 45–68.

Rieber, R., and K. Salzinger, eds. *Psychology: Theoretical-Historical Perspectives.* New York: Academic Press, 1980.

Rinard, Ruth. "The Problem of the Organic Individual: Ernst Haeckel and the Development of the Biogenetic Law." *Journal of the History of Biology* 14 (1981): 249–76.

Roget, Peter Mark. *Animal and Vegetable Physiology Considered with Reference to Natural Theology.* 2 vols. London: Pickering, 1834.

Romanes, Ethel. *Life and Letters of George John Romanes.* 4th ed. London: Longmans, Green, 1897.

Romanes, George. *Animal Intelligence.* 4th ed. London: Kegan Paul, Trench, 1886; unchanged from 1st ed. of 1881.

————. "Animal Life and Intelligence." *Mind* 16 (1891): 262–67.

———— [Physicus, pseud.]. *A Candid Examination of Theism*. London: Kegan Paul, Trench, Trübner, 1892; unchanged from the 1st ed. of 1878.

————. *Christian Prayer and General Laws, being the Burney Prize Essay for the Year 1873*. London: Macmillan, 1874.

————. *Darwin and After Darwin*. 3 vols. 2d ed. Chicago: Open Court, 1897.

————. "Definition of the Theory of Natural Selection." *Nature* 38 (1888): 616–18.

————. "Disuse as a Reducing Cause in Species." *Nature* 10 (1874): 164.

————. "The Fallacy of Materialism." *Nineteenth Century* (1882): 871–88.

————. "Mr. Herbert Spencer on 'Natural Selection.'" *Contemporary Review* 63 (1893): 497–516.

————. "Is There Evidence of Design in Nature?" *Proceedings of the Aristotelian Society* 1 (1891): 66–76.

————. *Jelly-Fish, Star-Fish, and Sea Urchins*. New York: D. Appleton, 1898; originally published in 1885.

————. "Mental Evolution." *Athenaeum* (January–June 1884): 312–13.

————. *Mental Evolution in Animals*. London: Kegan Paul, Trench, 1883.

————. "Mental Evolution in Animals." *Athenaeum* (January–June 1884): 411–12.

————. *Mental Evolution in Man*. London: Kegan Paul, Trench, 1888.

————. *Mind and Motion and Monism*. London: Longmans, Green, 1895.

————. "Mr. Mivart on the Rights of Reason." *Fortnightly Review* 45 (1886): 329–38.

————. "Natural Selection and Dysteleology." *Nature* 9 (1874): 361–62.

————. "Natural Selection and Natural Theology." *Nature* 27 (1883): 362–64, 528–29; 28 (1883): 100–101.

————. "Nature and Thought." *Contemporary Review* 43 (1883): 831–41.

————. "A Note on Panmixia." *Contemporary Review* 64 (1893): 611–12.

————. "Physiological Selection." *Nature* 34 (1886): 314–16, 336–40, 362–65.

————. "Physiological Selection." *Nineteenth Century* 21 (1887): 59–80.

————. "Physiological Selection: An Additional Suggestion on the Origin of Species." *Journal of the Linnean Society: Zoology* 19 (1886): 337–411.

————. "Physiological Selection and the Origin of Species." *Nature* 34 (1886): 407–8, 439.

————. "Professor Mivart on Instinct." *Fortnightly Review* 44 (1885): 90–101.

————. "Rudimentary Organs." *Nature* 9 (1874): 440–41.

————. "The Scientific Evidence of Organic Evolution." The Humboldt Library of Science, no. 40. New York: Humboldt, 1882.

————. "The Spencer-Weismann Controversy." *Contemporary Review* 64 (1893): 50–53.

————. "The Springs of Conduct." *Nature* 33 (1886): 436–37.

————. "The Struggle of Parts in the Organism." *Nature* 24 (1881): 505–6; 25 (1881): 29–30.

————. *Thoughts on Religion*. Edited by Charles Gore. Chicago: Open Court, 1895.

————. "Mr. Wallace on Physiological Selection." *Nature* 35 (1887): 247–48, 390–91.

Rorarius, Hieronymus. *Quod animalia bruta saepe ratione utantur melius homine.* Edited by Gabriel Naude. Amsterdam: Ravesteinium, 1654.

Rosenfield, Lenora. *From Beast-Machine to Man-Machine.* New York: Octagon Books, 1968.

Roule, Louis. *Lamarck et l'interprétation de la nature.* Paris: Flammarion, 1927.

Royce, Josiah. *Herbert Spencer: An Estimate and Review.* New York: Fox, Duffield, 1904.

Rudwick, Martin. "Charles Darwin in London: The Integration of Public and Private Science." *Isis* 73 (1982): 186–206.

————. *The Great Devonian Controversy.* Chicago: University of Chicago Press, 1985.

Ruse, Michael. "Charles Darwin and Group Selection." *Annals of Science* 37 (1980): 615–30.

————. *The Darwinian Revolution.* Chicago: University of Chicago Press, 1979.

————. "Karl Popper's Philosophy of Biology." *Philosophy of Science* 44 (1977): 638–61.

————. "The Morality of the Gene." *Monist* 67 (1984): 167–99.

————. *Sociobiology: Sense or Nonsense?.* Dordrecht: Reidel, 1979.

Russell, Bertrand. *The Autobiography of Bertrand Russell.* 3 vols. Boston: Little, Brown, 1967–1968; Simon and Schuster, 1969.

————. *A History of Western Philosophy.* New York: Simon and Schuster, 1945.

————. *Religion and Science.* New York: Holt, 1935.

Russell, Bertrand, and Patricia Russell. *The Amberley Papers.* 2 vols. London: Allen & Unwin, 1937.

Russell. E. S., *Form and Function: A Contribution to the History of Animal Morphology.* Chicago: University of Chicago Press, 1982; originally published in 1916.

Ryle, Gilbert. *The Concept of Mind.* London: Hutchinson, 1949.

Sahlins, Marshall. *The Use and Abuse of Biology.* Ann Arbor: University of Michigan Press, 1976.

Samelson, Franz. "J. B. Watson's Little Albert, Cyril Burt's Twins, and the Need for a Critical Science." *American Psychologist* 35 (1980): 619–25.

————. "Struggle for Scientific Authority: The Reception of Watson's Behaviorism, 1913–1920." *Journal of the History of the Behavioral Sciences* 17 (1981): 399–425.

Sarton, George. *The History of Science and the New Humanism.* Bloomington: Indiana University Press, 1962.

————. *Sarton on the History of Science: Essays by George Sarton.* Cambridge: Harvard University Press, 1962.

Schleicher, August. *Die Darwinsche Theorie und die Sprachwissenschaft.* Weimar: Bölau, 1863.

Schneewind, J. B. *Sidgwick's Ethics and Victorian Moral Philosophy.* Oxford: Clarendon Press, 1977.

Schneider, Georg. *Der menschliche Wille vom Standpunkte der neueren Entwick-lungstheorien*. Berlin: Dummlers, 1882.

———. *Der thierische Wille*. Leipzig: Abel, 1880.

Schneirla, T. C. "A Consideration of Some Conceptual Trends in Comparative Psychology." *Psychological Bulletin* 49 (1952): 559–97.

Schonewald, Richard. "G. Eliot's 'Love Letters': Unpublished Letters from George Eliot to Herbert Spencer." *Bulletin of the New York Public Library*. 79 (1976): 362–71.

Schweber, Silvan. "Darwin and the Political Economists: Divergence of Character." *Journal of the History of Biology* 13 (1980): 195–289.

———. "The Origin of the *Origin* Revisited." *Journal of the History of Biology* 10 (1977): 231–316.

Schwehn, Mark. "Making the World: William James and the Life of the Mind." *Harvard Library Bulletin* 30 (1982): 426–54.

Sebright, John. *Observations upon the Instincts of Animals*. London: Gossling & Egley, 1836.

Sellars, Wilfrid. "Being and Being Known." In *Science, Perception and Reality*. London: Routledge and Kegan Paul, 1963.

———. "Concepts as Involving Laws and Inconceivable without Them." *Philosophy of Science* 15 (1948): 287–315.

Sellars, Wilfrid, and John Hospers, eds. *Readings in Ethical Theory*. New York: Appleton-Century-Crofts, 1952.

Seward, A. C., ed. *Darwin and Modern Science*. Cambridge: Cambridge University Press, 1909.

Shaw, George Bernard. *Back to Methuselah*. London: Penguin Books, 1961; first published in 1921.

Sherrington, Charles. *Integrative Action of the Nervous System*. New Haven: Yale University Press, 1906.

Shields, Senator. *Congressional Record—Senate* 65, Part 7 (1924): 6461.

Sidgwick, Henry. *Lectures on the Ethics of T. H. Green, Mr. Herbert Spencer, and J. Martineau*. London: Macmillan, 1902.

———. *The Methods of Ethics*. 7th ed. Indianapolis: Hackett, 1981; reprint of last edition, 1907.

Simmel, Georg. "Über eine Beziehung der Selectionslehre zur Erkenntistheorie." *Archiv für systematische Philosophie* 1 (1895): 34–45.

Simpson, George G. "The Baldwin Effect." *Evolution* 7 (1953): 110–17.

Singer, Charles. *A Short History of Scientific Thought*. Oxford: Clarendon, 1962.

Sloan, Phillip. "Darwin, Vital Matter, and the Transformism of Species." *Journal of the History of Biology* 19 (1986): 369–445.

———. "Darwin's Invertebrate Program, 1831–1836: Preconditions for Transformation." In *The Darwinian Heritage,* edited by David Kohn. Princeton: Princeton University Press, 1985.

Smart, J. J. C. "Sensations and Brain Processes." In *The Philosophy of Mind,* edited by V. Chappell. Englewood Cliffs, N. J.: Prentice-Hall, 1962.

Smith, Adam. *The Theory of the Moral Sentiments*. Edited by A. Macfie and

D. Raphael. Glasgow ed. of the works of Adam Smith. Oxford: Oxford University Press, 1976.

Smith, C. U. M. "Evolution and the Problem of Mind: Part I: Herbert Spencer." *Journal of the History of Biology* 15 (1982): 55–88.

Smith, John Maynard. "The Concepts of Sociobiology." In *Morality as a Biological Phenomenon,* edited by Gunther Stent. Berkeley: University of California Press, 1978.

Sober, Elliott. *The Nature of Selection: Evolutionary Theory in Philosophical Focus.* Cambridge, Mass.: M. I. T. Press, 1984.

Spalding, Douglas. "On Instinct." *Nature* 6 (1872): 485–86.

———. "Instinct and Acquisition." *Nature* 12 (1875): 507–8.

———. "Instinct, with Original Observations on Young Animals." *Macmillan's Magazine* 27 (1873): 282–93.

"Douglas A. Spalding." *Nature* 17 (1877): 35–36.

Spencer, Herbert. *Autobiography.* 2 vols. New York: D. Appleton, 1904.

———. "The Developmental Hypothesis." *Leader* 3 (20 March 1852): 280–81.

———. *Education: Intellectual, Moral and Physical.* New York: D. Appleton, 1883; reprint of 1st ed. of 1860.

———. *Essays, Scientific, Political and Speculative.* 3 vols. New York: D. Appleton, 1892.

———. *The Factors of Organic Evolution.* London: Williams & Norgate, 1887.

———. *First Principles.* London: Williams & Norgate, 1862.

———. "The Inadequacy of 'Natural' Selection." *Contemporary Review* 63 (1893): 152–66, 439–56.

———. "Letters." *The Nonconformist,* 15 June to 23 November 1842.

———. "Lyell and Owen on Development." *Leader* 2 (18 October 1851): 996–97.

———. "Mental Evolution in Animals." *Athenaeum* (January–June 1884): 446.

———. "The Origin of Animal Worship." *Fortnightly Review* 13 (1870): 535–50.

———. *Principles of Biology.* 2 vols. New York: D. Appleton, 1884; reprint of first edition of 1864–1867.

———. *The Principles of Ethics.* 2 vols. Indianapolis: Liberty Classics, 1978; reprint of original ed. of 1893.

———. *Principles of Psychology.* London: Longman, Brown, Green, and Longmans, 1855.

———. *The Principles of Psychology.* 2d ed. 2 vols. London: Williams & Norgate, 1872.

———. "Progress: Its Law and Cause." *Westminster and Foreign Quarterly Review* 67 (1857): 445–85.

———. "A Rejoinder to Professor Weismann." *Contemporary Review* 64 (1893): 893–912.

———. *Social Statics: Or, the Conditions Essential to Human Happiness Specified, and the First of them Developed.* London: Chapman, 1851.

———. *Social Statics, Abridged and Revised: together with The Man versus the State.* New York: D. Appleton, 1904; originally published in 1892.

———. *Study of Sociology.* Ann Arbor: University of Michigan Press, 1961; first published in 1873.

———. "A Theory of Population, deduced from the General Law of Animal Fertility." *Westminster Review* 57 (1852): 468–501.

———. "The Universal Postulate." *Westminster Review,* n.s. 3 (1853): 513–50.

———. "Weismannism Once More." *Contemporary Review* 66 (1894): 592–608.

Spiegelberg, Herbert. *The Phenomenological Movement* 2d ed. 2 vols. The Hague: Nijhoff, 1965.

Staum, Martin. *Cabanis: Enlightenment and Medical Philosophy in the French Revolution.* Princeton: Princeton University Press, 1980.

Stein, Jay. *The Mind and the Sword.* New York: Twayne, 1961.

Stent, Gunther, ed. *Morality as a Biological Phenomenon.* Berkeley: University of California Press, 1978.

Stewart, Dugal. *Elements of the Philosophy of the Human Mind.* Vol. 3. Philadelphia: Carey, Lea & Carey, 1827.

Stocking, George. *Race, Culture, and Evolution.* 2d ed. Chicago: University of Chicago Press, 1981.

———. *Victorian Anthropology.* New York: Free Press, 1987.

Stout, G. F. *Analytic Psychology.* 4th ed. 2 vols. London: Allen & Unwin, 1914.

———. "Instinct and Intelligence." *British Journal of Psychology* 3 (1909–1910): 237–49.

———. *A Manual of Psychology.* 5th ed. London: University Tutorial Press, 1938; 1st ed. in 1899.

Strassen, Otto Zur. *Die Neuere Tierpsychologie.* Leipzig: Teubner, 1908.

Strout, Cushing. "William James and the Twice-Born Sick Soul." *Daedalus* 97 (1968): 1062–82.

Stuewer, Roger, ed. *Historical and Philosophical Perspectives on Science. Minnesota Studies in the Philosophy of Science,* vol. 5. Minneapolis: University of Minnesota Press, 1970.

Suarez, Francis. *Opera omnia.* 28 vols. Paris: Vives, 1856–1878.

Sulloway, Frank. "Darwin and His Finches: The Evolution of a Legend." *Journal of the History of Biology* 15 (1982): 1–53.

———. "Darwin's Conversion: The Beagle Voyage and Its Aftermath." *Journal of the History of Biology* 15 (1982): 325–96.

Swisher, Charles. "Charles Darwin on the Origins of Behavior." *Bulletin of the History of Medicine* 41 (1967): 24–43.

Tait, Lawson. "Has the Law of Natural Selection by Survival of the Fittest Failed in the Case of Man?" *Dublin Quarterly Journal of Medical Science,* n.s. 47 (1869): 102–13.

Terman, Lewis. "The Intelligence Quotient of Francis Galton in Childhood." *American Journal of Psychology* 28 (1917): 209–15.

Titchener, Edward. "Wilhelm Wundt" (1921). In *Wundt Studies,* edited by W. Bringmann and R. Tweney. Toronto: Hogrefe, 1980.

Tosti, Gustavo. "Baldwin's Social and Ethical Interpretations." *Science* 15 (1902): 551–53.

————. "The Sociological Theories of Gabriel Tarde." *Political Science Quarterly* 12 (1897): 490–511.

Toulmin, Stephen. *Human Understanding.* Oxford: Oxford University Press, 1972.

Trivers, Robert. "The Evolution of Reciprocal Altruism." *Quarterly Review of Biology* 46 (1971): 35–57.

Turner, Frank. *Between Science and Religion.* New Haven: Yale University Press, 1974.

Tylor, E. B. *Researches into the Early History of Mankind.* 2d ed. London: Murray, 1869.

Uexküll, Jacob von. *Strafzüge durch die Umwelten von Tieren und Menschen.* Berlin: Springer, 1934.

Valen, Leigh Van. "Two Modes of Evolution." *Nature* 257 (1974): 298–300.

Vartanian, Aram. "Trembley's Polyp, La Mettrie, and Eighteenth-Century French Materialism." *Journal of the History of Ideas* 11 (1950): 259–86.

Virey, Julien. "Animal" (1803). In vol. 1 of *Nouveau dictionnaire d'histoire naturelle.* 36 vols. Paris: Deterville, 1803–1819.

————. *Histoire des moeurs et de l'instinct des animaux.* 2 vols. Paris: Deterville, 1882.

————. "Instinct" (1817). In vol. 16 of *Nouveau dictionnaire d'histoire naturelle.* 36 vols. Paris: Deterville, 1803–1819.

Vonèche, Jacques. "An Interview Conducted with Piaget: Reflections on Baldwin." In *The Cognitive Developmental Psychology of James Mark Baldwin,* edited by John Broughton and D. Freeman-Moir. Norwood, N. J.: Ablex, 1982.

Vorzimmer, Peter. *Charles Darwin: The Years of Controversy.* Philadelphia: Temple University Press, 1970.

Waddington, C. H. "The 'Baldwin Effect,' 'Genetic Assimilation' and 'Homeostasis.'" *Evolution* 7 (1953): 386–87.

————. *The Ethical Animal.* New York: Athenaeum, 1961.

————. "Evolutionary Adaptation." In *Evolution after Darwin.* Vol. 1 of *The Evolution of Life,* edited by Sol Tax. Chicago: University of Chicago Press, 1960.

Wade, Michael. "A Critical Review of the Models of Group Selection." *Quarterly Review of Biology* 53 (1978): 101–44.

————. "An Experimental Study of Group Selection." *Evolution* 31 (1977): 134–53.

————. "Group Selection among Laboratory Populations of Tribolium." *Proceedings of the National Academy of Sciences* 73 (1976): 4604–7.

Wallace, Alfred Russel. *Contributions to the Theory of Natural Selection.* London: Macmillan, 1870.

————. *A Defense of Modern Spiritualism.* Boston: Colby and Rich, 1874.

————. "The Descent of Man." *Academy* 2 (1871): 177–83.

————. "How to Civilize Savages." *Reader* 5 (17 June 1865): 670–72.

————. "Modern Biology and Psychology." *Nature* 43 (1891): 337–41.

————. *My Life.* 2 vols. New York: Dodd, Mead & Co., 1905.

————. *Natural Selection and Tropical Nature.* London: Macmillan, 1891.

————. "The Origin of Human Races and the Antiquity of Man Deduced from the Theory of 'Natural Selection.'" *Anthropological Review* 2 (1864): clviii–clxxxvii.

————. "Review of *Principles of Geology* by Charles Lyell, 10th ed., and *Elements of Geology* by Charles Lyell, 6th ed." *Quarterly Review* 126 (1869): 359–94.

————. "Dr. Romanes on Physiological Selection." *Nature* 43 (1890): 79, 150.

————. "Romanes versus Darwin." *Fortnightly Review* 46 (1886): 300–316.

————. *Studies, Scientific & Social.* 2 vols. London: Macmillan, 1900.

Wasmann, Erich. "Zur Entwicklung der Instincte." *Verhandlungen der zoologisch-botanischen Gesellschaft in Wien* 47 (1897): 168–83.

————. *Die Gastpflege der Ameisen: ihre biologischen und philosophischen Probleme.* Berlin: Borntraeger, 1920.

————. *Instinct und Intelligenz im Thierreich.* Freiburg i. B.: Herder, 1897.

————. *Die moderne Biologie und die Entwicklungstheorie.* Freiburg i. B.: Herder, 1904.

————. *Der Trichterwickler.* Münster: Aschendorf, 1884.

————. *Vergleichende Studien über das Seelenleben der Ameisen und der höhern Thiere.* 2d ed. Freiburg i. B.: Herder, 1900.

Watson, John B. "Autobiography." In vol. 3 of *History of Psychology in Autobiography,* edited by Carl Murchison. Worcester, Mass.: Clark University Press, 1936.

————. *Behavior: An Introduction to Comparative Psychology.* New York: Holt, 1967; reprint of the 1st ed. of 1914.

————. *Behaviorism.* Rev. ed. Chicago: University of Chicago Press, 1930; originally published in 1924.

————. "The Behavior of Noddy and Sooty Terns." *Carnegie Publications* no. 103 (1908): 187–255.

————. *Psychological Care of Infant and Child.* New York: Norton, 1928.

Watson, John B., and Rosalie Rayner. "Conditioned Emotional Reactions." *Journal of Experimental Psychology* 3 (1920): 1–14.

Watson, R. *Chemical Essays.* 6th ed. London: Evans, 1793.

Webb, Beatrice Potter. *The Diary of Beatrice Webb: Volume One 1873–1892.* Edited by Norman and Jeanne Mackenzie. Cambridge: Harvard University Press, 1982.

————. *My Apprenticeship.* Cambridge: Cambridge University Press, 1979; first published in 1929.

Webb, R. K. *Harriet Martineau: A Radical Victorian.* New York: Columbia University Press, 1960.

————. *Modern England: From the 18th Century to the Present.* New York: Dodd and Mead, 1971.

Wedgwood, Hensleigh. *On the Origin of Language.* London: Trübner, 1866.

Weismann, August. "The All-Sufficiency of Natural Selection: a Reply to Herbert Spencer." *Contemporary Review* 64 (1893): 309–38, 596–610.

————. *Essays upon Heredity.* Edited and translated by E. Poulton et al. 2d ed. 2 vols. Oxford: Clarendon Press, 1891; 1st ed. in 1889.

———. *The Romanes Lecture, 1894: The Effects of External Influences upon Development.* Oxford: Clarendon Press, 1894.

Wells, Algernon. *On Animal Instinct.* Colchester: Longman, Rees, Orme, Brown, Green, and Longman, 1834.

Whewell, William. *History of the Inductive Sciences.* 3 vols. London: Parker, 1837.

———. *On Induction.* London: Parker, 1849.

———. *The Philosophy of the Inductive Sciences, Founded upon their History.* 2 vols. London: Parker, 1840.

Wiener, Philip. *Evolution and the Founders of Pragmatism.* Cambridge: Harvard University Press, 1949.

Wild, John. *The Radical Empiricism of William James.* New York: Doubleday Anchor, 1970.

Williams, George. *Adaptation and Natural Selection: A Critique of some Current Evolutionary Thought.* Princeton: Princeton University Press, 1966.

Willis, Thomas. *Opera omnia.* Edited by Geradus Blasius. Amsterdam: Westen, 1682.

Wilshire, Bruce. *William James and Phenomenology: A Study of "The Principles of Psychology."* Bloomington: University of Indiana Press, 1968.

Wilson, A., G. Bush, S. Case, and M. King. "Social Structuring of Mammalian Populations and Rate of Chromosomal Evolution." *Proceedings of the National Academy of Science* 62 (1975): 5061–65.

Wilson, Edward. *On Human Nature.* Cambridge: Harvard University Press, 1978.

———. *Sociobiology: The New Synthesis.* Cambridge: Harvard University Press, 1975.

Wiltshire, Bruce. *The Social and Political Thought of Herbert Spencer.* Oxford: Oxford University Press, 1978.

Wimsatt, William. "Reductionistic Research Strategies and Their Biases in the Units of Selection Controversy." In *Scientific Discovery: Historical and Scientific Case Studies,* edited by Thomas Nickles. Dordrecht: Reidel, 1980.

Woodward, William. "Introduction to William James's Essays in Psychology." In *The Works of William James: Essays in Psychology,* edited by Frederick Burkhardt. Cambridge: Harvard University Press, 1983.

Woodward, William, and Mitchell Ash, eds. *The Problematic Science: Psychology in Nineteenth-Century Thought.* New York: Praeger, 1982.

Wozniak, Robert. "Metaphysics and Science, Reason and Reality: the Intellectual Origins of Genetic Epistemology." In *The Cognitive-Developmental Psychology of Baldwin,* edited by John Broughton and John Freeman-Moir. New York: Ablex, 1982.

Wright, Chauncey. "Evolution of Self-Consciousness." *North American Review* 116 (1873): 245–310.

Wundt, William. *Ethik.* Stuttgart: Enke, 1886.

———. *Grundriss der Psychologie.* Leipzig: Engelmann, 1896.

———. *Vorlesungen über die Menschen- und Thierseele.* 2 vols. Leipzig: Voss, 1864.

————. *Vorlesungen über die Menschen- und Thierseele*. 2d ed. Leipzig: Voss, 1892.

Wynne-Edwards, V. C. *Animal Dispersion in Relation to Social Behavior*. New York: Hafner, 1962.

Yates, Francis. *Giordano Bruno and the Hermetic Tradition*. Chicago: University of Chicago Press, 1964.

————. "The Hermetic Tradition in Renaissance Science." In *Science and History in the Renaissance*, edited by C. Singleton. Baltimore: Johns Hopkins University Press, 1968.

Youatt, William. *Cattle: Their Breeds, Management, and Disease*. London: Library of Useful Knowledge, 1834.

Young, Robert M. *Darwin's Metaphor*. Cambridge: Cambridge University Press, 1985.

————. *Mind, Brain and Adaptation in the Nineteenth Century*. Oxford: Oxford University Press, 1971.

Ziegler, Heinrich. *Der Begriff des Instinktes einst und jetzt*. 3d ed. Jena: Fischer, 1920; originally published in 1904.

————. *Die Naturwissenschaft und die Socialdemokratische Theorie*. Stuttgart: Enke, 1893.

Ziehen, Theodor. *Introduction to Physiological Psychology*. Translated by C. van Liew and O. Beyer. London: Sonnenschein, 1892.

Zirkle, Conway. "The Early History of the Idea of the Inheritance of Acquired Characters and of Pangenesis." *Transactions of the American Philosophical Society*, n.s. 35 (1946): 91–151.

Index